The Bare Bones

Life of the Past

James O. Farlow, *editor*

THE BARE BONES

An Unconventional
Evolutionary History
of the Skeleton

Matthew F. Bonnan

Indiana University Press Bloomington & Indianapolis

This book is a publication of

Indiana University Press
Office of Scholarly Publishing
Herman B Wells Library 350
1320 East 10th Street
Bloomington, Indiana 47405 USA

iupress.indiana.edu

The paper used in this publication meets
the minimum requirements of the American
National Standard for Information
Sciences–Permanence of Paper for Printed
Library Materials, ANSI Z39.48–1992.

Manufactured in the
United States of America

Library of Congress Cataloging-in-
Publication Data

Names: Bonnan, Matthew F.
Title: The bare bones : an unconventional
    evolutionary history of the skeleton /
    Matthew F. Bonnan.
Description: Bloomington : Indiana
    University Press, 2015. | Series: Life of the
    past | Includes bibliographical references
    and index.
Identifiers: LCCN 2015020766| ISBN
    9780253018328 (cloth : alk. paper) |
    ISBN 9780253018410 (ebook)
Subjects: LCSH: Skeleton—Evolution.
Classification: LCC QL821 .B66 2015 | DDC
    599.9/47—dc23 LC record available at
    http://lccn.loc.gov/2015020766

1 2 3 4 5   21 20 19 18 17 16

TO MOM AND DAD—without your early encouragement,
this would not have happened.

MY CHILDREN, QUINN AND MAXWELL—may you always appreciate
and respect the vertebrate animals that form our greater family tree.

JESS BONNAN-WHITE—as always, for your love and support.

So have a toast and down the cup, and drink to bones that turn to dust.

OINGO BOINGO, *No One Lives Forever*

Everybody's got mixed feelings about the function and the form.

RUSH, *Vital Signs*

Contents

C

·   PREFACE   *x*

·   ACKNOWLEDGMENTS   *xv*

·   PART ONE: SETTING THE STAGE

1   Introduction: How Vertebrates and Cars Are (and Are Not) Similar   3

2   Evolution to Deep Time, Pedigree to Anatomy   8

·   PART TWO: THE ORIGIN AND EARLY EVOLUTION OF THE VERTEBRATE CHASSIS

3   Inferring the Basic Vertebrate Chassis   32

4   Evolution of a Bony Chassis   50

·   PART THREE: THE EVOLUTION OF THE JAWED VERTEBRATE CHASSIS AND SOMETHING FISHY

5   The Jawed Vertebrate Chassis: A Primer   72

6   Placoderms and Cartilaginous Fishes   90

7   The Fishlike Osteichthyes, Part 1   114

8   The Fishlike Osteichthyes, Part 2   136

·   PART FOUR: THE VERTEBRATE CHASSIS MOVES TO LAND

9   The Tetrapod Chassis: A Primer   156

10   The Tetrapod Chassis in Transition   178

11   The Amphibian Chassis   200

12   The Amniote Chassis: A Primer and the Lead-Up to True Amniotes   226

· PART FIVE: DEEP SCALY I: REPTILIAN CHASSIS
FROM EARLY REPTILES TO SEA MONSTERS

13   Modern Lizards and the Tuatara as an Introduction   251

14   Early Reptiles and Turtles   268

15   Snakes and Sea Dragons   290

· PART SIX: DEEP SCALY II: THE ARCHOSAUR CHASSIS,
THOSE RULING REPTILES

16   The Archosaur Chassis, Part 1:
Modern Archosaurs   314

17   The Archosaur Chassis, Part 2:
A Primer on Archosaur Posture and Diversity   344

18   The Archosaur Chassis, Part 3:
Pterosaurs, Dinosaurs, and the Origins of Birds   362

· PART SEVEN: OVERCOME BY FUR:
THE MAMMALIAN CHASSIS

19   The Mammalian Chassis: A Primer   392

20   The Evolution of the Mammal Chassis   420

21   Brains, Milk, and the Modern Radiations of Mammals   444

· APPENDIX: THE CARDS OF TIME   475

· REFERENCES   477

· INDEX   497

The author's first paleontological award.

Preface

I DON'T KNOW WHY IT STRUCK ME, BUT IT DID WITH SUCH FORCE that my life's trajectory was forever changed. I'm speaking, of course, about my passion for dinosaurs, big dead reptiles that have captivated me since the tender age of five. I was that annoying kid in the first-grade classroom who knew all the dinosaurs and corrected the teachers on their pronunciation. I was even given a "Paleontologist Award" by my first-grade teachers for my, ahem, "abundant knowledge about dinosaurs." I did have varied interests outside of dinosaurs, including everything from human anatomy to science fiction to role-playing games to music editing (I'm not suggesting these were cool interests). In fact, I even did a stint at a radio station in college (WDCB, Glen Ellyn, Illinois), where I played songs that sounded interesting (at least to me) in headphones. Yet, over time, the allure of fossils kept its pull.

I was never so much into the treasure-hunter aspect of paleontology, but rather the thrill of reconstructing long-dead animals and breathing life into old bones. In other words, I am a zoologist and anatomist at heart who happens to be fascinated by dinosaurs. I see dinosaurs as living animals, and I want to reconstruct how these animals moved and behaved when their bones were still pulsing with blood. As time has gone by, I have come to have a deep appreciation and fascination with all the backboned (vertebrate) animals, their collective natural history, and their evolution. I have come to realize that the questions about dinosaurs that I began to pursue in earnest in graduate school have a broader and more powerful context across our vertebrate family tree.

I was inspired to write this book when I began teaching my own vertebrate evolution and paleontology course for undergraduate students. What I found was that many of these students were fascinated by vertebrate evolution, but that few, if any, went on to careers in museums and academe. Instead, many of my students were future teachers, doctors, veterinarians, and perhaps even politicians. There are many excellent books available on vertebrate paleontology, many of which I consulted in writing this book, but their focus tends to be strongly taxonomic and linearly chronological: who is who, who is related to whom, and in what order do we find them. However, the books that had truly inspired me to become a paleontologist were those that tackled the issue of functional morphology and paleobiology: what does the skeleton tell us about how the animal moved, fed, and behaved? This is the type of questions that motivated me as a student to learn about vertebrate history.

During my formative years and into my undergraduate days, I read a number of books that inspired the more colloquial approach I take here. First and foremost, *The Dinosaur Heresies* by Robert T. Bakker, *After Man: A Zoology of the Future* by Dougal Dixon, and *Archosauria: A New Look at the Old Dinosaur* by John C. McLoughlin were the books that truly lit the fires of my imagination in junior high. The stories these books told about past life or possible future evolution, combined with engaging artwork, solidified my desire to be a professional scientist. I even corresponded with Dougal Dixon, and his letter of encouragement to a thirteen-year-old budding paleontologist was inspirational. However, during my undergraduate days, I stumbled upon a small book called *The Evolution of Vertebrate Design* by the late paleontologist Leonard Radinsky that would truly influence my approach to writing. Radinsky took a complex subject like vertebrate paleontology and, using cartoons and brief but informative language, distilled the essence of our evolutionary story into a format that was friendly and approachable. In fact, I initially used his book in my vertebrate paleontology and evolution courses because it served as a jumping-off point for exploring the rich tapestry of vertebrate life past and present.

Given that Radinsky passed away in 1985, his beautiful book was never updated. Despite its appeal to my students, with each passing year the stack of articles I was assigning to supplement the understandably dated material was becoming larger than the book itself! Simultaneously, as my research developed into understanding the evolution of dinosaur locomotion, I was beginning to question why I had never paid more attention to classical mechanics in my physics courses. When I took physics, I found the course to be absolutely dull and dry. However, if you can understand the way that the machines and tools that surround us in our daily lives work the way that they do, you can approach the skeleton the same way. And then I thought, what if I tried to write a book about the evolution of the vertebrate skeleton as if I were someone trying to teach my younger self about classical mechanics and physics? Using Radinsky's book as an inspiration and launch point, I began writing the book you now hold in your hands: what I hope is a friendly but somewhat unconventional introduction and exploration of the history of the skeleton, using machine metaphors, for those who want to learn but do not (yet) have the chops for anatomy.

My disclaimer is as follows. In attempting to write a generalized and friendly but unconventional introduction to vertebrate evolution, my goal was never to replicate the already excellent works available that cover this topic in the depth and scope it deserves. I want to be clear to my readers and colleagues alike that this book was intended to serve as an introduction, and must understandably truncate or simplify what is often a much more complex, sprawling picture of vertebrate life. Moreover, my selection of vertebrate examples past and present was designed to highlight the major pathways of skeletal evolution, and should not be interpreted as an exhaustive survey of vertebrate life's true diversity. I also emphasize that my illustrations are meant to serve as diagrams and abstractions that, while conveying the essentials of vertebrate anatomy, should not be misconstrued as scientifically rigorous reconstructions. In a nutshell, this was not intended to be a professional reference text, but I hope in some small way it encourages those who are interested to dig deeper. I especially hope that it will serve undergraduates as a text for their vertebrate paleontology and comparative anatomy courses.

To end this preface on a bit of a philosophical note, as I tell my children and students, nothing worth doing in life is easy. That is certainly true for the field of paleontology. I have been fortunate in having parents who supported my dreams even though they were not scientists or academics, and that helped tremendously. I was also fortunate to marry another academic who understands the quirkiness and obsession of this type of career. In fact, paleontology and academics in general tend to be less of an occupation than a vocation. You pursue this type of career because you love it, and many of my friends and colleagues in vertebrate paleontology and related fields are proof of that sentiment. There is a saying that effort expended at making your dreams reality takes the work out of the courage, and I certainly follow that philosophy. As with many of us who go into basic scientific research, there were and continue to be many personal and professional challenges to overcome. However, I wouldn't have it any other way: I feel lucky and grateful to be someone who has his dream job. My role in the discovery of three new dinosaurs has been one of the greatest recent rewards of this career, and my inner five-year-old very much approves.

Acknowledgments

WRITING A BOOK LIKE THIS IS A LABOR OF LOVE, AND I AM INDEBTED to all the people who have helped me realize this project in so many different ways. I must start by thanking the editors at Indiana University Press, Jim Farlow and Bob Sloan, who gave this book a chance and were more than patient and understanding when, in the midst of writing, my life turned upside down. As fate would have it, I was offered a position at the Richard Stockton College of New Jersey (now Stockton University), and in the process of moving our family from Illinois to New Jersey and settling into new jobs, writing a book became difficult. Thank you for standing by this project, Jim and Bob! I also express my thanks to Nancy Lightfoot and the staff at Indiana University Press as well as freelance copyeditor Carol Kennedy for fine-tuning the writing and style of my manuscript. I'm particularly grateful for the accommodation of last-minute edits to keep the book as up-to-date as possible.

As a working paleontologist, I am honored to know so many good scientists whose expertise I was able to consult through conversations over the phone, through e-mail and Facebook, and face-to-face at conferences. So many individuals helped clarify my understanding of various vertebrate groups, provided me with papers and books, and were often good and patient listeners and teachers. Among these people I would especially like to thank: Fernando Abdala, Paul Barrett, Willy Bemis, John Bolt, Elizabeth Brainerd, Juan Cisneros, Ted Daeschler, Peter Dodson, Peter Falkingham, Brooke Flammang, Stephen Gatesy, Lance Grande, Jason Head, David Hone, Thomas Holtz, Angela Horner, John Hutchinson, Michael Lague, Margaret Lewis, Zhe-Xi Luo, Tyler Lyson, Heinrich Mallison, Darren Naish, Robert Reisz, Olivier Rieppel, Michael Romano, Emma Schachner, Kenshu Shimada, Hans-Dieter Sues, Stig Walsh, Matt Wedel, and Adam Yates. The section title "Deep Scaly" is used with the blessing of John Wiens as it was inspired by his National Science Foundation grant of the same name. For answering seemingly unending questions about physics and mechanics, I thank Jim Rabchuck and especially Jason Shulman, who is a friend, a lab coconspirator, and the only physicist who tolerates me enough to collaborate on research. It goes without saying that any and all errors in regard to the diversity of vertebrates or the laws of physics discussed in this book are mine and mine alone.

I thank the Field Museum in Chicago for their permission to reprint several photographs of their display specimens as color plates in this book. I also thank my colleague Heinrich Mallison for the courtesy of providing me with additional photographs of fossils and animals. The staff at the Cape May Zoo in New Jersey were extraordinarily helpful in giving me close-up access to many of their animals. Finally, I would be remiss if I did not thank lab director Justine Ciraolo, as well as John Rokita and the animal lab staff at Stockton University for their assistance with obtaining and maintaining my research animals, as well as their help in photographing them. All photos are my work except where indicated.

I am indebted to all my former professors, both good and bad, for giving me valuable lessons in how to, and how not to, teach, which translates over to my writing style for this book. My approach to teaching was particularly influenced by Dan Olson of Northern Illinois University, who was my favorite human anatomy professor. Many pearls of anatomical wisdom were given to me by Virginia Naples at Northern Illinois University, who taught me to appreciate mammal anatomy during my PhD studies. Daniel Gebo and Neil Blackstone, also PhD committee members, kept me from going off on too many tangents and opened my eyes to a wider view of the animal world. Finally, my PhD advisor, J. Michael Parrish, is to be thanked for honing my skills as a scientist and a science writer, and for being a great mentor. A special thank you is necessary to Ron Toth, now professor emeritus at Northern Illinois University, who taught me as a graduate student about the process of science and the biological theory of evolution. The second chapter of this book is greatly influenced by his writings and discussions with me on this subject.

I could not rightly call myself a paleontologist if not for the opportunity to do field and prep lab work. As a high school student, I participated in the Dig-A-Dinosaur program at the Milwaukee Public Museum, and I thank Peter Sheehan, David Fastovsky, and those organizers for my initial experiences in field paleontology, which opened many doors. Lance Grande, Bill Simpson, and the Field Museum provided another opportunity just as I was entering college, to learn fossil preparation as a summer intern. This experience in turn opened other doors to field work. Jim Kirkland, Brooks Britt, Rod Sheetz, the Dinamation International Society, and the Museum of Western Colorado provided me with additional internship opportunities in the field during my undergraduate and graduate school years. Additional opportunities for fieldwork arose when I was approached by Scott Williams in 2007 to work with the Burpee Museum in Eastern Utah with my students at Western Illinois University to excavate Jurassic sauropods. My fieldwork in South Africa, and the privilege of naming new dinosaurs, is thanks to Adam Yates, Johann Neveling, and John Hancox, who invited me to work with them on drafting what turned out to be a successful National Geographic grant to explore the Early Jurassic.

To all my current and former undergraduate and graduate students, you should know that you have helped me become a better teacher, and a professor always learns more from his students than he imparts to them. Among the colleagues at my previous institution, Western Illinois University, I want especially to thank Susan Meiers for her friendship, support, and wisdom on teaching. I also thank my new colleagues at Stockton University for giving my career a new and exciting jumpstart.

Finally, I have to acknowledge the many people in my life who have contributed to this book through their love and support. My parents, Fred and Threse Bonnan, who were neither scientists nor academics, saw the spark in me to pursue this unusual career, and were always supportive of my dreams. Barrie Jaeger, one of my mother's college friends, was also an early influence on my growth as a budding scientist. She sent me a copy of *Archosauria: A New Look at the Old Dinosaur* by John McLoughlin, when I was but five years old! I still fondly recall her inscription to me, "Long live dino bones! May they never crumble or rot!" My siblings and extended family have always been a source of encouragement, even though, like my parents, many of them are probably convinced I'm crazy (but in a good way). My children, Quinn and Maxwell, are always an inspiration and keep me hopeful for the future. Last, but not least, I express my love and gratitude to my partner and soul mate, Jess Bonnan-White. Her patience, love, and understanding through the often tedious process of writing my first book made this possible.

Setting the Stage

The sands of time were eroded by the river of constant change.

GENESIS, *Firth of Fifth*

Introduction: How Vertebrates and Cars Are (and Are Not) Similar

I KNOW VERY LITTLE ABOUT CARS. I KNOW THAT I LIKE TO GET FROM point A to point B in a reliable vehicle. I know that I like my car to start when I turn the ignition. I know that whereas I am theoretically capable of changing a tire on the expressway, it is probably best for me and passing drivers that this hasn't occurred too often in my life. And it is probably obvious that manual transmissions and I have not made a proper acquaintance.

My wife and father-in-law, however, know very much about cars. They speak a foreign language to one another about gears, models, makes, and tire treads. Both have racing experience. They follow Formula One races religiously. The two of them even have a special bond with an old Porsche named "Helmut" – both can make that car do things that require special powers I just don't possess.

And if my father-in-law and I were dropped into a junkyard full of discarded automobile parts, you can bet that he would be in a better place than I to tell you the make, model, and year of the particular bits we stumbled across. Here a discarded spoiler, there a V-8 engine block. To the untrained eye, such as mine, these are pieces of junk. To the eyes of someone with knowledge of car mechanics, however, these pieces may be valuable salvage, or at least they very clearly show their functionality. The spoiler's shape causes the car to suck air downward, holding the vehicle more firmly on the pavement. The arrangement of the V-8 engine allows for both more horsepower and less vibration in a smaller space.

The appearance and shapes of the automobile parts not only highlight their function but often indicate their pedigree. Pin striping, the angle and orientation of a particular corner or side, or the size and shape of tail fins can inform the expert of whether he or she has a Chevy, a Ford, a Toyota. Furthermore, if you are a car enthusiast, you have probably seen the complete versions of many of these dismembered vehicles, and even know in great detail how the components have been modified over time by different manufacturers. In some cases, these changes have been functional – improving engine performance, controlling emissions, or damping vibrations. In other cases, the changes have been largely superficial and stylistic to attract new or returning car buyers.

Although I am about the farthest thing from a car enthusiast that one can be, it has often struck me that the skeleton and its associated soft tissues are perhaps best understood in machine metaphors. In fact, certain aspects of automobiles work well as concepts for explaining the evolution and functionality of the vertebrate body. I especially like the concept of the chassis, the framework that supports a car. The chassis in my mind is akin to the skeleton of the vertebrate animals I study. In cars, chassis shape, strength, and size can tell you a lot about a particular vehicle even when much else is missing. Likewise, even with the loss of soft tissues, the skeletons of vertebrate animals can reveal much about both the functionality and the pedigree of the particular individual you are studying.

The concepts of gear and torque are also useful. The size and shape of the gears in a car determine how fast the axle will spin and how much rotational force (torque) it will transfer to the tires. The arrangements of many skeletal muscles in the vertebrate body have similar mechanical consequences: muscles and bones act as gears producing torque around particular joints. Even concepts such as front-wheel and rear-wheel drive can be applied to understanding movement in vertebrate animals: for example, many land vertebrates utilize their hind limbs for propulsion (rear-wheel drive) and "steer" with their forelimbs.

A junkyard and a fossil-bearing layer of rock offer an intriguing comparison. My field paleontologist mentor, Jim Kirkland, once remarked that a vertebrate paleontologist in the field is equivalent to a car expert dumped into a mess of disassembled cars. As I now understand, a vertebrate expert can often identify the functionality and pedigree of a particular animal by examining the shape and structure of the disarticulated bones. Just as the features on a discarded engine can tell the car enthusiast the make, model, and size of the car, as well as its probable top speed, so can a femur (thigh) bone tell the vertebrate expert aspects of its possessor's common ancestry, pedigree, probable top speed, and overall body size.

Of course, cars and vertebrates are not the same. If a manufacturer wants to change the chassis or engine of a car, they can redesign it from scratch, improving its efficiency or style in the process, without being tied to problems or limitations of the previous car body. Vertebrates are living creatures—a heart or brain or skeleton cannot simply be replaced or modified wholesale. Unlike a car's engine, which can be shut off during modifications, the living engine of a vertebrate must continue to work throughout its life. Therefore, changes in form and function must occur while the animal continues to live. Often this means that the organs of vertebrate animals all begin from the same basic blueprint and are modified—many old parts are used or tweaked in ways different from their original functionality. It is as if an engineer was forced to redesign a car's engine while it was still running, using only the parts already making up the car. Some interesting and weird but perhaps elegant solutions to various problems would most likely result. Hence, vertebrate skeleton shape is often a bizarre mix of retooled but old features containing stamps of long-lost ancestors.

Perhaps the most important difference between cars and vertebrates is that we can't directly ask a particular animal lineage how and under what circumstances particular backbones fused, or limbs developed, or jaw power increased. Cars are made and designed by people, so we can ask them how and why they did what they did. Our approach to deciphering the skeletal evolution of vertebrates has to be much more indirect, more akin to reconstructing a crime scene or piecing together a family tree from old photos or lost memoirs. Fortunately, because the skeleton is the living, moving framework of the body, its shape, scars, articulations, and openings provide vital clues about the muscles, nerves, and guts of living and extinct vertebrates. Thus, if we can understand the shape and form of the skeleton, we purchase a window into the evolutionary history of the vertebrates, even if we can't directly ask our ancestors how the various backboned animals came to be.

It can be exceedingly difficult for the uninitiated to become acquainted with the skeleton, its evolution, and function from a purely anatomical approach. Whereas professionally there is no substitute for discussing vertebrate evolution anatomically, many beginning students and laypersons will quickly become lost when we speak of animals being dorsoventrally compressed, or having craniofacial prognathism, or possessing a manus with ulnar deviation. However, although vertebrates and cars are not the same, approaching the skeleton in a way similar to that of the car enthusiast approaching a vehicle provides us with a framework within which vertebrate evolution and function can be understood and in which certain key anatomical terms can be introduced.

If you want to know about the history and functional design of an automobile, you are generally interested in knowing the following basic information:

1. Manufacturer
2. Date of production
3. Specialties of the chassis
4. Intended use – e.g., racing, utility vehicle, etc.

In this book, I modify and apply this approach of investigation to vertebrate animals. In each chapter, we will explore the evolution and functionality of the vertebrate skeleton using the following framework:

1. Pedigree – relationship to other vertebrates
2. Date of first appearance in the fossil record
3. Specialties of the skeletal chassis
4. Econiche – e.g., herbivore, carnivore, etc.

I should emphasize that our consideration of vertebrate skeletal function will not be limited to automobile analogies. Although we will approach vertebrates as a car enthusiast might, the mechanical analogies will run the gamut from scissors to engines, from medieval armor to socket wrenches. Overall, I wish to convey the functional anatomy of the vertebrate skeleton and its historical changes using a simplified, mechanical perspective rather than the more typical approach of anatomical analysis rooted in complex evolutionary diagrams.

Most of the chapters in this book are arranged by pedigree and skeletal chassis. The exceptions to these criteria fall under chapters 2 and 3. Chapter 2 deals with subjects such as evolution, deep time, and phylogeny, subjects that help us reconstruct the pedigree and functional anatomy of vertebrate animals. Chapter 3 introduces the basic vertebrate chassis and predicts what major anatomical structures would be present in the earliest vertebrate animals. This chapter also introduces the best-known earliest vertebrate in light of our predictions of what structures should be present in our common ancestor. From chapter 4 onward, we cover the long and intriguing history of the vertebrate skeleton stretching back over 540 million years. My hope is that, by approaching this topic from a mechanical perspective, you will ultimately appreciate the fundamental role that the skeletal chassis has played in ensuring the survival and diversity of the vertebrate animals.

2.1. This is evolution. Note that a common ancestor is not one individual but rather a group of two or more, just like on your family tree.

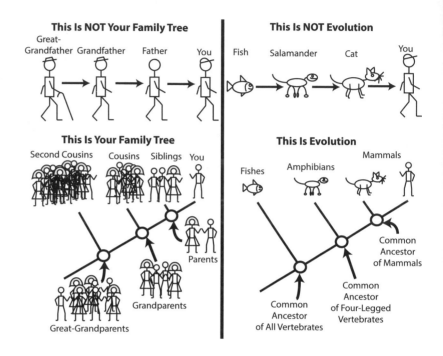

This Is NOT Your Family Tree

Great-Grandfather   Grandfather   Father   You

This Is NOT Evolution

Fish   Salamander   Cat   You

This Is Your Family Tree

Second Cousins   Cousins   Siblings   You

Parents

Grandparents

Great-Grandparents

This Is Evolution

Fishes   Amphibians   Mammals

Common Ancestor of Mammals

Common Ancestor of Four-Legged Vertebrates

Common Ancestor of All Vertebrates

8

Evolution to Deep Time, Pedigree to Anatomy

EVOLUTION IS MISUNDERSTOOD BECAUSE SCIENCE ITSELF IS misunderstood. Science has been conflated with atheism, social Darwinism, and the cold, inhuman march of progress that ultimately leads to the subjugation of humans by machines in many a science fiction novel. Science is no more any of these things than a mechanical engineer is the person with the striped hat in the locomotive engine. Science is a discipline narrowly focused on posing answerable questions about the physical universe (Popper, 1959). Science is simply a useful tool for understanding the physical world.

The confusion and maligning of science, and especially the theory of biological evolution, is unfortunate. Part of this confusion stems from the acceptance by some of a false choice: either you accept science and reject spirituality, or you accept spirituality and reject science. However, this misses the point: science is simply a tool for understanding the natural world—it cannot provide evidence for or against what is beyond nature.

To borrow an analogy from Eugenie Scott and Ron Toth, a scientist is no more qualified to show or falsify the existence of God than an auto mechanic. A vertebrate animal is (kind of) like a car, and an auto mechanic's job, like that of a scientist, is firmly planted in the physical. I don't know about you, but when I take my car to the auto mechanic, I expect that he or she will diagnose and fix the problems I have with the car physically. I expect to get a bill with items such as: low brake fluid, ruptured radiator, pinhole leak in the air conditioning compressor, or worn brake pad. I expect to see how the mechanic physically has altered my car so that it is once again safe to drive. I do not expect my auto mechanic to wax philosophic on the nature of good and evil, or to tell me that something supernatural is to blame for my car's troubles. I also don't plan to ask my auto mechanic for spiritual guidance or emotional counseling, or even for oboe lessons, because all these things fall outside his or her job description (unless the mechanic moonlights in oboe lessons, of course).

Science works similarly. There may be an ultimate purpose to the universe, and there may be a spiritual realm, but science is not the right tool to address those issues. On the other hand, understanding the pattern and functional evolution of vertebrates does fall under the purview of science, because it has occurred in the physical world and can be tested under the explanatory umbrella of the theory of biological evolution.

But what is the theory of biological evolution? Biological evolution is not simply "change over time" as is often claimed. Your watch changes over time, but it certainly does not evolve. It is telling that when Charles Darwin drew his first conception of evolution during his travels aboard

Evolution

the HMS *Beagle*, he drew a family tree of finches. In science lingo, Darwin drew a phylogeny—a branching diagram of common ancestry and descent of the finches on the Galapagos Islands (Darwin, 1859; Jones, 1999; Zimmer, 2001; Kelly and Kelly, 2009).

Darwin, and Alfred Russel Wallace independently, successfully united two concepts into one that provided a simple but effective explanation for the diversity of life: (1) common ancestry and (2) the passing of heritable traits via natural selection (Darwin, 1859; Zimmer, 2001; Kelly and Kelly, 2009). Put simply, the biological theory of evolution can be stated as descent with modification from a single, common ancestor. All life on earth is related through a great family tree. Different branches of the family tree have inherited modified characteristics unique to their portion of the pedigree through a process of natural selection.

Biological evolution is a theory, but that is not a weakness. A theory in science gets a specific definition: it is an overarching generalization that explains and makes sense of a given set of phenomena. A theory can be tested, has potential to be falsified, and has predictive power (Popper, 1959). Data ultimately decide whether or not a theory is rejected or modified.

Biological evolution is supported by data to be an effective explanation for many biological phenomena. For example, the groups-within-groups arrangement known as the hierarchy of life is the predicted end result of descent of modification from a single, common ancestor—we would predict there to be general traits shared by all organisms, followed by more and more exclusive traits shared by groups with more recent common ancestry, producing a groups-within-groups hierarchy. As another example, descent with modification over a long time would produce a sequential fossil record, such as the one we find around the world. The basic sequence of vertebrate evolution is the same everywhere you look in the rock record—birds never appear before fish in the fossil record, for example.

At its base, all of us recognize the family tree concept and its application in our own lives (Fig. 2.1). For example, I share common ancestors with my siblings—my mother and father. Further back in my family tree, my siblings and I along with our cousins share other, more distant common ancestors—our grandparents. Moving out to my second cousins, I share even more distant common ancestors with them—our great-grandparents. Take the common ancestry back several more generations, and we quickly approach a common ancestor for all of humanity. Beyond this, the family tree's scope enlarges to encompass more distant common ancestors, such as those shared with other animals, plants, and eventually all living organisms on earth. Every living thing on earth is related. All vertebrate animals, including us, have a common ancestor.

Natural selection is the motor behind evolution. It was the insight of both Darwin and Wallace, and it provides a mechanism for how descent with modification from a common ancestor may occur. Natural selection is a biological law that can be summarized as follows: (1) all populations vary; (2) all populations produce more offspring than can survive; and (3) individuals within populations with traits that allow them to successfully mate and reproduce viable offspring will be selected for (Darwin, 1859; Jones, 1999; Zimmer, 2001). Having viable offspring means having children that can themselves successfully reproduce. Natural selection is sometimes, unfortunately and mistakenly, referred to as "survival of the fittest" (Kelly and Kelly, 2009), a statement that ignores the chance involved in survival. Many of the great vertebrate extinction events have had less to do with "fitness" than with the effect of rapid environmental change on organisms that are well adapted to the previously stable environment.

Within a given population, different individuals possess different traits or characteristics. Those individuals with traits that allow them to survive to reproductive age in a given environment, and to therefore mate and pass on these inheritable traits, are "selected for." In other words, natural selection is simply the differential survival of individuals with any combination of traits that allow them to reproduce viable offspring.

We now understand, as Darwin, Wallace, and their contemporaries did not, that descent with modification is due to the inherited mutations (modification) of genes, the molecular units of heredity (Freeman and Herron, 2007). Genes ultimately produce the physical traits we see in animals. A brief review of genetics is in order here. We now know that DNA is the universal code stored in the nucleus of animal cells, and it acts as a library that can be read (Fig. 2.2). The books in the library are genes, and they must be read (transcribed) by messenger RNA (mRNA). The transcribed gene message is then translated from the mRNA by cellular machinery (such as ribosomes and other forms of RNA) into an amino acid sequence that becomes modified into a protein. When we say a gene is "expressed," we mean a protein has been generated from a gene's code. If you remember nothing else from this genetic lesson, remember this: it is the proteins that cause cells, tissues, and organs to take on their characteristic features. Mutations in genes, therefore, ultimately lead to the expression of different or modified proteins that can cause subtle to major changes in an organism's anatomy and future success in passing on its genes to the next generation. This means that the shape of the skeleton is impacted by which genes are and are not expressed.

Recently, research into embryonic development has given us an even better insight into how major structural changes might occur in a given population of organisms. We now understand that there are two major types of genes: developmental and "house-keeping" genes (Slack, 2013). Developmental genes are those that are expressed during embryonic development and growth, and their proteins control the symmetry, skeletal development, organ placement, and overall form of the developing animal (Wilt and Hake, 2003; Gilbert, 2010). In contrast, "house-keeping" genes are expressed during the animal's daily life to generate proteins that keep the cells, tissues, and organs in the body functioning properly. As you might suspect, mutations in developmental genes can have radical

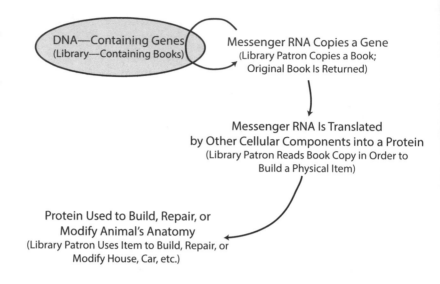

2.2. A basic schematic of gene expression. Genes within DNA are read (transcribed) by messenger RNA; this is sometimes called 'genetic dogma.' This gene copy (mRNA) is then 'read' to build a protein – the act of translation. Here, gene expression is compared to a human patron using a copy of a library book to help build, repair, or modify a house or car.

DNA—Containing Genes
(Library—Containing Books)

Messenger RNA Copies a Gene
(Library Patron Copies a Book;
Original Book Is Returned)

Messenger RNA Is Translated
by Other Cellular Components into a Protein
(Library Patron Reads Book Copy in Order to
Build a Physical Item)

Protein Used to Build, Repair, or
Modify Animal's Anatomy
(Library Patron Uses Item to Build, Repair, or
Modify House, Car, etc.)

consequences for body form and function, whereas mutations in "housekeeping" genes tend to affect the health and reproductive success of the postembryonic animal.

Let us reiterate natural selection, now with genes in mind. Within a given population, different individuals possess different gene combinations. Those with genes that, when expressed, allow them to survive to reproductive age in a given environment, and to therefore mate and pass on their genetic inheritance, are "selected for." Again, natural selection is simply the differential survival of individuals in a population with any combination of developmental and "house-keeping" genes that allow them to reproduce viable offspring. Only genes that are passed on in the gametes of vertebrates affect evolution: the sperm or eggs must carry the modified, duplicated, or mutated developmental or "house-keeping" genes.

It should be emphasized that evolution occurs at the level of the population, not the individual (Freeman and Herron, 2007). The individual is born into the world with a particular mixture of traits, but that individual does not and cannot evolve. For example, if a predator attacks a population of a particular species, the individuals that survive are those that already have traits to fight or escape – an individual without these traits cannot "evolve" them in response to this threat. In our hypothetical population attacked by a predator, faster animals with long legs may survive more often than slower animals with stubby legs (Fig. 2.3). Thus, more of the faster individuals will pass on their traits than the slower ones. However, the slower individuals cannot "evolve" longer legs or faster speeds – they will simply be selected against. Over time, the population will evolve so that the average individual is relatively faster simply because individuals with this trait tend to survive and reproduce, whereas slow individuals tend be eaten before reproduction. Thought of yet another way, biological evolution and natural selection boil down to sex and time against a given environment. In a particular environment,

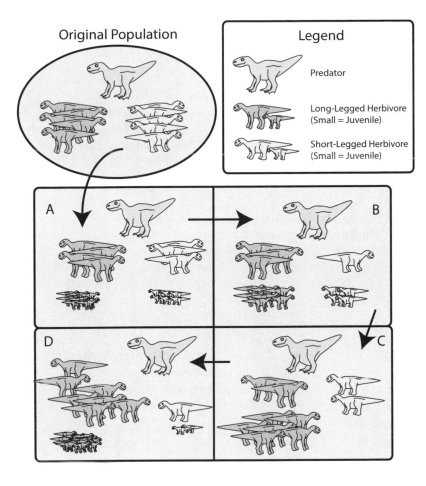

Original Population

Legend

Predator

Long-Legged Herbivore
(Small = Juvenile)

Short-Legged Herbivore
(Small = Juvenile)

A

B

D

C

2.3. Natural selection illustrated. In this hypothetical example, we start with a population of long-legged and short-legged herbivores and a predator. Initially, there are equal numbers of long-legged and short-legged herbivores, but as time goes on, the short-legged herbivores are more easily caught and eaten by the predator. Even though the reproductive output (number of offspring) is the same for each animal, there is a shift in the population toward the long-legged trait due to selection from the predator.

inherited combinations of genes allow some individuals to survive to reproduce viable offspring, whereas other individuals are less successful at these tasks. The shape and form of the vertebrate skeleton is a testament to all of these processes.

The sort of evolutionary changes we see in the vertebrate skeleton could not have occurred without an exceptionally long period of time for natural selection to work on thousands of generations. It is possible, of course, in the short term to produce startling changes in various animal breeds under human-guided, artificial selection. For example, we can select, in certain individual dogs, for traits we like, and then allow only those favored individuals to breed. This sort of human selection has resulted in the variety of dog breeds that we see today at pet stores, animal shelters, and various dog pedigree contests. But these are just variations on the domestic dog species – a Chihuahua and a Great Dane are both extremes of *Canis familiaris*. The domestication of dogs from wolves – literally the human-encouraged evolution of a new species (*Canis familiaris*) from a population of a wild species (*Canis lupis*) – took much longer (Wang and Tedford, 2008), as did the domestication of other animals for our

Deep Time

agriculture (Zimmer, 2001). Thus, much more copious amounts of time are required to explain the major structural changes we see in the vertebrate skeleton during their evolutionary history.

That the Earth is old has not always been appreciated or understood, even to this day (McPhee, 2000). Moreover, the history of the geological sciences has been replete with efforts to estimate and calculate the age of the Earth, a history that goes well beyond the scope of this book. Here, we will focus on two approaches to understanding what has become known as deep time. The first approach is relative dating, where clues in the rock record allow a paleontologist to determine the sequence in which various vertebrate groups have appeared in time. The second approach, called absolute dating, requires knowledge of radioactive element decay (Lambert, 1998; McPhee, 2000; Winchester, 2009), and it has allowed us to put numbers on the relative sequence of fossil vertebrate appearances.

The first scientists in "recent" history (late 1700s to early 1800s) to attempt to decipher the rock record gradually devised a number of laws that allowed them to develop a relative time scale based on sedimentary rocks (e.g., Winchester, 2009). Many of these laws seem obvious, but their simplicity and application was their power–these laws were basic rules that allowed for a more detailed understanding of the sequence of major events on Earth and (in our case) the vertebrate story.

First, there was the Law of Superposition (Fig. 2.4 and Plate 1). This law states that in a sequence of originally horizontal rock layers, the rocks on the bottom have to be older than the rocks on the top (Lambert, 1998). This makes a great deal of intuitive sense–if we have a stack of books, we realize that the books at the bottom of the pile had to be there first to support the books at the top of the pile. Unless the laws of physics as we know them were suspended, you cannot place books in thin air, and then later get around to placing books underneath them. It is harder still to imagine how whole sequences of heavy rock could float in air, waiting for the day they could rest on top of other rocks. Thus, if we have a sequence of undisturbed horizontal rock layers, we can be very confident that the lowest rocks are older than the upper rocks.

As you have probably noticed, rock layers are not always neatly arranged in horizontal layers. During processes of mountain building, earthquakes, and continental movement, layers of rock can be folded, buckled, tilted, and even overturned. How do we know which end is up, and therefore which way we are going (backward or forward) in time? Part of the solution to this problem involves another law, the law of original horizontality (Fig. 2.4). Simply stated, this law says that the sediments that formed the rock layers must have originally been deposited in horizontal sheets (Lambert, 1998). This follows from the observation of sediments forming in modern environments, where they are first deposited horizontally. This means that any folding, buckling, tilting, or overturning has to occur sometime after the sediment layers are deposited. A sequence with folded rock layers overlaid by horizontal rocks layers would suggest that whatever event bent the lower rocks had to have occurred before

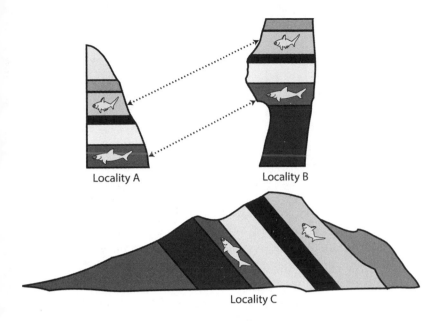

Locality A   Locality B

Locality C

2.4. Laws of superposition, original horizontality, and faunal succession illustrated. Three different localities are shown with two index fossils specific to a certain place in the geologic record. In localities A and B, the rock layers are still in their original horizontal orientations, and this means that rocks at the bottom of the sequence must be older than rocks at the top (law of superposition). Although the rocks in localities A and B preserve somewhat different portions of the rock record, we can fill in missing data from one or the other by matching up the index fish fossils from one location to the other (law of faunal succession). At locality C, we find tilted layers—these originally horizontal layers must have been tilted by some event, such as mountain building, after their deposition (law of original horizontality). The index fossils at locality C help us determine which direction in the tilted rock layers is toward older time periods and which direction is toward younger rock layers. The index fossils also help us 'stitch' this sequence of rocks into the sequences found at localities A and B.

the upper rock layers were deposited. Also, we would anticipate that the folded rocks were once horizontally deposited, and should determine their relative ages after reconstructing them as "unbent."

It was also recognized that the order and sequence of fossil animals and plants in the rock record was specific and reliable, even across great horizontal distances. No matter where you were in the world, fish fossils always preceded the appearance of amphibians, and mammals were never found before the appearance of reptiles. This meant that you could cross-correlate one region of the world's rock sequences with that of another (Fig. 2.4). From these observations came the law of faunal succession, which states that there is a specific, unique, and sequential order to the appearance of fossil animals in the rock layers (Lambert, 1998). As a side note, the same observations were made for fossil plants, and hence the full law is that of faunal and floral succession.

An even more precise method for cross-correlating rock units was through the recognition and use of index fossils. Certain fossil species, especially those of invertebrates, appeared only in a narrow band of rock units and nowhere else in the rock record. Because these fossils were associated only with a specific, short-term unit of rock, if you found these fossils in another locality somewhere else in the world, you could be sure that you were in the same specific interval in the rock sequence. Such index fossils allowed early geologists to further refine, match, and cross-correlate rock units around the world (Winchester, 2009).

By use of these geological laws, and several others, by the mid-1800s a relative geological time scale was constructed (McPhee, 2000; Zimmer, 2001). Now when fossil vertebrates were discovered, their appearance could be placed on a relative timeline. To many geologists in the 1800s, it appeared conceivable that the Earth was an old place—it had to be, given how much sedimentary rock had accumulated. But how old was old?

Numerical or absolute dating of rock wasn't possible until the 20th century, with the discovery that radioactive elements (such as uranium) change over time (decay) into stable elements (such as lead) at a predictable rate (Lambert, 1998). Deep within the Earth radioactive elements are abundant, and they travel with molten (igneous) rock to the surface. When igneous rock cools, its crystalline structure "traps" these radioactive elements inside, where they naturally decay into their stable daughter products at a constant, predictable rate. The ratio of how many radioactive elements versus stable daughter elements there are in a given igneous rock sample indicates how much time has passed since the rock cooled. For example, two varieties (isotopes) of uranium are present that decay at different but slow rates. In the case of uranium-238, half of the atoms in a given sample (the half-life) will decay to lead-206 within about 4.5 *billion* years. For uranium-235, the half-life (to decay to lead-207) is approximately 704 million years. The ratios of these products are measured using mass spectrometers and other sophisticated instruments that can literally count atoms in pulverized rock material (Lambert, 1998).

With rare exceptions, fossil vertebrates are not preserved in igneous rock–molten rock tends to incinerate animals. Instead, we date fossil vertebrates by obtaining the dates of igneous rock layers that occur above and below the sedimentary rocks in which our fossils of interest are preserved (Parker, 1991; Lambert, 1998). Hence, you often hear that such-and-such a fossil vertebrate is approximately 10.2–11.1 million years old–the upper and lower bounds of these numbers being based on the upper and lower igneous rock layer dates that sandwich the fossil in between.

But how do we know such dates are reliable? After all, the half-lives are calculated from extrapolations based on short-term observations of radioactive decay. How can we be confident that the dates we get are not random or overinflated? Our confidence in these numbers stems from two observations. The first is the heat produced by radioactive decay. When radioactive elements decay to their natural daughter products, they release a substantial amount of heat–the same heat that boils the water that turns the turbines in a nuclear power plant. In fact, the interior of the Earth is unbelievably hot, thanks in no small part to the decay of billions of radioactive elements generating heat so tremendous that it melts rock! It follows that if one were to somehow increase the speed of radioactive decay to give misleading dates, there would be so much heat that the Earth itself would be cooked (Isaak, 2007). So, we can be fairly certain that the rate of radioactive decay has not sped up at some point in the past.

However, perhaps the most telling evidence that radioactive decay rates are constant over long periods of time comes from the rock record itself. When geologists began to apply radioactive dating techniques to igneous rocks all over the world, it was found that the dates they computed matched in an uncanny way the relative time sequence established by the early geologists (Lambert, 1998). In other words, the relative ages of the rocks predicted merely on the basis of the geological laws were borne out. As an example, rocks of the Cambrian period around the world give radioactive decay dates of approximately 542 to 488 million years ago. The rock sequences above the Cambrian rocks, from the Ordovician period, give younger dates of approximately 488 to 433 million years ago. Every time igneous rocks of Cambrian age are dated, they are always and without exception older than the rocks above them, the Ordovician rocks. And the same can be shown with other rock units and time periods. If the rate of radioactive decay measured were not constant over long periods of time, we would expect to get all sorts of contradictory age estimates (Isaak, 2007). That this does not occur is compelling evidence that these dates are accurate.

We now know that the common ancestor of all vertebrate animals appeared on Earth sometime during or just before the Cambrian period, approximately 542 million years ago. Although this age may seem incredible, it is but a small percentage of the total age of the Earth, which has been estimated through radioactive dating methods of meteorites (the rocky bits left over from the formation of the planets) back to over 4.5 billion years ago. This means that the entire history of vertebrates, as grand as it is, spans only 12% of the entire history of the planet!

To put this in perspective, imagine if all of Earth history were contained in a stack of 100 index cards: in this scenario, each card would represent 45 million years of time. It is not until we get to Card 89 that the common ancestor of vertebrates makes their appearance. By Card 93 we have all major fish groups and the common ancestors of the tetrapods. By Card 95 the first dinosaurs appear, followed soon after by the first mammals. Card 97 welcomes in the first birds, and the great dinosaur radiations go extinct on the middle of Card 99. The last one and one half cards are devoted to the so-called Age of Mammals, and the entirety of human evolution and civilization, agriculture, and history are summarized on the razor-thin edge of the last card—certainly, this is a humbling view of our place in deep time. To keep our perspective, when geologic dates are given throughout the book, they will be indicated by the card number out of 100 in parentheses. The appendix on the Cards of Time at the end of the book illustrates this concept, placing the earliest appearance of each major vertebrate group we encounter on their appropriate cards.

Fossils

Fossils are the traces or remains of organisms preserved in the rock record (Carroll, 1988; Parker, 1991). Traces can include footprints, skin impressions, and burrows, whereas mineralized bones to soft tissues can constitute body fossils. That fossils exist at all is amazing. It is estimated that of all the organisms that have ever lived, probably fewer than 1% have been preserved as fossils (Prothero, 2007).

We can play a rather macabre game called "Who Wants to Be a Fossil?" In this game, the object is to die and then be preserved as a fossil for time immemorial . . . or until some paleontologist digs us out and establishes their career on us. If you were a contestant in such a game, where would you go to become fossilized?

Before we go off dying, let's first consider what happens when a vertebrate animal dies. Upon death, a vertebrate becomes a readily available source of nutrients for the still living (Lyman, 1994; Rogers, Eberth, and Fiorillo, 2007). Scavengers pick the bones clean, or even crush them up and eat them. Various insects use the carcass as a nursery for their larvae. In fact, one of the best ways to clean a skeleton is to let the larvae of dermestid beetles consume the rotting flesh, leaving behind gleaming white bones. Dermestid beetle larvae are one of a museum curator's best tricks for obtaining beautiful skeletons. Bacteria and fungi help along the process of decomposition, turning the soft tissues into a blackened soup, and plants, fungi, and assorted microbes also play a large role in skeletal destruction, leaching out the calcium and phosphorous salts stored in bones.

So, if we are to be a winner in the fossil game, we must make sure we die in such circumstances that it will be difficult for predators, scavengers, insects, plants, fungi, and microbes to use and disperse our remains. As you might imagine, losers of the fossil game are those vertebrates who have any kind of long-term exposure to the living world after death.

Let us also not forget the effect of simple environmental exposure on a carcass. Want to bleach bones using no chemicals? Stick them in the sun! Want to break bones into tiny pieces or reduce them to dust? Let the rain, frost, humidity, and heat have at them! Even flowing water itself can act to tumble, polish, but ultimately destroy even the hardiest of skeletal elements. Another way to lose the fossil game, then, is to be exposed to the elements for too long (Lyman, 1994).

If we are to be a winner in the fossil game, it is the combination of exposure to the living world and the particular environment in which we die that will initially make or break us. Certain environments clearly do not lend themselves well to fossilization. If we die in the rain forest, the plethora of opportunistic organisms combined with the stifling heat and humidity almost certainly ensures that we will be reduced to nothing but dirt very quickly. If we die up in the mountains, our chances of preservation are equally poor, given that these are places of great environmental exposure and erosion. Much of the sediments of rivers, beaches, and oceans are the transported remains of old mountains (Rogers, Eberth, and Fiorillo, 2007; Blakey and Ranney, 2008).

To win the fossil game, it is clear that we must be buried, and quickly. A flash flood might do it, the churning sediments of an engorged river smothering us in an instant, trapping us away from lots of living things and environmental exposure. Even a death on the bank of a river offers some hope—if the river is meandering back and forth, it may bury us under bar sediments after a rain storm. Some kind of landslide would do nicely as well—even in a desert, a collapsing sand dune could quickly and efficiently bury us, taking us out of the land of the living. Some quieter deaths are also possible under certain circumstances. In the quiet, anoxic backwaters of a swamp, the oxygen-depleted water would ensure that few organisms would survive long enough to dine upon us, and the slow, steady accumulation of fine sediments would entomb us nicely. Or we could fall to the bottom of some ocean bed where sediments gradually accumulate and the temperature, pressure, and oxygen content all work against the agents of decomposition.

Perhaps the biggest consideration in all of this is where sediments are most likely to accumulate over time. We avoid mountains if we wish to become fossils because sediment accumulation is minimal to negative in these geographies. Sediments collect reasonably well on flat floodplains when rivers overflow their banks and dump mud and silt in thick layers. Better yet are basins, regions set low relative to the surrounding areas that catch and store up sediments (Blakey and Ranney, 2008). Lake and lagoon bottoms can act as basins, as can the abyssal plain of the ocean floor. It should be no surprise, then, that most fossils are preserved in the sediments of floodplains and river deltas, of flash floods and collapsed desert dunes, of backwater lagoons and old ocean floor, of basins and not mountain peaks (Parker, 1991; Lambert, 1998; Blakey and Ranney, 2008).

Now that we are buried, our next challenge is to survive being part of the lithosphere (literally, the rock world). Even if we successfully escape from the biosphere, where we are buried in the lithosphere can be helpful or detrimental. If we are buried too deeply, the rock pressures may crush us into oblivion. Heat also increases with depth, so that if we are unfortunate enough to be buried in sediments being drawn beneath a crustal plate, we are doomed to warp and melt.

Even if we avoid the cruel fates of being crushed or melted, we still have to be lucky enough to be mineralized. For most skeletons, the soft tissues that give bones their resilience decompose fairly quickly after death. One such tissue is collagen – its ropelike fibers give strength and flexibility to bones. When collagen decays, the mineral portion of a bone, the calcium and phosphorus, becomes brittle, increasing the possibility that the bone will not survive long in the rock record (Chinsamy-Turan, 2005).

Many bones are preserved as fossils through a process of permineral-ization (Fig. 2.5) (Chinsamy-Turan, 2005). As we will see later, bones are not solid objects but porous materials. In life, blood vessels and collagen fibers course through bones, and many limb bones have a hollow center where marrow can be stored (Martin, Burr, and Sharkey, 1998; Carter and Beaupré, 2001). After death and the decay of these tissues, there are plenty of spaces to be filled in or patched up by other elements.

Sediments, like biological tissues, decompose into simpler elements (Boggs, 2011), and it is this process that frees certain minerals to be trans-ported and redeposited inside skeletons. Many of the rocks in the earth's crust, for example, are dominated by the mineral silica, the same mineral that makes up quartz, comprises glass, and is the main constituent of beach sand (Boggs, 2011). As the sediments that fossils are buried within decay over time, the free silica is leached out by ground water. When the ground water comes into contact with the porous skeletal elements buried in the sediments, the silica has somewhere to latch onto and fills

2.5. The process of fossilization illustrated. In this figure, three major types of fossilization are represented. Mineralization is illustrated in the shaft of a limb bone in cross-section. In mineralization, some of the original bone remains, but its spaces are infilled with minerals from the surrounding rock matrix. In the process of recrystallization, eventually all of the original bone cells (represented as bricks) are replaced with new materials that take up identical dimensions. Finally, natural molds and casts can form when the original bone is destroyed or dissolved but leaves a moldlike space in the surrounding sediments. If this natural mold is filled in with new minerals, a natural cast of the bone can form.

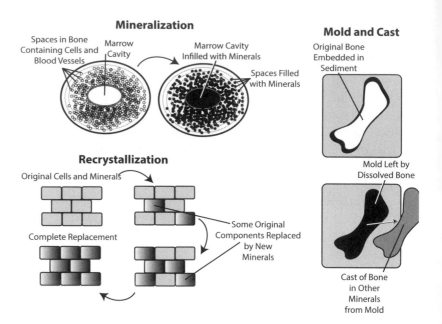

the skeletal spaces. Over time, the skeleton may become completely infilled with silica and other minerals (Chinsamy-Turan, 2005).

In other cases, the original skeletal material may be completely replaced by another mineral, a process called recrystallization (Fig. 2.5). During this process, materials at the cellular level are replaced bit by bit by other minerals (Chinsamy-Turan, 2005). An example of this is the replacement of the calcium phosphate material of the original bone with silica crystals. Eventually, we are left with what amounts to an exact replica of the original bone replaced by another mineral. One way to think about this is to imagine a house originally built of clay bricks. Now imagine that over time, each brick is replaced by an exact replica crafted from plastic. Eventually, the entire house would be composed of a new material, the plastic bricks. However, the dimensions and details of the house would remain exactly as they were when the original clay bricks were present. In the same way, although the minerals and cells of the original bone are gone, a recrystallized bone still retains all the dimensional information of the original bone.

Finally, we can have a case where the original fossil itself is completely destroyed, but it leaves a detailed space or mold in the sediments surrounding it (Fig. 2.5). This mold itself is a type of fossil, and can give at least some details of the original bone. The mold may even become filled with another mineral or materials that form a natural cast, which provides beneficial three-dimensional information about the surface structures of the original fossil as well (Parker, 1991).

If we have survived our trip from the biosphere to the lithosphere, and have become some kind of fossil, we are nearly there to winning the game. However, now we have to be discoverable, and this means we must get back to the surface somehow. Fossils are often brought up to the surface during mountain-building events where large piles of sedimentary rocks are thrust up and begin weathering, exposing their fossilized contents. Fossils may also be exposed during the weathering and erosion of rock layers by down-cutting rivers or even wind (Parker, 1991; Lambert, 1998).

Once you as a fossil are exposed to the elements, you again have the same problems that were equally detrimental just after death. Wind and rain can beat and break your fossil remains down to dust. Ice can invade microcracks in your fossilized skeleton and, through expansion and contraction during freeze-thaw cycles, split you into little pieces. Even the release of the great pressure of rock layers on top of you can result in the formation of expansion cracks. To finally win the fossil game, then, you have to be spotted by a human and collected before you are destroyed by Mother Nature.

In summary, becoming a fossil and being discovered is not an easy task. Most vertebrate animals that have ever lived have been unfortunate losers in the game "Who Wants to Be a Fossil?" The conditions under which a vertebrate dies, how quickly and in what situations it is buried, how it becomes mineralized, and whether it is returned to the surface world and given a narrow window of time to be discovered are all filters that result in our having just a glimpse at the larger picture of vertebrate evolutionary history.

During vertebrate evolution, the continents have not been fixed—they have been moving. That continents are not stable and have traveled across the surface of the earth is a relatively new discovery (McPhee, 2000; Redfern, 2003). In fact, the theory of plate tectonics did not gain wide acceptance among geologists and paleontologists until the 1970s, when overwhelming data from numerous sources showed that continents were mobile (Redfern, 2003). The history and volumes of data that support continental movement go well beyond the scope of this book. Here I very briefly summarize what we know and encourage interested readers to find more detailed accounts (of which there are many) elsewhere.

Continental land masses move because the continents themselves literally float on viscous and molten rock below their surfaces (Redfern, 2003). Inside the earth, the remaining heat from the planet's formation and the decay of radioactive materials collectively generate tremendous heat that melts rock. Like the heated gel inside a 1960s lava lamp, the heated, molten rock rises. By the same token, cooler rock tends to sink. The rising and sinking molten rocks together form a sort of conveyor belt that moves the more rigid continents above them closer together or pulls them farther apart (Redfern, 2003). Where continents are pulled apart, deep sea trenches and oceans may arise. Where continents are pushed together, mountains may form from the collision of one continent against the other. At times, denser oceanic crust can be pulled (subducted) under lighter, continental crust. The molten rock generated from subducted crust can rise to the surface of the denser continent, gushing out as lava

Plate Tectonics in Brief

from volcanoes. If you were to plot the major volcanic eruptions and earthquakes that have occurred in the past 100 years, you would see clearly that these events nearly all happen at continental plate boundaries (Redfern, 2003).

The positions of the continents during the 540-plus-million-year history of the vertebrates have been mapped with a fair degree of accuracy (Blakey, 2010). One way that past continental arrangements have been inferred is through the comparison of rock units and fossil faunas and floras across different parts of the world. For example, during the Late Permian and Early Triassic (~260–240 Ma; Card 95), all the continents had assembled into a single, gigantic continental landmass called Pangea (Blakey and Ranney, 2008; Blakey, 2010). Comparison of Permian and Triassic rock units on all the continents shows remarkably similar makeup and mineralogy, something difficult to reconcile with the continents in their current positions in different environments and climates. Also, a number of small plant and animal fossils are present in the rocks of such far-flung places as South America, Southern Africa, India, Australia, and Antarctica. It is difficult to imagine how small plants and animals could make a trans-Atlantic voyage to reach all of these continents. More importantly, the small Permo-Triassic plants and animals were adapted to a tropical climate, something difficult to resolve with the present-day position of Antarctica.

The major reason for us to briefly consider plate tectonics is that movement of the continents has had a significant effect on vertebrate evolution. As the continents have moved into new positions over the course of the past 540 million years, they have pushed animals into new climates and created geographic barriers in the form of mountains or oceans. Environmental changes and isolation of various populations provides the "fuel" for natural selection, which in turn leads to the evolution of new species. In other cases, the moving continents destroyed old barriers and allowed formerly isolated groups of vertebrates to intermix, which has also spurred on evolution in various ways.

One example of the long-term effect of continental movements on vertebrate evolution involves the dinosaurs. The earliest dinosaurs existed during the Triassic period, the last days of the supercontinent Pangea. Intriguingly, we find very similar dinosaur faunas on all the continents in rocks of this age (Fastovsky and Weishampel, 2012). Beginning in the middle of the age of dinosaurs 180 Ma (Jurassic period; Card 97) and continuing until their demise 66 Ma (Late Cretaceous; Card 99), Pangea began and continued to split and separate into different landmasses. Comparison of dinosaur faunas from Jurassic- and Cretaceous-age rocks in different parts of the world reveals quite different dinosaurs on different continents. As the supercontinent Pangea split into the major continental masses we know today, each continent began to act like a large island, isolating a particular population of dinosaurs there. Over time, due to different environments and selective pressures, different dinosaur species evolved on the continents of North America, South America, Africa,

Eurasia, Australia, and Antarctica (Fastovsky and Weishampel, 2012). This simple example shows the effect of geographic isolation on natural selection and the evolution of vertebrates over time.

As evolution is misunderstood, so is the concept of a hypothesis. It is common to have a hypothesis described as "an educated guess." Yet, a hypothesis is something much more specific in science, more than just a guess. A scientific hypothesis is a generalized statement intended to guide a researcher. More specifically, it has to have the qualities we have already touched on with a theory: it has to be testable, falsifiable, and able to predict certain outcomes (Popper, 1959).

It is in this sense that we talk about reconstructing the vertebrate family tree. A pedigree of relationships is known formally as a phylogeny. A phylogeny is a hypothesis of relationships, based on data, that is open to testing and that predicts the shared common ancestry of various groups (Schuh and Brower, 2009). Any phylogeny is not permanent – it remains supported so long as it is supported by the data. Why is vertebrate phylogeny so important to us? If our goal is to understand the function of the skeletal framework and how it has changed in different vertebrate animal groups, we have to understand the order and appearance of these skeletal features. To return to our automobile analogy, if we wanted to understand the development that led up to the modern sports car, we would need to know about general car mechanics, then about the first sports cars, and so on. We need to understand generalities, and then get more specific to see what has changed (what is special) and what has remained the same to appreciate why the particular car or cars we are interested in function the way they do.

For vertebrates, we typically reach an understanding of changing function by comparing the anatomy of several related groups to see what has changed (what traits are special to our group of interest) and what has remained the same (what traits are present in all vertebrates). This requires an understanding of the relationships of vertebrates. With automobiles, the history and pedigree of particular cars can be traced via access to the machines themselves, through written histories or photographs, and in interviews with surviving designers. We have no such luck with vertebrates – we cannot ask vertebrate animals to recall their ancient history.

We establish vertebrate phylogeny on traits, the attributes or features of particular animals. Simply put, a phylogeny is a hypothesis of relationships established using trait data (Schuh and Brower, 2009). But traits are not self-evident – they are not labeled neatly on fossils or written on the tags of pickled museum specimens. So, how do we select informative traits?

The simplest approach to reconstructing vertebrate phylogeny would be to survey as many animals as possible and then compare the distribution of similar traits. We could then tabulate all the similarities, and group

2.6. Convergent evolution in body form. In this illustration, the shark and dolphin have a streamlined body form with fins. Despite this superficial similarity, dolphins share more trait states in common with other mammals such as cats than they do with sharks. The streamlined form is due not to common ancestry, but to convergence on a form that allows the dolphin and shark to move quickly through the same medium, water.

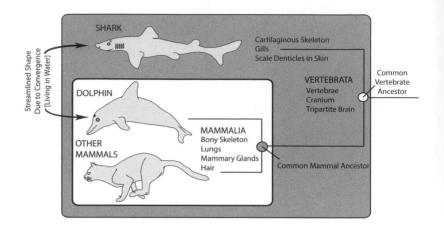

together vertebrates that share more traits in common. For example, we group all known vertebrates together because they all possess specialized back bones called vertebrae. We could continue comparing similar traits, divvying up the vertebrates into smaller subgroups possessing more exclusive shared traits. Fish have gills, amphibians have a moist skin, reptiles are scaly, birds have feathers, and mammals have hair. By doing this, we would eventually have a phylogeny, a hypothesis of relationships among the living vertebrates.

Yet, not all similar-looking traits are related to common ancestry. For example, a shark and a dolphin both have a streamlined body form with fins (Fig. 2.6). At face value, we might conclude that these traits were evidence that sharks and dolphins shared a recent common ancestor. However, on closer inspection, we would begin to notice some large discrepancies. The skeletal structure of the shark is cartilaginous whereas that of the dolphin is bone. A shark's skin is rough and covered in toothlike scales, yet that of a dolphin is smooth and overlies a layer of blubber. Sharks breathe using gills, but dolphins have lungs and must surface occasionally to take in fresh air. Dolphins nurse their young on milk from mammary glands while shark pups must fend for themselves.

Eventually, it would occur to us that, more likely, the similar shapes of the shark and dolphin were due not to common ancestry but instead to a common environment: water. Water is denser than air, and there are only so many "solutions" to swimming fast in it. The shark and dolphin have converged onto a similar functional solution, the streamlining of their bodies and the possession of fins, to move fast in a dense medium. So, overall similarity is not good enough: we must be able to distinguish between traits inherited through common descent and those that are due to convergent evolution (Fig. 2.6) (Schuh and Brower, 2009).

Traits are not static. Under the theory of biological evolution, traits are inherited and modified during descent from a common ancestor. Traits, the key features we are establishing vertebrate relationships on, are plastic and malleable – to wit, they change. Although the plasticity of traits seems at first to be a stumbling block to establishing vertebrate phylogeny, it is in fact a great asset. This is because traits have states – that is, they have an original form and one or more variant forms inherited and passed down to different descendants (Schuh and Brower, 2009).

The original form or state of a trait is referred to as 'primitive,' whereas a trait is considered to be in a 'derived' state if it has changed from this original condition. It is the change from the primitive to the derived trait state that reveals evolution and pedigree (Schuh and Brower, 2009). For example, one type of trait is that of possessing appendages. The appendages of the earliest jawed vertebrates were fins. Therefore, fins are the primitive appendage state, and modern jawed fishes would be said to retain the primitive appendage condition. In contrast, amphibians, reptiles, birds, and mammals have changed the original condition (a fin) into a new structure (a limb with digits). So, in this example, fins are the primitive (original) appendage type, and limbs with digits are the derived (changed from the original) appendage type. All vertebrates possessing the derived appendage state of limbs with digits are grouped together as tetrapods. We would hypothesize that all tetrapods share a closer common ancestry with each other than any of these animals would with fishes (Liem et al., 2001).

The word 'primitive' is often confused with something inferior or less developed. It is important to note that all living vertebrates today actually possess a complex combination of primitive and derived traits. Bony, ray-finned fishes are the most successful vertebrates ever, with over 25,000 known species alive today. Their evolutionary history did not require them to modify the primitive appendage type (a fin) to be successful. However, they are certainly not inferior or less-developed animals – if they were, they would be extinct! To cure you of equating a primitive trait with inferiority, imagine being thrown into a tank with a hungry shark. Yes, the shark has primitive appendages (fins), and you have derived appendages (limbs with digits), but which of you will swim well? Which of you will be the diner, and which the dinner?

But how does one determine whether or not a trait is in its primitive or derived state? The answer is something called polarity. Because trait states are changeable, they have polarity, or a direction of change from primitive to derived. Polarity is an arrow of change pointing in the direction that a particular trait evolved (Schuh and Brower, 2009).

Polarity is inferred in a number of ways. In an approach called outgroup comparison, the trait state of interest is noted over a large number of animals outside the ones being studied (Schuh and Brower, 2009). If one particular state is widely distributed over these so-called outgroups, the researcher may conclude that this is the primitive state. This reasoning would follow from evolutionary theory: all vertebrates should share certain general trait states, followed by different descendant groups that possess more exclusive trait states. Therefore, commonality may indicate that a particular trait is in its primitive state (Schuh and Brower, 2009).

Another approach to determining trait polarity would be to note when certain anatomical features appeared during embryonic development (Schuh and Brower, 2009). The idea here is that more general, foundational trait states should appear earlier in development, whereas more derived states of various traits should develop at a later time. For example, all vertebrate embryos develop throat (pharyngeal) slits during development, but these are retained as spaces for gills only in fishes; the slits anneal and contribute to other structures in tetrapods. Because all vertebrates develop these pharyngeal slits before these features transform into more specialized structures within each group, the retention of these slits in fishes would be considered primitive, and the annealed versions would be called derived.

The modern biologist or paleontologist thus collects trait state data and, using specialized software, analyzes the distribution of primitive and derived trait states. The relationships of the vertebrates are determined by how many derived trait states they share in common, and this is translated into a branching diagram of relationships: the phylogeny (Schuh and Brower, 2009).

The phylogeny is established on trait states that are independent of the chronological order of their appearance in the vertebrate fossil record. This means that the fossil record can serve as a check or test against the phylogeny (Schuh and Brower, 2009). Because the assigned polarities are hypothetical, the chronological sequence of when certain trait states appear in the fossil record can either support or call into question whether something is indeed primitive or derived. For example, the earliest jawed vertebrates in the fossil record possess fins, whereas it is not until much later that we see the first fossil vertebrates possessing limbs with digits. This observation independently bolsters the assignment of fins to the primitive appendage state. Through careful analysis of both traits and the fossil record, a reasonably consistent phylogeny of vertebrate animals has emerged. This phylogeny will be elaborated on throughout the book.

Functional Morphology

Functional morphology is the study of how anatomical shape and operation are correlated (Liem et al., 2001; De Iuliis and Pulerà, 2011; Kardong, 2012). Just as we would study the size, tooth-count, and shape of a gear to predict its effect on turning a car's axle, so too can we study the size, shape, and orientation of limb bones to predict their contribution to movement in a vertebrate animal. Inferring function in fossil vertebrates involves two major approaches. The first is form-function analysis, where the shape or form of the skeleton or a part of it is used to infer its function. Another approach is that of biomechanical analysis, where the principles of physics and engineering are applied to understanding skeletal structure (Radinsky, 1987).

As a simple example of form-function analysis, consider the teeth of a given vertebrate. Vertebrates with sharp, conical teeth are typically flesh-eaters, whereas those with squared-off, blunted teeth are more likely to be omnivorous or strictly vegetarian. These conclusions follow from the form of the teeth—sharp, conical teeth are better for slicing meat than blunted ones. Thus, fossil vertebrates that we find with sharp teeth

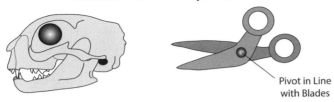

Carnivore—Scissors-Style Jaws

Pivot in Line
with Blades

Herbivore—Nutcracker-Style Jaws

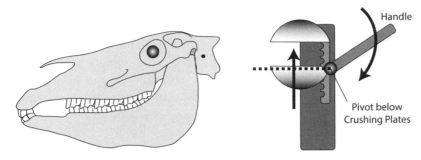

Handle

Pivot below
Crushing Plates

2.7. Simple diagram showing the application of biomechanical knowledge to inferring the functional morphology of vertebrate skulls. Carnivorous mammals, such as a cat, tend to have a jaw joint in line with their sharp, shearing teeth, much as the handles of a pair of scissors align with the blades. This puts the best cutting surface toward the back of the jaws. In contrast, herbivorous mammals such as horses have a jaw joint located above the tooth row, allowing their teeth to simultaneously contact one another like a nutcracker.

are likely to have been carnivorous, whereas those with blunted teeth are more likely to have been omnivorous or herbivorous.

Applying biomechanics to our simple tooth example, we could measure the mechanical advantage of the jaw joint in carnivorous mammals versus those of herbivores. In particular, we could measure the distance between the jaw joint and the tooth row. In carnivorous mammals, slicing is important, and we find that the tooth row is in line with the jaw joint (Liem et al., 2001). This is analogous mechanically to what we see in a pair of scissors—the handles through which you exert force to cut paper are in line with the cutting blades (Fig. 2.7). As you intuitively know, you get greater slicing force near the back of the scissors blades, and cutting proceeds from the back of the blades to the front. Not surprisingly, the sharpest, bladelike teeth of a carnivorous mammal such as a cat are located closest to the jaw joint, and these engage first, followed by more forward teeth (De Iuliis and Pulerà, 2011). In herbivorous mammals, such as a horse, we find that the jaw joint is located far above the tooth row. This is analogous to the situation in a nutcracker where, unlike scissors, the broad crushing surfaces contact the nut all at once, applying their force across the whole nut simultaneously (Fig. 2.7) (Radinsky, 1987; Liem et al., 2001; Fastovsky and Weishampel, 2012). In similar fashion, in a horse all the tooth surfaces come together at once, providing a single, broad crushing surface for vegetable matter.

These approaches work reasonably well on their own to help us establish an initial testable hypothesis on the functional morphology of a fossil vertebrate. However, a more thorough test of our functional hypothesis involves placing these inferences within the context of a vertebrate phylogeny. In such an approach, the relationship of our fossil vertebrate to living vertebrates is determined. Most importantly, we want to know which living vertebrates are the most closely related to our fossil. In a phylogeny, these would be the closest living relatives that fall on either side of our fossil vertebrate. In other words, we want to find the nearest living relatives that surround or bracket our fossil animal. In such an approach, the living vertebrates and their phylogenetic relationships are our control group – they help us delimit and constrain the possible scope and function of a particular anatomical system in our fossil vertebrate.

This approach, known as the extant (living) phylogenetic bracket (EPB), is the comparative method of choice among paleontologists (Witmer, 1995). Returning to our tooth example, we could apply our knowledge of form-function and biomechanics to that of a fossil mammal. Perhaps we have found a new carnivorous mammal, and we find through analysis of its derived trait states that it falls between dogs and cats. In both dogs and cats, the jaw joint is in line with the tooth row. In cats, the back teeth called carnassials are narrow and bladelike, and lack crushing areas (Radinsky, 1987; Pough, Janis, and Heiser, 2002). In contrast, although dogs have carnassials, theirs retain both a blade and a crushing area (Wang and Tedford, 2008). This is why dogs occasionally eat some vegetable matter. In our hypothetical new carnivorous mammal, we find that the jaw joint is in line with the tooth row, and this goes along with what we find in both cats and dogs in terms of the scissors-like action of their jaws. However, we find that the carnassial teeth do retain a small, flattened crushing region in our fossil carnivore. Thus, based on what we see in the living relatives, we would conclude that this carnivore had some ability to consume some vegetable matter along with a diet of meat.

EPB can even be used to infer and constrain possible soft tissues in fossil vertebrates that typically do not preserve (Witmer, 1995). For example, let's say we construct a hypothesis that says dinosaurs had four-chambered, double-pump hearts. Although we have no preserved dinosaur hearts, we could still test this hypothesis using EPB. First, we know through phylogenetic analyses that the closest living relatives of dinosaurs are birds and crocodylians and that both of these living relatives have four-chambered hearts (Fastovsky and Weishampel, 2012). Second, we also know that, physically, a four-chambered, double-pump heart functions to pump low-pressure blood to the lungs and high-pressure blood to the head. In animals as large as the largest dinosaurs, high-pressure blood would need to reach the head, but low-pressure blood would still need to be pumped to the lungs to prevent their capillaries from bursting (Fastovsky and Weishampel, 2012). Therefore, the data from dinosaur relatives and the physical properties of hearts support our hypothesis that dinosaurs probably had four-chambered, double-pump hearts. We could falsify this hypothesis if it could be shown either that dinosaurs do not share a close common ancestry with birds or crocs, and thus may not share their heart shape, or that separation of high- and low-pressure blood could be accomplished another way. So far, the data derived from the EPB approach best support the four-chambered heart hypothesis.

In summary, the skeletal diversity of vertebrates is the result of evolution via the process of natural selection within the space of deep time. Vertebrate evolutionary history resulted from sex and time, coupled with selection and adaptation. We interpret and understand this diversity by studying the sequence of vertebrate fossils and their anatomy. Fossil vertebrate anatomy, in turn, is understood in light of the vertebrate evolutionary tree (the vertebrate phylogeny) and in combination with studies of functional morphology.

Now we can at last turn our attention to the vertebrate story, and learn how the skeleton has changed to adapt these amazing animals to air, land, and sea.

Where Do We Go from Here?

The Origin and Early Evolution of the Vertebrate Chassis

2

You feel it running through your bones.

THE CAESARS, *Jerk It Out*

3.1. Family tree (phylogeny) of deuterostome animals, emphasizing chordates. See text for more specific details. Phylogeny based on Putnam et al. (2008).

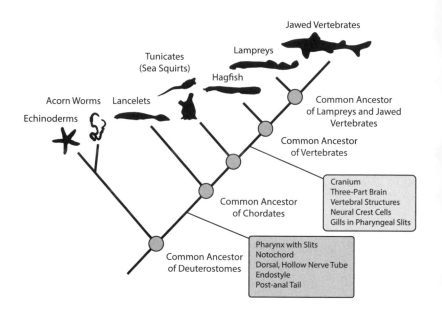

Jawed Vertebrates

Lampreys

Tunicates
(Sea Squirts)

Hagfish

Acorn Worms   Lancelets

Echinoderms

Common Ancestor
of Lampreys and Jawed
Vertebrates

Common Ancestor
of Vertebrates

Cranium
Three-Part Brain
Vertebral Structures
Neural Crest Cells
Gills in Pharyngeal Slits

Common Ancestor
of Chordates

Common Ancestor
of Deuterostomes

Pharynx with Slits
Notochord
Dorsal, Hollow Nerve Tube
Endostyle
Post-anal Tail

THE THEORY OF BIOLOGICAL EVOLUTION PREDICTS THAT ALL VERtebrate animals alive today and those contained in the fossil record are descendants of a single common ancestor. This implies that across all vertebrate animals we should see a deeper pattern, a hidden chassis from which all other derived traits have been built or modified.

We could simply jump to the fossil record to look for the ancestral body plan, but we need a search image to know that whatever fossil we find is indeed an early vertebrate. To understand what truly constitutes the original vertebrate blueprint, we must first turn to the living relatives of vertebrates and the relationships of vertebrates to other animals. After establishing both the undergirding and the overall blueprint of the basic vertebrate chassis, we can then turn to the fossil record to test our hypothetical model of the ancestral vertebrate.

Before we go further, it is time to introduce a few directional terms to our vocabulary. Directional terms are very useful in that they always mean the same thing no matter the orientation of the animal we are talking about, and we avoid convoluted ways of saying back-to-front or head-to-tail. The head end of a vertebrate is considered to be 'cranial,' whereas the tail end is known as 'caudal.' For example, the vertebral column runs from the back of the skull to the end of the tail. We would therefore say that the vertebral column of vertebrates runs craniocaudally. The back and belly sides of a vertebrate are also given directional names. The back is known as 'dorsal.' One way to remember this is that the triangular fin of a shark that projects above its back is called the dorsal fin. The belly side of a vertebrate is referred to as 'ventral.' As an example of how this term is used, in vertebrates like stingrays that are flattened back-to-belly, we would say these animals are dorsoventrally compressed. For the remainder of the book, we will use these four directional terms to simplify and clarify our descriptions of the vertebrate chassis and its movements.

Comparisons of embryonic, genetic, and anatomical trait states place vertebrates among an intriguing group of animals known as the deuterostomes ('second mouths') (Liem et al., 2001; Kardong, 2012) (Fig. 3.1). In these animals, the digestive tract literally develops from the bottom up, with the anus appearing first and the mouth appearing much later in development. This is different from the development of many other animals, such as insects, in which the mouth develops first. Among the deuterostomes are the echinoderms, which include the familiar sea stars,

The Deuterostome-Chordate Undergirding

3.2. The tunicate, or sea squirt, in its mobile larval stage (above) and after metamorphosis into its immobile adult stage (below). Note that the larval sea squirt shares many features in common with other chordates, including vertebrates. Figures based on Liem et al. (2001).

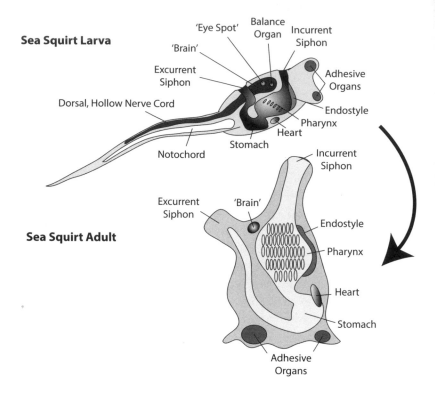

Sea Squirt Larva

'Eye Spot'
Balance Organ
Incurrent Siphon
'Brain'
Excurrent Siphon
Adhesive Organs
Dorsal, Hollow Nerve Cord
Endostyle
Pharynx
Heart
Stomach
Notochord

Sea Squirt Adult

Incurrent Siphon
Excurrent Siphon
'Brain'
Endostyle
Pharynx
Heart
Stomach
Adhesive Organs

sea urchins, and sea lilies, and the hemichordates, strange worm-like creatures that live in burrows on the ocean floor (Fig. 3.1).

As astounding as it seems, we as vertebrates share a closer common ancestor with a sea star than with an insect. Despite our radically different body plans and lifestyles, our underlying development is remarkably similar. It is highly unlikely that such a fundamental developmental trait as the order and pattern in which the digestive system forms would evolve multiple, independent times in such diverse animals as echinoderms and vertebrates. A simpler explanation for this underlying pattern of development is that it evolved once and was inherited from a common ancestor shared by echinoderms, vertebrates, and all other deuterostome animals. This inference is supported by the presence of shared developmental genes and other anatomical features found only among deuterostomes.

More exclusively, vertebrates collectively share a number of derived trait states with other living deuterostome animals called chordates. These chordate animals are an odd bunch. On the one hand, we have curious bag-like animals called sea squirts (or tunicates), immobile, brainless filter feeders that spend most of their adult lives attached to various underwater surfaces (Fig. 3.2). Yet, as larvae these animals are radically different – they have a neural enlargement in their head reminiscent of a brain and a tadpole-like body that they wriggle vigorously to move about until they find a permanent place to settle down and transform into their sponge-like adult form (Fig. 3.2) (Liem et al., 2001).

On the other hand, there are the small, eel-like lancelets, most no longer than the last segment of your pinky finger (Fig. 3.3 and Plate 2).

Lancelet—External View

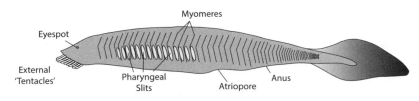

Lancelet—See-through View

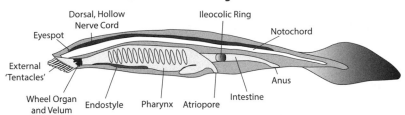

3.3. The lancelet shown in external and see-through views. This chordate spends most of its time in a sandy burrow with its head just poking out to filter particles and zooplankton passing over its tentacles. A wheel organ and velum help to capture and shunt food toward mucus in the endostyle, which is then passed through the pharynx to the esophagus and intestine. A small, muscular ileocolic ring is present to control the passage of food to the guts. An atriopore drains water sieved out through the pharynx. Figures based on Hildebrand and Goslow (2001) and Liem et al. (2001).

These animals are shaped like a lance (hence their names), being broad in their middles and tapering to pin-like points at head and tail. In fact, it is not too off base to call lancelets pin-headed. Where a brain would be located in us, they possess only a small nerve cluster with an eyespot, and light detecting spots line their flanks. Despite their eel-like bodies, they spend most of their time tucked away in sandy burrows in shallow marine waters, waving tentacles on the ends of their tiny heads to capture and filter feed on various particles and microscopic animals and plants (Pough, Janis, and Heiser, 2002).

What do vertebrates have in common with sessile sea bags and wriggly lancelets? It turns out, quite a lot, actually. During development tunicates, lancelets, and vertebrates generate five structures in a combination seen nowhere else. Some of these structures do appear in deuterostomes generally and also in some hemichordates, but not all five together simultaneously. These five structures together form the basic chassis of the chordate animals, and were the undergirding upon which the vertebrate blueprint was established (Fig. 3.1).

The pharynx is a specialized organ for chordates (Liem et al., 2001; Kardong, 2012). In an adult human, the pharynx is a tiny region of the throat, squeezed between the mouth and windpipe (trachea). However, as embryos, humans all sport a large pharynx region, so large in fact that initially it is one of the largest organs of the tiny developing body. In tunicates, lancelets, and embryonic vertebrates, the pharynx is a large, barrel-like organ pierced by vertical slits.

In both tunicates and lancelets, the pharynx is not a respiratory organ (Pough, Janis, and Heiser, 2002). This is because the tiny bodies and thin skins of these chordates automatically allow gases to be absorbed into or expelled out of the bloodstream. These chordates are literally skin-breathers. Were you to subject a tunicate or lancelet to high-powered microscopic scrutiny, you would quickly realize that there are no gills inside their vertical pharyngeal slits. Instead, the slits are part of the filter-feeding apparatus of these little animals. When a tunicate or lancelet sucks food particles into its mouth, they come along on a stream of water (Liem et al., 2001). Just as you would drain water from your spaghetti noodles in a colander prior to eating dinner, so too the pharyngeal slits allow excess water to drain from the tunicate or lancelet pharynx, concentrating the food bits. The pharyngeal slits are lined with hairlike cilia that beat rhythmically to sweep excess water out of the pharynx (Pough, Janis, and Heiser, 2002).

Keeping the food bits on the right track, so to speak, involves another chordate trait we as vertebrates develop as embryos. A groove-like organ called the endostyle lies on the floor of the pharynx (Liem et al., 2001). As water drains through the pharyngeal slits of tunicates or lancelets, the settling food becomes trapped in the endostyle, which is coated liberally with sticky mucus. Special cilia push this string of food-laden mucus into the gut. (It should be noted that the tentacles surrounding the mouths of lancelets are also mucus covered. Snot, it turns out, has been a chordate's best friend from the beginning.) The endostyle also does something very interesting—it secretes proteins linked to iodine that regulate, among other things, growth and reproductive behavior (Liem et al., 2001). The endostyle of embryonic vertebrates is eventually transformed during development into the more familiar thyroid gland, which among other things regulates our growth, metabolism, and the maturation of our testes or ovaries (Gilbert and Raunio, 1997; Gilbert, 2010).

Little chordates and vertebrate embryos are softies with no bones, and they would be very easy to squish or permanently kink were it not for the next unique chordate trait, the notochord. The notochord develops dorsally and functions as a long, incompressible rod (Gilbert, 2010). In automobile terms, the notochord is the frame of the chassis providing support and attachment points for different components. Close inspection of the notochord reveals that it is composed of chambers filled with a watery, gel-like substance that allow it to resist compression. It is the 'backbone' of a chordate or vertebrate embryo in that it supports and straightens the body, and prevents the head from being smashed accordion-style into the tail. Unlike a telescope, a chordate or vertebrate embryo (or car for that matter) does not benefit from head-to-tail collapse. We should note that although adult sea squirts have lost their notochord, it is prominent in their free-swimming larvae.

The notochord is also excellent at resisting side-to-side bending of the body trunk, much like a car frame. Before we continue, we are at good juncture to introduce another directional term: 'lateral.' Lateral refers to the side of the body, and so side-to-side bending is more simply defined as lateral bending or lateral undulation. Most deuterostomes possess segmented blocks of muscle (myomeres) that contract in a lateral sequence so that the right and left halves of the body bend rhythmically in opposite ways (Liem et al., 2001; Gilbert, 2010). Myomeres are powerful in chordate animals, and this creates an undulating or eel-like body movement that pushes against the water, propelling chordates forward. Even in vertebrate embryos, these myomeres generate powerful contractions that bend and flex the trunk sideways, and they are modified into most of the major body muscles later in development.

Myomere contractions would overflex the chordate body, causing it to flop against itself, were it not for the resistance and springlike nature of the notochord. Unlike in a car frame, some lateral bending in the notochord is a good thing. As myomeres on one side contract, they pull on and bend the notochord toward them, in much the same way the pull on a bowstring will bend a bow toward an archer. When the muscles on the pulling side relax, the tension in the notochord is released and it springs back to its straightened shape, imparting its stored elastic energy to propulsion (Liem et al., 2001). This same physical interaction is what causes an arrow to fly when released by the archer: the stored elastic energy in the bow is imparted through the string to the arrow shaft (Vogel, 2003).

The nervous system of chordates lies above the notochord and is coupled with this organ in development. The notochord acts as a powerful signaling center that triggers gene expression related to the pattern of the nervous system, which becomes a dorsal, hollow, fluid-filled tube (Wilt and Hake, 2003; Gilbert, 2010). The nervous system of developing chordates begins as a plate of specialized cells that change their shape and bend like wave crests toward one another on receiving various signals from the notochord. The two 'neural crests,' as they're called, anneal together at the midline, and a hollow space is left inside the newly formed nerve cord. This space still persists today in adult vertebrates, including humans, and can be observed without a microscope when a spinal cord is sectioned during an autopsy (Hildebrand and Goslow, 1995).

The nervous system is somewhat akin to the electrical system of an automobile. In fact, the nervous system is chemoelectric, sending its messages through chemical interactions between nerves and muscles, and along nerves through electrical impulses. In a car, wires carry electrical signals sent by the driver to automate power windows, power steering, and the all-important audio player. In many modern cars, a small computer is on board that monitors the engine's performance, and also gets feedback from other portions of the car. In general, though, the electrical system is one-way, sending activation signals initiated by the driver or on-board computer to the windows and windshield wipers, to the steering and stereo.

Much more feedback is required for functionality in a chordate animal, and the nervous system cannot simply be a one-way street of signals to actuators (muscles). In chordates, major trunks of nerves, correlated with the segmented myomeres, flow out from the main, hollow spinal cord. Within each pair of outgoing nerve trunks are two paths, one that takes activator signals out to muscles and glands and one that takes sensations such as pressure, temperature, and pain back to the spinal cord. The outgoing nerve trunks activate the myomeres in regular patterns on alternating sides, allowing for the relatively smooth eel-like movements discussed earlier (Radinsky, 1987; Liem et al., 2001).

3.4. Diagram of major nerve pathways between the brain and spinal cord, and spinal cord and body (in this case, a fin). Sensations from the sensory nerves in the fin are sent to the spinal cord, which then relays some of them to the brain for further processing. However, some sensory signals are automatically looped through a reflex pathway to an outgoing series of motor nerves, which in turn are capable of activating muscles in response to the sensed stimulus. Sensory signals to the brain are processed, and outgoing motor commands return to the spinal cord and are passed to the fin as well. Aspects of spinal cord outflows modified from Liem et al. (2001).

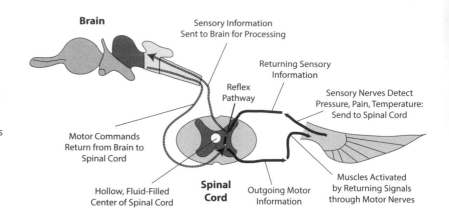

Vertebrate Appendage (Fin)

The spinal cord itself relays signals to and from the brain (if it is there), and also handles and reroutes numerous signals on its own (Fig. 3.4). The ingoing and outgoing nerve tracks are not insulated entirely from one another, and special nerve tracks within the spinal cord shunt messages from the incoming to the outgoing tracks or from one side of the body to another (Hildebrand and Goslow, 1995; Liem et al., 2001). All children quickly learn that touching hot items falls under the category of 'don't try this again.' Even as adults we have all had the misfortune of touching or brushing against a hot plate or stove. The reaction we have is unconscious and seemingly instantaneous – we jerk our hand or body away from the source of the hot pain. You may have realized at moments like this that the reaction generally occurs before you register the pain. This is because of your spinal cord's reflex connections. The incoming tracks are sensory and carry the pain signals to the spinal cord, where part are sent to the brain for processing. However, special interneuron tracks cross the pain signal directly over to the outgoing or motor tracks that activate muscles. Because the pathway from the sensory to motor nerves through the spinal cord is shorter than the pathway of sending pain signals to the brain and back, you feel the pain just fractions of a second after you remove your hand from the source (Liem et al., 2001). Such a basic system is present in all embryonic or larval chordates, but is secondarily lost in the adults of sea squirts – these animals retain only a simple nerve net embedded in their skins after metamorphosis (Liem et al., 2001).

The fifth and final trait of chordates may not seem remarkable but it nonetheless is rare among other animals: the possession of a post-anal tail. The tails of chordates extend beyond the anus, and thus beyond the end of the digestive system. In many animals, the anus terminates at the end of the tail. The tail also extends to some degree beyond the anus in the weird hemichordate animals discussed previously, but this feature by itself may be another holdover from the common ancestor of all deuterostomes (Liem et al., 2001). In chordates, the post-anal tail is typically muscular and enhances the laterally undulating locomotion enabled by the notochord and myomeres (Liem et al., 2001). In fact, the post-anal tail commonly possesses a tail fin that provides an expanded area for pushing against the water. Again, in adult sea squirts, the post-anal tail disappears during their metamorphosis.

In summary, the five chordate traits are: (1) a pharynx with vertical slits; (2) an endostyle organ; (3) a notochord; (4) a dorsal, hollow nerve tube; and (5) a post-anal tail. The presence of these traits in all chordates suggests that their common ancestor (and thus that of vertebrates) was a small, wormlike, filter-feeding animal. It probably lacked gills and respired through a thin skin, and was probably capable of at least some lateral wriggling or undulations from time to time. Even if a brain was present, it probably was not very enlarged. The chordate ancestor was likely supported by a rigid notochord, possessed a muscular post-anal tail, and was controlled by an integrated series of incoming and outgoing tracks to and from a hollow spinal cord. It is upon this body plan that vertebrates have built their chassis.

The Vertebrate Chassis

Modern vertebrate animals are very diverse in their body forms and habits, and so it takes some doing to get underneath hundreds of millions of years of evolutionary history to see the core chassis on which they are all built. Paring down the major trait states of the vertebrates has involved painstaking studies of comparative anatomy, vertebrate paleontology, and embryology, but a consensus of the basic features has emerged. Here, we look to what the anatomy of the two most primitive living vertebrates, as well as information gleaned from vertebrate embryology, has to tell us about the ancestral vertebrate chassis.

Hagfish and lampreys, the most primitive living vertebrates, are the sort of animals that do little to inspire affection (Fig. 3.5 and Plate 2). Both are jawless, both are alien-like and eel-shaped, and both have poor table manners. Hagfish possess a multitude of snotty slime glands (for which they are sometimes called slime hags). When threatened, they exude a specialized mucus that swells exponentially when it comes in contact with sea water (Jorgensen et al., 1998; Lim et al., 2006). In fact, a single hagfish can turn a bucket of sea water into a jellylike soup in minutes (Jorgensen et al., 1998; Pough, Janis, and Heiser, 2002)! The slime coat is a predator-deterrent (it can clog gills) (Lim et al., 2006), and when attacked, the hagfish will generate a great gob of the stuff and then literally twist its body into a knot through which it pulls itself free of the mucus (Jorgensen et al., 1998; Pough, Janis, and Heiser, 2002). Hagfish make a living feasting upon the putrid carcasses of fish and whales that sink to the bottom of the ocean. Lampreys are parasites on other fishes. Their jawless mouths are sucker-shaped with toothlike projections that allow them to adhere to the sides of their prey. A piston-like tongue is used to scour through the scales and skin of the hapless fish to which they have attached, and the victim's blood is drained until the lamprey has had its

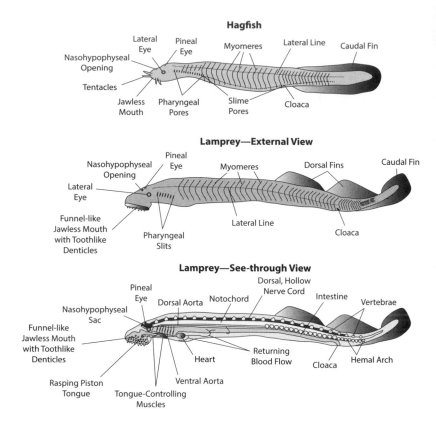

3.5. Hagfish and lamprey illustrated. See text for more details. Figures based on illustrations in Liem et al. (2001).

fill, resulting in death for its host (Liem et al., 2001; Pough, Janis, and Heiser, 2002).

Hagfish and lampreys are much larger animals than their nearest chordate relatives, with an average length of 1 to 3 feet (approximately 0.5–1 meters). Their body size reflects a broader trend among vertebrates, and larger body size has driven vertebrate evolution. It is often said in zoological circles that body size is one of the most important characteristics of an animal because how large or small you are affects so many aspects of your diet, your movement, and your metabolism.

If you are an American citizen, then you are familiar with the tradition of Thanksgiving wherein a large bird, usually a turkey, is the centerpiece of dinner. If you simply cook the bird for the recommended time without any dressing or marinade, you generally end up with a bland and dry meal, which your guests then liberally coat with gravy. I have 'discovered' (by which I mean stolen from other chefs) that the secret to a juicy but well-cooked bird is to immerse it for several days prior to cooking in a brine of various ingredients, such as cranberry juice. Common sense dictates that if you want cranberry flavors to infuse the muscle fibers of the bird, the bigger the bird, the more time it must soak. It comes as no surprise that smaller birds brine faster, whereas large birds must soak much longer . . . but why?

Diffusion of the tasty flavors into your turkey is affected by two factors—the surface area of the bird and its volume. Surface area is simply

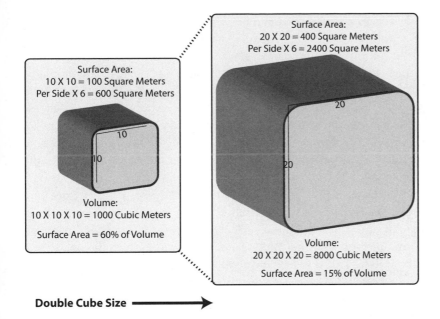

3.6. The scaling of surface area and volume illustrated. Here, a cube that is 10 meters in all dimensions is doubled in size. Note that as the size of the cube increases, its volume increases faster than its surface area. This results in the larger cube having relatively less surface area compared to its volume than the smaller cube.

the area exposed to the environment of a given object, in this case the skin and exposed muscle surfaces of the turkey. As you may recall from your math classes, increasing the dimensions of an object increases its area (Vogel, 2003). Area increases as the square of length and width, so if you could cut off and spread a turkey skin flat, the larger the bird, the greater the area in two dimensions. So, no surprise, larger turkeys have more skin and muscle surface area than smaller ones. However, the volume or space within the muscles and the body cavity of the turkey are much greater in big birds than in small ones, because volume increases in three dimensions with increasing length (Fig. 3.6) (Vogel, 2003). In other words, the volume of a turkey increases as the cube of its length. This means that larger turkeys have far more volume compared to their surface areas than smaller ones, and so it takes exponentially more time to diffuse your special brine into a truly huge holiday bird.

There are tricks for getting around this surface area to volume problem when cooking. The tricks all involve creating a greater surface area while diminishing the volume of the turkey. I am certainly not suggesting I have special turkey-bending laws of physics. Instead, I manually increase the surface area on the bird in ways familiar to most kitchen chefs. I stab the skin and muscles repeatedly with a large fork and knife, creating new surface areas and entrances for the briny fluids to find. I also ensure that the giblets are removed and that I do not place anything into the turkey (such as stuffing) that would create a greater volume for the brine to penetrate. With the innards removed, the meat of the turkey is now exposed inside and out to the brine, essentially doubling its surface area. Compared to a whole turkey with no poking or removal of giblets, the stabbed and gutted bird more quickly takes up and becomes saturated with the cranberry brine.

What do turkeys and cranberry brine have to do with vertebrates? Because vertebrates are generally larger animals than their fellow chordates, diffusion of respiratory gases into and out of their bodies is no longer feasible. As with the large turkey in our example, the available skin surface area for gas exchange is nowhere near sufficient to soak up enough oxygen. Unlike a turkey dinner, a live vertebrate cannot increase its surface area for respiration by removing its guts and poking itself full of holes. Instead, the solution involves increasing the surface area compared to volume in a particular location of the body where gases can diffuse into and out of the blood and be transported throughout the body.

The slits in the pharynx are an ideal place for such structures, and here, instead of cilia, we find the reddish-orange tissues we call gills in hagfish and lampreys. Gills consist of microscopic loops of capillaries that develop within delicate, folded tissues suspended inside the pharyngeal slits (Liem et al., 2001). The fine blood vessels and ultra-thin tissues of gills present a large surface area for gas exchange, and their low volume ensures the rapid uptake of oxygen and release of waste gases such as carbon dioxide. Given the larger size of jawless fishes, possession of gills is a physical necessity. Thus, the originally water-straining slits of the chordate pharynx have become co-opted for respiration in the most primitive living vertebrates.

In other chordate animals, the pharyngeal slits are held up by collagen, a soft but strong material that gives our skin its elasticity and strengthens our bones. However, these slits are relatively passive structures lined with beating cilia that gently sweep excess water from their pharynx. In contrast, the pharyngeal slits of hagfish and lampreys are suspended from segmented cartilaginous arches to which specialized pharyngeal muscles anchor. This means that the pharynx in these animals is a rigid but flexible and collapsible organ, and its gill-containing pharyngeal slits can be compressed and opened by direct muscular action (Liem et al., 2001; Pough, Janis, and Heiser, 2002). This enhances respiration in hagfish and lampreys by improving water circulation across their gills. This internal cartilaginous basket additionally acts to hold the jawless oral cavity open.

Respiration is also dependent on steady, pressurized blood flow through the gills and transport of the oxygen-saturated blood in the gills back to the body organs. In other chordate animals, blood moves through vessels and body cavities via body motions and the pulsations of large arteries. In adult sea squirts, for example, an enlarged tubular artery, sometimes called a 'heart,' will pump blood in one direction for a few minutes, and then pump blood in the reverse direction (Kardong, 2012). For most chordate animals, no true heart exists because most respiration occurs by diffusion. Because larger vertebrates like hagfish and lampreys require steady, pressurized blood flow, a distinct, muscular heart develops that pumps blood returning from the body on toward the gills. Blood then leaves the gills and flows either to the head or through a large vessel along the notochord called the dorsal aorta, whose branches supply most of the major body organs. In all embryonic vertebrates and throughout

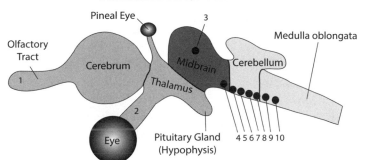

Generalized Vertebrate Brain

3.7. Diagram of a generalized vertebrate brain, based on the brains of lampreys, bony fish, and sharks. Vertebrate brain schematic based on illustrations in Hildebrand and Goslow (2001) and Liem et al. (2001).

Legend
Olfactory Tract = Conveys Odor Information to Brain
Cerebrum = Thinking, Memory, Emotional Integration
Pineal Eye = Light Detection and Regulation of Day-Night Cycles
Thalamus = Emotional Integration, Regulation of Metabolism
Pituitary Gland (Hypophysis) = Regulation of Metabolism
Midbrain = Visual and Auditory Interpretation; Equilibrium
Cerebellum = Motor Coordination Center
Medulla Oblongata = Basic Life Functions (Heart Rate, etc.)

Numbers = Cranial Nerves
1 - Odor Detection
2 - Vision
3, 4, 6 - Control of Eye Muscles
5 - Jaw Muscles and Face Sensation
7 - Taste, Pharynx Muscles
8 - Auditory Sensation and Equilibrium
9 - Taste, Pharynx Muscles
10 - Taste, Body Organ Monitoring

life in fishes, this one-way muscular heart ensures the proper circulation of blood and gases throughout the body (Gilbert and Raunio, 1997; Gilbert, 2010). As we will find later, this pattern changes in air-breathing vertebrates.

If you were to watch any given vertebrate embryo develop, you would notice that, unlike other chordates, the head is enlarged and distinct. Vertebrates really use their heads, and the development of an enlarged cranium is correlated with an expansion of the head-end of the neural tube into a brain. All vertebrate brains are tripartite; that is, they are arranged in three segments craniocaudally: forebrain, midbrain, and hindbrain (Fig. 3.7) (Liem et al., 2001; Gilbert, 2010). The forebrain contains the cerebrum and diencephalon. The cerebrum is the seat of somatosensory integration: it does the thinking, decision making, and emotional integration for vertebrates. Sprouting from the cerebrum are odor detectors (olfactory tracts) that lead to the nose. The diencephalon, or caudal forebrain, is the origin of the optic tracts for the eyes and a median light-detecting eye called the parietal, or pineal, eye. The diencephalon also contains the thalamus, a brain region that conveys information from the mid- and hindbrain to the cerebrum so that the vertebrate animal is 'aware' of sensations and information gathered from the body. The thalamus also regulates hunger, thirst, sex, and metabolism. The midbrain segment integrates and interprets auditory, balance, and visual signals. The hindbrain segment contains two major regions known as the cerebellum and the medulla oblongata. The cerebellum coordinates muscle movements, and the medulla oblongata regulates basic life functions such as heart beat and respiration. Radiating from the brain are specialized tracts of cranial nerves that supply sensation and motor control to the head, to the pharynx, and even to many internal organs (Liem et al., 2001).

Our noses and those of most vertebrates have a pair of openings called 'nostrils.' In hagfish and lampreys, there is but a single, central nostril connected to a dead-end sac that contains thousands of sensitive olfactory nerve endings. Odors are detected when they are carried into the single nostril by water. Unlike the situation in ourselves, the nostril and olfactory sac have no connection with respiration – a hagfish or lamprey cannot channel the water drawn in to detect odors on to its gills. Water is simply pumped into and out of the nostril and olfactory sac, and the nose of these primitive vertebrates is solely for odor detection (Liem et al., 2001; Pough, Janis, and Heiser, 2002). We will see a somewhat similar but more complex situation in cartilaginous and bony fishes later.

A median, light-detecting eye, often called the pineal organ, is present in all vertebrates in some form or another (Lutterschmidt, Lutterschmidt, and Hutchison, 2003; Falcón et al., 2009). In hagfish and lampreys, it protrudes through a dorsal opening in their braincase, where it lies beneath a patch of translucent skin. The pineal organ detects light and dark cycles, and uses these cycles to regulate the sleep and waking cycles of hagfish and lampreys. Technically, the pineal organ commonly develops as an asymmetrical set of two structures. One is a hormonal portion that uses light signals to regulate body cycles (Lutterschmidt, Lutterschmidt, and Hutchison, 2003; Falcón et al., 2009). The other portion, often called the parietal eye (because it pokes through a parietal bone in various vertebrates), is the actual light-detecting portion of the pineal organ (Falcón et al., 2009). However, the parietal eye is not an image-forming eye.

In the water, the body of an animal is approximately the same density as water, so sounds pass directly through the body with little to no disruption (Liem et al., 2001). We would say vertebrate animals are transparent to water-borne sounds. The inner ear, which develops in all vertebrates from the midbrain, is filled with cilia-like nerve 'hairs' surrounded by a viscous gel. In water, sound vibrations pass directly through the cranium and into the midbrain, disturbing the gel and causing nerve ends to vibrate. These signals are then passed on to the midbrain, where they are interpreted as sound (Liem et al., 2001). However, in the water, the head, and usually the entire body, can act as a sounding board, and horizontal tubes on the surface of the skin, called lateral line canals, are present that contain gel and nerve 'hairs' sensitive to vibrations. The nerve endings in lateral line canals detect the direction and origin of sounds and other vibrations in three-dimensional space (Liem et al., 2001; Pough, Janis, and Heiser, 2002). Differences in the timing of when lateral line nerves on one side of the body first detect vibrations compared to the opposite side tell the vertebrate whether the pressure waves are coming from the right or left. Special lateral line nerves follow several tracts back to the midbrain, to an area called the otic capsule, where these signals are interpreted and acted upon. The lateral line system, though lost in us, is prominent in most fishes, and assists in the detection of prey movements and schooling.

In hagfish, the lateral line system is rudimentary and present solely on the head (Liem et al., 2001).

The inner ear region also possesses hairlike nerve endings that are attuned to movements of the head in three-dimensional space. These nerve endings are commonly capped by what are called otoliths (literally, ear-stones). The otoliths of most vertebrates are composed of calcium carbonate, the same material that makes up clam shells and many antacids. A weird exception to this is found in hagfish, where the otoliths are composed of the mineral amalgamation called apatite, normally found in bones (we will discuss more about apatite in later chapters) (Liem et al., 2001). When the head turns, gravity pulls on the otoliths, which slide and bend the sensitive inner ear nerve endings. These signals are used to interpret the position of the head (Liem et al., 2001).

Additionally, rotational movements of the head are detected by collections of sensitive nerve endings that cap the ends of U-shaped inner ear tubes called semicircular canals. As the head rotates, the fluid inside the semicircular canals is forced against the opposite end of these tubes, much as you are pressed back against your seat when a car suddenly lurches forward. The fluid pressure on the nerve endings indicates to the brain that the vertebrate animal is turning in a particular direction (Liem et al., 2001). All jawed vertebrates have three semicircular canals to detect rotational movements in three dimensions. Hagfish possess only a single semicircular canal, and lampreys have two (Liem et al., 2001; Pough, Janis, and Heiser, 2002; Kardong, 2012). This lack of three semicircular canals is probably related to the lifestyles of these primitive jawless fishes, which don't participate in a lot of active swimming or pursuit of fast-moving prey.

Protection of the specialized brain is accomplished in hagfish and lampreys with the development of a cartilaginous braincase that both shields the brain and provides exits and entrances for cranial nerves, lateral line nerves, and the spinal cord. The braincase is commonly trough-shaped in developing vertebrate embryos, with the brain nestled inside, and covered by a cartilaginous canopy (Liem et al., 2001). As a consequence of the intimate relationship between the braincase and the brain, fossils of vertebrate crania often yield insights into the basic structure of the brain and the distribution of the cranial nerves.

Much of the enlarged cranium, its associated cranial nerves, and the pharyngeal skeleton develop from special cells in vertebrate embryos called neural crest cells (Wilt and Hake, 2003; Gilbert, 2010). These special cells bud off the crests of the developing neural tube (hence their name) and migrate like little slugs to all corners of a vertebrate's body, contributing to many anatomical structures, such as a great deal of the peripheral nerves that exit the spinal cord, the adrenal glands on the kidneys, and even skin pigment (Wilt and Hake, 2003; Gilbert, 2010). The neural crest cells that migrate into the head region form the cranium, parts of the cranial nerves, and the pharyngeal skeleton during a complex series of interactions with each other and surrounding tissues. The mineral constituents of teeth, enamel and dentine, also develop directly or indirectly from neural crest cell interactions (Gilbert, 2010). It has been hypothesized that the advent of neural crest cells and the subsequent development of a cranium with specialized sense organs gave vertebrates a 'head' start over their chordate cousins (Gans and Northcutt, 1983; Northcutt, 2011).

The larger size of vertebrates such as hagfish and lampreys was accompanied by the development of more complex and powerful myomeres capable of propelling the body through the viscous medium of water. The notochord is still prominent in hagfish and lampreys, but new segmented arches of cartilage form around the notochord and the overlying spinal cord in lampreys. These arches are the vertebrae, the segmented back bones that in most living vertebrates both anchor and control the pull of the trunk muscles and protect the spinal cord from injury. Such structures are absent in the hagfish, which, among its other odd features, has called into question whether these animals are indeed vertebrates. For simplicity, here we will consider them to be a very primitive but strange vertebrate.

With larger bodies and extended periods of swimming, there was also selection for additional body fins in vertebrates. Embryonic and primitive living vertebrates retain a cylindrical body form, somewhat akin to a torpedo. Cylindrical objects pushed through water tend to be unstable, wobbling and spinning aberrantly. In mechanical engineering, a cylindrical object is often stabilized using fins placed in parallel with the item (Bloomfield, 2006). Following these basic physical principles, the appearance of dorsal (back) and ventral (belly) fins in primitive vertebrates such as hagfish and lampreys stabilizes their bodies during bouts of active swimming. Cartilaginous rays have infiltrated the body and tail fins in hagfish and lampreys as well, allowing these structures to hold their shape against the water rushing around them (Liem et al., 2001).

The Inferred Basic Vertebrate Chassis – A Hypothesis Tested

By comparing and contrasting the anatomy of chordates and primitive living vertebrates, we have come away with a general hypothesis of how the chassis of the earliest vertebrates preserved in the fossil record should be constructed (Fig. 3.8). These animals will have a deuterostome-chordate undergirding comprised of a large pharynx with slits, an endostyle, a dorsal notochord surrounded by segmented myomeres and overlain by a hollow nerve tube, and a muscular post-anal tail. The vertebrate chassis should consist of gills suspended within the pharyngeal slits and an internal cartilaginous pharyngeal skeleton. Vertebral arches should cover the spinal cord and perhaps partly surround the notochord. An enlarged head with a cartilaginous braincase should be present, as should some evidence of a three-part specialized brain and cranial nerves. There might be evidence of a lateral line system, and body fins in addition to a tail fin supported by cartilaginous rays should make an appearance.

Hypothetical Ancestral Vertebrate—External View

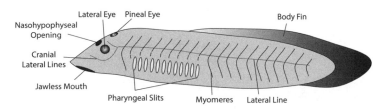

Lateral Eye Pineal Eye
Nasohypophyseal Opening
Cranial Lateral Lines
Jawless Mouth
Pharyngeal Slits
Myomeres Lateral Line
Body Fin

Hypothetical Ancestral Vertebrate—See-through View

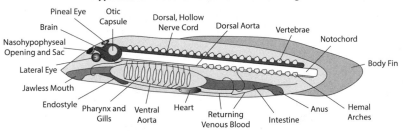

Pineal Eye Otic Capsule
Brain
Dorsal, Hollow Nerve Cord
Dorsal Aorta
Vertebrae
Notochord
Nasohypophyseal Opening and Sac
Lateral Eye
Jawless Mouth
Endostyle
Pharynx and Gills
Ventral Aorta
Heart
Returning Venous Blood
Intestine
Anus
Hemal Arches
Body Fin

3.8. A hypothetical, stylized vertebrate ancestor based on comparisons of outgroup traits with other chordates and the jawless lamprey and hagfish. See text for more details. Modeled after Radinsky (1987).

Armed with our hypothesis of what we should expect to see in the earliest vertebrates, we now turn our attention to what the fossil record has yielded in terms of the earliest vertebrate animals. Until recently, our best evidence and information on the earliest vertebrates were bone fragments from approximately 480 Ma (Card 90), and complete fish fossils dating to about 420 Ma (Card 91). Yet we know from an examination of living chordates and primitive vertebrates that the earliest vertebrate animals must have predated all of these fossils, not least of all because our earliest ancestors would not have had bony skeletons.

Around the world, remarkable new discoveries from rocks nearly 540 Ma (Card 89) have offered up well-preserved soft-bodied animal specimens that were part of an evolutionary event called the Cambrian Explosion. During this time, animals underwent an incredible radiation in body forms, and most of the major animal groups alive today can trace their ancestry back to this time (Foster, 2014). A dizzying legion of invertebrate and deuterostome creatures with otherworldly body forms are known from the Cambrian Explosion (Gould, 1989; Shu et al., 2003; Foster, 2014), but their anatomy and relationships go well beyond the scope of this book. Because the probable prevertebrate deuterostome animals are soft-bodied and somewhat squished in preservation, there has been considerable debate about their relationships to one another and to vertebrates. For reasons of simplicity, we will focus here on the best example we currently have of an early vertebrate from this time, named *Haikouichthys* (Shu et al., 1999, 2003). This little animal is known from hundreds of fossils discovered in southern China, and so the information we have about its body form is robust and sheds much light on what the early vertebrate chassis was like.

Haikouichthys is a small animal, just over an inch (~2.5 cm) long (Fig. 3.9). The body has a tapered oval shape with the head and tail forming the narrow end points. The head is small and tapered, but more

3.9. One of the earliest known vertebrates, *Haikouichthys,* from Cambrian Period rocks (540 Ma) in China. This diagram is based on composites of several different specimens. Although vertebrae are visible in some *Haikouichthys* specimens, their number and precise distribution cannot be known. Here, vertebrae are speculated to cover the dorsal, hollow nerve cord from head to tail. Illustrations based on Shu et al. (1999, 2003).

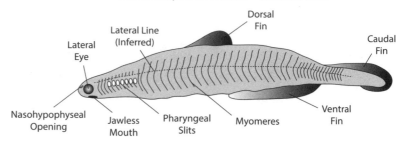

Haikouichthys **Restoration—External View**

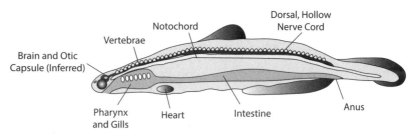

Haikouichthys **Restoration—See-through View**

developed than in chordates like the lancelet. Two large eyes adorn the head, and in some specimens impressions of the olfactory stalks from the brain can be discerned (Shu et al., 1999, 2003). A series of pharyngeal arches with gill pouches lie caudal to the head, and in some specimens even the feathery outlines of the gill tissues themselves can be seen (Shu et al., 1999, 2003). A jawless mouth is present, supported by internal cartilaginous arches as in hagfish and lampreys. Extending much of the length of the body is a notochord that is surrounded by small arch-like vertebrae. Clearly defined, segmented impressions of myomeres are present across the trunk and tail of the body. Dorsal, ventral, and tail fins are present with what appear to be fin rays embedded within them, outlines of the gut tract can be observed, and the tail appears to extend beyond the anus (Shu et al., 1999).

One of the trait states that cannot be confirmed in *Haikouichthys* is the presence of a lateral line system. Of the nearly 500 specimens of this fossil discovered, none shows anything that can definitively be considered a lateral line canal (Shu et al., 1999, 2003). This does not necessarily mean that there was no lateral line system – it may have been there but the decay of the carcass failed to preserve it. It seems highly likely that a lateral line was present, considering all the other vertebrate characteristics that are present in this little animal. In some specimens, there are small structures preserved behind the eyes that may indicate the presence of otic capsules, the portion of the braincase that houses inner ear structures for balance and hearing that develop from the midbrain (Shu et al., 2003). If these are indeed otic capsules, this would be indirect evidence for some sort of lateral line system to detect water-borne vibrations, but the evidence is still somewhat ambiguous.

These incredible fossils support our predictions of what the earliest vertebrate chassis was like. They also show us that the earliest vertebrates were already somewhat predatory. You will recall that in filter-feeding chordates, the pharyngeal slits acted in a sieve-like fashion to pump out excess water and concentrate food particles. The presence of gills in the pharyngeal slits of hagfish and lampreys precludes such a function in these primitive living vertebrates because it would interfere with respiration. Instead, hagfish and lampreys scavenge and parasitize other organisms. The large eyes, well-developed notochord and vertebral arches, and segmented myomeres of *Haikouichthys* suggest this early vertebrate was more active than filter-feeding chordates such as the lancelet. Its jawless mouth could not have crushed or chewed up large prey, but it certainly would have been capable of pursuing and engulfing tiny invertebrates and other zooplankton. Another more recently discovered early vertebrate similar to *Haikouichthys*, *Metaspriggina* (Morris and Caron, 2014), shows more developed pharyngeal arches that may presage jaws, a subject we return to in chapter 5.

In *Haikouichthys*, other less well known early vertebrates from the Cambrian Explosion, and lampreys and hagfish, a well-defined internal skeleton is lacking, and more significantly, there is little to no calcification of the cartilage or the presence of anything like bone. At this point, you may well wonder why other vertebrates possess calcified cartilage or bone when early vertebrates, hagfish, and lampreys do not. You may also wonder if these primitive vertebrates once possessed bone and then lost it. Come to think of it, why don't these animals have teeth and jaws? All of these features, while present in a majority of living vertebrates, were apparently absent in the earliest vertebrates, and would be predicted to appear piecemeal in the various descendants of the vertebrate common ancestor.

We now turn to the fossil record to see when, how, and in what vertebrate groups bones and teeth came into being. More significantly, we want to discover which group of early vertebrates shared the closest common ancestor with the diverse, jawed vertebrate groups of today.

4.1. Diagrams demonstrating compression versus tension (top) and composite materials (bottom). Bone is a composite tissue made of both hard minerals and soft but strong collagen fibers.

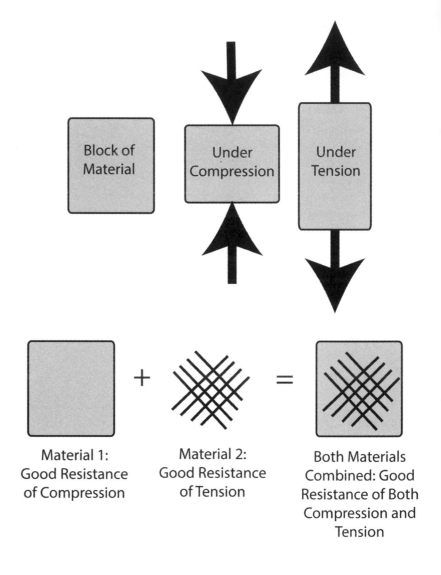

Block of Material

Under Compression

Under Tension

Material 1: Good Resistance of Compression

+

Material 2: Good Resistance of Tension

=

Both Materials Combined: Good Resistance of Both Compression and Tension

Evolution of a Bony Chassis

4

FILLETING ANY FISH INVOLVES SKILL. I AM CERTAINLY NOT THE person to be in charge of fish fillets, unless you like the extra crunch of little bones in your salmon. But this culinary aggravation is also one of the most obvious features of many vertebrate animals – possessing hardened, internal skeletons made of bone. As we have seen, the earliest vertebrates were naked-skinned and lacked an internal, bony skeleton, which is one of the reasons why the preservation of any of these animals is so remarkable.

The major hard parts of a vertebrate animal that typically fossilize are bones and teeth, and these are all composed from derivatives of apatite, a crystalline mineral consisting of interlinked calcium-phosphate (Pough, Janis, and Heiser, 2002). The presence of apatite, however it is modified in bones and teeth, is a unique chemical signature that only vertebrates possess (Liem et al., 2001; Pough, Janis, and Heiser, 2002). Therefore, if we find the presence of apatite in the hard tissues of a fossil animal, this strongly suggests that we have a vertebrate on our hands. Cartilage, although not composed of apatite, may also calcify on occasion (Currey, 2002), and can be preserved in fossils under certain circumstances.

Let us first consider bone (Plate 3) – what is it, and why has it become so important to vertebrates? Bone is a special, mineralized tissue that contains a mixture of minerals (calcium and phosphorus) coupled to soft tissues such as collagen (Carter and Beaupré, 2001; Currey, 2002). Calcium and phosphorus act like concrete in the floor and walls of a building: they are great at resisting compressive stresses. Compressive stresses are squeezing forces that press in opposite directions on either side of an object (Fig. 4.1). An example of compressive stress from everyday life occurs when you sit on a foam cushion on your couch. The cushion experiences compressive stress, and is flattened (compressed) because its top and bottom are squeezed closer together. Concrete is a material that resists this squeezing force well, which is why it is used to hold up heavy, vertical buildings. Likewise, many bones in the arms and legs of land-living vertebrates are vertically oriented for similar physical reasons.

Collagen fibers, on the other hand, act to increase the flexibility of the bone, much as steel rebar embedded in the concrete of buildings reduces the chances that the cement will crack or break by providing extra give (Fig. 4.1) (McGowan, 1999). Collagen fibers give bones a 'springiness' they would otherwise lack. Scurvy is a disease related to collagen fibers best known among the pirates and mariners of old (Brown, 2004).

Of Bone, Cartilage, and Teeth

Bone

4.2. Diagram of the arrangement of bone cells and minerals at the microscopic level. Bone cells deposit calcium-phosphate salts around themselves, and eventually trap themselves in a self-imposed chamber. Blood vessels and special canals close to the bone cells ensure their survival by providing them with oxygen and nutrients and by taking away carbon dioxide and other waste products. Diagram based on figures in Liem et al. (2001) and Chinsamy-Turan (2005).

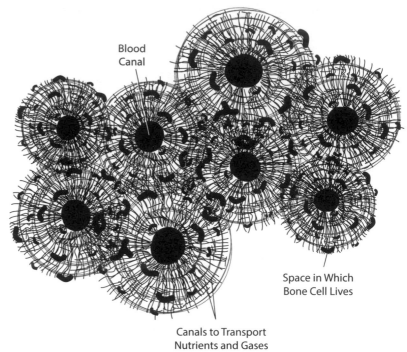

Blood Canal

Space in Which Bone Cell Lives

Canals to Transport Nutrients and Gases to and from Bone Cells

Collagen fiber architecture is constructed in part from vitamin C, and lack of vitamin C results in weaker collagen fibers (Currey, 2002). Weaker collagen fibers, in turn, result in weakened bones—people with scurvy, who usually do not get enough vitamin C in their diets, tend to develop deformed bones and lose their teeth (which are held by ligaments built on collagen fibers) (Brown, 2004). Hence, the British navy enacted regulations that required their servicemen to transport and drink lemon juice and eat limes. This in turn led to a derogatory term for British sailors, 'limeys' (Brown, 2004).

Bone is a living and dynamic tissue—its shape is sculpted both by genetics and by the environment. The genes that encode for bones generate proteins that supply the basic shape, but the nuanced shape of a bone with all of its pockets and fissures, its condyles and tubercles, is generated by the interaction of this tissue with the environment (Carter and Beaupré, 2001). How is this possible?

Bone is made up of living cells that deposit bony matrix around themselves, eventually creating a tiny space in which they live out the rest of their days. A rich blood supply finds its way into living bones, and this allows locked-in bone cells to survive their self-imprisonment by providing them with necessary oxygen and nutrients (Fig. 4.2) (Carter and Beaupré, 2001). When a bone is exposed to stress from body weight or the movements of the body muscles, a remarkable series of events is set into action (Fig. 4.3). When a bone experiences stress, new bone cells are generated within the bone itself—this occurs in our bone marrow and in a flexible outer sheath called the periosteum. These new bone cells migrate

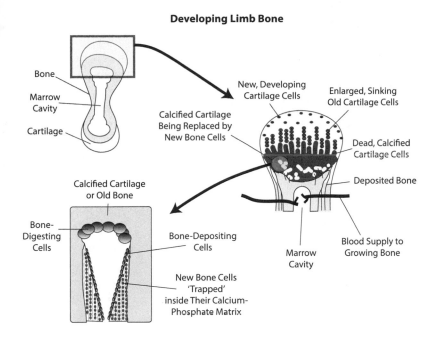

Developing Limb Bone

Bone

Marrow
Cavity

Cartilage

Calcified Cartilage
Being Replaced by
New Bone Cells

New, Developing
Cartilage Cells

Enlarged, Sinking
Old Cartilage Cells

Dead, Calcified
Cartilage Cells

Deposited Bone

Calcified Cartilage
or Old Bone

Bone-
Digesting
Cells

Bone-Depositing
Cells

Blood Supply to
Growing Bone

Marrow
Cavity

New Bone Cells
'Trapped'
inside Their Calcium-
Phosphate Matrix

4.3. Development of bone and cartilage in a growing limb bone. See text for more details. Diagram based on figures in Liem et al. (2001) and Chinsamy-Turan (2005).

to the regions of greatest repetitive stress and begin to lay down calcium and phosphorus salts, entangling collagen fibers with these minerals. Eventually these cells surround themselves with bony matrix and remain as so-called mature bone cells. The mature bone cells, although trapped and immobile, still play a role in bone chemistry–they can absorb and recycle the calcium salts around them. Calcium and phosphorus are minerals important for many life-sustaining functions in vertebrates, as we will see. Other cells, related to the white blood cells that are part of our immune system, digest and break down bone where stress is minimal or absent. In this way, an interaction between bone-building and bone-destroying cells related to the stresses placed upon the bones sculpts the skeleton (Carter and Beaupré, 2001).

Why Is Bone Based on Calcium-Phosphate?

Muscle contractions depend heavily on calcium to jump-start the complex microscopic sliding of their filaments past one another (Liem et al., 2001). It is the movements of these sliding filaments that ultimately cause muscles to contract. Therefore, a reliable source of calcium is essential for the rapid, powerful muscle contractions vertebrate animals rely on to propel themselves through life. In fact, the cells living within the calcium-phosphate matrix of bone can and do absorb and send stored calcium to muscles and other organs requiring this precious material (Liem et al., 2001). It has often been suggested that the origin of bone in vertebrates was related, in part, to this need for a calcium supply (Donoghue and Sansom, 2002).

Vertebrate hard parts differ from those of most other animals, which tend to form their hard parts from calcium carbonate. Snail and clam shells, for example, are a rich source of calcium carbonate, and the carbon is drawn out of the carbon dioxide gases dissolved in the water in which these animals live. Vertebrate bones, in contrast, are composed of calcium-phosphate salts. Why should phosphorous and not carbon be the element associated with calcium in vertebrate skeletons? Sustained muscle contractions of the type used for locomotion in vertebrates generate lactic acid as a by-product (Romer, 1962). As muscles work, they quickly use up readily available oxygen and begin to extract energy through reactions that do not require oxygen. Lactic acid is the end result of these oxygen-poor reactions (Romer, 1962), and is one source of muscle soreness after strenuous workouts (Miles and Clarkson, 1994). Before the lactic acid can be soaked up and recycled, the acidity of the surrounding tissues increases (Donoghue and Sansom, 2002).

Unlike the exoskeleton of a snail or clam, the skeleton of a vertebrate is buried deep within its muscles. Thus, as lactic acid builds up after bouts of exercise, the bones are directly exposed to a corrosive environment. When we suffer from heartburn or other types of acid-induced indigestion, we often turn to antacids to soothe our troubled stomachs. The main ingredient in most over-the-counter antacids is (you guessed it) calcium carbonate, which readily reacts with stomach acids to neutralize them. Imagine a vertebrate skeleton built on calcium carbonate being routinely subjected to pulses of corrosive lactic acid. The skeleton would soon corrode and dissolve—a very bad outcome in animals that rely on the skeleton for movement and protection.

A common buffer in many a chemistry experiment is the addition of phosphorus. Phosphorus is less reactive with acids, and it acts to buffer or protect other minerals, such as calcium, that would otherwise be much more easily or rapidly dissolved. We infer that the phosphorus contained with the apatite of vertebrate skeletons acts as a buffer against the inevitable lactic acid buildup that goes with active muscle metabolism (Ruben and Bennett, 1987).

However, nice as this hypothesis sounds, it is important to note that the earliest vertebrates had an exoskeleton of bones and not an internal bony skeleton. Therefore, lactic acid buildup may not have been the primary driver behind the evolution of a calcium-phosphate skeleton. Perhaps the phosphorous was a serendipitous evolutionary quirk that later allowed for the evolution of an internal, bony skeleton. It should be mentioned that the main energy currency of organisms, ATP (adenosine triphosphate), contains and utilizes phosphorus to propagate most metabolic processes. Because vertebrates are typically energetic animals, a ready supply of phosphorus for the generation of ATP is probably also beneficial. It may be that the phosphorous component was useful first as a source for physiological mechanisms but later became crucial to the evolution of the internal skeleton as we know it.

You will recall that cartilage is a tissue that has been present since the earliest vertebrates made their appearance nearly 540 Ma (Card 89). In many cases, the internal bony skeleton of living vertebrates develops by replacing a cartilaginous framework laid down during development (Carter and Beaupré, 2001; Liem et al., 2001). However, bone completely replaces cartilage when this occurs, and so cartilage is not simply an 'inferior,' rubbery version of bone waiting to be mineralized. Cartilage itself is a superior tissue in its own right. It can be hard, but it is also flexible. Cartilage cells receive their nutrients and exchange gases through diffusion with the surrounding matrix (Carter and Beaupré, 2001; Liem et al., 2001). Therefore, it does not depend, as bone does, on a vast blood supply. In fact, cartilage is an excellent starter tissue for the vertebrate skeleton because it does not require a vast supply of blood during a time when the circulatory system itself is developing (Gilbert, 2010). Cartilage is lighter than bone, and for aquatic animals this is definitely a plus – there is less body mass to fight against when swimming. Finally, cartilage can be calcified or mineralized to a point where it is a very strong tissue – the jaws of sharks, which exert some of the greatest pressures of any animal on earth, are comprised entirely of mineralized cartilage (Liem et al., 2001).

Like bone, cartilage cells surround themselves in a matrix of materials. Unlike bone, the cartilage cells are surrounded by a gel-like amalgamation of long-chain proteins, sugars, and collagen fibers (Carter and Beaupré, 2001). The gel-like matrix of cartilage tissue is what gives it its springiness and flexibility. The matrix inside the cartilage is fluid, much like water inside a water balloon, and shifts and moves within the cartilage tissue when it is placed under stress. This property is why cartilage is so common in joints between vertebrae and limb bones, areas where sudden shocks of pressure must be softened to cushion the blow to the harder bones. Movement of the fluid matrix dissipates the energy of the stress, spreading out its impact across the skeleton rather than concentrating it in a local region (Simon, 1970).

Where bone replaces cartilage tissues, a particular sequence of events usually plays out (Fig. 4.3). First, the cartilage cells enlarge and sink within the gel-like matrix, making room for the addition of new, smaller cartilage cells. The old, enlarged cartilage cells eventually stack up in columns and begin to die, their cells being invaded and replaced by calcium salts such that they become calcified cartilage. In the meantime, new bone cells penetrate into the dead, calcified cartilage cells by hitching rides on invading blood vessels. The bone-destroying cells mentioned earlier are also excellent at digesting calcified cartilage, and this process creates nooks and crannies within the dead cartilage that new bone cells can invade to begin laying down their calcium-phosphorus matrix. In this way, the original internal framework of cartilage is replaced by a bony skeleton (Haines, 1942a; Carter and Beaupré, 2001; Liem et al., 2001).

Cartilage

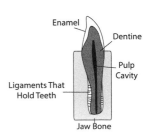

4.4. A section through a mammal tooth in a jawbone to show the basic arrangements of the dentine, enamel, and pulp cavity. Based on Liem et al. (2001).

Enamel and Dentine

The mineral components of teeth make a very early appearance in the history of vertebrates. Enamel, enamel-like minerals, and dentine first appear in the fossil record some 540 Ma (Card 89). Enamel is the diamond of the vertebrate skeleton, being the strongest mineral component. Its mineral composition is typically inert apatite crystals, and unlike bone it contains no collagen or blood vessels. Dentine is softer than enamel, and it contains a pulp cavity where nerves and blood vessels reside (Fig. 4.4) (Liem et al., 2001; Pough, Janis, and Heiser, 2002). Several small tubes radiate from the pulp cavity to the surface of the dentine, where they relay sensations to the dentine-bearing structure.

Your teeth are composed of an outer veneer of enamel and a deeper, thicker layer of sensitive dentine (Fig. 4.4). Your enamel, like that of most vertebrates, is insensitive. The pressure, temperature, and pain that you do feel in your teeth are detected by the nerves of the pulp cavity within the dentine (Liem et al., 2001). Most of us have had the unpleasant experience of tooth cavities where holes form in the enamel, exposing the sensitive dentine directly to the environment, and generating an awful amount of pain. People with no cavities but sensitive teeth are those whose dentine is exposed near the gum line, where the enamel is thinnest.

The First Bony Vertebrates

Pedigree Earliest Apatite-Bearing Vertebrates

Date of First Appearance ~540 Ma (Card 89)

Specialties of Skeletal Chassis Toothlike Elements Arranged within the Pharynx; Otherwise Soft-Bodied

Eco-niche Small, Eel-like Carnivores

Conodonts

The first hard parts to develop in vertebrates seem to have been strange toothlike structures that appear in rocks approximately 500–254 Ma (Cards 89–95). Called conodonts, they are common fossils for the first 160 million years of vertebrate history, but until 1983 it was uncertain what type of animals they belonged to (Benton, 2005; Knell, 2013). Conodonts are made of apatite-derived dentine and enamel-like minerals, and many are fossilized in groups, arranged and articulated in particular ways. This led researchers prior to the discovery of conodont animals to surmise that conodonts were part of the feeding apparatus of some animal, but which animal was anyone's guess (Benton, 2005).

Remarkably preserved soft-bodied fossils from coal-age (~350–300 Ma; Cards 93–94) rocks in Scotland (others were later discovered in much older rocks in South Africa and the U.S.A.) showed clearly that conodonts belonged to an extinct line of vertebrate animals now called conodonts (Fig. 4.5) (Pough, Janis, and Heiser, 2002; Benton, 2005). The conodont animals were small, being approximately 5–7 cm long. The only hard parts of these animals were the conodonts themselves, bony toothlike elements that were arranged in the mouth and pharynx in such a way that captured prey could be sliced up and directed into the gut cavity. The preserved soft tissues show that conodonts had a well-developed cranium with large eyes, a notochord, what appears to be a dorsal, hollow nerve tube, and clearly defined myomere segments. A short tail fin is present, as are what appear to be gill cartilages in the pharynx.

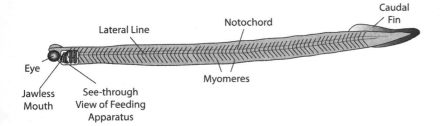

4.5. A conodont-bearing animal, one of the earliest vertebrates with apatite-based elements (in this case, the conodonts, toothlike structures in its mouth and pharynx). This reconstruction is a composite of several known fossils. Based on figures in Janvier (1996) and Benton (2005).

Here was an animal, somewhat like a small modern lamprey, that could move about in eel-like fashion, using its large eyes to locate prey. Although no vertebrae are known for any conodont animal, the apatite-based dentine-enameloid tooth structures leave little doubt that these are early vertebrates. Hagfish lack vertebral structures, which is part of the reason why controversy still remains whether or not hagfish are part of the vertebrate group or whether they lie just outside 'true' vertebrates (Pough, Janis, and Heiser, 2002; Benton, 2005). Unlike conodont animals, hagfish do not possess dentine or enamel-like minerals in their bodies, although you will recall that small bits of apatite are present in their inner ear. Even while lacking evidence for vertebrae, the mineral composition of conodont tooth elements places their owners squarely among the early vertebrates. As dentine and enamel-like tissues can arise only from the neural crest cells unique to vertebrates, this further strengthens the contention that conodont animals are early vertebrates.

'Ostracoderms'

We are accustomed to thinking of our skeletons as an internal framework, so it comes as a surprise to find that bone first appears not as an endoskeleton but as an external shell in vertebrates. Not only that, but the earliest bone fragments we have from the fossil record are actually sandwiched together with toothlike enamel and dentine elements (Benton, 2005). The bony parts themselves are acellular, meaning that unlike the bone of most living vertebrates (with some exceptions in modern fish groups) living bone cells were not present inside the calcium-phosphorus matrix. These acellular bones, called aspidine, grew much like trees, with new bone added to the outside as they aged (Benton, 2005).

The first truly bony vertebrates comprise an odd collection of animals called 'ostracoderms' (meaning shell-skinned). Collectively, these vertebrates were all small, jawless creatures with a protective outer casing of hard but sensitive bone-enamel-dentine armor. We place quotes around the word 'ostracoderm' because it is a grouping of convenience—we are purposely excluding the jawed vertebrates that shared a common ancestor with these animals. Technically, we would call them the stem group to jawed vertebrates because, just as the leaf of a plant develops from its stem, so too did the earliest jawed vertebrates diverge from a common ancestor (stem) last shared most closely with an 'ostracoderm' group (Janvier, 1996, 2008a).

Pedigree Outgroups to the Ancestor of Jawed Vertebrates

Date of First Appearance ~480 Ma (Card 90)

Specialties of Skeletal Chassis External Skeletal Armor with a Cartilaginous Internal Skeleton

Eco-niche Carnivore / Omnivore / Detritivore

The most numerous and successful vertebrate species alive today are the jawed vertebrates. Jawed vertebrates include all familiar fishes as well as amphibians, reptilians (reptiles and birds), and mammals. 'Ostracoderms' are significant relative to jawed vertebrates in two ways. First, the various species of 'ostracoderms' represent closer and closer evolutionary relatives of the jawed vertebrates. Second, the functional anatomy of 'ostracoderm' skeletons and their inferred soft tissues provides a window into the changes that eventually set the stage for the successful evolution of the jawed vertebrates (Janvier, 1996, 2008a). Several different lines of 'ostracoderms' existed, but their interrelationships go beyond the scope of this book. Instead, we will focus on the shared features of these shelled vertebrates and concentrate on the osteostracans, the branch of 'ostracoderms' with the closest relationships to the jawed vertebrates.

Let's begin our investigation of the 'ostracoderm' chassis by first focusing on the bony and 'toothy' exoskeleton of these animals. A typical chunk of 'ostracoderm' armor developed directly in the skin. An anatomist would call this dermal bone, meaning skin-bone, to differentiate it from the bone of the internal skeleton, which is often called endochondral, meaning that it develops within cartilage. The bony exoskeletons of 'ostracoderms' did not develop from a model of cartilage or replace cartilage. They developed directly inside the skin of these animals (Janvier, 1996). The bones of our skull and face, as well as certain chest elements such as our collar bones, still develop as dermal bones (Moore and Dalley, 1999; Liem et al., 2001).

The base of the 'ostracoderm' exoskeleton consisted of compact acellular bone layers, overlaid by porous or 'spongy' acellular bone, and was capped by sensitive dentine projections reminiscent of the toothy scales found on modern sharks (Fig. 4.6) (Janvier, 1996). In most vertebrates, layered bone is called laminated or compact bone, and this bone is tough, stiff, and resistant to compression. The compact acellular bone of 'ostracoderms' may have served such a functional role, resisting compressive stresses such as those from the claws or jaws of predators (Pough, Janis, and Heiser, 2002). Lest you wonder, 'ostracoderms' were probably first prey items for larger invertebrate animals (Foster, 2014), but also survived into later times when vertebrates with jaws were present. Spongy bone in most vertebrates is laid down in regions where some give is necessary. For example, underneath the joint cartilage of limb bones you will find spongy bone, and this bone in turn grades into compact bone (Carter and Beaupré, 2001; Currey, 2002). Perhaps the spongy bone in 'ostracoderms' acted as a cushion against various forces that might impinge upon these animals. Regarding the toothlike outer layer, a cross-section through pieces of fossilized 'ostracoderm' bone shows that, as with your teeth, pulp cavities were present from which small tubes ran to the surface of the dentine. These tubes relayed signals to the nerves within the pulp cavity, alerting such early armored vertebrates to environmental cues. In some cases, a thin layer of tough enamel or enamel-like material covered the dentine (Janvier, 1996).

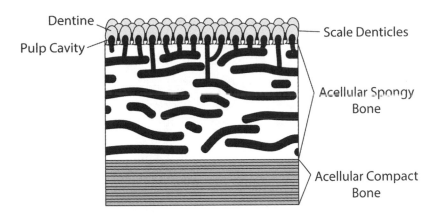

Early Vertebrate Exoskeletal Armor

Dentine

Pulp Cavity

Scale Denticles

Acellular Spongy Bone

Acellular Compact Bone

4.6. 'Ostracoderm' exoskeletal armor piece in cross-section. Note that the scales are very toothlike in having dentine and sensitive pulp cavities. The bone itself is acellular, meaning that bone salts are laid down without trapping bone cells in the matrix. Based on figures in Janvier (1996) and Pough, Janis, and Heiser (2013).

The heads of many, though not all, 'ostracoderms' are typically encased in a bony head shield (Janvier, 1996). Like the helmet of a well-armored medieval knight, the construction is solid but jointed, and sparse openings are present for the eyes, a median olfactory organ, and the pineal organ of the brain. The braincase itself was commonly bony, and its close association with the brain and its cranial nerves has revealed a lot of important information about these soft tissues in 'ostracoderms' (Janvier, 2008b). The body itself was less well-armored, and many 'ostracoderms' possessed thinner apatite scales that may have allowed bending of the trunk and tail for more active swimming. Overall, the exoskeleton appears to have served two functional roles in these vertebrates – protection and environmental sensitivity. An internal skeleton comprised of cartilage was also present in these animals. Although not usually preserved in detail, the tail shape and structure of fins in many 'ostracoderms' attests to there being cartilaginous rays and vertebrae to hold up these elements (Janvier, 1996; Benton, 2005).

Based on fragments of fossilized armor, the earliest 'ostracoderms' appeared nearly 480 Ma (Card 90), and by 420 Ma (Card 91) we have good body fossils of them in the fossil record (Benton, 2005). Most 'ostracoderms' were small, only 5–10 cm in total length (Fig. 4.7). Large, sometimes scalloped plates of jointed exoskeleton cover the head, pharynx, and mouth regions. The mouth was usually small and surrounded by several so-called mouth plates, and there were openings in the head armor for eyes, a nostril (or nostrils), and the pineal organ. A string of pores containing gill filaments ran down both sides of the pharynx, somewhat reminiscent of the condition seen in modern hagfish and lampreys. In some species, the eyes and what might be paired nostrils were situated on the most cranial portion of the head, directly in front of the jawless mouth. The earliest 'ostracoderms' already show pores in defined canals flanking the sides of the body, giving us our first evidence of the lateral line system (Janvier, 1996).

4.7. The body form and anatomy of an early 'ostracoderm' and two heterostracans. See text for details. Based on figures in Radinsky (1987), Janvier (1996), and Benton (2005).

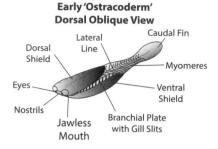

Early 'Ostracoderm' Dorsal Oblique View

Dorsal Shield · Lateral Line · Caudal Fin · Myomeres · Eyes · Ventral Shield · Nostrils · Jawless Mouth · Branchial Plate with Gill Slits

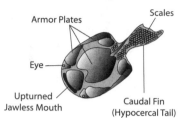
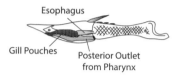

Heterostracan 'Ostracoderm' Dorsal View of Dorsoventrally Flattened Species

Armor Plates · Scales · Eye · Upturned Jawless Mouth · Caudal Fin (Hypocercal Tail)

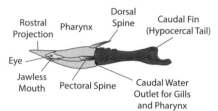

Heterostracan 'Ostracoderm' External View

Rostral Projection · Pharynx · Dorsal Spine · Caudal Fin (Hypocercal Tail) · Eye · Jawless Mouth · Pectoral Spine · Caudal Water Outlet for Gills and Pharynx

Heterostracan 'Ostracoderm' See-through View

Esophagus · Gill Pouches · Posterior Outlet from Pharynx

The earliest 'ostracoderms' were tadpole-shaped and apparently lacked body fins, suggesting, along with their heavy head shields, that these were slow, intermittent swimmers. Later 'ostracoderms' (420–400 Ma; Cards 91–92) built upon and enhanced this basic chassis in various ways. One group, called the heterostracans, more fully developed the head shield so that large plates completely covered the first third of their bodies (Fig. 4.7) (Janvier, 1996, 2008a; Pough, Janis, and Heiser, 2002; Benton, 2005). Some of these 'ostracoderms' developed a dorsal spine that may have provided stability while swimming (we will discuss how fins and spines provide stability in chapter 5). Two openings for eyes and an opening for the pineal organ were present, but nostril openings are not to be found. Grooves running along the insides of the oral opening of some heterostracans suggest that access to olfactory cues was obtained through the mouth (Janvier, 1996; Benton, 2005). In many heterostracans the large head shield plates completely cover the gill slits, and water exited through paired openings at the back of the head. Although most heterostracans were small animals 5–12 cm in length, one group contained species measuring over 1 meter across (Janvier, 1996; Benton, 2005)! These large heterostracans were flattened dorsoventrally (Fig. 4.7). We see similar adaptations in certain fishes today such as stingrays. The flattening of the body was so great in the head region of these heterostracans that the normally downward-facing mouth was pitched up (Janvier, 1996). What advantage does such a body shape confer to its owner, and what does this tell us about dorsoventral flattening in these large heterostracans?

We learn as children helping our parents or grandparents garden that much fun can be had with a water hose. The diameter of the hose determines what volume of water can move through it, and at what speed.

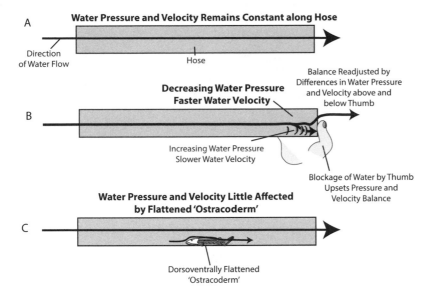

A   Water Pressure and Velocity Remains Constant along Hose

Direction
of Water Flow

Hose

B   Decreasing Water Pressure
Faster Water Velocity

Balance Readjusted by
Differences in Water Pressure
and Velocity above and
below Thumb

Increasing Water Pressure
Slower Water Velocity

Blockage of Water by Thumb
Upsets Pressure and
Velocity Balance

C   Water Pressure and Velocity Little Affected
by Flattened 'Ostracoderm'

Dorsoventrally Flattened
'Ostracoderm'

4.8. Water hoses and changing water pressure. When a garden hose is completely unblocked (A), the water pressure and velocity remain constant from one end to the other. If, however, you use your thumb to block a portion of the hose's exit (B), this upsets the pressure and velocity balance. This imbalance is corrected by differences in the water pressure and velocity above and below your thumb, resulting in the water speed increasing and the water shooting farther out from the hose as it exits over your thumb. For animals living in the water, if we imagine the animal's body creating a 'blockage' in a 'hose of water,' we can see that dorsoventrally flattened animals cause less disruption of the water pressure and velocity, allowing them to more easily stay put without expending much energy.

If you wanted to drink from the water hose, you simply turned it on and a slow-moving column of water would emerge from the hose and almost immediately drop. Normally, you would hold the hose above your mouth to take a drink. If you wanted to spray plants on the far side of the garden (or your parent, grandparents, or other bystanders) you quickly learned a cool trick. By placing your thumb over the opening to the hose, you would block some of the water. You could feel the water pressure quickly build behind your thumb, and a thinner but faster stream of water would shoot out, moistening both garden soil and your human targets (Fig. 4.8).

The reason that blocking a portion of the water going through a hose causes its speed to increase has to do with the physics of fluid flow (Fig. 4.8). The hose is connected to a spigot on the house, and that metal or plastic spigot is sending out water of a certain volume, speed, and pressure. If the hose is the same diameter as the spigot, the same volume, speed, and pressure of water will exit the hose. There are physical laws that dictate why this occurs. First, as a fluid moves along a system (in this case our garden hose), energy is conserved (Bloomfield, 2006). This means that as water travels through the hose, its potential energy (pressure) and its kinetic energy (speed) must balance out. Second, there is a law of continuity, meaning that the volume of water going into the hose must match the volume of water coming out of the hose (Vogel, 2003; Bloomfield, 2006).

If a part of the hose becomes narrowed, say by you placing your thumb over the opening, you have just altered the previously balanced energy and continuity relationships of the water in the hose. This imbalance is corrected in the following ways. First, pressure is redistributed by the blockage (in this case your thumb): pressure decreases above the blockage and increases behind it (Bloomfield, 2006). This is why you feel increased water pressure on your thumb. Second, the volume must be redistributed because the same volume of water going into the hose must

4.9. Other 'ostracoderms' with more streamlined shapes. Some anaspids were covered in elongate body scales and were elongated with a hypocercal tail in which the lower lobe was larger and more prominent than the upper lobe. Some thelodonts looked very fishlike, even having fine scales, a dorsal fin, and a symmetrical caudal tail. Based on figures in Janvier (1996).

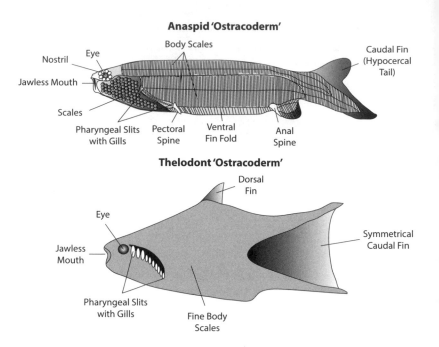

pass through any given point in the hose at a given time (Bloomfield, 2006). What this means is that if your thumb blocks half of the hose's opening, only half the volume of water can make it through at the previously slower speed. Therefore, to balance out the incoming and outgoing volume, the water's speed must increase until the same volume of water is pushed through a given region of the hose. All of this has to occur to achieve what seemed to you to be a simple trick.

If we now imagine an 'ostracoderm' animal lying on the sea floor, there will be a current of water moving around it (Fig. 4.8). We can pretend, in a sense, that the body of water in which the 'ostracoderm' lives is like a garden hose–there is conservation of energy and continuity. As a given volume, speed, and pressure of water moves over an animal's body, it will cause an imbalance that must be corrected. Like a thumb blocking the opening to our garden hose, the blockage of our 'ostracoderm' in the water column will cause water pressure to build up in front of it, water pressure to drop over the top of it, and the speed of the water traveling over the top of the animal to increase. An animal whose body projected well above the sea floor sediment would struggle to prevent being lifted. Continuous struggling to stay put would be bad news for various reasons–muscular energy would be expended fighting the current, or a previously hidden animal would now be exposed to predators and prey. As it turns out, dorsoventral flattening is a common way for bottom-dwelling animals to get around these problems (Pough, Janis, and Heiser, 2002; Vogel, 2003). When the body is flattened as much as possible, energy conservation and continuity of water flow are only slightly perturbed, and this in turn creates fewer pressure and speed differences to lift the animal off the sea bottom. The large, dorsoventrally

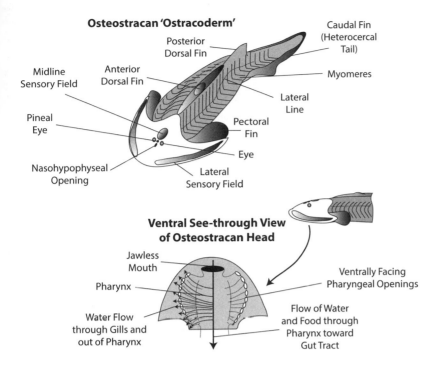

Osteostracan 'Ostracoderm'

Posterior Dorsal Fin

Caudal Fin (Heterocercal Tail)

Midline Sensory Field

Anterior Dorsal Fin

Myomeres

Pineal Eye

Lateral Line

Pectoral Fin

Nasohypophyseal Opening

Eye

Lateral Sensory Field

Ventral See-through View of Osteostracan Head

Jawless Mouth

Pharynx

Ventrally Facing Pharyngeal Openings

Flow of Water and Food through Pharynx toward Gut Tract

Water Flow through Gills and out of Pharynx

4.10. An example of an osteostracan 'ostracoderm.' These vertebrates had large, shield-like heads with a downward-facing mouth and gills. Two modestly developed dorsal fins were present in some species, and a heterocercal tail with a larger upper lobe was prominent. Sensory fields along the sides of the skull and just caudal to the eyes may have detected various combinations of pressure, temperature, pain, and even taste. Based on figures in Janvier (1996).

flattened heterostracans were most likely bottom-dwellers, based on these physical considerations.

Still other groups of 'ostracoderms' took on more streamlined, fish-like shapes. Anaspids and thelodonts were small animals up to 17 cm in length that had elongate bodies covered in small, interlocking scale denticles (Janvier, 1996) (Fig. 4.9). These animals had dorsal and ventral body fins in addition to their tails, and some even had spiny pectoral fin-like projections lying just caudal to the head. These features suggest that they were more maneuverable than heterostracans (Janvier, 1996; Pough, Janis, and Heiser, 2002).

Osteostracans and several closely related species are considered the closest fossil relatives to the jawed vertebrates (Benton, 2005; Janvier, 2008a). Osteostracan 'ostracoderms' are characterized by their large, single-piece head shields and heavily armored bodies (Fig. 4.10). The head shields are reminiscent of an insect's carapace in overall form and were so large and extensive that they completely covered the tops and sides of the head. This head shield configuration forced the mouth and gill slits in these 'ostracoderms' to face ventrally. Openings for eyes, a single nasal opening (called the nasohypophysial opening), and an opening for the pineal organ were present dorsally. In addition, there were strange, thinly scaled regions along the lateral edges of the head shield and in the midline of the head just behind the eyes that may have been sensory fields (more on this later) (Janvier, 1996; Benton, 2005). Well-developed, paired pectoral fins, covered in tiny scales, are present in these animals. These fins were possibly quite flexible and may have been used actively in swimming. Broad, rectangular scales protected the flanks of the body,

4.11. Heterocercal tail function illustrated in an osteostracan 'ostracoderm.' In the upper panel, the classic hypothesis is shown, in which the tail creates a downward force partially balanced by uplift from the pectoral fins. The lower panel shows what is known from more recent studies, suggesting that the tail provides mostly forward thrust and that overall body movements allow for changes in direction of the swimming animal.

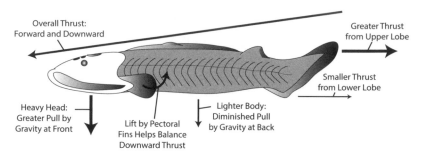

Classical Model of Heterocercal Tail Function

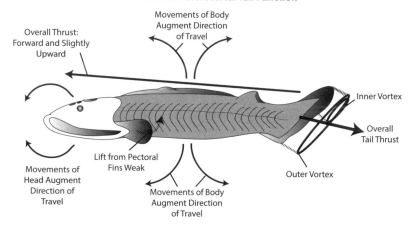

Modern Model of Heterocercal Tail Function

and several scales fused onto the back, forming a dorsal fin. The tail of osteostracans was asymmetrical: the notochord and vertebral column of the tail bent upward, creating a large upper lobe and a small, thinner lower lobe (Janvier, 1996; Benton, 2005). This sort of tail shape is called heterocercal (Fig. 4.11).

What are the functional advantages of a heterocercal tail? At first consideration, it would seem that this sort of tail would put an animal at a distinct disadvantage. In many modern fish, the tail is symmetrical and provides a forward-directed thrust when beating against the water (Lauder et al., 2003). In contrast, the large upper lobe and small lower lobe of a heterocercal tail would be predicted to create an imbalanced force. You might predict that the upper lobe should create greater force than the lower lobe, generating a lot of downward thrust with each beat of the tail. This would result in the head of the osteostracan tilting downward, and eventually leading to a situation where the poor animal would flip end over end or bury its head into the sand. But sharks and certain bony fishes like paddlefish have a heterocercal tail, and both of these animals are active swimmers. How can this be?

Early experiments on amputated shark tails seemed to confirm what we have predicted: the upper lobe of the tail generates more force than the lower lobe (Fig. 4.11) (Alexander, 1965). It was then argued that sharks and other animals with a heterocercal tail rectified this unbalanced force equation using their large pectoral fins. The pectoral fins of sharks and osteostracans are relatively flat, but slightly rounded on their dorsal sides. It was thought that as they swim through the water, their fins act in much the same way as wings on an airplane. In aircraft, the flat but slightly rounded wing cross-section causes air to rush over the top of the wing faster than the wind underneath it (Bloomfield, 2006). Just like with our hose analogy earlier, the difference in wind speed above and below the wing causes a difference in air pressure. Air pressure builds underneath the wing and lessens above it, causing lift (Bloomfield, 2006). This is why very large and heavy airplanes, if they can get enough speed, are able to fly. Imagine now a shark or osteostracan swimming through the water with its pectoral fins out like airplane wings. Water moves faster over their tops, slower underneath them, and causes lift, which tilts the head end of the animal up. Thus, it was suggested that the forward but downward thrust from the heterocercal tail was counteracted by the wing-like pectoral fins, allowing sharks, osteostracans, and other animals with similarly shaped tails to swim in a straight line (Alexander, 1965). This is all well and good, except that modern experiments don't quite support this balancing scenario.

In recent experiments, sharks and other fishes are placed in flow tunnels in which they swim in place against a current. Dyes are sometimes injected around the animal to show water flow patterns, and now there are even laser-scanning methods where reflective particles in the water can be traced and analyzed as they move around a shark's body. These experiments have shown that, despite the difference in size, the upper and lower tail lobes beat together and create forces that push the animal mostly forward, not downward (Fig. 4.11) (Wilga and Lauder, 2000, 2004; Lauder et al., 2003; Flammang et al., 2011). Moreover, Brooke Flammang discovered a specialized tail muscle in sharks, the radialis, that stiffens the tail fin and enhances forward propulsion (Flammang, 2009). However, it is not clear whether such a muscle was present in osteostracans. Such experiments have also shown that the pectoral fins of sharks do not create much lift. Instead, the whole body of the shark is used to steer—by changing the overall angle of the body relative to the tail, a shark can orient itself up, down, or sideways as it swims (Wilga and Lauder, 2000). Certainly, the pectoral fins still play an important role in steering, but their role as glorified airplane wings is not supported by the new data.

Returning to osteostracans: their heterocercal tails combined with their large, flattened heads probably worked together well. It is possible that these 'ostracoderms' maneuvered quite well through the water, tilting their heads into various positions in the water currents while their heterocercal tails propelled them forward (Janvier, 1996; Pough, Janis, and Heiser, 2002; Benton, 2005). Osteostracan fossils are common in sandy siltstone rocks that strongly suggest these animals lived in turbulent waters near or at the mouths of rivers—this is where we find brackish estuaries and deltas in modern environments (Benton, 2005). Living in such sediment-rich environments would provide ample nutrients and planktonic food for these vertebrates to feed upon. Their domed heads, dorsal and pectoral fins, and heterocercal tails all suggest that these animals had a chassis built for swimming against the current. However, their flattened chassis also would allow them to lie flat and rest on the sediment without fighting lift. A similar thrust-and-lift scenario may have played out in other 'osteostracans' in which the asymmetry of the tail was reversed (large lobe ventral, small lobe dorsal).

Found along with the carcasses of 'ostracoderms' are the exoskeletal remains of fearsome sea scorpions, called eurypterids, distant early relatives of modern scorpions and spiders (Benton, 2005). Although not true scorpions, these animals had flattened, horseshoe-shaped heads with compound eyes, armored and segmented bodies and legs, and some species possessed large pinchers. Some eurypterids were large, much larger than most 'ostracoderms,' with some reaching the size of a crocodile! Although eurypterids themselves were not present when the first 'ostracoderms' made their appearance, other large predators such as anomolacarids were (Foster, 2014), and it is not beyond the pale to hypothesize that bony exoskeletons were, in part, an evolutionary response to invertebrate predation on early vertebrates (Benton, 2005).

'OSTRACODERM' BRAINS

The spaces and perforations in the braincases of fossil vertebrates can provide some evidence for what general regions of the brain were present, and for the distribution of the cranial nerves. As it so happens, osteostracan braincases provide us with excellent data for reconstructing their brains. This is because osteostracan brain cases were composed of a paper-thin bony layer inside and out with cartilage sandwiched in between (Janvier, 2008b). In modern vertebrates, the brain itself floats in cerebral spinal fluid and is supported and protected by tissues called meninges (Liem et al., 2001). This means that there is a space, sometimes a considerable amount of space, between the brain itself and the interior of the skull. In many cases, a vertebrate endocast (a mold of the inside of the braincase) is subtle and shows mostly general regions of the brain (Benton, 2005). Not so in osteostracans—the brain appears to have been very closely associated with the inside of the bony braincase (Janvier, 2008b). Evidence for this close association comes from impressions of a structure called the choroid plexus, a tissue closely associated with the brain itself that delivers fluid and nutrients (Janvier, 2008b).

Extensive studies of osteostracan endocasts reveal that the major brain regions present in all jawed vertebrate groups were already developed in these animals (Fig. 4.12) (Janvier, 2008b). The brain was typically a straight tube divided into the three primary brain sections we discussed in chapter 3, each expanded slightly like little water balloons. The forebrain shows evidence of a cerebrum and diencephalon, and olfactory tracts from the former and an organ known as the hypophysis from the latter extend into the so-called nasohypophysial opening

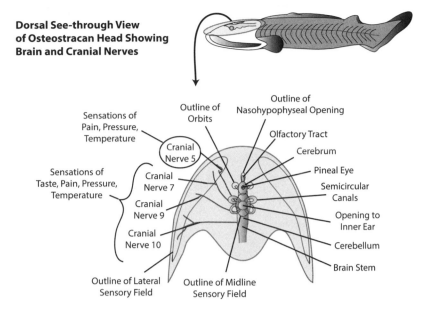

Dorsal See-through View of Osteostracan Head Showing Brain and Cranial Nerves

Sensations of Pain, Pressure, Temperature

Outline of Orbits

Outline of Nasohypophyseal Opening

Olfactory Tract

Cranial Nerve 5

Cerebrum

Sensations of Taste, Pain, Pressure, Temperature

Cranial Nerve 7

Pineal Eye

Semicircular Canals

Cranial Nerve 9

Opening to Inner Ear

Cranial Nerve 10

Cerebellum

Brain Stem

Outline of Lateral Sensory Field

Outline of Midline Sensory Field

4.12. Schematic of an osteostracan brain based on composites of various endocasts. Notice that the cranial nerves that supply the lateral sensory fields not only detect pain, pressure, and temperature, but also may have conveyed sensations such as taste. Based on information and figures in Janvier (2008b).

mentioned earlier. The hypophysis is better known in human anatomy circles as the pituitary gland (Liem et al., 2001). The pituitary regulates growth and reproduction, and its association with the odor-detecting olfactory tracts in osteostracans suggests it closely 'monitored' odors or pheromones (species-specific chemicals that can trigger behavioral changes) to regulate important aspects of these vertebrates' lives. Large optic nerves branched to the eyes from deep within the diencephalon, and a pineal organ projected dorsally to an opening in the head shield (Janvier, 2008b). The midbrain was modestly developed and contained regions for visual and audio interpretation (Fig. 4.12). The inner ear and semicircular canals were fairly well developed, and the openings for the cranial nerves related to the audio and lateral line signals were significant. These findings suggest that the osteostracans could have been reasonably good swimmers, especially because their well-developed inner ears would have supplied ample data on body position in the water (Janvier, 2008b). The hindbrain shows a cerebellum and the medulla oblongata (Fig. 4.12) (Janvier, 2008b).

Remarkably, bone surrounded most of the cranial nerve tracts themselves, so we can actually trace the major pathways these important nerves took in osteostracans (Janvier, 2008b). Not only were the cranial nerves present and well developed in these early, bony vertebrates, but they reveal intriguing if not well understood information about the sensory fields in the head shield (Fig. 4.12). The sensory fields that ring the edges of the head shield received neural input from the cranial nerves numbered 5, 7, 9, and 10. Cranial nerve 5 is a complex nerve that controls jaw muscles but also provides sensation to the face (Liem et al., 2001; Kardong, 2012). It is therefore fascinating that the sensory branches of cranial nerve 5 supply part of these sensory fields in osteostracans. Cranial nerves 7, 9, and 10

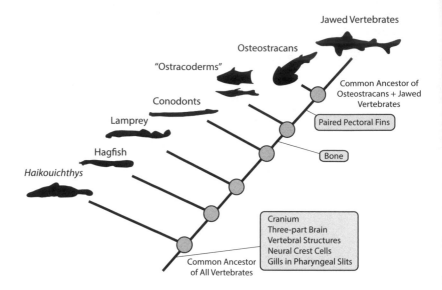

4.13. Evolutionary relationships of the early vertebrates. See text for more details.

Jawed Vertebrates

Osteostracans

"Ostracoderms"

Conodonts

Lamprey

Hagfish

Haikouichthys

Common Ancestor of Osteostracans + Jawed Vertebrates

Paired Pectoral Fins

Bone

Cranium
Three-part Brain
Vertebral Structures
Neural Crest Cells
Gills in Pharyngeal Slits

Common Ancestor of All Vertebrates

do a huge variety of things, but intriguingly these three nerves together supply vertebrates with their sense of taste (Liem et al., 2001; Kardong, 2012). So, the sensory fields on the edges of the head shield were supplied with nerves of general sensation and the special sense of taste. Of course, we cannot resurrect osteostracans to test whether this is indeed what they were sensing, or whether they were also detecting other combinations of sensation. However, we might tentatively speculate that the sensory fields on the edges of the head shield were sensitive to pain, pressure, and temperature, and might have 'tasted' the environment around these animals.

The midline sensory field directly behind the eyes may be more interpretable based on some modern fishes. This sensory field lies directly above the center of the inner ear region. In modern sharks and other cartilaginous fishes, a thinly covered opening exists in the same region of the head. In these modern fishes, a direct connection is made through this opening so that pressure waves and sound are directly channeled into the inner ear (Liem et al., 2001). It is possible that osteostracans were utilizing something similar, using this sensory field to direct sound and pressure waves into their inner ear region. More research waits to be done on these intriguing but poorly understood regions of osteostracan anatomy.

Onward to Jaws

The 'ostracoderms' were a remarkable and fairly diverse group of early, bony vertebrates, and many of them coexisted with early jawed vertebrates for quite some time. The latest known surviving members of the 'ostracoderms' come to us from fossils approximately 360 Ma (Card 93). However, these animals eventually went extinct, leaving no direct modern descendants. The distantly related hagfish and lampreys discussed in chapter 3 remain the only jawless vertebrates left on earth. See Figure 4.13 to review the relationships of these early vertebrates.

Despite their diversity, a jawless mouth restricted the diets and body sizes of the earliest bony vertebrates. The development of jaws sparked a diverse evolutionary bush of vertebrate animals, and all but one of these lineages have surviving, thriving members with us today. The chassis of the 'ostracoderms,' especially that of the osteostracans with its bone, paired fins, and well-developed tail, set the stage for greater things to come once jaws made their first appearance.

The Evolution of the Jawed Vertebrate Chassis and Something Fishy

3

You know when that shark bites with his teeth, babe, scarlet billows start to spread.

BOBBY DARIN, *Mack the Knife*

5.1. A stylized jawed vertebrate, based on placoderms and sharks, showing the positions and orientations of the jaws, head, and pharynx. Note the location of the spiracle, a 'trapped' gill pouch caught between the palatoquadrate and Meckel's cartilage cranially and the hyoid arch caudally. In the exploded view, note how the vertebrae coalesce and fit over both the spinal cord and the notochord.

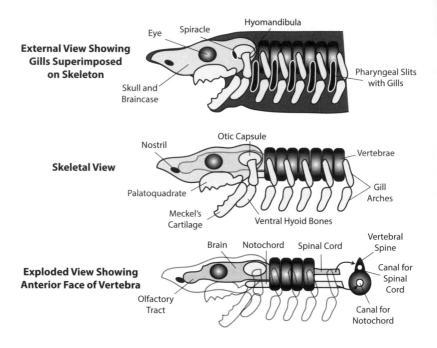

External View Showing Gills Superimposed on Skeleton

Eye
Spiracle
Hyomandibula
Skull and Braincase
Pharyngeal Slits with Gills

Skeletal View

Nostril
Otic Capsule
Vertebrae
Palatoquadrate
Meckel's Cartilage
Ventral Hyoid Bones
Gill Arches

Exploded View Showing Anterior Face of Vertebra

Brain
Notochord
Spinal Cord
Vertebral Spine
Canal for Spinal Cord
Olfactory Tract
Canal for Notochord

The Jawed Vertebrate Chassis: A Primer 5

WE TAKE OUR JAWS AND THEIR ASSOCIATED TEETH FOR GRANTED. I was reminded of this with my children when both were less than a year old and only beginning to sprout their first teeth. When you begin to feed a baby real food and wean them from a strictly milk- or formula-based diet, food has to be thoroughly mashed up and puréed. As my wife and I have learned, it takes quite a while before a child masters eating larger pieces of food. Thanks to our jaws and teeth, we can eat quite a variety of food items of almost unlimited size, provided we have the time and inclination (and preservatives!) to take the necessary number of bites. In a strange and abstract sense, as we grow from infant to child we move from the micro-particle food consumption of our jawless ancestors to the larger food items of our jawed ancestors.

But, in our following discussions on the evolution and glory of the jaws and teeth, let us not belittle our conodont and 'ostracoderm' friends. Fossil fragments of the earliest jawed fishes go back to nearly 450 Ma (Card 91). It is important to emphasize again that jawed fishes and the jawless conodonts and 'ostracoderms' described previously coexisted for at least 150 million years. So, the evolution of jaws and teeth, while an important event in vertebrate history, was not the immediate death knell for conodonts and 'ostracoderms,' who apparently went on with the business of life just fine for a long time.

Before we continue, we must also have an aside on the word 'fish' and how it will be used throughout this book. Technically, we name natural, related groups of animals based on the following criterion: they must include the last common ancestor and all of the descendants (Schuh and Brower, 2009). For example, mammals are a natural, related group because they include the common ancestor and all its descendants: the egg-laying mammals, the pouched mammals, and the placental mammals. Even though a duck-billed platypus lays eggs and does other weird things, it is not excluded from mammals. In contrast, the general term 'fish' is familiar but misleading: it groups together several different vertebrate groups because of their fishy bodies and excludes others who don't fit the bill. But, based on shared trait states, such as a bony skeleton and jaws, we humans share a common ancestor with bony 'fishes' such as a trout. To be accurate, we would have to call humans and all the land-adapted vertebrates 'fishes' for the grouping to be considered natural! Also, whereas we humans do share a distant common ancestor with animals like trout, we are not direct descendants from trout-like animals.

What to do? To keep things simple and less confusing, we will apply the term 'fish' or 'fishes' to any group we discuss whose living relatives are still, well, fishy. In other words, if the living descendants of a particular jawed vertebrate group still possess fins, gills, and so on, we will call them fishes. But, to be more precise and less confusing, we will add a descriptive term in front of 'fish.' For example, sharks and their kin will be identified as cartilaginous fishes, and trout, salmon, and other such familiar animals will be called ray-finned fishes. In this way, we can avoid long phylogenetic abstractions but not suggest that something like a trout gave rise to the land-living vertebrates, including humans.

Jaws

Jaws are advantageous to have, and two major functional reasons stand out above the rest. First, jaws can help you catch larger prey by directly 'grabbing' them with the mouth. Teeth in a variety of sizes and shapes can help you take prey that is, when whole, too big to swallow, and break it up into manageable bits. Second, jaws, by virtue of their surrounding the mouth and being just in front of the pharynx, can help you actively pump water across your gills, improving your respiration (Liem et al., 2001; Pough, Janis, and Heiser, 2002; Kardong, 2012). Unlike jawless vertebrates with their smaller mouths that cannot completely open and close, jawed vertebrates can swing open and slam shut their jaws in rapid succession, forcibly pumping water across their oxygen-hungry gills.

But jaws had to come from somewhere. They were not simply made to order as a new-fangled part for some luxury vertebrate chassis. Where did jaws come from? What part of the original vertebrate chassis could be modified in such a way that life could continue and jaws, with all their advantages, could come into being? The embryonic development of jawed vertebrates provides a compelling answer.

The cartilaginous pharyngeal arches that support the mouth and gill pouches are the source of vertebrate jaws (Liem et al., 2001; Pough, Janis, and Heiser, 2002; Gilbert, 2010). A pharyngeal arch itself is not a single element but a series of jointed struts, divided into a dorsal and ventral portion, that allow the pharynx to collapse and open, pumping water over the gills. These arches are closely associated with muscles and nerves that can be actively controlled by the animal (Liem et al., 2001). You may recall that the initial source of the pharynx and its associated arches is from the unique neural crest cells that create much of what is special about the vertebrate chassis. Remarkably, during jawed vertebrate development, we can clearly see that the first pharyngeal arches are modified into the upper and lower jaws (Gilbert, 2010).

The development of jaws and their evolution is complex, and new embryonic and genetic evidence provides an intriguing but complicated picture of how jaws form (Kimmel, Miller, and Keynes,, 2001; Kuratani et al., 2001; Cohn, 2002; Cerny et al., 2004; Kuratani, 2004; Mallatt, 2009). Here, I knowingly simplify the development of jaws down to the basics, acknowledging that our current understanding of this phenomenon is more detailed than our treatment and the scope of this book allow. What I present is called the neoclassical hypothesis of jaw evolution and development (Liem et al., 2001). This hypothesis does gel in a broad way with newer embryonic and genetic data. Because the intricacies of the interactions that lead up to the development of jaws are still debated, I direct interested readers to the primary literature for more information.

In the neoclassical hypothesis, the first two pharyngeal arches (and possibly one or two smaller elements) develop into the jaws and their associated anatomical structures (Fig. 5.1). The dorsal portion of the first pharyngeal arch becomes the upper jaw, and the ventral portion becomes the lower jaw. Some technical language is necessary at this point so that we can avoid overly long descriptions such as 'the back portion of the upper jaw' or 'the cranial connecting point of the lower jaws.' The upper jaw element is formally known as the 'palatoquadrate.' We call the roof of our mouth our palate, and this might help you remember that the palatoquadrate is the upper jaw element. In addition, the 'quadrate' part of 'palatoquadrate' refers to the generally rectangular shape of its caudal portion. The upper jaw joint is located at the caudal corner of the palatoquadrate. The lower jaw element is called the Meckel's cartilage, after the anatomist who first named this structure (Liem et al., 2001). Most people call the lower jaw bone the 'mandible,' so it may help you to remember that the Meckel's cartilage is the lower jaw element because, like the 'mandible,' it begins with the letter 'M.' An articular process at the caudal end of the Meckel's cartilage connects with, and allows it to rock against, the quadrate portion of the palatoquadrate. I emphasize that both the palatoquadrate and the Meckel's cartilage are formed from the dorsal and ventral parts, respectively, of the first pharyngeal arch (Liem et al., 2001).

The Meckel's cartilage has always been a relatively free element—this makes mechanical sense: you want to be able to open and close your mouth! However, in the first jawed vertebrates, the palatoquadrate was not directly fused to the skull or braincase, but was instead suspended underneath the head via ligaments, tough, ropelike tissues that link bone to bone (Liem et al., 2001; Benton, 2005). This means that there was some movement of the palatoquadrate against the skull when the jaws were opened and closed. This is difficult for humans to imagine because our upper jaws are firmly fused with our skull—only our lower jaws can move. However, based on conditions noted in living jawed vertebrates, the somewhat mobile and loose connection of the palatoquadrate with the skull in our jawed ancestors allowed the mouth to protrude while it was opened (Liem et al., 2001; Pough, Janis, and Heiser, 2002; Kardong, 2012). This motion would have propelled the mouth forward toward prey when the animal was attacking or feeding. The loose connections may have also allowed the palatoquadrate to swing upward, increasing the size of the mouth's gape. As we will see, how loose or how attached the palatoquadrate is to the skull varies among different vertebrates.

The pharyngeal arch caudal to the jaws also plays an important but diverse role in jawed vertebrates (Fig. 5.1). To find a portion of this arch in ourselves, find your so-called Adam's apple. This feature is well developed in, and is characteristic of, sexually mature male humans. This protrusion in the throat is not actually a bone but instead a prominent ridge of cartilage (the thyroid cartilage), positioned superior to the thyroid gland (the old endostyle, as you may recall) (Moore and Dalley, 1999). It is above the Adam's apple that we find the hyoid, a small U-shaped bone suspended entirely by muscles, cartilage, and ligaments. This odd little bone develops from the ventral portion of the second pharyngeal arch (Liem et al., 2001; Gilbert, 2010), and in humans it currently serves a variety of functions related to swallowing (Moore and Dalley, 1999). However, the second pharyngeal arch from which the hyoid develops is a critical anatomical structure that has played a large and varied role in jawed vertebrates.

The second pharyngeal arch of jawed vertebrates is technically called the hyoid arch. Its dorsal portion is given the technical name hyomandibula. The hyomandibula spans the space between the jaws and the otic capsule (where the inner ear of the brain is housed) of the braincase (Liem et al., 2001). The ventral portion of the hyoid arch is composed of one or several bones collectively called the hyoid bones, and these commonly link to the hyomandibula through ligaments (Liem et al., 2001). As we will see, the hyoid arch serves many different functions in various vertebrate groups, with a lot of variation seen in the function of the hyomandibula.

We will later discuss specific functions of the hyomandibula in the context of different vertebrate groups, but here we briefly discuss the role of the ventral hyoid bones, which are usually more conservative in their functional role. The ventral hyoid bones generally serve as a site of attachment for muscles that pull open the jaws, aid in swallowing, and/or give purchase for the attachment of the tongue (Liem et al., 2001). Based on the condition observed in most living jawed fishes, the hyoid bones of the earliest jawed vertebrates were most likely an anchoring spot for some of the jaw-opening muscles. These strap-like muscles span from the hyoid to the Meckel's cartilage (or its bony derivatives) in many living fishes. When these muscle straps contract, they act like the chains on an inverted drawbridge, lowering the Meckel's cartilage and opening the jaws (Liem et al., 2001). Other muscles also play important roles in opening the jaws, and these will be discussed in the next section.

In hagfish, lampreys, and extinct jawless vertebrates, there is a gill pouch located between the first and second pharyngeal arches. One quirk that occurred when the first two pharyngeal arches became the jaws and their associated support structures was the formation of a small, 'trapped' gill pouch, caught between the jaws and the hyoid arch. In the most primitive living jawed vertebrates, sharks and their allies, this remnant gill opening is called the spiracle (Fig. 5.1) (Liem et al., 2001; De Iuliis and Pulerà, 2011). In these animals, the spiracle appears as a small, ear-like opening on either side of the head behind the eyes and nearly in line with the otic capsule. The spiracle retains a small gill pouch in many of these fishes. Many of the bony fishes we will discuss later lose the external opening of the spiracle, but all jawed vertebrates in one way or another hold on to this old jaw evolution relic.

Even we humans hold on to the spiracle as our so-called Eustachian tube, the small, angled tunnel that connects our inner ear to the back of our throats (Moore and Dalley, 1999; Liem et al., 2001; Gilbert, 2010). We become acutely aware of our remnant spiracles when we go up quickly in the elevator of a tall building or take off in an airplane – the change in air pressure outside our ear drums differs from the higher pressure behind it. We swallow, cough, or chew gum, and this uncomfortable pressure goes away (we say our ears 'pop'), thanks to our old spiracle relieving the pressure behind our ear drums, allowing the excess air to escape into the back of our mouths. Another role our spiracle (Eustachian tube) continues to play for us is drainage of fluid from the inner ear, especially during a head cold or ear infection (Moore and Dalley, 1999).

In sum, vertebrate jaws probably evolved from the first several pharyngeal arches. The main elements of the jaws themselves are the upper palatoquadrate and the lower Meckel's cartilage. Behind the jaws proper, another pharyngeal arch developed into the hyoid arch, comprised of an upper hyomandibula and a lower series of hyoid bones. These bones serve a variety of purposes, but the lower hyoid bones are typically involved in swallowing, tongue anchoring, and the attachment of jaw-opening muscles. Along with jaws, the jawed vertebrates developed a suite of modified or new anatomical adaptations that resulted in their increased diversity and success.

Teeth, Tongue, and Stomach

Elements of the Basic Jawed Fish Chassis

The teeth that line the jaws (and roof of the mouth in some cases) of most vertebrates probably trace their origins to scales or scalelike structures that invaded the mouth (Kawasaki and Weiss, 2006; Soukup et al., 2008). It is perhaps no coincidence that teeth, like scales, develop from the special neural crest cells that form the pharyngeal arches, jaws, and facial skeleton in vertebrate embryos. More specifically, teeth and scales form from a complex interaction between the neural crest cells and surrounding mouth tissues (Kawasaki and Weiss, 2006; Soukup et al., 2008). The neural crest cells themselves give rise to the dentine of the teeth (or scales), and the interaction of the surrounding tissues with the dentine-forming neural crest cells results in their becoming the outer enamel (Gilbert, 2010).

As mammals, we are accustomed to thinking of our teeth and their various shapes as the norm for vertebrate animals. In fact, as we will see later in the book, we are the oddballs among the vertebrates. Most jawed vertebrates have similarly shaped teeth of different sizes. This is very apparent if you have ever peered into the jaws of a fish on the end of a hook or watched slow-motion photography of sharks feeding. The teeth of most jawed vertebrates vary in size craniocaudally along the mouth, but their shapes remain fairly constant. The bladelike, serrated teeth of a great white shark in the front of its jaws are not much different (except, maybe, for size) from those residing at the back of its mouth.

We mammals chew our food, which (as we will later see) created selective pressures to develop a variety of tooth shapes capable of nipping, shearing, shredding, and crushing food as it was passed from the front of the mouth to the back. In contrast, most jawed vertebrates don't bother very much with chewing. If you have ever watched fishes feeding, you know that they simply gulp down smaller items of prey, or they may break off chunks of larger organisms. The teeth function mostly to grab, rend, and direct chunks of food back toward the pharynx and gut tube (Liem et al., 2001). In fact, a close look at the teeth of many fishes will reveal that they have a recurved shape – that is, the crown of the tooth is curved somewhat like a banana held upside down. The sharpest, pointiest part of the tooth in most fishes points, not directly downward, but backward toward the pharynx. This ensures that when struggling prey is caught in the jaws, its thrashing will direct it further down the mouth and pharynx (Liem et al., 2001).

Struggling prey or hard items can break or loosen teeth. As mammals, we worry about our teeth because we get only two sets. Once the second set has worn down (or, these days, been eroded by sweets and soda), we must rely on false teeth to get us through the rest of our eating lives. Most vertebrates, however, have a nice tooth replacement plan. As an old tooth breaks, wears down, and falls out, a new tooth emerges to take its place (Liem et al., 2001; Kardong, 2012). As we will see, most jawed vertebrates are continually replacing their teeth throughout their lives, and this process occurs in numerous and fascinating ways.

Most animals fight not to be eaten, and struggling prey often present problems to the predator. Keepers of snakes, for example, may feed their reptiles with dead or unconscious mice for this very reason. Many a snake keeper has watched their reptilian charges lose eyes, scales, or even their tail to panicking rodents fighting to survive. At least in the wild, snakes subdue or kill their prey through constriction or venom prior to swallowing. Most fishes, however, do not have this 'luxury,' and must contend with uncooperative prey bent on saving its life at any cost. If struggling prey punched through the pharyngeal slits and gills of a predator, the predator's own life would be cut short. Many jawed vertebrates solved this problem through the development of structures called gill rakers. These are cartilaginous or bony extensions of the gill arches with recurved and pointed processes that both block access to the gills and help to direct prey down the pharynx to the gut (Liem et al., 2001). Some gill rakers are even studded with toothlike denticles that further restrict prey to the center of the pharynx and put them on a one-way path to digestion.

The earliest jawed vertebrates probably had a tongue, but not the mobile, fleshy organ we know in ourselves. Based on outgroup comparisons with the most primitive living jawed vertebrates, the cartilaginous fishes (Chondrichthyes), the original tongue of jawed vertebrates was likely a stiffened and relatively immobile projection residing in the floor of the mouth (Iwasaki, 2002). In humans, the tongue is important in smashing and swallowing our food. In living aquatic vertebrates and presumably in our earliest jawed ancestors, water is forcibly pumped into the mouth, pushing food items down the pharynx into the gut tube. Therefore, there is no need for a muscular organ to assist with swallowing—water pressure, teeth, jaws, the hyoid bones and their associated swallowing muscles, and a pumping pharynx all push food to the gut. In many fishes, there may even be denticles on the tongue itself, which may aid with directing prey toward the gut (Liem et al., 2001).

Speaking of guts, here is a question I use to stump my anatomy students: where is most food digested? Many students (and many of you reading this) immediately say, "The stomach!" However, whereas the stomach does absorb water, alcohol, and some vital nutrients, this is not where most of the digestion and absorption of food takes place. Most digestion in vertebrates takes place in the intestinal tract, specifically in the middle of the small intestines (Liem et al., 2001; Kardong, 2012). So what good is a stomach?

With the evolution of jaws came the evolution of a stomach. The living jawless vertebrates, the lampreys and hagfish, have no stomach. Both of these vertebrates are feeding either on a blood meal (lampreys) or softened, decayed flesh (hagfish), foodstuffs that are easily chemically broken down and absorbed in the intestines. With jaws, an animal can break off and gulp down large pieces of food. You will anticipate from our previous discussion of diffusion and surface area that larger food items will take longer to break down and absorb. And there is a limit to how much and how big a food piece one can swallow before certain problems will result.

Let's think about a backyard barbeque and that all-American dish, the hamburger. No matter how clean or how thoroughly cooked your food is, you are always swallowing uninvited guests when you eat. These travelers include fungal spores, bacteria, and other microscopic organisms all trying to tap into the energy of your plate of hamburgers. If the meal could be made into something finer than the finest purée (however disgusting the idea), the food could go right to the intestines and the powerful enzymes there would quickly kill and break these invaders down. However, when you chew up a hamburger, the mashed up pieces are actually quite large, so large in fact that if this material went straight to your intestines it would take quite a while to break down (and probably cause a blockage). Any invaders that came along for the ride would have ample time to begin feeding on the food itself, releasing their toxic wastes into your body. Simply put, large chunks of food going right to the intestines would often rot before they could be digested.

As a scientist in a biology department, I must often use a variety of chemicals to clean and prepare anatomical material for my research and classes. I am no stranger to caustic acids, such as hydrochloric acid, which will quickly cause chemical burns on any exposed skin. Some of my students have initially been resistant to wearing pants and lab coats in the lab until they begin to read the safety sheets on these various chemicals and what they do to humans who inadvertently come into contact with them. It is perhaps no surprise, then, that the stomach of most jawed vertebrates secretes copious amounts of hydrochloric acid – any foreign invaders that hitch a ride on our food soon encounter a hostile, caustic environment (Liem et al., 2001). Very few living creatures are resistant to the corrosive power of hydrochloric acid.

An acidic environment can also be beneficial for the work of enzymes. A number of stomach-specific enzymes that exist in vertebrates rely on an acidic environment to speed up the process of food breakdown (Liem et al., 2001; Kardong, 2012). Like all enzymes, these proteins act like catalysts, substances that participate in chemical reactions by speeding them up, but do not themselves get broken down in the process. Thus, the larger food chunks are more quickly broken down into small bits, and any invaders are quickly destroyed. Stomachs are also good for continuing the work teeth initially put into smashing food into smaller pieces. The stomach can expand and accommodate large chunks of food, and its powerful, muscular walls will eventually smash and grind the ingested food into a paste called chyme (Liem et al., 2001; Kardong, 2012).

Based on these considerations, the stomach probably evolved in concert, or close in time, with the evolution of jaws. Animals that are swallowing large chunks of food must be able to temporarily store and physically alter them so that they are ready to be properly digested in the small intestines. An acidic environment, combined with enzymes, ensures that a harsh and deadly environment is present during a period of time when large food pieces that may possess detrimental invaders are present. This ensures that none of these potentially dangerous hitchhikers get into the intestines.

I end this discussion by pointing out that one jawless vertebrate appears to have evolved a stomach in parallel with that of jawed vertebrates. You will recall the very fishlike 'ostracoderms' called thelodonts. In some specimens of these animals, a distinct gut tract is preserved (Janvier, 1996). Between what is most likely the esophagus (the tube that brings food from the mouth to the guts) and the intestines is a large, in some cases barrel-shaped, pouch structure. This structure has been interpreted as a stomach (Wilson and Caldwell, 1993). It is intriguing to note that thelodonts have toothlike denticles lining their jawless mouths, and it is possible that these 'ostracoderms' were capable of swallowing larger prey items than most other jawless contemporaries (Janvier, 1996; Benton, 2005). If so, perhaps the independent evolution of a stomach was necessary in these vertebrates as well. Future research on these odd 'ostracoderms' may illuminate the function and role of this probable stomach in a jawless vertebrate.

Buoyancy, Drag, Vertebrae, and Fins

Buoyancy is a very critical issue for aquatic vertebrates. Simply put, buoyancy is an upward force generated by the water displaced by an object (Vogel, 2003; Bloomfield, 2006). As you may remember from your experiments as a child in the bathtub or kitchen sink, an object submerged in a cup or bucket of water full to the brim will cause the water to overflow. This is displacement–if you could measure the volume of water displaced, you would find that it is the same volume as the object (McGowan, 1999). In other words, if your rubber ducky has a volume of 12 cubic cm, then 12 cubic cm of water will be displaced or pushed aside when you submerge it. In your bathtub, if you completely submerge the rubber ducky, you will feel the force of the water trying to push it back up to the surface–this is the upward force of buoyancy generated by the water the bath toy displaced (McGowan, 1999; Vogel, 2003; Bloomfield, 2006).

Not everything that is placed or thrown, or that willingly jumps, into water will completely submerge. As we know, people float (some better than others), ice floats, boats float, and even some types of wood float. This is related to a property called density. Density is the mass of an object per unit volume (Vogel, 2003; Bloomfield, 2006). Put another way, density is governed by the mass of each particle of an object within a certain volume. For example, let's say we have two glass containers that we fill up with marbles. Each marble has its own mass, and the combined number of marbles in a given volume (the glass container) would give us a particular density. In one glass container, we place heavier marbles. We fill another container of equal size with the same number of lighter marbles. Logically, the second container would be less dense than the first container. If we placed these two containers in water, the denser (first) container would sink or submerge further into the water than the lighter (second) container.

The flesh and bone of vertebrates are very close to the density of water–all things being equal, the body of a vertebrate will be mostly submerged but float just at the surface of the water (Fig. 5.2) (McGowan, 1999; Vogel, 2003). For an aquatic vertebrate, staying submerged and controlling when to sink to the bottom or come to the surface are tasks critical for survival. Also, if you can avoid it, you don't want to be fighting to stay beneath the waves–that uses energy one could spend chasing prey or hiding from predators.

In the water, there are two opposing forces that act on an aquatic vertebrate (Fig. 5.2). The first is a downward pulling force from gravity. Just like on land, a greater mass will result in a greater gravitational pull on a given animal. Unlike on land, the animal is displacing a volume of water equal to its own volume, and that generates the upward force of buoyancy. If a vertebrate wishes to remain stable below the waves, the trick is to balance out the gravitational force pulling down on it with the upwardly buoyant force (Liem et al., 2001; Pough, Janis, and Heiser,

5.2. Vertebrate buoyancy illustrated. Because the shark, like all vertebrates, has a body that is naturally denser than water, it will tend to sink in the water. Although the shark displaces a volume of water equal to its volume that pushes up against it (buoyancy), the pull of gravity on its dense body will overcome the buoyancy force and it will sink.

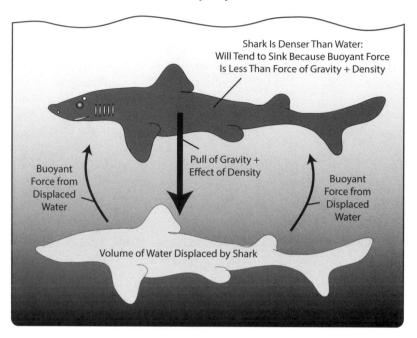

Vertebrate Buoyancy Illustrated

Shark Is Denser Than Water: Will Tend to Sink Because Buoyant Force Is Less Than Force of Gravity + Density

Pull of Gravity + Effect of Density

Buoyant Force from Displaced Water

Buoyant Force from Displaced Water

Volume of Water Displaced by Shark

2002; Vogel, 2003). This is more difficult than it first sounds: although the water is the same density around an animal, a vertebrate is not of a uniform density. Think of your own body—the organs are distributed in different regions, in different configurations, and at different densities. Some spaces are packed with organs, whereas others are filled with air or fat. This can mean that gravity pulls down on a vertebrate in a different region of the body than the force of buoyancy pushes up on it, creating instability (McGowan, 1999; Vogel, 2003).

If we added up all the high- and low-density spaces of a vertebrate body in three dimensions, we would arrive at a point called the center of mass. The center of mass can be thought of as an axis point through which gravity pulls (Vogel, 2003). Think of the center of mass like the pivot point on a seesaw—if forces are balanced on either side of the pivot, the seesaw remains horizontal. If a heavier force is placed on one side or the other, the seesaw will become unbalanced and tip toward the heavier side. By the same token, imagine if we had an aquatic vertebrate with more mass concentrated near its head than its tail—its center of mass (pivot) would be located closer to the head. This means that the animal would tend to rotate about its center of mass and dip forward. Just like on a seesaw, we would want to generate more force on the lighter side of the center of mass or decrease the force on the heavier side.

As we will later see, some aquatic vertebrates can actively control their balance in the water using an air-filled organ called a swim bladder. However, fins can greatly assist with balancing the opposing forces of gravity and buoyancy. It is probably no surprise, then, that the earliest jawed vertebrates developed more fins than the common ancestor they

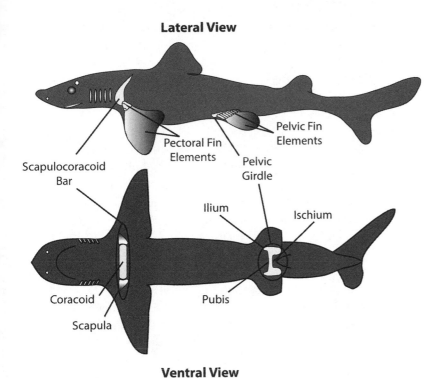

Lateral View

Pectoral Fin Elements

Pelvic Fin Elements

Scapulocoracoid Bar

Pectoral Fin Elements

Pelvic Girdle

Ilium

Ischium

Coracoid

Pubis

Scapula

Ventral View

5.3. Pectoral and pelvic girdles of a stylized jawed vertebrate, patterned after a shark. The pectoral and pelvic girdles provide expanded area and anchoring for the larger muscles necessary to control and move fins in the water.

shared with the 'ostracoderms.' The large dorsal fins, the expanded pectoral and pelvic fins, the presence of an anal fin, and the expansion of the heterocercal tail fin are all indicators that the earliest jawed vertebrates were fighting to stay balanced while pursuing their prey. As discussed previously in chapters 3 and 4, fins serve to stabilize animals in the water in a variety of ways. The vertically projecting dorsal fins and ventral anal fins act to prevent the body from rolling on its long axis, much as do the vertical stabilizers on torpedoes (Vogel, 2003; Bloomfield, 2006). The winglike pectoral and pelvic fins assist with steering but may also help with lift as the animal swims. The heterocercal tail provides an overall forward thrust, and adjustments of the entire body help to balance out the animal as it swims. In these ways, aquatic jawed vertebrates could actively overcome balance issues—many living fishes, such as sharks, still spend a majority of their lives swimming to keep balanced and oriented right-side up.

Having the muscle strength and force to move and orient the pectoral and pelvic fins during swimming requires a strong anchor point (Fig. 5.3). This led to selective pressure for the development of girdles, parts of the endoskeleton (and dermal skeleton in some parts of the pectoral girdle) into which the fins could connect and to which larger, more powerful fin-moving muscles could anchor into the body wall. The pectoral girdle develops mainly from two bones in jawed vertebrates, the scapula and the coracoid (Liem et al., 2001). The scapula is better known in humans as our shoulder blade, and it still remains the major girdle element into which our forelimbs attach. Humans and other mammals generally do not have a prominent coracoid bone, but in most other jawed vertebrates it is a large element. The coracoid establishes a connection between the scapula and fin on one side and the center of the chest region on the other. In the earliest jawed vertebrates, the scapula and coracoid on each side of the body fused to each other, and the left and right scapulo-coracoid (as they are now called) fused together at the midline to form a single, brace-like element called the scapulocoracoid bar (Pough, Janis, and Heiser, 2002; Benton, 2005; De Iuliis and Pulerà, 2011). This structure is still present in modern sharks and other cartilaginous fishes. As we will see, in different groups of jawed vertebrates additional bones comprise the pectoral girdle, and these will be highlighted where necessary.

The pelvic fins of most aquatic vertebrates tend to be smaller and less well developed (Fig. 5.3) (Liem et al., 2001). The pelvic girdle of nearly all vertebrates is generally comprised of three bones: the ilium, the pubis, and the ischium (Liem et al., 2001). The ilium forms the dorsal portion of the pelvic girdle, the pubis is the forward-facing portion, and the ischium is the caudal element. All three of these elements meet in the middle of the pelvic girdle, and here they form a recessed socket called the acetabulum into which the pelvic fin attaches and rotates. In some cases, unlike the scapulocoracoid bar, the right and left pelvic elements do not fuse together (but there are many exceptions), and the pelvic girdle bones are suspended within the trunk musculature.

The vertebral column was also modified in jawed vertebrates. No longer were there simple, arch-like segments that hovered above and below the spinal cord and notochord. Instead, vertebrae developed in two to three pieces that fused in various species. Typically, the main body of the vertebra encircled the notochord, much like a spool threaded onto a string (Fig. 5.1) (Liem et al., 2001). Vertebrae occupying the trunk developed neural spines and arches that protected the spinal cord from above, whereas vertebrae in the tail commonly developed another series of arches (hemal arches) below the notochord that supported and protected tail nerves and blood vessels (Fig. 5.1) (Liem et al., 2001; Pough, Janis, and Heiser, 2002). The segmented vertebral column both was stronger than the notochord by itself and, as a bonus, protected the spinal cord. In larger, faster-swimming active predators, the early jawed vertebrates also developed larger and more powerful myomeres. The strong contractions of these trunk muscles were resisted most effectively by the vertebral column, which also provided additional anchorage for the myomeres (Liem et al., 2001; Benton, 2005).

Finally, we turn to the problem of drag. As an object pushes through a fluid such as water, it must overcome the fluid's resistance to forward travel. It turns out there are two distinct but intertwined types of drag. The first is related to friction. As layers of water are separated and move around the traveling object, this causes friction both between the layers of water and between the water and the object (Bloomfield, 2006). This friction acts to slow the forward progress of a moving object in the water. As speed increases, the trail of disturbed water behind a moving object,

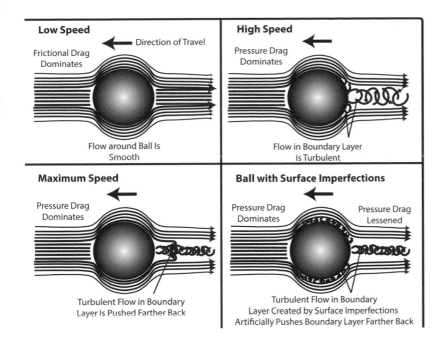

5.4. Examples of drag on a ball moving through the water. At low speeds, the friction of the water in the boundary layer around the ball is the stronger drag force. When speed increases, the boundary layer water cannot reach the back of the ball, and is pushed backward, creating drag and turbulence behind the ball. However, at faster speeds the turbulent boundary layer is pushed far enough behind the ball that less turbulence is generated. A ball with surface imperfections creates a physical situation similar to what occurs at maximum speed – it pushes the boundary layer farther back and decreases the relative drag and turbulence. Diagrams patterned after Bloomfield (2006).

the wake, becomes more turbulent, and that causes the second form of drag, known as 'pressure drag' (Bloomfield, 2006). Pressure drag occurs because the turbulent water behind the moving object has a lower pressure than the water in front of the object. This pressure difference 'sucks' the object back, requiring more energy to keep moving forward. You can think of pressure drag like a growing invisible weight towed behind a moving object. The faster the object travels, the more difficult it becomes to pull the ever-growing invisible weight. Thus, at low speeds frictional drag dominates, but pressure drag becomes dominant and more severe as speed increases (Vogel, 2003; Bloomfield, 2006). How do aquatic vertebrates manage to overcome frictional and pressure drag?

As you might imagine, an object's shape determines how well it travels through water and how well it reduces its drag. Let's start by considering a spherical object like a ball (Fig. 5.4). As the ball moves through the water, it must push the layers of water aside. If the ball is traveling slowly, the water in front of the ball separates smoothly and the layers of water travel over or under the ball's surface, connecting back together behind the ball. Behind the ball a smooth and gentle wake forms that will not cause much pressure drag. In this case, the only significant drag on the ball will be frictional. The water layers must bend around the top or bottom of the ball, which changes their pressure and speed, much as these physical properties of water changed inside our garden hose when obstructed by our thumb in the previous chapter. In this case, the water flowing over and under the ball travels at a higher speed and lower pressure than the water in front of and behind the ball. The slower and higher-pressure water in front of and behind the ball balances out the lower-pressure and higher-velocity water traveling around it. This balancing of forces is why significant pressure drag does not form around the slow-moving ball (McGowan, 1999; Vogel, 2003; Bloomfield, 2006).

If you could look closely at the surface of the ball as it moves through the water, you would notice something called the boundary layer. This is an area next to the surface of the ball where frictional forces slow down the passing water (Bloomfield, 2006). At low speeds, this slowing is not a problem because the difference in speed between the water in the boundary layer and those layers traveling over it is minimal. Because the boundary layer is low-pressure water, it needs an 'assist' from the overlying water layers to travel all the way to the back of the ball where the water pressure is greater (Bloomfield, 2006).

However, as speed increases, the differences in velocity between the boundary layer and overlying water become greater–the water over the boundary layer begins to move much faster. Although the overlying water layers have enough momentum to reach the back of the ball, they can no longer impart enough energy to the boundary layer water to assist it. There comes a point where the boundary layer water does not reach the back of the ball any longer. The boundary layer then begins to be pushed in the reverse direction, toward the front of the ball and in opposition to the water flowing over it. This causes the wake behind the ball to become extremely turbulent, which further pushes the water flowing around the boundary layer even farther away. This is the pressure drag, and it can eventually stop the ball dead in its tracks (Bloomfield, 2006).

This brings us back to the importance of shape. Certainly, having a spherical body is not ideal for efficiently swimming through the water. It is not necessarily the front of the ball that is the problem, because here the water layers separate smoothly. It is on the back half of the ball where we encounter turbulence and pressure drag. If we could keep a relatively spherical front end but stretch out the back end into a tapered tail, this would certainly change things. Such a teardrop-shaped form is called 'streamlined' (Fig. 5.5) (Vogel, 2003; Bloomfield, 2006). If something is streamlined, the shape is carefully tapered so that the pressure of the water in the boundary layer gradually rises from front to back (McGowan, 1999). This resolves much of the problem associated with pressure drag–the water in the boundary layer has enough pressure on its own, regardless of speed, to make it to the back of the object without separating from the water above it or reversing its direction. In essence, a streamlined shape greatly cuts down on turbulence.

Not surprisingly, many fishes and other aquatic vertebrates, such as dolphins and whales, and even some invertebrates such as squid and shrimp, have a streamlined shape (McGowan, 1999; Liem et al., 2001; Pough, Janis, and Heiser, 2002; Vogel, 2003). We know, then, when we see a fossil vertebrate with a streamlined shape that we have an animal that can move through the water without much resistance from pressure drag. We can also infer that fossil aquatic vertebrates without streamlining led lives that did not require them to be active swimmers.

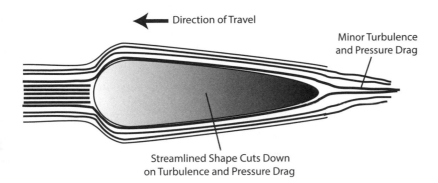

Direction of Travel

Minor Turbulence
and Pressure Drag

Streamlined Shape Cuts Down
on Turbulence and Pressure Drag

5.5. How a streamlined shape cuts down on turbulence and pressure drag. Unlike a ball or sphere, a streamlined shape allows the water layers to flow back together behind it much more gradually, creating less turbulence and pressure drag. The boundary layer of water remains smooth for much longer, and the region of turbulence is small and pushed well back beyond the object. Diagrams patterned after McGowan (1999) and Bloomfield (2006).

We still, however, have the problem of frictional drag. With increased speed, the frictional force of the water traveling over the body of a fish still rises, and this acts to slow the animal. One way that many living fishes may reduce their frictional drag is through the secretion of a mucus coat across their body surface (Liem et al., 2001; Pough, Janis, and Heiser, 2002; Kardong, 2012). The slippery snot-like coating would theoretically allow water in the boundary layer to more easily slip over a fish, and in some experiments this has indeed been shown to (slightly) decrease frictional drag. However, other experiments have shown no significant difference arising from the mucus coat in other fishes (Vogel, 2003). Also, we must bear in mind that the mucus coat serves many nonlocomotor functions in fishes, such as bacterial resistance (Liem et al., 2001). More research is required to determine just what, if any, advantage the mucus coat of many fishes gives to decreasing frictional drag.

Another 'trick' for decreasing pressure drag, at least in the balls of many sports, is the presence of small imperfections, dimples, or fuzz. These dimples and projections interfere with the typically smooth-flowing boundary layer, causing it to become turbulent (Bloomfield, 2006). To understand why turbulence in the boundary layer acts to decrease pressure drag, we must first return to our fast-moving ball. When we left off, our speeding ball's boundary layer did not have enough energy to fight the pressure gradient and ended up reversing on itself, causing turbulence. It turns out that if you can get an object like a ball moving very fast, the boundary layer itself becomes turbulent. This boundary layer turbulence allows it to mix with, and be pushed backward by, the overlying layers (of air in the case of sports balls) (Bloomfield, 2006). The boundary layer still separates from the ball and causes turbulence, but this occurs much further back on the ball. As a consequence, the wake behind the ball is much smaller, leading to less relative pressure drag (Bloomfield, 2006). A similar scenario would occur with our ball in the water. As counterintuitive as it may seem, a turbulent boundary layer actually allows a ball to go faster and farther!

Returning to the imperfections present in many sports balls, you may now realize that these dimples and projections cause the boundary layer to become turbulent much sooner than it normally would, allowing that particular ball to travel farther than one that was completely smooth. If you imagine our ball in the water covered in small bumps, you can begin to appreciate how these would quickly cause the boundary layer to become turbulent. Such turbulence would decrease the relative pressure drag. Now imagine a jawed vertebrate with bumpy skin, and you can infer that such projections might cause boundary layer turbulence. Sharks, for example, have skin studded with tiny, toothlike scale denticles (Liem et al., 2001; Pough, Janis, and Heiser, 2002). It has been suggested that this feature of shark skin increases the boundary layer turbulence around the animal, allowing it to swim faster with less pressure drag (Dean and Bhushan, 2010). Presently, it remains unclear how much these denticles actually contribute to reducing pressure drag. However, it is significant to note that the exoskeletal bones of many early vertebrates were also covered in denticles, which may have created somewhat similar imperfections in the skin surface. By extension, the presence of such projections in an aquatic fossil vertebrate might suggest that the animal was an active swimmer. Certainly, more research into this is required.

Next Stop – Jawed Vertebrate Diversity

In summary, the evolutionary modification of the cranial pharynx into jaws marked a turning point in the vertebrate story. The possession of jaws with teeth, a stomach, strongly anchored pectoral and pelvic fins, and a body form that could balance out problems associated with buoyancy and drag heralded the beginnings of more active and predatory lifestyles. Body form and function in the jawed vertebrates quickly diversified, and one group of jawed vertebrates built upon this chassis to move onto the land. What follows in the next three chapters is a general survey of the three major branches of jawed vertebrates, the common ancestor of which arose nearly 450 Ma (Card 91). Although some of these lineages are completely extinct, others continue to be successful and adaptive to this day.

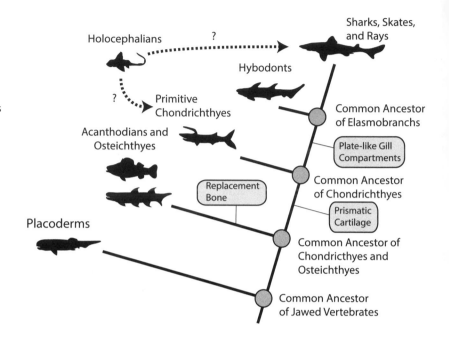

6.1. Family tree (phylogeny) of jawed vertebrate relationships. Note that the position of the holocephalians (ratfish) is uncertain: they may be an off-branch of the primitive chondrichthyans, they may be odd descendants of the sharks and rays, or they may be their own derived group. Derived from Pough, Janis, and Heiser (2002, 2013) and Benton (2005).

Sharks, Skates, and Rays

Holocephalians ?

Hybodonts

? Primitive Chondrichthyes

Common Ancestor of Elasmobranchs

Acanthodians and Osteichthyes

Plate-like Gill Compartments

Replacement Bone

Common Ancestor of Chondrichthyes

Prismatic Cartilage

Placoderms

Common Ancestor of Chondricthyes and Osteichthyes

Common Ancestor of Jawed Vertebrates

Placoderms and Cartilaginous Fishes

THE EARLY HISTORY OF AUTOMOBILES IS REPLETE WITH A VARIETY of internal combustion engines, but certain forms were eventually more successful than others. After a time, the most efficient and most economical versions remained, and other variants, even ones that may have had certain distinct advantages, were discontinued. Much as in the history of automobiles, when the new jawed chassis made its appearance, it came in diverse forms, but eventually a much narrower variety survived. Why some of these forms persisted and others went extinct remains difficult to resolve, but changes in the environment and access to food resources were almost certainly the primary causes.

The earliest jawed vertebrates appeared 450 Ma (Card 91) and had already diversified into several lineages around the same time as the jawless 'ostracoderms' were populating the globe (~420 Ma; the tail end of Card 91). Traditionally, the early jawed vertebrates have been divided into four major groups on the basis of trait states (placoderms, acanthodians, chondrichthyans, and osteichthyans), but current research suggests a more complex picture. New fossil and trait state data suggest that, as with the 'ostracoderms,' a number of jawed fishes have been grouped together for convenience, and not because they form a natural group (Friedman and Brazeau, 2010; Zhu et al., 2013; Zhu, 2014; Giles, Friedman, and Brazeau, 2015). For our purposes, we are interested in the mechanical developments in the early jawed vertebrate chassis. To keep things simple, we will follow a somewhat more traditional scheme and focus on three major groups: the placoderms, the cartilaginous fishes, and the bony vertebrates (bony fishes and the land-adapted vertebrates, the tetrapods) (Fig. 6.1). According to this arrangement, the placoderms are extinct, whereas the other two jawed vertebrate groups still populate the earth to this day (Carroll, 1988; Benton, 2005; De Iuliis and Pulerà, 2011). This scheme allows us to more easily outline the accumulating adaptations in the skeleton from the most primitive jawed vertebrates to modern forms.

The geological period known as the Devonian has been called the 'Age of Fishes' and for good reason: during this time (~416–360 Ma; Cards 91–93) a huge diversity of jawed vertebrate groups with fishlike body forms coexisted. This time period is excellent for our purposes for several reasons. First, it is a time when whole-body fossils of many early jawed vertebrates appeared, giving us a nice window into the basic jawed vertebrate chassis and its diversity. Second, during this time we see the rise and fall of the now totally extinct placoderms. Third, it is the environmental setting under which the first land-living vertebrates evolved. If we could travel back in time to this period, any paleontologist worth his or her salt would take a boat (let's call it *Devonian Destiny*) and troll the ancient waves across the globe to net some early jawed vertebrates. Some jawed vertebrates had external bony armor; some developed an internal, bony skeleton; and others lost bone altogether. In this chapter, we turn first to the extinct placoderms and the cartilaginous fishes. Although members of these groups possess or possessed bone tissues in some form, they do not possess the fully bony internal skeleton we will later see in the various bony fishes and the tetrapods.

Placoderms: Armored Attitude

Pedigree Most-Primitive Jawed Vertebrates

Date of First Appearance ~420 (?450) Ma (Card 91)

Specialties of Skeletal Chassis Armor-Plated Head with Shear-like Jaws

Eco-niche Active Carnivores, Scavengers, Filter and Deposit Feeders

Placoderms have traditionally been regarded as an extinct, early branch of jawed vertebrates with no modern descendants. More recent analyses suggest they are not a natural group but instead the basal members of all jawed vertebrates (Giles, Friedman, and Brazeau, 2015). Thus, they are probably both the most primitive jawed vertebrates and closer and closer cousins of the other jawed vertebrates in our family tree (Janvier, 1996; Benton, 2005; Friedman and Brazeau, 2010; Zhu et al., 2013; Giles, Friedman, and Brazeau, 2015). Whatever their exact relationships, their place at the root of the jawed vertebrate family tree makes them an ideal model from which to compare more derived jawed vertebrates. Bits of scale and skeletal fragments suggest placoderms may have first appeared as long ago as 450 million years (Card 91) (Benton, 2005). However, to see these animals in their full glory, we first set sail to the northwest of what will one day be Australia, approximately 375 million years ago (Card 92). At this time, there was a large reef system that, like similar environments in today's oceans, was teeming with schools of fishes and other oceanic life (Long, 2006). To swim among these environments would have been a spectacular experience, and we would have encountered an astounding diversity of placoderms. Some were small bottom scavengers that looked like a cross between a stingray and a crab, whereas others were large, fearsome, and downright intimidating sharklike animals (Long, 2006). In fact, the jaws of some placoderms were probably as strong and lethal as those of living sharks. Placoderm diversity was truly remarkable, but goes far beyond the scope of this book. For our purposes, it is most instructive to look at the two most polar-opposite groups in the placoderm family tree: the arthrodires and the antiarchs.

Were they alive and patrolling the oceans today, the arthrodire placoderms, especially the largest ones, would have made the shark in *Jaws* appear laughable by comparison. These animals had thick, bony, armor-plated skulls with large, sideways-facing eyes (Fig. 6.2 and Plate 4). We see in these animals the first examples of sclerotic bones, little bones that ringed and support the eyeball. We don't have such bones, but many vertebrates do. In aquatic vertebrates such as placoderms, sclerotic bones are thought to hold the shape of a large eye constant as the water rushes over it (Young, 2008). In this way, the animal's ability to focus on prey will

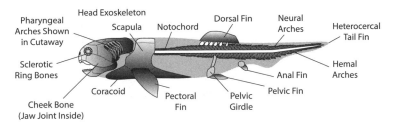

Lateral View

Pharyngeal Arches Shown in Cutaway

Head Exoskeleton

Scapula   Notochord

Dorsal Fin

Neural Arches

Heterocercal Tail Fin

Sclerotic Ring Bones

Hemal Arches

Anal Fin

Pelvic Fin

Coracoid

Cheek Bone (Jaw Joint Inside)

Pectoral Fin

Pelvic Girdle

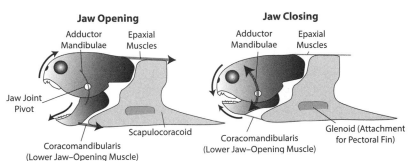

Jaw Opening

Adductor Mandibulae

Epaxial Muscles

Jaw Joint Pivot

Scapulocoracoid

Coracomandibularis (Lower Jaw–Opening Muscle)

Jaw Closing

Adductor Mandibulae

Epaxial Muscles

Coracomandibularis (Lower Jaw–Opening Muscle)

Glenoid (Attachment for Pectoral Fin)

6.2. Anatomy and skeletal function of placoderms. A generalized arthrodire placoderm based on the predators *Compagopiscis, Eastmanosteus,* and *Dunkleosteus.* The inset shows the basic mechanics of jaw opening and closing in an arthrodire placoderm. The major muscle groups that were probably involved are indicated with dashed lines and arrows to show their lines of action. Generalized arthrodire placoderm based on specimens illustrated in Carroll (1988), Benton (2005), and Long (2006).

not be distorted when swimming. The head armor was so extensive that it covered the pharyngeal arches that housed the gills, and thick bony plates also surrounded the connection between the body trunk and the back of the skull. Likely, movement of the bony plates over the gills was necessary for their aeration. The jaw bones were equipped with odd, bladelike teeth. So odd are these structures that there has been considerable debate among the paleontologists working on these animals as to whether or not they were teeth at all (Burrow, 2003; Smith and Johanson, 2003; Young, 2003; Rücklin et al., 2012; Donoghue and Rücklin, 2014). It is now recognized that these bladelike extensions from the jaws were indeed teeth thanks to powerful X-ray techniques that can show the internal structures necessary for such an identification (Rücklin et al., 2012). Odd as they are, one thing is certain – they would have caused substantial, nasty bites! In some arthrodire placoderms there is even evidence that these teeth acted like a pair of self-sharpening scissors, refining their bladelike surfaces with each bite (Benton, 2005; Long, 2006).

There were two nasal openings, and in many arthrodire placoderms canals of the lateral line system for detecting water-borne sound waves are preserved, etched into the exoskeleton. The head armor and trunk armor were not plain, but were instead covered by thousands of toothlike bumps. Whether these bumps were for display, protection, reduction of drag while swimming, or all of these functions remains unclear. Beneath the head skeleton were cartilaginous pharyngeal arches that held up the gills. Paired pectoral and pelvic fins were present, and these were anchored to a scapulocoracoid bar and pelvic girdle, respectively. It should be noted that several additional bones comprise the pectoral girdle in addition to the scapulocoracoid in placoderms. While not important to our story here, these additional bones become significant in the Osteichthyes, or the bony vertebrates, discussed in more detail in later chapters. A long dorsal fin and small anal fin were present in some arthrodire placoderms, and the tail was heterocercal (Benton, 2005; Long, 2006).

Two well-known arthrodire placoderms from the Devonian of Australia allow us to explore how the jaw mechanics of these early jawed vertebrates worked. One is a smallish arthrodire called *Compagopiscis* that measures in at just over 30 cm, and another is a large, apex predator called *Eastmanosteus* that probably stretched over 2 meters long from head to tail (Fig. 6.2) (Long, 2006). We will also make some reference to *Dunkleosteus*, a well-known, colossal, 6-meter-long apex predator we find across the world in Cleveland, Ohio (Fig. 6.2) (Anderson, 2009). Despite their great differences in size, the heads of these animals have been preserved as exquisite three-dimensional fossils, and a wealth of information can therefore be gleaned from their careful study.

You will recall that, in jawed vertebrates, the palatoquadrate forms the upper jaw and the Meckel's cartilage forms the lower jaw. These elements form first as cartilaginous struts in jawed vertebrates, and then are fused with overlapping skull bones later in development (except for cartilaginous fishes, as we will later see). In arthrodire placoderms, the palatoquadrate and Meckel's cartilage are covered and incorporated into a series of large, platelike jaw bones. A cleaver-like bony mandible engulfs the Meckel's cartilage, and it bears an articular process that connects it to the upper jaw joint in the skull. This region of the jaw, called the articular, is an ancient and common lower jaw joint that we will see again and again in nearly all jawed vertebrates. In arthrodire placoderms, the palatoquadrate, being surrounded by large bony plates, has shrunk to a few small pieces of cartilage that do not articulate (connect) directly with the lower jaw. Instead, platelike cheek bones in these animals have an inwardly facing knob-like process that the articular of the lower jaw connects with. In addition, because the lower jaw connects to the skull inside the cheek bones, this had the effect of eliminating any lateral motions of the jaws and enhanced their shear-like rotation in a vertical plane.

To appreciate the terrifying force behind the jaws of these armored predators, we take our first foray into lever mechanics. Levers are simple machines that translate and amplify power (Bloomfield, 2006). In its most basic form, a lever is a board that, when balanced over a wedge, can be used to transmit force and move objects (Fig. 6.3). If I want to lift a heavy object (for example, a refrigerator) off the ground, I could simply bend my legs, grab both sides, and lift with all my might. I might be able to lift a heavy refrigerator, but probably not for long, not very well, and not without rupturing something in my back. If I instead slide a board under the refrigerator, prop the board on a block of wood, and push down on the opposite end of the board, I will find that it is much easier to lift the refrigerator. Unlike my muscles on their own, the rotation of the board translates and amplifies my relative strength.

More specifically, there are three parts to any lever system: the so-called in-lever and out-lever, and the fulcrum (McGowan, 1999; Vogel,

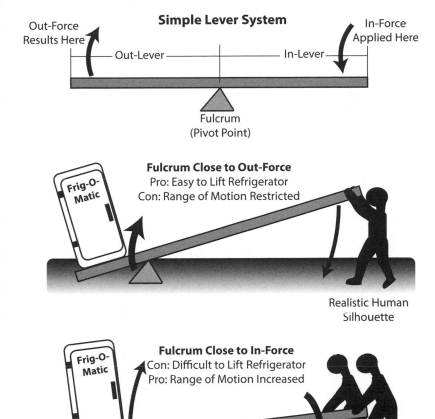

Simple Lever System

Out-Force Results Here

Out-Lever

In-Lever

In-Force Applied Here

Fulcrum
(Pivot Point)

Fulcrum Close to Out-Force
Pro: Easy to Lift Refrigerator
Con: Range of Motion Restricted

Frig-O-Matic

Realistic Human Silhouette

Fulcrum Close to In-Force
Con: Difficult to Lift Refrigerator
Pro: Range of Motion Increased

Frig-O-Matic

Realistic Human Silhouettes

6.3. How a simple lever system works. Changes in the length of the in-lever and out-lever relative to the fulcrum affect the so-called mechanical advantage of lifting or moving an object (in this case, a Frig-O-Matic refrigerator). See text for further explanation.

2003; Bloomfield, 2006). In our refrigerator-lifting system, the board is the lever and the fulcrum is the wood block that acts as the point around which the board rotates (it is a pivot). The part of the board that runs from underneath the refrigerator to the fulcrum is the out-lever, the part that applies the force coming 'out' of the lever to move the refrigerator. The portion of the board opposite the fulcrum and refrigerator that we push on is the in-lever, the part that transmits the force we put 'in' to the out-lever. In any lever system, if you want to amplify power and transmit more force at the out-lever, you increase the size of the in-lever relative to the out-lever. In other words, a longer in-lever and a shorter out-lever yield more applied force. However, in any system there are bound to be trade-offs, and lever systems are no exception. The trade-off in lever systems is related to the range of motion at the out-lever. Although a short out-lever coupled with a long in-lever will provide a great deal of force, its arc of rotation is limited because the out-lever is short. In contrast, if your goal is to increase the range of motion at the out-lever by making it longer relative to the in-lever, you will sacrifice force. As with everything in life, you can't have your cake and eat it too.

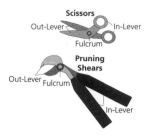

6.4. Application of lever system mechanics to understanding the mechanical advantage of scissors versus pruning shears. As with a simple lever system, there is a trade-off between range of motion and force applied to an object. See text for more details.

You intuitively understand levers even if you don't know the physics behind them. For example, if you have ever used various pruning shears to trim bushes, you have used levers to your advantage. By way of comparison, let's say you were unsure as to whether to use scissors or pruning shears to trim your bushes. In these two tools, the handles are the in-levers, the screw around which the blades rotate is the fulcrum, and the blades themselves are the out-levers (Fig. 6.4). Thus, in scissors with short handles (in-levers) and long blades (out-levers) you will get a wide range of motion but not very much cutting power. This doesn't matter so much with something thin and as easily cut as paper. To cut thick plant branches, the pruning shears have an advantage over scissors because they have handles (in-levers) that are longer than their short blades (out-levers). Certainly, with their shorter blades, the pruning shears have a shorter range of motion than do the scissors, but as long as the blades can gape wide enough to seize the offending branch, this is good enough.

Now imagine the mouth of a 6-meter-long armored *Dunkleosteus* as a giant pruning shear, and you begin to get the picture. *Compagopiscis*, *Eastmanosteus*, and *Dunkleosteus* specimens show that arthrodire placoderm jaws worked on the same basic principle as a pruning shear with a so-called anvil design (Anderson, 2009; Anderson and Westneat, 2009). In these tools, there is essentially one movable upper blade that squeezes and cuts branches by pressing them against a nonmovable lower platform called an anvil. It turns out that in arthrodire placoderms much of the jaw movement happened in the skull, whereas much less movement occurred at the Meckel's cartilage and surrounding lower jaw elements. As with an anvil-design pruning shear, opening the mouth required the upper jaw and skull to rock dorsally, while the lower jaw dropped and jutted cranially (Anderson, 2009; Anderson and Westneat, 2009). To bite prey, the upper jaw and skull were rotated down and into contact with the lower jaw, which retracted and rose to act as the anvil in these formidable mouth shears (Fig. 6.2).

How could the arthrodire skull rock dorsally, especially if placoderm skulls were rigid and covered in heavy dermal armor plates? The answer lies in a chink in the armor, a gap that formed between the head and trunk armor that gave the skull the necessary room to rotate dorsally (Janvier, 1996; Benton, 2005; Long, 2006; Anderson, 2009; Anderson and Westneat, 2009). To be specific, the trunk armor forms dorsal to the scapulocoracoid. The gap between the skull and trunk armor, technically called the nuchal gap, is similar to the space between your car's body and the engine hood that allows you to lift it up to check the engine block (Plate 4). In placoderms, as in all vertebrates, an ancient series of muscles, the epaxials, pull on the back of the skull and rotate it dorsally (Liem et al., 2001). If you place your hand on the back of your neck and then tilt your head up, you will feel these cord-like muscles squeezing and shortening. It has been determined that the nuchal gap in placoderms was spanned by epaxial muscles, and that their contraction pulled the skull up and back (Anderson, 2009; Anderson and Westneat, 2009). Unlike the hinge joint in your car's hood, ball-and-socket joints between the trunk and skull allowed these movements.

But there is more to this jaw-opening story. The skull, lower jaw, trunk, and scapulocoracoid were all interconnected to one another: when one part of the system moved, it caused other parts to move as well. In short, jaw opening in arthrodire placoderms was a simultaneous, linked action (Fig. 6.2) (Anderson 2009; Anderson and Westneat, 2009). The skull linked to the trunk armor by the epaxial muscles. The lower jaw linked to the scapulocoracoid through a jaw-dropping muscle, and, of course, there is the joint between the cheek bones of the skull and the lower jaw. As the epaxial muscles pulled back on the skull, it rotated upward. Because the cheek bones were attached to the lower part of the skull, such a movement rotated these bones forward. This action in turn pushed the lower jaw down and forward, a movement likely aided by the jaw-dropping muscle. What did this system do for arthrodire placoderms? Simply put, it allowed them to quickly open and at the same time project their jaws toward potential prey. Because water is a dense medium, 800 times more dense than air (Pough, Janis, and Heiser, 2002; Vogel, 2003), such rapid mouth opening would have caused water and any potential prey in front of the jaws to be sucked into the gaping maw of these predators at terrifying speed.

This same system, worked in reverse, slammed the jaws shut with immense power. To explain jaw closing, it is now time to introduce another muscle. All jawed vertebrates share a muscle that closes the jaws, named the adductor mandibulae. The name tells precisely what it does: an adductor is a muscle that brings bones together, and mandibulae is the Latin word for 'mandibles.' The adductor mandibulae closes the jaws by bringing them together. This muscle always anchors itself across the side of the skull just behind the eye (what you would call your temple and cheek bone), and pulls powerfully on the jaw bones. In fact, this muscle is most always associated with the same bones, and often leaves calling cards of its presence in the form of bony scars. In other cases, as in many fishes and the placoderms, the adductor mandibulae was sandwiched in a canal between these temple and cheek bones.

Returning once more to our shears analogy, you will recall that the longer the handles and shorter the blades, the more force could be delivered when cutting. In arthrodire placoderms, the adductor mandibulae was the handles and the skull and lower jaws were the blades. Compared to the size of the jaws, the region for attachment of the adductor mandibulae was huge. Recently, a computer model was developed that took into account a range of adductor mandibulae sizes and contraction speeds to calculate the force behind placoderm jaw closing. In scientific studies we measure force in newtons (N), where the force (F) equals the mass (m) times the acceleration (a). For example, an object that weighs 1 kilogram (~2.2 pounds) has a force of 9.8 N (mass of 1 kilogram times acceleration due to gravity, or 9.8 meters per second per second = 9.8 N). For the 6-meter-long *Dunkleosteus*, the computations revealed an astonishing

6000–7400 N of bite force for this animal (Anderson and Westneat, 2009). For reference, the bite of great white sharks has been estimated with bite forces in excess of 18,000 N (Wroe et al., 2008). Still, this is an impressive amount of force and would certainly have caused substantial damage to whatever animal was unlucky enough to end up between those jaws.

Recently, discoveries of fossil embryos inside adult arthrodires and their close relatives show that some of these animals gave live birth and had internal fertilization (Long, Trinajstic, and Johanson, 2009). Some of these placoderms even have special bones that project off the pelvic girdle that look reminiscent of claspers, organs of sperm transfer seen today in sharks and their kin (Long et al., 2014). We will discuss claspers with the cartilaginous fishes, but it is interesting to note that this would be the first evidence of this mode of reproduction in vertebrates (Long et al., 2014). It would also mean that we would be able to sex some placoderms, a task usually difficult or impossible with most fossil vertebrates.

Before we leave placoderms, there is a kinder, gentler side to these creatures, best exemplified by the odd antiarchs (Fig. 6.5). Unlike the arthrodires, the antiarch placoderms were slow-moving, clay-eating bottom-dwellers that were flattened dorsoventrally like some the 'ostracoderms' we encountered earlier. The best known example of an antiarch is a species known from fossil sites throughout the Devonian world called *Bothriolepis*, an animal with a body length approximately 30-cm-long (Benton, 2005). Their skulls were rounded at the front and adorned with two small, sideways-facing eyes perched dorsally. Unlike in the arthrodire placoderms, the skull armor and trunk armor were solidified into a single unit. The mouth pointed straight down, and the jaws of these animals were just plain weird. The original upper jaw bone of *Bothriolepis* was reduced and bore no teeth, but the cheek bones, with toothlike projections, had migrated down to serve as mud graters (Janvier, 1996; Benton, 2005). The lower jaw was a pathetic, thin, twisted piece of bone, and unlike its arthrodire cousins, *Bothriolepis* obviously did not require great jaw strength. Then again, how much force does one need to chew clay?

The pectoral fins of *Bothriolepis* are just plain bizarre for a vertebrate: they look like crab legs! The pectoral fins were completely encased in jointed armor, and the muscles, nerves, and blood vessels found their way inside via a large opening in the trunk armor (Fig. 6.5) (Carroll, 1988; Benton, 2005; Long, 2006). This is very strange indeed, as this is the typical anatomy of limbs of arthropods, the invertebrates that include spiders, scorpions, insects, lobsters, and crabs. The shoulder joint of these pectoral fins was a complex series of pits and ridges that allowed the appendage to move and orient itself in various positions. Based on the presence of fin spines and somewhat similar looking fins in some modern fishes, it is hypothesized that the pectoral fins were used to anchor and hold *Bothriolepis* in the muddy bottoms of lakes and oceans (Long, 2006). The remainder of the body was fairly unremarkable, consisting of a single dorsal fin and a tail that tapered into a small heterocercal fin.

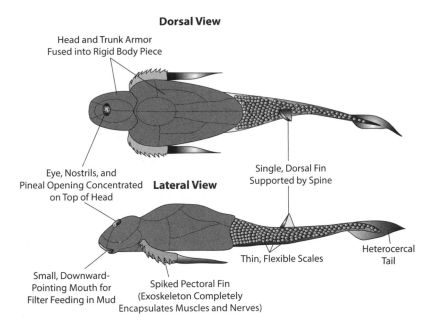

Dorsal View

Head and Trunk Armor
Fused into Rigid Body Piece

Eye, Nostrils, and
Pineal Opening Concentrated **Lateral View**
on Top of Head

Single, Dorsal Fin
Supported by Spine

Small, Downward-
Pointing Mouth for
Filter Feeding in Mud

Spiked Pectoral Fin
(Exoskeleton Completely
Encapsulates Muscles and Nerves)

Thin, Flexible Scales

Heterocercal
Tail

6.5. Dorsal and side views of the antiarch placoderm, *Bothriolepis*. Note the resemblance to modern stingrays in the dorsoventrally compressed body form. The crab-like pectoral appendages have an exoskeleton that covers fin-moving muscles within. Reconstruction based on specimens illustrated in Carroll (1988), Benton (2005), and Long (2006).

We sometimes associate the term 'bottom-dweller' with something inferior. However, based on their diversity and distribution across the world, *Bothriolepis* shows that such labels can be deceiving. This little animal is found in all sorts of Devonian environments and probably was so successful because the bottom muds and clays of rivers, estuaries, and continental shelves tend to be organic rich (Benton, 2005). *Bothriolepis* and other antiarchs are the first vertebrates to truly appreciate the adage "slow and steady wins the race."

The Chondrichthyes: The Cartilaginous Fishes

The cartilaginous fishes, technically the Chondrichthyes, are one of the major success stories among vertebrate animals. Still going strong today, the cartilaginous fishes include sharks and their relatives, the skates and rays, along with an oddball group, the chimaeras (Liem et al., 2001; Pough, Janis, and Heiser, 2002; Benton, 2005). Given their relationship to the more primitive placoderms, it is somewhat mysterious why they would have done away with bone altogether in their skeletal chassis. Certainly, we will point out how wonderfully constructed their skeletons are, but no one has yet figured out why bone was dispensed with in the first place. We do know that modern Chondrichthyans lack genes present in other jawed vertebrates that produce endochondral (internal) bone (Venkatesh et al., 2014; Zhu, 2014), but the evolutionary factors that led to this loss of bone remain unclear. To understand the evolution of the cartilaginous fish chassis and appreciate it in all its glory, we first turn to modern sharks to give us some perspective on what we are looking at in the fossil record.

Pedigree Cartilaginous Jawed Fishes

Date of First Appearance 450 Ma (Card 91)

Specialties of Skeletal Chassis Cartilaginous Skeleton Lacking Bone

Eco-niche Carnivores, Scavengers, Filter Feeders

For many years I have had the opportunity to get to know shark anatomy intimately. I teach a course called comparative vertebrate anatomy that strikes fear into the hearts of many students. It is a rite of passage for many future zoologists, medical doctors, veterinarians, dentists, physical therapists, and paleontologists in which they learn the overwhelming introductory details of the vertebrate body and its organs while honing dissection skills that will serve them later in their chosen fields. It is not an easy course, and even the hard-core 'A' students find they must study for this class intently. After some brief introductions to the chordate animals and the lamprey, the students become immersed in the anatomy of a smallish shark (80–100 cm long), the spiny dogfish (known to my colleagues as *Squalus acanthias* or the 'spiny squaloid shark') (De Iuliis and Pulerà, 2011).

Without dermal bones covering the skull and jaws, these regions in the spiny dogfish can be perplexing to the novice anatomist because they look so strange (Fig. 6.6 and Plate 5). The skull is technically called the chondrocranium (meaning 'cartilage skull'). We can divide the chondrocranium into the braincase (the part that encapsulates the brain) and the facial skeleton (the region cranial to the braincase). Viewed laterally, the braincase is rectangular whereas the facial skeleton tapers into a triangular nose cranially. Large orbits are present, surrounded by numerous small openings. Dorsally, the triangular nose is seen to be scooped out and hollow. A pineal opening lies just caudal to the nose, and additional tiny openings run along each side of the facial skeleton. There is also a large hole over the inner ear, similar to what we have seen previously for osteostracans. The ventral region of the facial skeleton has nostril openings piercing bulbous pieces of cartilage caudal to the triangular nose. However, there are two nostril openings on each side, giving us a grand total of four nostrils. To top it all off, ventral to the orbits along the length of the facial skeleton are rail-like tracks. I told you: weird city.

Of course, when we begin to associate the soft tissues with these odd features, one begins to appreciate that they are signposts on a skull chassis that tell you what goes where. Some obvious signals are the size of the orbits and the bulbous regions of cartilage-bearing nostrils caudal to the nose, indicating, respectively, the presence of large eyes and large olfactory bulbs for detecting odors. After all, sharks are known for their excellent eyesight and keen sense of smell. The dorsally scooped-out nose is filled in life with a gel that is a good conductor of electrical currents. Sharks have an electroreceptive system that allows them to detect minute electrical signals from any living thing moving in the water. Above the gel-filled nose are special electrical-detecting organs (ampullae of Lorenzini) that use this gel to amplify and send these signals to the brain for interpretation (Liem et al., 2001; De Iuliis and Pulerà, 2011). As you might expect, a pineal eye pierces the pineal opening to detect light, and the series of tiny openings that run dorsally along the orbits are conduits for branches of cranial nerve 5 that detect sensations.

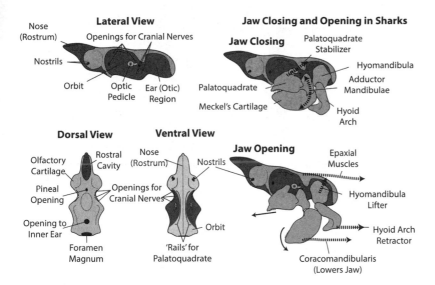

Lateral View

Nose (Rostrum)
Openings for Cranial Nerves
Nostrils
Orbit
Optic Pedicle
Ear (Otic) Region

Jaw Closing and Opening in Sharks

Jaw Closing
Palatoquadrate Stabilizer
Hyomandibula
Adductor Mandibulae
Palatoquadrate
Meckel's Cartilage
Hyoid Arch

Dorsal View

Olfactory Cartilage
Rostral Cavity
Pineal Opening
Openings for Cranial Nerves
Opening to Inner Ear
Foramen Magnum

Ventral View

Nose (Rostrum)
Nostrils
Orbit
'Rails' for Palatoquadrate

Jaw Opening
Epaxial Muscles
Hyomandibula Lifter
Hyoid Arch Retractor
Coracomandibularis (Lowers Jaw)

6.6. Anatomical details of a dogfish (*Squalus acanthias*) skull, showing the odd anatomy of these chondrichthyans and their jaw-opening mechanics. In the jaw movement diagrams, note that both epaxial (back) and jaw muscles play important roles in mouth movements, as discussed previously for placoderms. See text for more details. Based on Liem et al. (2001) and De Iuliis and Pulerà (2011).

Now we come to that odd opening dorsal to the inner ear—it turns out that this provides a direct route for water-borne vibrations and sounds to be channeled right into the head! Of course, this opening is covered by skin in life and filled with a special gel called endolymph (the same substance in the inner ear), but because the density of shark's body and that of the water are almost the same, sound and pressure waves have no trouble finding their way into this conduit (Liem et al., 2001; De Iuliis and Pulerà, 2011). This feature is unique to cartilaginous fishes, and when you see it in a fossil skull there is no doubt as to what you have (Liem et al., 2001; Pough, Janis, and Heiser, 2002). The rail-like structures running ventral to the orbits act as train tracks that keep the palatoquadrate centered on the head when the mouth is opened (more on this in a moment). But why should there be four nostrils? As we've seen before, like many fishes sharks do not breathe through their nose: its only purpose is that of odor detection. In sharks, water enters the cranial set of nostrils, passes across huge, convoluted olfactory bulbs that are highly sensitive to water-borne odors, and then exits through the caudal set of nostrils (Liem et al., 2001).

The jaws of modern sharks like the spiny dogfish are loosely suspended underneath the skull (Fig. 6.6). The palatoquadrate has two projections that jut upward and contact the rail-like 'tracks' on the skull with the help of ligaments. At the jaw joint, where the palatoquadrate and Meckel's cartilage join together, the upper part of the hyoid arch (the hyomandibula we spoke of back in chapter 5) attaches and suspends the jaw underneath the skull. As you may recall, the hyomandibula is firmly connected to the otic region of the skull (where the inner ear resides), and in sharks it hangs down and bends cranially under the skull from this attachment point with the jaws perched on its end.

If you think about it, sharks don't have a mouth that opens at the end of the head like we saw in the placoderms. Instead, modern sharks have a long nose, and the jaws are tucked underneath the head. Why should this be? It all comes back to our pruning shears analogy. Sharks have some of the most powerful bites among vertebrates, and they achieve this power by keeping their jaws short and their adductor mandibulae muscle huge (Radinsky, 1987; Liem et al., 2001). The trouble is, this has resulted in jaws that are so short that they are tucked underneath the skull and ventral to the nose (Moss, 1977). But prey doesn't just come along, swim under the shark's nose, and give up the ghost. Now think about the jaws perched on the end of that hyomandibula bone that can swing back and forth. When a shark goes to bite something, the hyomandibula swings cranioventrally, and the gaping jaws are thrust toward the prey (Liem et al., 2001). As the jaws spring toward their intended victim, the palatoquadrate glides on the aforementioned rail-like 'tracks.' If you have ever watched nature shows in which great white sharks are baited with hooks of meat, you have seen this: the shark comes up for the bait, its huge jaws seem to pop out of its mouth, clamp down with incredible force, and then pop back into the closing mouth. So, not only do sharks have powerful jaws, but they can throw them at their prey rather than give up some mechanical advantage.

The teeth are probably the most fantastic thing about sharks. In modern species, the teeth generally form as pointed triangles with serrated edges and broad bases inserted into the rind of the mouth (Liem et al., 2001; Pough, Janis, and Heiser, 2002; Kardong, 2012). Teeth of this shape are good for gouging out flesh. The teeth develop in long rows, known as tooth families, across the palatoquadrate and Meckel's cartilage, and each tooth in a family is stacked on top of another in a conveyor belt pattern. Most teeth are unseen, hidden and upside down behind the jaws, and only the teeth being used for their intended purposes are right side up and visible. As each tooth at the front of the mouth breaks, wears, and fall outs, the tooth family member hiding behind it replaces it by rotating upright into the vacant spaces left by the former toothy tenant (Liem et al., 2001). In this way, sharks continuously replace their teeth throughout their lives.

Before we leave the head, a few words on the brain are necessary. Modern cartilaginous fishes, and especially sharks, have relatively large brains for their body size (Liem et al., 2001). In fact, some sharks have larger brains than many other vertebrates (including some birds and mammals!) of similar body mass. Unlike in the 'ostracoderms,' the brain is not as tightly associated with the inside of the skull in sharks. However, the braincase is sufficiently close to this organ that it preserves the general outline and major features nonetheless. Given the importance of odor detection for prey capture in sharks, the portion of the cerebrum given over to odor detection and interpretation is large. In fact, this observation led early anatomists to surmise that most of a shark's thinking revolved around odor and little else. However, more recent comparisons of shark brains with those of other vertebrates, including other fishes, show that other regions of the cerebrum devoted to thinking and integration are equally well developed (Lisney and Collin, 2006). The cerebellum, a region of motor coordination noted in chapter 4, is also quite large in sharks, which is not surprising given their need as active predators to have well-coordinated movements.

Caudal to the shark's head is the spiracle (the 'trapped' pharyngeal slit we spoke of in chapter 5) and a series of bony, jointed pharyngeal arches that support the gills. In living sharks, there are wall-like flaps of tissue between each gill slit. If you have ever watched a shark at rest at the bottom of an aquarium you have probably noticed that the gill slits seem to open and then close over and over, with a piece of skin-like tissue temporarily filling each gill opening when it closes. The wall-like tissue flaps between each gill slit are causing most of the action you see. When a shark takes in water through its mouth (and its spiracle, as it turns out), muscles around the flap-like tissue walls press them against the slits, allowing oxygen-rich water to surround the gills at high pressure, forcing this gas into the bloodstream (Liem et al., 2001; De Iuliis and Puler à, 2011). When the pharyngeal muscles relax, the flaps are opened, the gill slits expand, and the now-stale water goes rushing out, taking with it wastes such as carbon dioxide and ammonia that have built up in the blood. As a side note, the spiracle has a tiny but functional gill apparatus buried inside of it.

Moving into the shark's body: a notochord is still present but diminished and surrounded by an interlocking series of vertebrae (Fig. 6.7). The vertebrae take the basic form described in chapter 5 and are threaded spool-like onto the notochord. Ball-like cartilaginous remnants of the notochord allow the vertebrae to move in relation to one another, helping the shark throw itself into a series of lateral undulations (Pough, Janis, and Heiser, 2002). However, between the neural and hemal arches are additional blocks of cartilaginous tissue called intercalary plates (Fig. 6.7). These plates fill the spaces between the vertebral arches, and because the vertebral column is cartilaginous, these blocks of tissue are flexible. In fact, the intercalary plates show us once again the amazing adaptability of sharks: by spanning the spaces between the vertebrae, they provide extra support and resistance to the bending forces on the vertebral column imposed when swimming, yet are flexible enough to allow the vertebral column to bend and sway as necessary (Liem et al., 2001). Nerves leaving the spinal cord can still get out from inside the vertebrae thanks to small, paired openings along the vertebral column that pierce the vertebrae and intercalary plates.

Spiny dogfish have two dorsal fins and a large heterocercal tail. What is interesting to the student of paleontology is that spiny dogfish sharks have prominent spines at the base of both dorsal fins, a very primitive feature found in early sharks and another primitive jawed vertebrate group, the acanthodians (more on this in chapter 7). The pectoral fins are large, triangular, and winglike, but are narrow where they attach to the scapulocoracoid that spans across the chest just behind the gills. This allows the fins to have a large paddle-like surface to push against the water, but the narrow base gives the fin a wide arc of rotation. The skeleton of the fin consists of three upper cartilage pieces arranged in a row, followed by

6.7. Diagram of spiny dogfish shark (*Squalus acanthias*) anatomy. Note the presence of dorsal fin spines and spool-like vertebrae strung on the notochord. As shown in the close-up view of the vertebral column, intercalary plates are sandwiched between adjacent vertebrae. These structures provide further resistance and power to the shark's swimming movements. Based on Liem et al. (2001) and De Iuliis and Pulerà (2011).

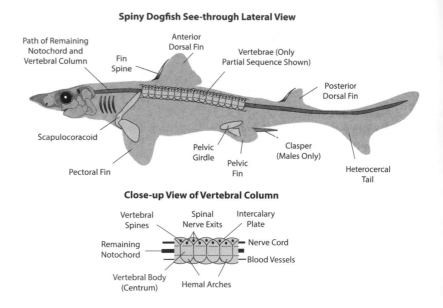

smaller and smaller rays of cartilage that extend to the fin's edge and give it a flexible border (Liem et al., 2001; Pough, Janis, and Heiser, 2002; De Iuliis and Pulerà, 2011; Kardong, 2012). The pelvic fins of female sharks are much smaller than the pectoral fins, but have the same basic skeletal arrangement. The pelvic fins connect via the acetabulum to the small pelvic girdle, and surround the reproductive and excretory opening called the cloaca (De Iuliis and Pulerà, 2011).

In male sharks paired, specialized structures known as claspers extend from the base of the pelvic fins all the way past the last fin rays, and usually terminate in a point or hook. Claspers allow sharks to have internal fertilization: during mating, the male shark inserts one of these structures into the female's cloaca and literally shoots what amount to sperm 'bullets' (sperm packed into spherical bundles) into her reproductive tract (Jones and Jones, 1982). At the base of the male's clasper is a sac that collects seawater and sperm. The sperm are protected within a gelatinous bubble of nutrients (a 'bullet') that supplies them with energy for their journey into the female's oviduct (Jones and Jones, 1982). When the male is ready to fertilize the female, the clasper sac squeezes the water and sperm down and out of his clasper and into her. Because claspers are present only in males, we can sex fossil cartilaginous fishes (Lund, 1985a; Benton, 2005).

The cartilage of cartilaginous fishes is not quite the same as the cartilage in the tip of your nose or your external ears. It is technically called prismatic cartilage, which means it has a bunch of calcium-rich bone-like fibers that radiate out in prism-like patterns inside it (Liem et al., 2001; Pough, Janis, and Heiser, 2002). These make the cartilage of sharks much stronger than your nose tip, and leave hollow spaces that lighten the skeleton. This is somewhat similar to the lightweight fiberglass bodies of race cars: the spindles of glass embedded in the plastic add strength but help keep the overall chassis weight low. The most calcified part of the shark skeleton is, as you might guess, the jaws.

The myomeres of sharks (the segmented muscle blocks that throw the body into lateral undulations) are more than just the simple, backwardly pointing V-shaped muscle blocks we saw in lancelets. Instead, the myomeres of modern sharks are W-shaped in side profile, and the fibers of one muscle interdigitate with the next in line (Liem et al., 2001). In other words, the W-shaped myomeres of modern sharks overlap one another a great deal and have fibers that insert into one another. This change in shape and relationship of the myomeres allows greater force to be generated and spread across the body, leading to more powerful swimming movements (Hildebrand and Goslow, 1995; Liem et al., 2001). We should note that the W-shaped pattern of myomeres in sharks is present but less well-developed in jawless fishes.

One peculiar aspect of spiny dogfish anatomy (one that will not appear in chondrichthyan fossils but that was almost certainly present) is a large, oily liver. Woe is the student who cuts open the already smelly liver of their spiny dogfish: the viscous oils in the liver ooze out, and the already fishy stench increases dramatically. (I might also add that this makes the offending student a very unpopular lab partner for the rest of the class.) Unlike the situation we will encounter in the ray-finned fishes, cartilaginous fishes do not have a swim bladder to help them maintain their buoyancy. Instead, sharks and their relatives employ the large, oily liver, which is much less dense than the water, to make them buoyant (Liem et al., 2001; Pough, Janis, and Heiser, 2002; De Iuliis and Pulerà, 2011).

Sharks are literally full of it: urea, that is. This is very unusual because urea, which contains toxic ammonia, is something most vertebrates try to get rid of. Oddly, sharks seem very tolerant of urea, and it is present in high concentrations in their muscles. It turns out that the large amount of urea in sharks has two profound effects. One is that urea is less dense than the surrounding water, and this seems to contribute to a shark's buoyancy (Withers et al., 1994). Even more odd is that by concentrating urea (which is salty) in its body, the shark becomes closer to the salt concentration of the water around it for reasons related to osmosis. Osmosis said another way is water diffusion (and no, you can't learn by osmosis). Just as a dye dropped into water diffuses from a highly concentrated point to a lowly concentrated tint, so water molecules diffuse from high to low concentrations. For example, freshwater contains more water molecules than seawater, which has more spaces occupied by salts. If you put a tank of freshwater next to a tank of saltwater and place a filter between them that allows water to move from one tank to another, the freshwater will move toward the saltwater. By the same token, a saltier shark doesn't have to fight osmosis so much (the water concentration in its body being close to that of the surrounding water), and it will therefore lose less water to the environment than other fishes (Liem et al., 2001; Pough, Janis, and Heiser, 2002).

The body of the spiny dogfish, as for most sharks, is covered in small toothlike scales or denticles. These scales, just like teeth, have a

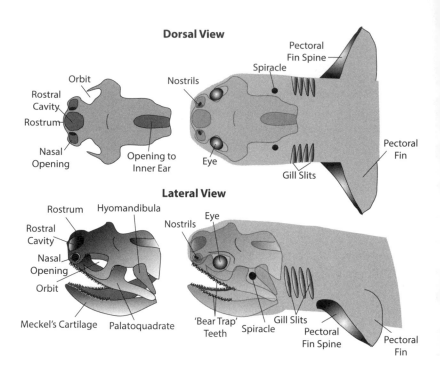

6.8. Diagram of one of the earliest chondrichthyans, *Doliodus,* showing the relationship of the skull and the odd opening to the inner ear. Note the 'bear trap' teeth and the location of the 'trapped gill slit,' the spiracle. Based on Maisey, Miller, and Turner (2009).

pulp cavity and dentine and enamel-like proteins (Liem et al., 2001). As discussed previously in chapter 5, these scales probably contribute to decreasing frictional drag in these animals (Dean and Bhushan, 2010). Although the skin itself in fossil cartilaginous fishes is rarely if ever preserved, the tough, toothlike scales occasionally are.

Having completed our tour de shark, we now turn to reviewing all-too-briefly the evolutionary diversity of the chondrichthyan chassis. As with placoderms, the earliest fossils of chondrichthyan fishes come to us as isolated scales and body fragments. It is not until approximately 407–370 Ma (Cards 91–92) that we finally get our first glimpse at good skulls and complete skeletons. Two early cartilaginous fishes, *Doliodus* (Miller, Cloutier, and Turner, 2003) and *Cladoselache* (Carroll, 1988), give us our first good look at the early chassis of shark ancestors (Figs. 6.8 and 6.9). We have only the head and trunk of *Doliodus* (Miller, Cloutier, and Turner, 2003; Maisey, Miller, and Turner, 2009), but it is preserved well enough to show us intimate details of this animal. On the other hand, we have many full-body specimens of *Cladoselache*, some so well preserved that impressions of muscle fibers and kidney anatomy are discernible (Carroll, 1988; Benton, 2005)!

Unlike modern sharks, these early cartilaginous fishes had short noses, longer jaws, and a mouth at the front of the head (technically, a terminal mouth), and the palatoquadrate was firmly attached and unmoving to the underside of the skull (Radinsky, 1987; Pough, Janis, and Heiser, 2002). These animals were the Model T of cartilaginous fishes and could not throw their jaws at their prey. As in modern sharks, the teeth form tooth families and have the conveyor-belt replacement pattern

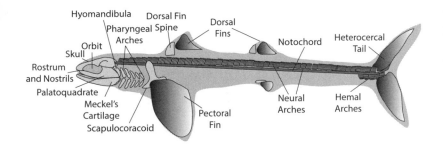

6.9. Diagram of *Cladoselache*, another primitive chondrichthyan from the Devonian period. Note the large dorsal fin spines and large heterocercal tail. Based on specimens illustrated in Carroll (1988) and Benton (2005).

we spoke of earlier. In *Doliodus*, each tooth consisted of two sharp cusps separated by a gap, and the gap accepted a sharp cusp from one of the opposing teeth (Fig. 6.8), allowing the tooth cusps to interlock like the blades of a bear trap (Maisey, Miller, and Turner, 2009). The teeth of *Cladoselache* had a tall central point that was flanked on either side by smaller, pointed crowns, all anchored to a broad base, and these cusps probably also interconnected when the mouth was closed (Carroll, 1988). Such teeth, unlike those of many modern sharks, seem best for grabbing smaller prey that can be swallowed whole. The backbone of these early cartilaginous fishes was still dominated by a notochord flanked by neural and hemal spines as in the placoderms.

Tough, toothlike spines were present at the bases of the pectoral fins, and, where known, also projected from the base of the dorsal fins as well. Unlike in modern sharks, the pectoral fins had broad bases and were not very flexible. Pelvic fins are not preserved for *Doliodus*, but they are in *Cladoselache* and show us something surprising: among all known cartilaginous fishes for which we have good body fossils, no specimens of *Cladoselache* have claspers (Carroll, 1988; Pough, Janis, and Heiser, 2002; Benton, 2005)! Such data might suggest that these early cartilaginous fishes did not have internal fertilization, but it could be we are simply getting a sample of all females. This is possible, given that many modern shark species travel in same-sex groupings, with mixing of sexes occurring only during times of mating (Jacoby, Busawon, and Sims, 2010).

The tail in *Cladoselache* is heterocercal, but the upper and lower lobes are similar in size (Fig. 6.9) (Carroll, 1988). In fact, the half-moon tail shape resembles that of a tuna, a modern ray-finned fish that is a fast and powerful swimmer. In fishes like tuna, most of the body remains rigid during swimming except for the tail: this cuts down on drag and makes these sorts of fishes very efficient long-distance swimmers (McGowan, 1999; Vogel, 2003). *Cladoselache* probably did not swim as fast as tuna, but these sorts of convergences in tail shape, along with teeth that seem ideal for grabbing small prey whole, collectively suggest an active, predatory lifestyle in these animals.

By the start of the Carboniferous period some 360 Ma (Card 93), the cartilaginous fishes had begun to diversify in both salt- and freshwater environments. Were we to drop anchor in what is now the Bear Gulch Limestone in Montana about 318 Ma (Card 93), we would find ourselves

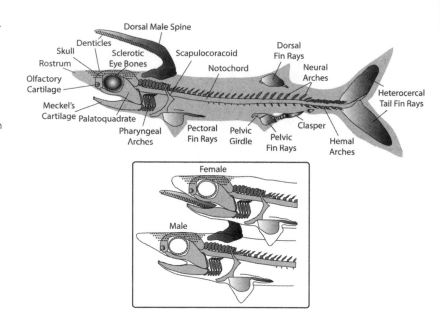

6.10. The primitive chondrichthyan, *Falcatus,* from the Bear Gulch Limestone in Montana. Only specimens with claspers (males) have the large 'hood ornament' of a dorsal fin spine. Remarkable preservation of *Falcatus* specimens from the Bear Gulch formation shows a female of the species (lacking claspers) biting onto the dorsal fin spine of the clasper-possessing male, presumably to help with copulation. Based on Lund (1985b).

on a tropical estuary. In these brackish waters we would find more than 65 species of cartilaginous fishes, many of fantastic shapes and dispositions (Lund, 1985a). One of the best known is *Falcatus*, a sharklike fish only about 30 cm long, whose males possessed quite the hood ornament: a large, denticle-covered spine that extended from over their shoulders to bend like a shade over their heads (Fig. 6.10) (Lund, 1985b). We know that this was a male-only structure because the only *Falcatus* specimens that possess the spine have claspers. In contrast, all specimens of this species without claspers lack the head spine. What function did the denticle-covered spine serve in male *Falcatus*? The fossil record from the Bear Gulch Limestone provides a stunning answer: there are exquisitely preserved fossils where female *Falcatus* are literally biting and holding on to the spine of the males (Fig. 6.10) (Lund, 1985b)! Biting may seem like a poor way to get romantic juices flowing, but it is actually a common behavior among modern sharks when they are mating: the male and female hold on to one another with their mouths as they thrash and spin while the male positions and inserts his clasper into her (Pratt and Carrier, 2001). Not that the Bear Gulch specimens as preserved were mating at the time (the male is on the bottom and couldn't possibly have positioned his claspers to mate with the female), but they could have been setting the mood.

Falcatus, unlike more primitive cartilaginous fishes, has a longer, larger nose, which suggests the electrical- and olfactory-detection systems were more enhanced, similar (but perhaps not quite as developed) to what we encountered in the spiny dogfish. Although the tail skeleton technically makes the tail fin heterocercal, long spines in the lower lobe make it approximately the same size as the upper lobe (Lund, 1985b). As a consequence, the tall, tuna-like tail of this chondrichthyan suggests it, too, was a fast and mobile predator.

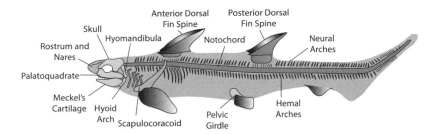

6.11. The hybodont shark, *Hybodus*. Note the short head and powerful jaws, and the large fin spines that are anchored into the fin bodies. The teeth of these primitive sharks were varied and had both shearing and crushing functions. These animals were the major chondrichthyan predators of Mesozoic seas, but were eventually replaced by the modern sharks and rays. Based on specimens illustrated in Carroll (1988) and Benton (2005).

Between 318 and 66 Ma (Cards 93 to 99), a number of chondrichthyan fish families appeared that replaced earlier forms like *Cladoselache* and *Falcatus* and began to show the modern anatomical trait states we saw in the spiny dogfish. These cartilaginous fishes, including their modern descendants (sharks, skates, and rays) are called elasmobranchs ('plate gills' because each gill set is embedded in a platelike gill flap) (Carroll, 1988; Benton, 2005). Some of these early elasmobranchs were eel-like and inhabited the freshwater, vegetation-choked environments of the Carboniferous period some 318 Ma (Card 93). As the Age of Dinosaurs (Mesozoic Era) dawned (245 Ma; Card 95), more modern-looking marine predators called hybodonts, some nearly 2 meters long, appeared (Fig. 6.11). These were very successful predators, and one species, *Hybodus*, was present for most of the time dinosaurs roamed the earth (235–65 Ma; Cards 95–99) (Benton, 2005). The anatomy of these elasmobranchs is so similar to those of modern sharks like the spiny dogfish that at first glance you would assume you were looking at a typical living shark. However, closer inspection would reveal that *Hybodus* and its kin had a short skull with a terminal mouth, and the teeth were varied and odd. Unlike the bear trap–style teeth of early cartilaginous fishes or the gouging teeth of modern sharks, the teeth of hybodonts were specialized for a variety of functions. The front teeth were sharp and pointy and appear ideal for piercing prey. The rearward teeth were broad, blunt, and arranged in sets that appear well suited to crush and mash hard-bodied prey like shelled mollusks and crustaceans (Benton, 2005). The notochord of these elasmobranchs was still prominent, and only neural and hemal arches were present as vertebral elements (Fig. 6.11) (Benton, 2005).

Like the hybodonts, the modern elasmobranchs made their first cameos around 245 Ma (Card 95) (Benton, 2005). Although the hybodonts were successful for a long time, their chassis was gradually retired by the modern radiation of elasmobranchs, the sharks, skates, and rays, by the end of the Mesozoic (65 Ma; Card 99). In all modern elasmobranchs, the mouth has become underslung and the rostrum longer and more prominent (Pough, Janis, and Heiser, 2013): recall the jaw mechanics of the dogfish shark. Physiologically, modern elasmobranchs run the gamut from being ectothermic ('cold-blooded' animals whose body temperature fluctuates with their environment) to being endothermic ('warm-blooded' animals that can internally control their body temperature irrespective of the environment) (Bernal et al., 2012). In fact, some sharks can selectively regulate the temperature of their eye and fin muscles (Bernal et al., 2012)–this prevents their eyes from 'locking up' or their fins from becoming sluggish while chasing down prey in cold water. Given that these ancestral modern elasmobranchs appeared during a time when the ray-finned fishes (the actinopterygians) (see chapter 7) were undergoing their own evolutionary radiations (Benton, 2005), the ability to rapidly project the mouth at prey may have been advantageous. Moreover, the modern sharks diversified during the Jurassic and Cretaceous periods (200–66 Ma; Cards 96–99) alongside the hybodont sharks (Benton, 2005; Pough, Janis, and Heiser, 2013). Therefore, it appears unlikely that the appearance of the modern shark radiations would have directly caused the demise of the hybodonts (Benton, 2005).

During their evolution, modern sharks developed the pointy, serrated teeth we discussed for the dogfish shark, as opposed to the more bear trap–like teeth of earlier chondrichthyans. With their expandable mouths and gouging teeth, the modern sharks (technically, the Neoselachii) successfully radiated into a variety of habitats and eco-niches. We have already discussed shark anatomy in some detail, so what follows is a brief overview of their recent diversification. Current molecular trait states suggest that modern sharks radiated into two major groups, the Galeomorphii and the Squalimorphii (Vélez-Zuazo and Agnarsson, 2011). The galeomorphs are the more diverse group and contain the largest known sharks alive today. For example, there are the giant filter-feeding basking and whale sharks: some of these elasmobranchs reach over 12 meters in length (Benton, 2005; Pough, Janis, and Heiser, 2013). Among the galeomorphs we also find the great white sharks (*Carcharodon carcharias*) of *Jaws* fame, the Carchariniformes. Females of this species can reach lengths of 6 meters, but this pales in comparison to the largest known carcharinform shark of all time, *Carcharocles megalodon*, the 'Megalodon' of recent media fame. This shark is known from only its teeth (after all, a cartilaginous skeleton rarely preserves), but what teeth they are: over 14 cm long and studded with flesh-slicing serrations (Benton, 2005). Body estimates derived from shark teeth of these sizes suggest some *C. megalodon* specimens may have approached 18 meters in length (Pimiento and Clements, 2014)! These giant sharks probably functioned as apex predators during their reign 16–3 Ma (Card 100), likely preying on the large marine mammals of their day (Pimiento and Clements, 2014). All recent 'documentary' claims to the contrary, this immense predator likely swam its last about 2.6 million years ago (Pimiento and Clements, 2014), and no credible evidence of its present-day existence has surfaced.

The other branch of the modern sharks, the squalimorphs, comprises a number of species. Among the squalimorphs we find our now-familiar friends the dogfish, as well as many deep-sea sharks, the so-called angel sharks, and saw sharks (Vélez-Zuazo and Agnarsson, 2011). Amid the squalimorphs we find odd and primitive 'six-gill' sharks, such as so-called cow sharks, some of which live in deep ocean waters near the continental shelves, where they feed on everything from crustaceans to fishes, and which lack the specializations found in most other modern sharks (Adnet and Martin, 2007; Soares and de Carvalho, 2013).

Before leaving the Chondrichthyes, it is necessary to kick the tires of the other modern elasmobranch chassis, the skates and rays. Skates and rays, which make their first appearance approximately 190–180 Ma (Cards 96–97), essentially take the shark chassis and flatten it dorsoventrally (Fig. 6.12) (Pough, Janis, and Heiser, 2002; Benton, 2005). In fact, their bodies are so flattened that both their mouths and gills open downward. This body shape, along with large flap-shaped pectoral fins, aids these animals in hugging the ocean bottom, where they spend their lives hunting for hard-shelled invertebrates (Liem et al., 2001). Their eyes are positioned dorsally on their flattened heads to keep a look out for predators and prey alike. Some skates and rays bury themselves in seafloor sediments with only their eyes protruding, hiding from would-be attackers and disguising themselves from potential dinner. They have large tooth plates lining their mouths that act like a crushing 'pavement' that allows them to crack open shells and hard-bodied invertebrates (Pough, Janis, and Heiser, 2002). The tails of many skates and rays are elongated and whiplike, and some rays have developed tails embedded with venomous barbs. As an aside, one might wonder how these animals breathe with both their mouths and gills stuck in the sediment. This would be like having the radiator, air intake valves, and exhaust covered on a car! The old 'trapped' gill pouch, the spiracle, comes to the rescue here. Skates and rays can suck water into the spiracle, which faces upward (dorsally), and force oxygen over their gills in this way, almost like sticking a hose through the hood of car to force air into the engine (Pough, Janis, and Heiser, 2002). It should be noted that the largest sharks and rays have become filter-feeders, giving up their toothy grins for a sieve-like apparatus (Pough, Janis, and Heiser, 2002).

Before we leave this all-too-brief overview of the cartilaginous fishes, we must acknowledge some of the oddest living jawed vertebrates around: the holocephalians or ratfish (Fig. 6.12). It remains unclear as to whether these cartilaginous fishes are actually weird elasmobranchs or some other branch of the cartilaginous fish family tree that split before their appearance (Pough, Janis, and Heiser, 2002). One very strange basal chondrichthyan radiation that is apparently related to holocephalians are the eugeneodonts. Eugeneodonts had a whorl of teeth situated between their lower jaws that was arranged like a spiraling Ferris wheel: new teeth rotated in from the back of the mouth, protruded in the center of the mouth, and then were shed at the front (Benton, 2005; Pough, Janis, and Heiser, 2013). The central whorl of teeth apparently acted to crush prey against a tooth 'pavement' on the upper jaw (Benton, 2005). Whatever their relationships to the eugeneodonts, holocephalians appeared with the early cartilaginous fishes at least as far back as the Devonian, and have been strange ever since. A typical ratfish, if there is such a thing, has a squat face with short but powerful jaws and large eyes. Some males have head claspers (!), the tail is long and whiplike (as it is in many

6.12. Diagrams of a generalized skate/ray and ratfish (holocephalian). Notice that for skates and rays the body is so flattened dorsoventrally that the mouth and gills open beneath (ventrally) the animal's head rather than at its terminal front end. These chondrichthyans can suck water through their dorsal spiracle to force oxygenated water over their gills. For ratfish, an odd combination of features is evident, including a venomous spine, crushing jaw parts, and whiplike (ratlike) tail. This odd mixture of traits has made it difficult to determine the precise placement of ratfish (holocephalians) among the chondrichthyan fishes. See Figure 6.1 as well. Based on figures in Liem et al. (2001), Pough, Janis, and Heiser (2002, 2013), and Benton (2005).

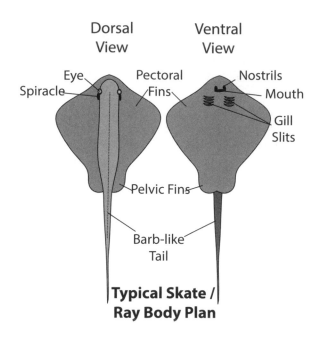

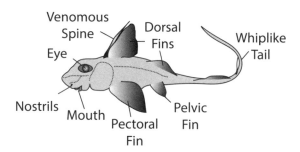

deepwater fishes), and a prominent spine is associated with their dorsal fin. The spine is also usually hollow and venomous (Pough, Janis, and Heiser, 2002). Ratfish live in deep water, and it has been difficult to study them alive and in their natural habitats. From the little we do know it appears that they eat hard-shelled invertebrates, and some have been known to consume spiny sea urchins. Their upper and lower jaws contain ever-growing tooth plates that adjust to wear by changes in their height within the mouth, which is similar in many ways to the teeth of the whorl-toothed eugeneodonts. Their evolutionary history remains an enigma, but this will surely change with continued study and the collection of new fossils in the future.

While the placoderms and cartilaginous fishes were busy establishing their fortunes in the Devonian seas, another group of jawed vertebrates was gaining a foothold (finhold?). These animals, the bony jawed vertebrates, include two living groups and possibly an extinct group. One of the living groups has gone on to become the most successful vertebrates ever. The other living group, although not a large part of aquatic ecosystems today, gave rise to the tetrapods, the group of land-living vertebrates that pioneered the invention of hands, feet, fingers, and toes from fishy fins.

7.1. Family tree (phylogeny) of the major actinopterygian fish groups. Note that kinetic (movable) skulls are primitive for the group, and that much of actinopterygian evolution has involved the increasing mobility of the skull. Phylogeny adapted from Liem et al. (2001) and Benton (2005).

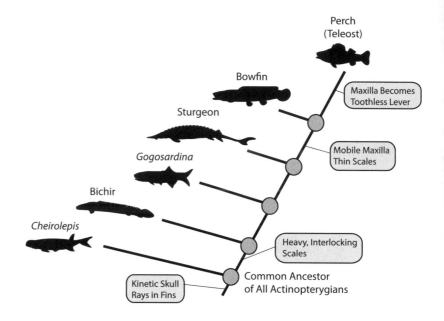

Perch
(Teleost)

Bowfin

Maxilla Becomes
Toothless Lever

Sturgeon

Mobile Maxilla
Thin Scales

Gogosardina

Bichir

Heavy, Interlocking
Scales

Cheirolepis

Kinetic Skull
Rays in Fins

Common Ancestor
of All Actinopterygians

JUST AS THERE ARE SEVERAL DIFFERENT WAYS TO GET ACROSS TOWN, whether by bike, car, or train, so there have been several different ways in which vertebrates have evolved the means to transport themselves with a skeleton. As we have discussed, bone as a tissue (apatite in particular) is one of the defining traits of vertebrates. In placoderms, the skeleton was cartilaginous, yet dermal bone covered the body. In cartilaginous fishes, bone was lost and the skeleton was made entirely of prismatic cartilage, but the teeth and scales contained apatite. Yet, it turns out that only one group of vertebrates develops a complex, internal bony skeleton: the bony jawed vertebrates, or osteichthyans as they are technically known. In osteichthyans, the internal skeleton develops from a cartilage model, and is technically known as endochondral bone. Essentially, the skeleton first develops in cartilage, and as development proceeds it is invaded by blood vessels carrying bone-making cells that eventually replace it with bone. Thus, the presence of an endochondral skeleton is what separates osteichthyans from all other vertebrates (Liem et al., 2001). Here, as everywhere in the study of the vertebrate family tree, there are disagreements about the interrelationships of these animals. For simplicity, and because we are attempting to follow the mechanical evolution of the vertebrate chassis, I have chosen to treat an odd and extinct group of fishes, the acanthodians, as the most primitive members of Osteichthyes, followed by the well-known living groups, the ray-finned fishes (actinopterygians) and the lobe-finned vertebrates (sarcopterygians) (Fig. 7.1). The fishlike sarcopterygians are covered in the next chapter, and their descendants, the tetrapods, get their full due starting in chapter 9 and throughout the rest of this book.

Acanthodians: Spiny Osteichthyes (?)

Acanthodians would have stuck out, literally, in any group of fishes because these animals were adorned with numerous fin spines (Plate 6). To be clear, the relationships of these fishes are in a state of flux, with recent analyses aligning them with Osteichthyes (Friedman and Brazeau, 2010), Chondrichthyes (Giles, Friedman, and Brazeau, 2015), or even in a basal position between both of these groups (Brazeau, 2009). Despite these uncertainties, these animals possess a number of interesting features that are likely holdovers from the common, bony ancestor of chondrichthyans and osteichthyans, and thus may shed light on the origins of the chassis in Osteichthyes (Friedman and Brazeau, 2010). Unfortunately, acanthodians are still poorly understood, partly because only a handful of species are

Pedigree Early Bony Jawed Vertebrates

Date of First Appearance ~450 Ma (Card 91)

Specialties of Skeletal Chassis Spiny Fins

Eco-niche Active Carnivores, Scavengers, and Filter Feeders

known in any great detail. As in the placoderms and cartilaginous fishes before them, bits of scale and bone that may belong to acanthodians are found as far back in the fossil record as 450 Ma (Card 91). However, it is not until the Devonian period (~416–360 Ma; Cards 91–93) that we get our first good look at whole-body fossils of these animals. Some of these fishes were carnivorous, whereas many of the last acanthodians were exclusively filter feeders. Most were not very large, with few ever growing beyond 20 cm or so, although a few species may have grown up to nearly 2 meters long (Pough, Janis, and Heiser, 2002; Benton, 2005).

The following description is based on two well-known acanthodians from approximately 400 Ma (Card 92), *Climatius* and *Ischnacanthus*, animals or their relatives that we might be able to catch and pull up onto our boat *Devonian Destiny* if we were sailing ancient seas in North America, Australia, and Asia (Fig. 7.2) (Radinsky, 1987; Carroll, 1988; Benton, 2005). Typically, the heads of acanthodians were bony and sported large eyes (as indicated by the sclerotic bones) near the terminal end of the mouth. The braincase and upper and lower jaws (palatoquadrate, Meckel's cartilage) were encased in bone, and some acanthodians bore teeth (Pough, Janis, and Heiser, 2002; Benton, 2005). Specialized gill support rays seen elsewhere only in other bony fishes project behind the skull, and the pharyngeal arches themselves were covered with one or more bony plates collectively called the operculum (Carroll, 1988; Pough, Janis, and Heiser, 2002; Benton, 2005). In modern bony fishes, the operculum is a movable, bony flap that circulates fresh water over the gills. In acanthodians, as well as other bony fishes, the gills are all contained in one open chamber hidden behind the operculum (Liem et al., 2001). You will recall that separate gill chambers were present in placoderms, cartilaginous fishes, and the 'ostracoderms,' and that water was forced through each of these openings to allow these animals to breathe. Because there was a wall of tissue between each gill arch, there is a certain amount of 'dead space' that cannot be used for respiration. In contrast, the operculum-covered gills of acanthodians and later bony fishes have a greatly increased surface area for gas exchange by virtue of their being exposed on all sides in one open chamber. Much like the air circulating around a lazy Roman emperor being fanned with palm fronds by his slaves, so the operculum circulates oxygen-rich water over the gills in these fishes (Liem et al., 2001).

The vertebral column, like that of many placoderms, consisted only of bony neural and hemal arches, and the notochord was still prominent (Benton, 2005). The bodies of most acanthodians were slender and bore one or two dorsal fins, a set of pectoral and pelvic fins, and one or more intermediate fins between the pectoral and pelvic girdles. These fins were anchored to stout spines in front of them and, as described in chapter 5, probably increased the stability of these fishes as they swam in search of prey or plankton. The pectoral girdle consisted of a sturdy but slender scapulocoracoid, and the pelvic girdle contained splinter-like pelvic bones. The tail, as with so many of the fishes we have discussed, was again

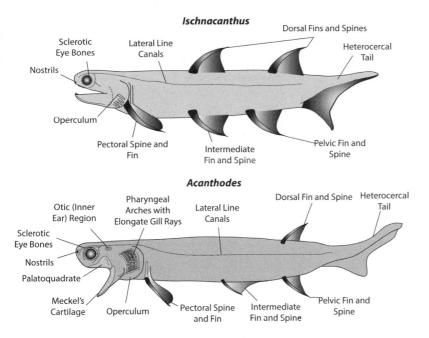

Ischnacanthus

Sclerotic Eye Bones
Nostrils
Operculum
Lateral Line Canals
Pectoral Spine and Fin
Intermediate Fin and Spine
Dorsal Fins and Spines
Heterocercal Tail
Pelvic Fin and Spine

Acanthodes

Otic (Inner Ear) Region
Sclerotic Eye Bones
Nostrils
Palatoquadrate
Meckel's Cartilage
Operculum
Pharyngeal Arches with Elongate Gill Rays
Lateral Line Canals
Pectoral Spine and Fin
Intermediate Fin and Spine
Dorsal Fin and Spine
Heterocercal Tail
Pelvic Fin and Spine

7.2. Two varieties of acanthodians (spiny Osteichthyes) showing to good effect the spines that are their namesake. Originally, most acanthodians were predators, such as *Ischnacanthus,* but the last lineages of these animals, such as *Acanthodes,* were specialized filter feeders. Note that for *Acanthodes* elongate gill rays were present that probably functioned as strainers, much like similar structures in modern filter-feeding fishes do. Skeletal diagrams based on illustrations in Radinsky (1987), Benton (2005), and Beznosov (2009).

heterocercal. The scales of acanthodians were small and in some species fit together with peg-like teeth that allowed the body to move while still providing protection (Pough, Janis, and Heiser, 2002; Benton, 2005). In many acanthodians, the scales had a base of bone (which could be acellular or cellular) topped with sensitive dentine. The scales were small, being mostly concentrated along the lateral line canals, eyes, and nostrils (Carroll, 1988; Pough, Janis, and Heiser, 2002; Benton, 2005). For the acanthodians *Climatius, Ischnacanthus,* and their kin, all of these anatomical features suggest that these animals were rather active predators.

As time went on, filter-feeding acanthodians (the best known of which is *Acanthodes*), which appeared around the same time as *Climatius* and *Ischnacanthus,* became successful and widespread, and eventually were the only type of acanthodian left (Fig. 7.2). Unlike *Climatius* and *Ischnacanthus, Acanthodes* and its relatives lacked teeth and had longer, eel-like bodies (Beznosov, 2009). How do we know *Acanthodes* and its close relatives were filter feeders? The specialized gill supports in these animals are very elongate and resemble the condition seen in other filter-feeding fishes, such as manta rays and the whale shark. The long gill supports apparently acted as a strainer for small invertebrates, and in some specimens of an *Acanthodes* relative, *Mesacanthus,* there are traces of many small crustaceans that were apparently swallowed as prey (Radinsky, 1987; Benton, 2005; Beznosov, 2009).

Acanthodians started as marine fishes, but as time went on they became mostly freshwater in their habits (Pough, Janis, and Heiser, 2002; Benton, 2005). By 270 Ma (Card 95), the acanthodians disappear completely from the fossil record. It remains unclear why these spiny fishes went extinct, although changes in the environment and competition from other bony fishes cannot be ruled out. Given our current understandings of their relationships, it is possible they are survived by their distant, living relatives in the form of the chondrichthyans, osteichthyans, or both. However, despite their sadly short but spiny tenure on planet Earth, the common ancestor that spawned the acanthodians also brought forth some of the most successful branches of the vertebrate family tree, including our own. So let's raise a glass, toast our long-lost spiny friends, and sleep well in the knowledge that we share a portion of their genome. If you were so inclined, you might even say that a part of the acanthodians has stuck with us. You may now groan.

Modern Osteichthyes—The Basic Chassis

Pedigree The Common Ancestors of Actinopterygians and Sarcopterygians

Date of First Appearance 420 Ma (Card 91)

Specialties of Skeletal Chassis Well-Developed Internal Bony Skeleton

Eco-niche Nearly Every Conceivable Lifestyle Available in Water or on Land

Before we can speak about the two major living branches of the modern Osteichthyes, the actinopterygians (the ray-finned fishes) and the sarcopterygians (the lobe-finned vertebrates), we must briefly consider the common chassis from which both groups diverged. Sometime around 450 Ma (Card 91), the same time the placoderms and perhaps the first chondrichthyans appeared, the modern Osteichthyes went their separate ways (Fig. 6.1 and 7.1). By 420 Ma (Card 91), the early ancestors of the ray-finned fishes and the lobe-finned vertebrates were already on the scene. The trouble is, few fossils of the ancestral modern Osteichthyes are known, and most are known from scales, teeth, and assorted body fragments. More difficult is the fact that disagreement and difficulty surround how to interpret various anatomical traits that would put any of the potential candidate fossils squarely among the original ancestors.

It is therefore instructive to start with a modern bony fish to highlight basic features of the osteichthyan body plan, and then to look for these commonalities in the earliest known fossil Osteichthyes. Such an example is the bichir (*Polypterus*): an odd freshwater subtropical fish native to the African continent (Fig. 7.3 and Plate 7). These fishes look like something you would fish out of a prehistoric river: they have a reptilian-looking face with a toothy grin, their bodies are densely covered in layers of interlocking scales, the dorsal fin has become divided into flag-like finlets, the pectoral and pelvic fins are squat and somewhat muscular, and the tail fin is homocercal in its appearance (although it is technically heterocercal inside). Phylogenetic studies of bichirs consistently place them near the base of the actinopterygians, so although they are one among the many thousands of ray-finned fishes, their skeleton retains primitive hallmarks inherited from the common ancestor of all Osteichthyes.

Let's start with the bichir skull. Gazing upon any bony fish skull, one gets the distinct impression that these animals share something in common with a hard-boiled egg struck by a spoon: like the fractalized eggshell, many of the bones are tiny and intricate, and worst of all, have names like preoperculosubmandibular. It is enough to frighten and discourage students and professionals alike, let alone someone learning vertebrate anatomy for the first time. Yet, it is still necessary to introduce some more technical bone terms at this point. I promise to keep the

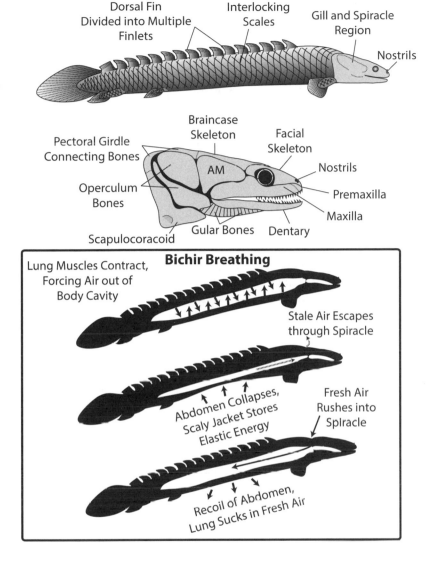

7.3. The bichir, an odd and primitive actinopterygian fish from Africa. Note the lobe-like fins and 'straight jacket' of its body scales. The inset is a series of diagrams that illustrate how the bichir uses the elastic recoil of scaly armor to breathe air into its lungs; based on illustrations and data from Brainerd, Liem, and Samper (1989). AM indicates the location of the adductor mandibulae group beneath the palatopterygoid arch (temple) region.

science lingo to a minimum, and it is comforting to note that these same bones will appear over and over again in all the other vertebrates we discuss. I emphasize that I am not detailing all of the bones of the skull, just ones that become important to our story.

The skull of the bichir is well-fortified with bone. Although a braincase, palatoquadrate, and Meckel's cartilage are present, they are now encased in strong, internal bony plates. The bony skull itself can be subdivided into a braincase region and a facial skeleton. For our purposes throughout the book, the braincase region consists of the bones caudal to the face that surround the brain and house the jaw muscles, whereas the facial skeleton encompasses the orbits, jaws, and nose. Let's start with the business end of the things, the jaws. The upper tooth-bearing jaw bones consist of the premaxilla and maxilla, whereas the lower tooth-bearing jaw bone is called the dentary. These three bones almost always bear teeth, and their shapes and dispositions tell us about important aspects of feeding in all bony vertebrates. In the bichir, the teeth are small and pointed, and there are rows of teeth along additional places on the roof (palate) of the mouth. Teeth of this shape tend to be ideal for catching small prey and piercing the exoskeletons of invertebrates. Not surprisingly, bichir diets commonly consist of small vertebrates and invertebrates. As in sharks, the teeth have a limited lifespan, and fall out to be replaced by new teeth regenerated in special stem cells in the jaws. There are many more jaw bones, of course, but detailing these causes the brains of most people to freeze up like a poorly maintained engine, so we will just acknowledge their presence and move along to the jaw joint itself. In the upper jaw, there is a bone called the quadrate, and in the lower jaw there is a bone called the articular. These two bones fit together and form the movable joint that allows the mouth to open and close. Now and for always (with one glaring exception described in later chapters) know that the quadrate and articular bones form the jaw joint. You can remember this by thinking of Q & A.

The facial skeleton consists of numerous bones, some of which will become important to our story in later chapters. Caudal to the orbits and in the braincase region are bones that surround the adductor mandibulae muscle (a region you would call your temple). As you will recall from chapter 6, this muscle is the main jaw closer, and the space contained by the bones of the temple region give some indication of the size of this muscle mass. Another, deeper muscle group in this same region consists of the pterygoids (the 'p' is silent), and these muscles work with the adductor mandibulae to close the jaws. Here, we will refer to the bones that form the temple region of the skull as the palatopterygoid arch. Finally, the bichir skull has an operculum comprised of several bones, and these bones (and thus the gill apparatus) lie just caudal to the palatopterygoid region of the skull.

The opening and closing of the jaws in bichirs operates in some ways that are similar to those we discussed for placoderms and sharks (see Figs. 6.2 and 6.6). In fact, what we discuss here in bichirs is modified to varying degrees in all other osteichthyans. Without becoming bogged down in muscle and bone minutiae, suffice it to say that the bones of the skull and jaw are interconnected in such a way that opening and closing the mouth is dependent on their combined movements. As we saw for placoderms and sharks, mouth opening involves rotating the skull dorsally and simultaneously depressing the lower jaw. In concert with these movements, the opercular bones rotate laterally. The combined upward rotation of the skull, downward thrust and rotation of the lower jaw, and the outward rotation of the operculum generate suction that pulls prey into the bichir's waiting mouth (Lauder, 1980). Closing the mouth likewise involves our good friend the adductor mandibulae, assisted by deeper pterygoid muscles, which forcefully rotates the skull and upper jaw down and into contact with the upwardly rotating lower jaw (Lauder, 1980). As the jaws are closing, the operculum is closed, squeezing out excess water that was sucked into the mouth with the prey (Lauder, 1980). Finally, a broad muscle, the intermandibularis, raises the floor of the mouth, allowing the fish to swallow its prey. So, when a bichir or any other osteichthyan closes its jaws, you can bet that the adductor mandibulae, the pterygoids, and the intermandibularis are all the usual suspects.

Bichirs, like all fishes, lack a neck! This adaptation is useful when swimming because it ensures that head and body are bound together and act as a single unit, thus cutting down on drag. But there's something weird and wonderful about this arrangement in bony fishes: it means that the pectoral girdle is attached directly to the back of the skull! A special series of bones that connect the pectoral girdle directly to the head are present, and among these two deserve special recognition: the cleithrum and the clavicle. The clavicle is still present even in ourselves: we call it our collar bone, and it is still attached to our scapula. The cleithrum and clavicle tend to be very large bones in most fishlike Osteichthyes, and may be larger than the scapulocoracoid itself. Typically, a scapulocoracoid (usually smaller than what we see in the bichir) is present, to which the pectoral fins attach through a shoulder socket joint called the glenoid.

Bichirs, in contrast to the chondricthyans we have previously considered, have well-developed internal skeletons and a robust vertebral column. The notochord in bichirs is small and surrounded by spool-like vertebrae; neural spines protect the spinal cord, and hemal arches are present that surround blood vessels and nerves. Small, horizontal ribs are also present in this fish, and they apparently act to brace the scaly body wall (Brainerd, Liem, and Samper, 1989). As previously mentioned, robust pectoral and pelvic fins are present, as well as dorsal, anal, and caudal fins. The robust pectoral fins are so strong, in fact, that bichirs are capable of walking on them for brief periods of time out of the water (Standen and Larsson, 2012). The interlocking scales that encase the bichir's body are composed of bone and topped with dentine and enameloid compounds, and these collectively give bichirs protection and flexibility. Finally, distinctive lateral line canals run across the head and body.

One of the fascinating things about the bichir is its ability to breathe through lungs that occupy much of the length of its body. In fact, a bichir will drown if it is denied access to air (Liem et al., 2001). The possession of lungs in something so clearly a 'fish' may seem odd, but in fact having lungs is a distinctive soft tissue trait found only in osteichthyans (Liem et al., 2001; Benton, 2005). During embryonic development in both actinopterygians and the sarcopterygians, a pocket of tissue surrounded by a rich blood supply develops from the esophagus. As development proceeds, this pocket of tissue eventually enlarges to become the lungs (Gilbert and Raunio, 1997; Liem et al., 2001; Gilbert, 2010). Among osteichthyans, lungs come in two flavors: a primary or supplemental respiratory organ, or a swim bladder that acts as a buoyancy organ (Liem et al., 2001). To emphasize that lungs are unique structures, there is no definitive evidence for their presence among cartilaginous fishes and acanthodians. Although some antiarch placoderm specimens have hollow areas that some paleontologists have suggested may have contained lungs, this inference is disputed by others (Benton, 2005). So far as we can tell at this time, the best evidence suggests lungs developed only once, in the common ancestor of the actinopterygians and sarcopterygians (Liem et al., 2001; Benton, 2005).

Given that bichirs are primitive and possess a chassis that retains hallmarks of the earliest osteichthyans, the mechanics underlying how bichirs ventilate their lungs is of great interest (Fig. 7.3). When a bichir gets ready to take a breath, it comes to the surface of the water and exhales (Brainerd, Liem, and Samper, 1989). Much as you would exhale through your nose, the bichir exhales with its mouth closed, but, of course, because it is a fish, its nose is not connected to its mouth. Instead, the air leaves through the spiracle behind the operculum bones, and this exhaling continues to a point where the air pressure in its lungs and body cavity drops significantly (Brainerd, Liem, and Samper, 1989; Graham et al., 2014). The rigid and interlocking scales on its body then allow something remarkable to happen. The body cavity temporarily caves in, and the scaly straitjacket collapses as well, but it puts a great deal of elastic tension on the body wall (Brainerd, Liem, and Samper, 1989). This is much like what happens when you squeeze a rubber bulb, a handy tool in the arsenal of those who raise small children that cannot seem to blow their own nose. As any parent knows, if you want to suck the snot out of your toddler's nose, you squeeze the rubber bulb, you distract the child, and then you quickly insert the end of the bulb into the child's nose. Just as quickly, you allow the bulb to spring back to its normal shape, and with gargling grossness, the offending mucus slurps into the bulb. Like our rubber bulb, the compressed body of the bichir will spring back because the scales pull on the body wall, causing it to quickly rebound (Fig. 7.3) (Brainerd, Liem, and Samper, 1989). As a consequence, fresh air rushes into the bichir's lungs through its spiracles (Graham et al., 2014).

If you have gills, why also have lungs? The answer seems to be related to safety in redundancy: by having multiple ways to get oxygen, you are not painted into a lethal corner if the water you are living in begins to dry up or becomes stagnant. Warm water is wonderful for relaxing and bathing, but it tends to be oxygen-poor. Most of the fossils we find from the time when the initial evolution and diversification of the Osteichthyes was occurring (420–375 Ma; Cards 91–92) are contained in sedimentary rocks associated with fresh and marine waters near the equator (Benton, 2005). The correlation between warm water, low oxygen, and the equatorial environments of the earliest osteichthyans suggests lungs initially evolved to allow the earliest osteichthyans an opportunity to gulp air when their aquatic environments became stagnant (Liem et al., 2001). It is significant to note that modern lungfish (which are sarcopterygians) typically inhabit warm, tropical waters where oxygen levels are low and fluctuate greatly (Liem et al., 2001). Moreover, goldfish owners know that these actinopterygians will gulp air (Burggren, 1982), especially if the tank water has not been changed for a few days. Although the lungs of goldfish have been modified into a swim bladder, they can hold air bubbles in their blood-rich esophagus (Burggren, 1982). It is possible that such a behavior, combined with any mutation to enlarge pockets of the esophagus, may have been the initial driving force behind the evolution of lungs. Eventually, the derived sarcopterygians known as tetrapods would come to rely on the backup lung system full-time in their new terrestrial habitats.

The odd bichir has provided us with a search image for the earliest Osteichthyes, and with these commonalities established for the general osteichthyan chassis, we can now turn our attention to one of the major groups to descend from this common framework: the actinopterygians.

Actinopterygians: The Ray-Finned Fishes

By any measure, whether by sheer numbers, reproductive success, diversity, or global distribution, the actinopterygians have been the most successful vertebrates ever. Well over 25,000 species of actinopterygians are known to inhabit the waters of the world today, and there are surely thousands more waiting to be discovered and described. Nearly every person on earth has eaten these animals, either directly (fish fillets) or indirectly (many barbeque sauces contain anchovy paste, for example), and so they are nutritionally and economically important as well. If scales and bony fragments are to be trusted, long before these fishes were being gobbled up by hungry humans they were swimming around almost as soon as the modern Osteichthyes separated from their last common ancestor with other jawed vertebrates some 450 Ma (Card 91). However, the first good body fossils of these animals and the beginnings of their diversity didn't appear until sometime in the Devonian (400–375 Ma; Card 92), but then the diversity and evolution of these fishes took off, and it has continued in unstoppable fashion ever since.

All modern actinopterygian fishes show why they are called ray-finned: delicate bony segmented rays radiate out from the bases of their fins. The base of the rays are embedded in muscles that can finely control the form and position of these fins, making them supple and capable of assuming a number of mechanically efficient shapes for swimming (Liem et al., 2001). Most modern actinopterygians show another characteristic developed to various degrees: a flexible, kinetic skull (Pough, Janis, and Heiser, 2002; Benton, 2005). In other words, various bones in the jaws and skull are loosely connected through a series of ligaments and can be slid past one another to project the mouth forward for suction feeding. The teeth themselves are unique to these fishes: they are long posts of dentine with a cap of specialized enamel (called acrodine) at their tips (Benton, 2005). Thus, if you find a fossil fish with these sorts of teeth, you are guaranteed to be holding an actinopterygian. One other characteristic we see on the skull is the presence of distinctly separated cranial and caudal nostrils, somewhat reminiscent of those in the cartilaginous fishes and placoderms (Liem et al., 2001).

Pedigree Bony Fish with Thin Fin Rays

Date of First Appearance ~450 Ma (Card 91)

Specialties of Skeletal Chassis Suction Feeding, Fins Composed of Thin Rays, Incredibly Diverse Body Forms

Eco-niche Nearly Every Conceivable Lifestyle Available in Water

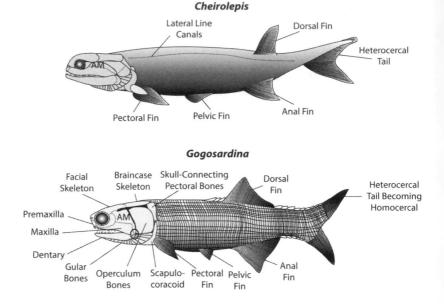

7.4. The early actinopterygians, *Gogosardina* and *Cheirolepis*. *Gogosardina* and *Cheirolepis* possess specialized actinopterygian traits including peg-like teeth with a cap of specialized enamel (acrodine) and delicate, bony rays embedded in their fins. In each diagram, the region of the adductor mandibulae muscle group, which lies beneath the temple bones (palatopterygoid arch), is indicated by AM. The quadrate-articular jaw joint is circled in *Gogosardina*. Based on specimens illustrated in Carroll (1988), Pough, Janis, and Heiser (2002, 2013), Benton (2005), and Choo, Long, and Trinajstic (2009).

We have already encountered one of the most primitive living actinopterygians, the bichir, and so we now turn to the fossil record to search for the beginnings of the actinopterygian chassis. Two fossils are informative: (1) *Cheirolepis*, an actinopterygian from the Old Red Sandstone of Scotland some 395–385 Ma (Card 92) (Radinsky, 1987; Carroll, 1988; Benton, 2005); and (2) *Gogosardina*, an actinopterygian (ray-finned fish) about 13–18 cm long from western Australia some 375 Ma (Card 92) (Choo, Long, and Trinajstic, 2009) (Figs. 7.4 and 7.5). The skulls and jaws of *Cheirolepis* and *Gogosardina* are reminiscent of the bichir, and this suggests they employed a similar suction-based means of feeding through mouth gaping and the lateral movements of their opercula (Fig. 7.5). The vertebral column of these fishes, unlike that of the bichir, was still comprised mostly of neural and hemal arches that encapsulated a prominent notochord. A dorsal fin, pectoral and pelvic fins, an anal fin, and a tail fin were all present as well, and these fins were more delicately constructed than those of the bichir. Although the vertebral column leading into the tail fin was still upturned and technically heterocercal, changes were beginning to occur in the outward shape of this region of the anatomy. The tail is still heterocercal in *Cheirolepis*, but in *Gogosardina* the tail fin has two equal-sized triangular lobes (technically, a homocercal tail) (Choo, Long, and Trinajstic, 2009). Unlike heterocercal tails, those that are homocercal create a series of vortices that propel the fish directly forward (Lauder et al., 2003; Lauder and Madden, 2006). This configuration also appears to enhance the ability of the rayed fins to more precisely control movement and directionality (Radinsky, 1987; Liem et al., 2001; Benton, 2005). Throughout actinopterygian evolution, the old heterocercal tail pattern has typically given way to the homocercal pattern.

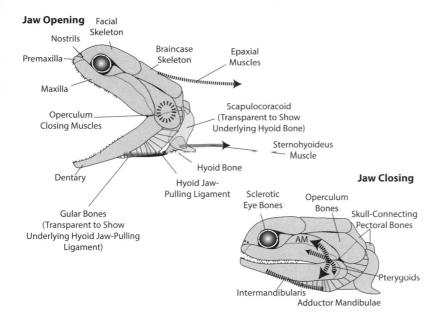

Jaw Opening

Facial Skeleton

Nostrils

Braincase Skeleton

Premaxilla

Epaxial Muscles

Maxilla

Operculum Closing Muscles

Scapulocoracoid (Transparent to Show Underlying Hyoid Bone)

Sternohyoideus Muscle

Dentary

Hyoid Bone

Hyoid Jaw-Pulling Ligament

Jaw Closing

Gular Bones (Transparent to Show Underlying Hyoid Jaw-Pulling Ligament)

Sclerotic Eye Bones

Operculum Bones

Skull-Connecting Pectoral Bones

AM

Pterygoids

Intermandibularis

Adductor Mandibulae

7.5. Inferred feeding mode in the early actinopterygian, *Cheirolepis,* from the Old Red Sandstone of Scotland in the Devonian period. The basic mechanics of jaw opening and closing in this fossil fish are inferred from studies on modern actinopterygians. Dashed lines and arrows indicate the major muscle groups that contribute to opening and closing the jaws. AM again indicates the location of the adductor mandibulae group beneath the palatopterygoid arch (temple) region. Diagram after Benton (2005).

Rows of thin, bony scales that interlock with peg-and-socket joints cover the bodies of *Cheirolepis* and *Gogosardina* (Carroll, 1988; Benton, 2005; Choo, Long, and Trinajstic, 2009). As you might anticipate from our previous analysis of lung-breathing in bichirs, this jacket of interlocking scales in *Gogosardina* and *Cheirolepis* is suggestive of a similar breathing mechanism in these animals – at least if their lungs were used to any significant degree for respiration. This inference is supported, in part, by the phylogenetic relationship of bichirs to other actinopterygians. Specifically, when fossils are added to the mix, we see that bichirs are sandwiched between the more primitive *Cheirolepis* and slightly more derived but still primitive actinopterygian fish variations on the scaly jacket chassis such as *Gogosardina* (Fig. 7.1).

As time ticked on, a variety of heavily scaled primitive actinopterygian members related to *Gogosardina* traditionally called 'paleoniscoids' continued to evolve and diversify, but some primitive actinopterygian groups took another route. The Chondrostei are primitive actinopterygians, more derived than the bichir and 'paleoniscoids,' which inhabit the earth today as sturgeon and paddlefish, and which provide us with our main source of caviar. The Chondrostei seem to have decided to do away with a lot of their bone and scale and assume some sharklike characteristics. Sturgeons, actinopterygians distributed throughout most of the Northern hemisphere (Fig. 7.6), can reach huge sizes, with the largest known individuals recorded as exceeding 6 meters in length and weighing over 450 kg (Bemis and Kynard, 2002; Bemis, Findeis, and Grande, 2002)! The skeleton of a sturgeon is composed of a great deal of cartilage, and the vertebrae consist of neural and hemal arches strung along a still-prominent notochord. Their body scales are tiny, and are actually close in

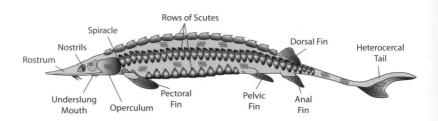

7.6. Diagram of a modern sturgeon. These strange actinopterygians parallel sharks and other chondrichthyans in many ways, having lost much of their bone and sporting toothlike scales. Based on a preserved specimen.

structure to the denticles we have discussed previously in sharks (Bemis, Findeis, and Grande, 2002). One exception to this scaling pattern is the five rows of large scales or scutes that run along the backs of these fishes. The tail is heterocercal, but the lower lobe is fairly large, making the tail more homocercal in appearance.

The head of a sturgeon is weird and fascinating. These fishes have lost parts of their jaw bones (including the premaxilla and maxilla), and use components of their palatoquadrate and lower jaw elements to capture prey (Bemis, Findeis, and Grande, 2002; Carroll and Wainwright, 2003), which in many cases is everything from clams and other mollusks to fishes. The jaws of many sturgeon species act like those of a shark in that the palatoquadrate and Meckel's cartilage regions are projected forward and downward by movements of the hyoid arch (Bemis, Findeis, and Grande, 2002; Carroll and Wainwright, 2003). What has actually happened in sturgeons, especially in their skulls, seems to be a case of what evolutionary biologists call pedomorphosis, when juvenile traits are retained into adulthood. Basically, adult sturgeons seem to hold on to the embryonic pattern of the skull: lots of cartilage and few bones (Bemis, Findeis, and Grande, 2002).

Paddlefish are another living member of the Chondrostei, close relations of the sturgeon (in fact, some species of paddlefish are now sought after for their caviar), and have a mostly cartilaginous skeleton. Their elongated rostrum is paddle-like (hence their common name) and has electric-detecting organs in it similar to those we encountered with sharks (Bemis and Grande, 1992). These odd fishes are filter feeders and strain tiny animals out of the water using their pharynx (Bemis and Grande, 1992). The odd parallels among sturgeon, paddlefish, and sharks suggest that loss of bone and development of a largely cartilaginous skeleton has been a frequent trend in various fish groups. In fact, in some of the most derived living actinopterygians, the deep-sea fishes, much of the skeleton has been given over to cartilage as well (Liem et al., 2001).

At some point toward the end of the Permian period (approximately 260 Ma; Card 95), more derived actinopterygians appeared that had more kinetic skulls, smaller body scales, and more symmetrical tail fins. From this radiation of fishes only two living forms exist today, the long-nosed gars and the bowfin. Gars still have the thick scaly jacket of primitive osteichthyans and have long, tweezer-like jaws ideal for quickly snapping up unsuspecting prey (Kammerer, Grande, and Westneat, 2006). However, due to their odd and derived body forms, they give us little insight

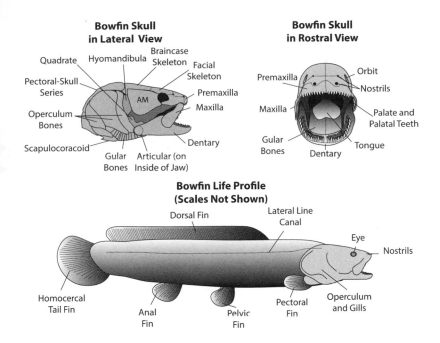

Bowfin Skull in Lateral View

Quadrate
Hyomandibula
Braincase Skeleton
Facial Skeleton
Pectoral-Skull Series
Premaxilla
AM
Maxilla
Operculum Bones
Dentary
Scapulocoracoid
Gular Bones
Articular (on Inside of Jaw)

Bowfin Skull in Rostral View

Premaxilla
Orbit
Nostrils
Maxilla
Palate and Palatal Teeth
Gular Bones
Tongue
Dentary

Bowfin Life Profile (Scales Not Shown)

Dorsal Fin
Lateral Line Canal
Eye
Nostrils
Homocercal Tail Fin
Anal Fin
Pelvic Fin
Pectoral Fin
Operculum and Gills

7.7. Diagrams of the bowfin fish *Amia*. For the skull, note that the maxilla and premaxilla bones are more loosely connected to the head, and this allows them to rotate forward toward prey. In the head-on view, numerous rows of teeth can be seen, including palatal teeth, and the tongue. The life profile shows why the bowfin gets its name: the large, bow-like dorsal fin that stretches across much of the length of its body. After the tail fin, the dorsal fin is the main propulsive organ of this fish. AM again indicates the location of the adductor mandibulae group beneath the palatopterygoid arch (temple) region. Life drawings based on Grande and Bemis (1998).

into changes in jaw and body structure that led to the modern radiations of actinopterygians. Instead, we will focus on the bowfin, a common freshwater fish with a skull intermediate between the early 'paleoniscoids' and the more derived and successful actinopterygian groups (Grande and Bemis, 1998).

Bowfins get their name from the long, bow-like dorsal fin that stretches across their backs (Fig. 7.7 and Plate 8). In fact, the bowfin's dorsal fin is flexible and mobile, and is the main organ of movement in these fishes after the tail (Grande and Bemis, 1998). In bowfins, skull mobility is more enhanced than what we saw in the bichir. Although the pattern of opening and closing remains familiar, two new functionalities have been added. First, the premaxilla and maxilla are independently movable, and when the jaws of the bowfin are opened, the maxilla slides into a more vertical and outward-angled position that expands the mouth cavity laterally (Liem et al., 2001; Benton, 2005). At the same time, just as in the bichir, the operculum bones are pulled by operculum muscles laterally and caudally, creating additional suction during mouth opening. However, in addition to suction, the operculum bones are now attached to the lower jaw through a series of ligaments. This means that when the operculum bones are rotated caudally, they pull on the lower jaw, further enhancing and coordinating jaw opening. The opened mouth of a bowfin is more circular when viewed mouth-on than in bichirs, and reveals additional rows of teeth along its roof (the palate), and toothlike projections along the hyoid and pharyngeal arches at its throat (Carroll, 1988; Liem et al., 2001; Benton, 2005).

Other anatomical details are worth noting here. The vertebral column of bowfins is robust. Ribs are present and associated with W-shaped

7.8. Teleost actinopterygian anatomy and skull function, based on a perch. Note that the pectoral fins have shifted upward directly behind the skull and the pelvic fins have moved forward to take their place. In the skull, the maxilla has now become a toothless lever that acts in concert with the opercular bones to throw the toothed premaxilla at prey. Jaw opening and closing are accomplished by muscles previously introduced in other fishes. Note that hyomandibula (H) and quadrate (Q) bones are recessed into the skull, flanking the palatopterygoid arch bones and the location of the adductor mandibulae (AM). Illustrations based on Liem et al. (2001), Westneat (2004), and Benton (2005).

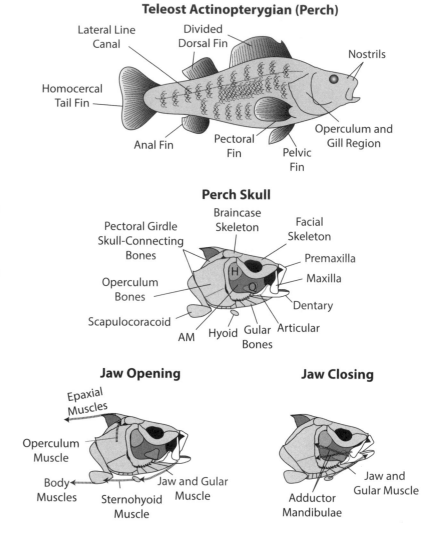

Teleost Actinopterygian (Perch)

Lateral Line Canal

Divided Dorsal Fin

Nostrils

Homocercal Tail Fin

Anal Fin

Pectoral Fin

Pelvic Fin

Operculum and Gill Region

Perch Skull

Braincase Skeleton

Pectoral Girdle Skull-Connecting Bones

Facial Skeleton

Premaxilla

Operculum Bones

Maxilla

Scapulocoracoid

Dentary

AM

Hyoid

Gular Bones

Articular

Jaw Opening

Epaxial Muscles

Operculum Muscle

Body Muscles

Sternohyoid Muscle

Jaw and Gular Muscle

Jaw Closing

Jaw and Gular Muscle

Adductor Mandibulae

myomeres that correlate with enhanced and more powerful swimming actions. The tail of bowfins is mostly symmetrical, although a small part of the vertebral column continues to extend into the upper lobe of this fin (Grande and Bemis, 1998). A quick beat of the mostly homocercal tail fin allows the bowfin to quickly charge and overtake prey. The body scales are also thinner than those of more primitive actinopterygians, being composed of slender bases of acellular bone capped with enamel, and embedded within folds of skin. In fact, the scales of the bowfin, and other more derived actinopterygian fishes, do not have the peg-and-socket arrangement we saw earlier, but instead are arranged in angled layers such that scales overlap and slide past one another (Grande and Bemis, 1998). This increases the flexibility of the body while still providing excellent protection. The skin in which the thin scales are embedded secretes a copious amount of mucus, which decreases friction and forms a full-body antibiotic barrier. An extinct group of actinopterygians related to the bowfin and gar, called the semionotids, utilized these adaptations to diversify in freshwater lakes in the Late Triassic to Mid-Jurassic (200–168 Ma; Cards 96–97). The semionotids looked a lot like teleosts, the most successful actinopterygians (described next), and thousands of fossils of these animals testify to their 'flocking' (closely related, coexisting species traveling together) in large and varying groups just as many cichlid fishes in African freshwater lakes do today (Benton, 2005).

By the end of the Triassic period some 204 Ma (Card 96), the most derived and successful actinopterygians, the teleosts, had appeared. These are the ray-finned fishes everyone knows, and they make up a bulk of the fishes eaten, and pets traded, by humans. Speaking of good eating, we will focus here on the perch as our example of a teleost actinopterygian (Fig. 7.8). Teleosts took skull kinetics one step further than other actinopterygians. In addition to all of the skull opening and closing mechanisms we have already discussed, teleosts have modified their maxilla into a toothless lever that throws the premaxilla cranially and upward toward prey (Westneat, 2004). As a teleost fish like the perch opens its mouth, the maxilla slides from a more horizontal to an almost vertical position, and this movement projects the premaxilla forward (Liem et al., 2001; Westneat, 2004; Benton, 2005). Because the maxilla is no longer used for biting prey, it is the premaxilla that has become enlarged and lined with teeth. In perch and most teleosts, the open mouth is jutted forward and is circular in cross-section, much like an extensible straw. This mouth-opening mechanism is what makes many teleost fishes, such as goldfish or perch, look like they're making kissy faces when they feed.

Just as we saw for sharks, the size of premaxilla, maxilla, and dentary bones compared with skull length gives some indication of jaw mechanics and leverage. In teleost actinopterygians that scrape algae from rocks and corals, you can bet that the three main jaw bones are very short: the short out-lever (jaws) and long in-lever (adductor mandibulae and friends, housed in the skull) generate great mechanical advantage in applying force (Westneat, 2004). In teleosts that must quickly open their mouths to catch fast-moving prey, the premaxilla and dentary (and sometimes the maxilla) are relatively long compared with the skull, and this gives the best mechanical advantage for rapid gaping (Liem et al., 2001; Westneat, 2004). The variation on this in-lever to out-lever situation in teleost jaws is remarkably diverse.

Let's summarize the evolution of jaw mechanics in actinopterygian fishes (Fig. 7.9). First, in the earliest actinopterygians (our bichir example), the jaws were opened by tilting the head back, forcing the lower jaw ventrally and forward, and rotating the operculum bones laterally (420 Ma; Card 91). Bowfins show us an intermediate step in this jaw-opening mechanism that appeared nearly 260 Ma (Card 95). At this step, the operculum becomes intimately connected with the lower jaw, and as the operculum bones rotate laterally and caudally, they pull on a ligament that, in turn, pulls the lower jaw downward. The premaxilla and maxilla also become more mobile, and the maxilla rotates into a more vertical position and flares laterally when the mouth is opened (Fig. 7.9).

7.9. Summary of the evolution of jaw mechanics in actinopterygian fishes. In primitive actinopterygians, a generally kinetic skull allowed a large gape well-suited for overtaking prey. In transitional or 'intermediate' stage actinopterygians like the bowfin, the maxilla and premaxilla become independently mobile units that help increase suction. In teleosts, the maxilla has become a toothless lever that projects the toothed premaxilla forward toward prey, creating suction. In bowfins and teleosts, the opercular bones are thrown out sideways to increase suction. In each diagram, the tooth-bearing jawbones are colored white and the opercular bones are black. Based on illustrations in Liem et al. (2001), Westneat (2004), and Benton (2005).

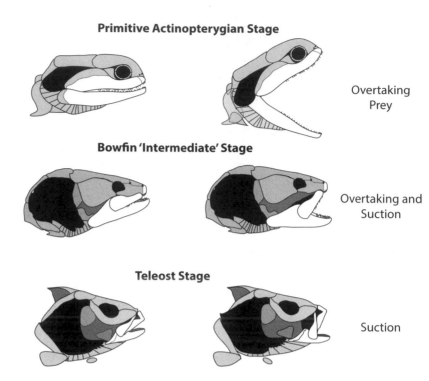

Primitive Actinopterygian Stage

Overtaking Prey

Bowfin 'Intermediate' Stage

Overtaking and Suction

Teleost Stage

Suction

Finally, in the teleost actinopterygians (our perch), the final icing on the cake was changing the maxilla into a toothless lever that projects the premaxilla forward and upward with great speed (Fig. 7.9).

But there is more to teleost success than simply a better fish-trap. As we did for sharks, we should peer inside the skull for a moment and note changes that have occurred in the brains of these most successful vertebrates. The brain of a teleost is largely given over to processing areas for sight and motor coordination (Fig. 7.10) (Butler and Hodos, 2005). Interpretation of images happens in the optic lobes of the midbrain, and these areas are huge in most actinopterygians. In fact, some evidence from studies on the brains of living teleosts points to the optic lobes as being almost as important as the cerebrum of the forebrain (Butler and Hodos, 2005). The cerebellum in the hindbrain is also impressively well developed, presumably to enhance motor coordination. These brain adaptations make sense given the generally active lifestyles of many teleosts and the need to visually find prey and then coordinate movements toward it.

Compared to many other vertebrates, the cerebrum and other areas of the forebrain are modest or even small in teleosts and other actinopterygians. Because the olfactory tracts in these fishes are nearly as big as the whole cerebrum, it was traditionally assumed (as with sharks) that most of an actinopterygian thinking was given over to odor detection and little else (Butler and Hodos, 2005). This has led to various myths, including the famous five-second rule: that a teleost like a goldfish can hold onto memories for only five seconds. The joke used to be that a goldfish

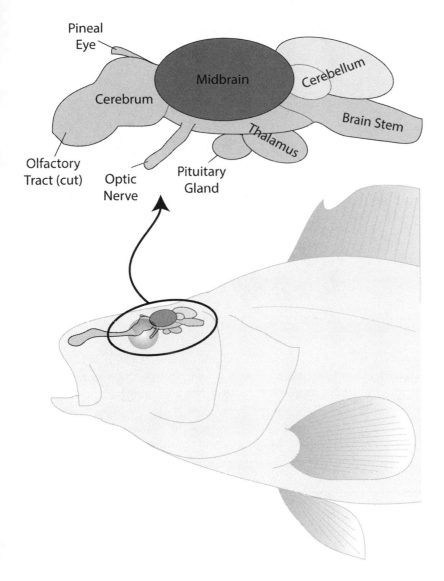

7.10. Diagram of a teleost actinopterygian brain. The largest parts of the brain are areas dedicated to visual and auditory interpretation (midbrain) and motor coordination (cerebellum). The relatively small higher thinking and integration center (cerebrum) has led to the myth that these animals are unintelligent or lack long-term memory, claims unsubstantiated by behavioral studies. Based on illustrations in Liem et al. (2001) and Butler and Hodos (2005).

couldn't remember when it had last rounded the fish bowl, and so every turn was a new adventure! It turns out that this was an oversimplification, and I'm surprised more scientists who have kept teleost fishes as pets didn't say something sooner: it is a common observation that the fishes learn to swim to the top of tank when you reach for the flake food. If they forget every five seconds, how are these hopelessly absent-minded fishes anticipating when they will be fed? More to the point, all vertebrates possess the same brain structures, including areas in the cerebrum for memory storage and processing (Butler and Hodos, 2005). Although the size and complexity of these regions varies dramatically among different vertebrate groups, there is no such thing as a fish without a memory. True, teleost fishes have some of the smallest brains for their body mass among most known vertebrates (Liem et al., 2001; Northcutt, 2002; Butler and Hodos, 2005). However, teleosts have been so successful that their brains

7.11. Teleost actinopterygian vertebrae, based on a perch. The vertebrae of many teleosts sport long ribs and form most of the vertebral structure, with the notochord becoming a smaller element. Illustrations based on Liem et al. (2001).

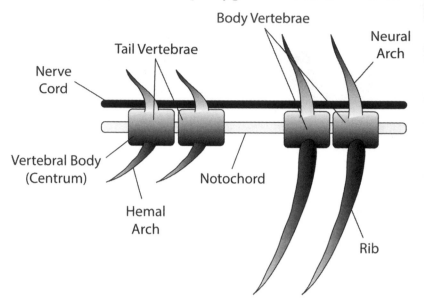

Teleost Actinopterygian Vertebral Column

and behavior must be incredibly adaptive or they would be extinct. It just goes to show that, like in Hollywood or Washington, you don't need to have a large brain to be successful.

The vertebral column of many teleosts is well developed, with prominent vertebrae, neural spines, and hemal arches surrounding and mechanically replacing most of the notochord (Fig. 7.11). In several lines of these fishes, the vertebrae actually develop joints that span the gaps between them. These specialized vertebral joints act both to strengthen the body wall and to resist side-to-side bending. In fact, in certain teleosts such as tuna, the vertebral joints stiffen the body and allow only the tail to be flexible (Hildebrand and Goslow, 1995; Liem et al., 2001). As you might imagine, this adaptation cuts down on drag because the body of the fish remains straight while the tail provides powerful forward thrust.

Not all teleosts are adapted for a high-speed lifestyle. Another fascinating aspect of teleosts is that many can hover and maneuver in complex three-dimensional environments such as coral reefs. As a consequence, pectoral, pelvic, and dorsal fins in many teleost fishes have become the major organs of locomotion, rather than the tail fin or body undulations (Lauder and Madden, 2006). The fin rays themselves are a marvelous adaptation in teleosts. These flexible, segmented bones can stiffen fins to provide propulsion and rapidly change fin shape to finely control movement and balance (Flammang et al., 2013). Remarkably, the flexible fin rays even allow the fins to quickly deform and recover their shape when perturbed, allowing passive damping of forces that might otherwise compromise efficient locomotion in complex environments (Flammang et al., 2013).

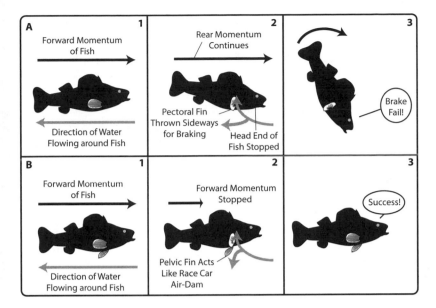

7.12. The mechanical advantage of having your pelvic fins where your pectoral fins used to be. In teleost fish, the pectoral fins act in braking. The pelvic fins (in the original place of the pectoral fins) can now act like an air dam or splitter in a race car, preventing the body from being lifted while braking. The sequence in (A) shows what happens to a teleost fish without the pelvic fins in their new place (result: brake fail!) whereas the sequence in (B) shows what actually happens in modern teleosts (result: brake success!).

The need to quickly accelerate, brake, change direction, accelerate, brake again, and so on, has led to the evolution of changes in the positions of the pectoral and pelvic fins in teleosts. Imagine how odd we would all look not only if our pectoral girdles and forelimbs were directly attached to the back of our skulls, but if our pelvis and hind limbs migrated forward to take the former place of our pectoral girdle and limb! In other words, can you picture your arms growing out of the back of your head and your pelvis and legs extending out of your chest? Well, this is precisely what many teleost fishes have done, and probably for the sake of maneuverability. Many teleost fishes have the ability to swim fast, and there has been selective pressure to ensure they can stop quickly as well. If you were rowing a boat and wanted to stop fast, your best bet would be to stick the paddles outward horizontally, which would create resistance and drag. In many teleost fishes, the pectoral fins have become employed for the same purpose. Instead of projecting ventrally, the pectoral fins have been rotated into a position just caudal to the head (Liem et al., 2001; Lauder and Madden, 2006). When a teleost fish wants to stop quickly, it throws the pectoral fins out laterally just like the paddles on our rowboat (Fig. 7.12).

All well and good, but there is a catch: unlike our rowboat, which is gliding along the surface of the water, the teleost fish is surrounded by water. Throwing out those pectoral fins when the fish is traveling fast is equivalent to hitting the front brakes on a motorcycle. Those unlucky enough to have locked up their front wheels on any bike know that one is thrown over the handles as the rear portion of the bike continues to move forward at high speed. Now imagine a teleost fish throwing out its pectoral fins only to have its tail rotate over its head. This is where the odd position of the pelvic fins comes into play. By moving the pelvic fins to where the pectoral fins used to be, teleost fishes now have what is equivalent to the splitter and air dam of a racecar at the front of the chassis. Splitters and air dams direct air under the car, causing suction that helps to hold the car body close to the ground and prevents it from being lifted off the track. The pelvic fins, in their place directly ventral to the pectoral fins, serve a similar purpose, directing water under the teleost fish as it rapidly slows and creating a downward force to counteract the upward force generated near the tail (Fig. 7.12) (Drucker and Lauder, 2003).

Finally, many teleosts have exquisite control over their swim bladders, the modified lungs used for buoyancy. For some, air can be gulped and stored in the swim bladder, but for many others a specialized gas gland controls how inflated or deflated their swim bladders become (Kardong, 2012). The swim bladders of teleosts and other actinopterygians can create a center of buoyancy quite different from the center of mass. In most teleosts, denser back muscles and vertebrae overlie the less dense air-filled swim bladder, making the fish unstable about its long axis (Hildebrand and Goslow, 2001). This is why a dead fish will go belly up. As my colleague Brooke Flammang has remarked, this is like trying to sit still on a beach ball in a swimming pool. But instability can be a good thing if controlled: because the body so easily tips side to side, this difference in the center of buoyancy and mass means most teleost fish are great at rapidly changing direction and maneuvering three-dimensional obstacles.

Overall, the teleost chassis has proven itself to be incredibly versatile. The first teleost fossils from the Late Triassic and Jurassic periods (204–145 Ma; Cards 96–97) were typically elongate animals, and some had already specialized into giant filter feeders that were nearly 9 meters long! From here, the diversity exploded into various chassis types. By the close of the Jurassic (145 Ma; Card 97), the first examples of true eels make their appearance, followed shortly thereafter in the Cretaceous period (145–65 Ma; Cards 97–99) by the appearance of carps and their relatives, and the beginnings of the radiations that led to salmon, trout, perch, cod, and other tasty fish groups. The bizarre anatomy and lifestyles of many deep-sea teleosts are a further testament to the flexible teleost design. One glimpse at a gulper eel, a deep-sea teleost with a tiny head perched on the end of absurdly large jaws, should be enough to convince anyone that the teleost chassis is the clear winner in the game of vertebrate diversity and success.

As the actinopterygians began their ascent to their currently diverse status in the Devonian, the sarcopterygians were also coming to prominence, especially in shallower bodies of water. In fact, sarcopterygians were once as diverse as the early actinopterygians were, but rather quickly became a smaller part of aquatic ecosystems. However, one line off the sarcopterygian family tree ended up moving toward a new frontier free of vertebrate competitors: the land.

8.1. Evolution of pectoral fin into limb elements in sarcopterygians. In actinopterygians and most fish, three proximal fin elements attach to the pectoral girdle at the scapulocoracoid. In turn, multiple thin fin rays radiate out from the proximal elements in a more or less symmetrical pattern. In sarcopterygians, one of the proximal elements becomes a larger bone called the humerus, whereas the other two bones either become very reduced in size (primitive sarcopterygians) or disappear altogether (modern sarcopterygian groups). The fin rays that do radiate from the distal end of the humerus tend to have an asymmetrical distribution. A similar pattern is present in the pelvic fin. Based on data and illustrations in Zhu and Yu (2009).

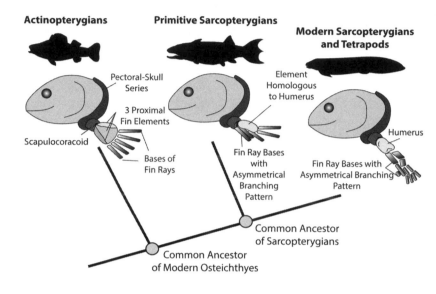

The Fishlike Osteichthyes, Part 2

AS CAR ENTHUSIASTS KNOW, DIFFERENT TIRES ARE IMPORTANT FOR different types of driving. The tires that work well for a racecar are a poor substitute on the family minivan. Bulky off-road tires that provide excellent traction over rough terrain are much less efficient on the highway. Ideally, a driver wants to match the right tire with the right environment. Like tires, fins vary in their biological construction for the conditions most often encountered by the animal in question. As we have seen for actinopterygians, flexible fins with thin, spindly rays are excellent for maneuvering through water. But our next group of vertebrates, the sarcopterygians, began to change the nature of their fins from wispy orienting devices to larger, bulky, muscular appendages.

All of the non-fishy vertebrates that you know and love are tetrapods, including us. The name 'tetrapod' means exactly what you might suspect: 'four-footed.' We know from careful study of the fossil record, anatomy, and genes that the modern tetrapods share a common ancestor with the living sarcopterygians that remain fishy: lungfish and two to three species of coelacanth. Hindsight is 20/20, and it would be tempting to conclude that the bulky, muscular appendages of our fishlike ancestors evolved to solve the problem of moving across land. However, it appears instead that these bulky fins developed to overcome problems associated with specialized aquatic conditions. It just so happens that, as an unintended consequence of developing such strong appendages, chance favored our common ancestor when a landlubbing existence presented itself.

But we are getting a little ahead of ourselves here. First, we want to establish some basic characteristics of the sarcopterygian chassis. As we have done in previous chapters, we will then introduce the living fishlike sarcopterygians, followed by a tour of the fossil diversity that once represented these fishy forms. Ultimately, we will get a glimpse at the beginnings of what would become the tetrapod chassis.

Fishlike Sarcopterygians: Fleshy-Finned Swimmers

Sarcopterygians, whether fishy or not, are characterized in part by the skeletal pattern of their appendages: only one large bone articulates with their pectoral or pelvic girdle, followed by two bones and then multiple bones in an asymmetrical branching pattern (Liem et al., 2001; Benton, 2005). This is in contradistinction to what we have seen in actinopterygians: a short base of bones from which radiate numerous thin, segmented rays. Therefore, an astute paleontologist can always spot a sarcopterygian, whether fossil or living, by its asymmetrical fins and bone-branching pattern.

Pedigree Vertebrates Close to the Common Ancestor of Tetrapods

Date of First Appearance ~450 Ma (Card 91)

Specialties of Skeletal Chassis Fleshy Fins with Fewer, Sturdier Bones

Eco-niche Some Deep and Shallow Aquatic Environments

137

To discuss changes to the appendages more precisely, we need to add a few more anatomical terms to our vocabulary. In anatomy-speak, the word 'proximal' refers to the part of an appendage or other body structure closest to the body wall or girdle. 'Distal,' the opposite of proximal, refers to the part of an appendage or body part farthest from the body wall or a girdle. For example, your upper arm bone is called your humerus, and it is the most proximal element in your forelimb due to its *proximity* to your shoulder girdle. In contrast, your fingers are the most distal of the bones in your forelimb due to their *distance* from your shoulder girdle. In all sarcopterygians, including us, the proximal-most bone in the pectoral appendage is the humerus, whereas the femur (thigh bone) assumes this position in the pelvic appendage (Fig. 8.1).

Before we take to the waves on our ship the *Devonian Destiny* once again, it is instructive to plunge into modern lakes, rivers, and oceans and get a look at some living fishlike sarcopterygians. Let's start with the modern lungfish, which is represented today by three species lurking in rivers and lakes in South America, Australia, and Africa (Fig. 8.2). The average lungfish species ranges in size from 1 to 1.5 meters long. As you might anticipate, lungfish get their name because they have functional lungs that they rely on for some of their oxygen needs. These animals live in regions where the water becomes stagnant from time to time, and their lungs help them make it through these trying periods. All lungfish can pull themselves from the water and move across land to another source of water (Liem et al., 2001; Pough, Janis, and Heiser, 2002).

A common misconception about land travel in vertebrates is that it requires limbs, but limbless lizards and snakes are the ultimate case in point for a terrestrial existence without pesky appendages. Among the actinopterygians, certain species of eels can cross from one pond to another simply by slithering over the ground (Moyle and Cech, 1996). So it should be no surprise, then, that the South American and African lungfish have spindly fins and very muscular bodies that they throw across the ground to make overland travels. Only the Australian lungfish has substantial, fleshy limbs that can be used to prop it up and propel it toward fresher sources of water (Moyle and Cech, 1996). More intriguingly, the South American and African lungfish will, like the bichir, drown if not given adequate access to air, whereas the Australian lungfish respires mostly through its gills (Moyle and Cech, 1996). Again, we see here that the evolution of muscular limbs capable of supporting the body took place beneath the waves rather than on terra firma. All these animals serve as a warning about our assumptions regarding the transition from water to land: fleshy appendages probably did not evolve initially for land travel.

Another warning has come in the form of new research on our (now) old friend, the bichir (*Polypterus*). As it turns out, the bichir can walk on land using its pectoral fins (Standen and Larsson, 2012; Standen, Du, and Larsson, 2014). Using a combination of lateral body undulation and alternating rotation of the pectoral fins, the bichir can propel itself along the ground until it reaches another pool of water. In an ingenious

African Lungfish (*Protopterus*)

Lateral Line

Pectoral Fin

Paired Nostrils inside Lips and Mouth Cavity

Pelvic Fin

Australian Lungfish (*Neoceratodus*)

Lateral Line

Diphycercal Tail

Pectoral Fin

Pelvic Fin

**African Lungfish (*Protopterus*)
Skull Detail**

Orbit   Braincase Skeleton

Remnant Facial Skeleton

Neural Arches

Palatal Bones

Notochord

Crushing Teeth

Prearticular

Quadrate   Hyold Arch Bones   Pectoral Girdle   Cranial Rib

8.2. Diagram of two living species of lungfish from the African and Australian continents. Note that the paired nostrils are present, but are hidden behind the nose in a chamber just cranial to the mouth. Curiously, it is the spindly-finned *Protopterus* lungfish from Africa, and not the lobe-finned *Neoceratodus* from Australia, that more frequently takes journeys across land. The bottom inset shows a diagram of the African lungfish skull. Note the reduced head skeleton, especially around the mouth, where the functional jaws are composed of the palatal bones and the prearticular, rather than the typical premaxilla, maxilla, and dentary, which are all absent. The short jaws and large teeth give excellent mechanical advantage in crushing invertebrate shells. Note also the odd cranial rib against which jaw-opening muscles pull when the lungfish feeds, and large notochord with small vertebrae. Lungfish drawings after Liem et al. (2001); lungfish skull after Bemis and Lauder (1986).

experiment, researchers created damp environments for bichir to live in without water deep enough to swim in (Standen, Du, and Larsson, 2014). Compared with typical bichirs, experimental fishes that were forced to walk developed more efficient patterns of support and movement (their fin movements were faster and more effective in propelling them forward), and their pectoral bones became more robust in relation to their new roles in locomotion and weight support (Standen, Du, and Larsson, 2014). Here again is an actinopterygian that is capable of walking on land.

What do we see in the modern fishlike sarcopterygians? The modern lungfish chassis is eel-like in overall appearance, very muscular, and has small, overlapping scales (Fig. 8.2 and Plate 7). Despite differences in the fleshiness of the pectoral and pelvic fins, the pattern of bones in all lungfish reflects their sarcopterygian pedigree: they all possess a humerus or femur proximally, followed by two and then multiple bones more distally (Liem et al., 2001; Clack, 2002). Living lungfish have a fairly cartilaginous skeleton (including their braincase) and they retain a prominent notochord (Clack, 2002; Pough, Janis, and Heiser, 2002; Kardong, 2012).

As in many actinopterygian fishes, the skulls of lungfishes are composed of numerous small bones. However, lungfish skulls are oddly constructed, and there has been the loss of many bones that are familiar to

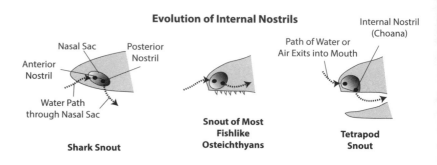

Evolution of Internal Nostrils

Internal Nostril (Choana)

Nasal Sac   Posterior Nostril

Anterior Nostril

Water Path through Nasal Sac

Path of Water or Air Exits into Mouth

Shark Snout

Snout of Most Fishlike Osteichthyans

Tetrapod Snout

8.3. Diagram showing a simplified evolution of the nostrils in jawed vertebrates. In both sharks and fishlike Osteichthyes, there is a pair of cranial and caudal nostrils. Odors carried on water pass through the cranial nostrils into an olfactory canal and then into the caudal nostrils, where the water exits. In tetrapods, the caudal nostrils have shifted into the mouth and point downward from the palate. These internal caudal nostrils are called the choanae. In tetrapods, these internal nostrils transfer air from the nose into the mouth and pharynx for breathing, but this does not appear to have happened in some sarcopterygians with choanae. Instead, water simply entered the nose and mouth for odor detection as in other fishlike Osteichthyes. Diagrams after Janvier (1996).

us now. For example, lungfish skulls lack the premaxilla, maxilla, and the dentary bones. The functional jaws are actually composed of the palatal bones in the roof of the mouth and a remnant lower jaw bone (technically, the prearticular) (Bemis and Lauder, 1986). The palatal and prearticular bones are fused and sport a closely packed array of tooth plates that grind against one another. Weirder still, a specialized cranial rib attaches to the back of lungfish skulls, and assists in jaw opening (Bemis and Lauder, 1986)! Overall, the short jaws and crushing tooth batteries of lungfishes are ideal for eating shelled invertebrates (Fig. 8.2) (Bemis and Lauder, 1986; Liem et al., 2001). Due to the loss of the premaxilla and maxilla in lungfishes, the position of the nostrils has moved. Both pairs of nostrils (cranial and caudal) now lie on the upper portion of the mouth inside a fleshy lip (in many fossil lungfishes, the lip was actually bony) and point downwards. However, lungfish do not breathe air through their nostrils, and these openings still work as they do in sharks and actinopterygians: water flows into them for odor detection but not for respiration (Atz, 1952) (Fig. 8.3).

Like the bichirs, lungfish can obtain oxygen both through their gills and lungs. However, their respiratory and circulatory systems are a bit more complex. The heart of a typical actinopterygian fish is a one-way pump: it takes oxygen-poor blood returning from the head and body and pushes it toward the gills. The gills, in turn, extract oxygen from the surrounding water where this gas diffuses into the blood, and from there the blood is shunted to either the head or the body (Liem et al., 2001; Kardong, 2012). In a lungfish, both the heart and gills are modified from this primitive one-way system. The heart of a lungfish is now a double-pump: it is divided internally into a right and left portion. The right portion acts like the primitive fish heart by receiving oxygen-poor blood from the head and body and pumping it toward the gills (Johansen and Lenfant, 1968; Szidon et al., 1969; Burggren and Johansen, 1986). However, the left portion of the heart now receives oxygen-rich blood from the lungs and pumps that directly to the head and body. This is possible because in lungfish, some of the old gill arches no longer bear gills. Whereas some of the blood vessels still dutifully pass through arches where gills reside, others follow arches devoid of gills to take oxygen-poor blood directly to the lungs (Johansen and Lenfant, 1968; Szidon et al., 1969; Burggren and Johansen, 1986). Within the lungs, tiny clusters of capillaries tucked into pockets called alveoli extract oxygen from each breath, and then pass their precious cargo back to the left side of the heart. Much of the system works on differences in pressure that trigger the closing and opening of various valves. For example, when the lungfish is breathing air, more blood naturally flows to the lungs. This causes a pressure drop in the gills, and a valve closes that shunts blood around the gills and directly to the head and body (Johansen and Lenfant, 1968; Szidon et al., 1969; Burggren and Johansen, 1986; Liem et al., 2001).

Analysis of trait states shows that lungfish are relatively primitive and near the base of the sarcopterygian family tree (Clack, 2002; Brinkmann et al., 2004). This suggests that their heart-lung-gill arrangement may be a holdover from the common ancestor that eventually gave rise to the tetrapods. Thus, a double-pump heart and flexible blood distribution system may have been present in many of the fossil fishlike sarcopterygians and tetrapodomorphs (close relatives of the tetrapods) we will soon encounter.

The African lungfish has an additional trick up its lobe fin: it can bury itself in mud, coat itself liberally in mucus (there's that magic snot again), and put itself in a state of suspended animation called aestivation (Liem et al., 2001; Pough, Janis, and Heiser, 2002; Chew et al., 2004). When normally wet areas dry up, the African lungfish goes through this routine to wait out the bad times until the rains and rivers return. That African lungfish can stay in this state of suspended animation for a long time was discovered in the early twentieth century when specimens were collected and shipped back to various museums across the world. According to one account, an African lungfish that was kept dry in its dirt cocoon for two years (in a metal coffee can, no less) resumed life as usual when a curious scientist placed it in a tank of water (Smith, 1931)!

The only other living fishlike sarcopterygian, long thought extinct until it was fished out of the Indian Ocean in the late 1930s, is the coelacanth (Fig. 8.4). Although they were more diverse in the past, there is only one living form (but at least two species), called *Latimeria*, around today, and it lives at depths of up to 180 meters in the Indian Ocean and Indonesian waters (Moyle and Cech, 1996; Fricke and Hissmann, 2000). The modern coelacanth is a large fish, with the largest known specimens weighing close to 90 kg and stretching over 1.5 meters in length. The coelacanth has a thick, deep body covered in interlocking scales and a short and robust head, and nearly every fin—the exception is one of its dorsal fins—is muscular. The fleshy fins are moved in a way eerily similar to those of land-living tetrapods, and previous generations of paleontologists suggested that this pattern of fin movements mimicked those of the earliest tetrapods. Alas, it turns out that these fin movements likely have little to do with those of tetrapods because they are used to row the coelacanth slowly through the water, and this fishy sarcopterygian does not walk on the ocean floor. However, if the pattern of fin movements is the same as limb movements, it may at least be an evolutionary precursor to terrestrial locomotion. The tail fin is diphycercal, being composed

8.4. Diagram of the modern coelacanth, *Latimeria*. Nearly all of its fins are muscular and lobed. Although the coelacanth moves the fins like a walking tetrapod, it rarely uses them to walk on the ocean floor. Instead, its slow fin movements position it in the water column, where it drifts within striking distance of prey such as small fishes. The inset shows the reduced skull of *Latimeria* and the intracranial joint that allows the snout to bend forward to help crush and impale its prey. An intracranial joint is a primitive trait state of sarcopterygians, but *Latimeria* is the only living form to retain this feature. A large basicranial muscle spans the skull beneath the braincase and causes the snout bending along the intracranial joint. Coelacanth and skull diagrams based on Liem et al. (2001).

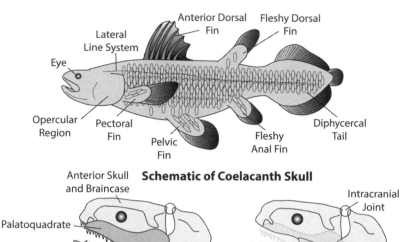

of two spade-shaped lobes converging around a central point, and this broad organ can be swept in powerful strokes to quickly launch the normally slow-moving coelacanth at potential fish prey. Modern coelacanths have nonfunctional lungs surrounded by small calcified plates, and a fatty organ that may aid buoyancy (Moyle and Cech, 1996; Fricke and Hissmann, 2000; Cupello et al., 2015).

Coelacanths have heavy, reinforced skulls with an intracranial joint that allows the snout to flex ventrally when the animal is feeding on other fish (Fig. 8.4) (Fricke and Hissmann, 2000; Liem et al., 2001; Pough, Janis, and Heiser, 2002). Coelacanths are the only surviving sarcopterygian with such a joint in their heads, but this joint was very common among the most primitive sarcopterygians, and had important evolutionary implications for feeding and eyesight. A broad, strap-like basicranial muscle spans this joint in coelacanths (and presumably other primitive sarcopterygians with this same anatomy), connecting the cranial and caudal ends of the skull, and its powerful contractions drive the teeth of the upper jaw bones deep into prey (Liem et al., 2001). This basicranial muscle would later become modified in some tetrapods to help them blink their eyes by pulling the eyeball back into the socket—more on this in later chapters.

With this basic overview of living fishlike sarcopterygians, we can now sail out on the *Devonian Destiny* and briefly explore the former diversity of these animals prior to their decline and the rise of their tetrapod descendants. The Devonian saw an explosive radiation of fishlike sarcopterygians. It turns out, from analyses of the trait states in these fossil forms, that the coelacanth and lungfish are some of the more primitive

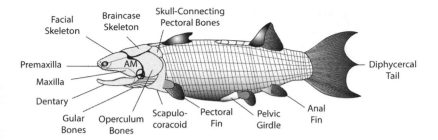

8.5. Diagram of the early fishlike sarcopterygian, *Guiyu*. The region of the adductor mandibulae muscle group, which lies beneath the temple bones (palatopterygoid arch), is indicated by AM, and the quadrate-articular jaw joint is circled. Skeleton diagram after Zhu et al. (2009).

members of the fishlike sarcopterygians (Clack, 2002). Therefore, with few exceptions, most of the fishlike sarcopterygians described next are actually more closely related to organisms like ourselves than to the lungfish or coelacanth.

Two examples of sarcopterygians more primitive than lungfish and coelacanths are the animal *Guiyu* and the onychodontiforms. *Guiyu* appears to be one of the earliest sarcopterygians, living as it did 419 Ma (Card 91) in what is now China (Fig. 8.5) (Zhu et al. 2009). Like coelacanths, it possessed an intracranial joint that was likely spanned by the basicranial muscle, which suggests that *Guiyu* had a powerful bite (Zhu et al., 2009). The pectoral and pelvic fins of *Guiyu* were not very robust and were probably not very muscular, and several of the fins were supported by spines. Unlike in modern sarcopterygians, the pectoral fins of *Guiyu* articulated with the pectoral girdle in much the same way as seen in actinopterygians: there are several proximal bones, rather than a lone humerus, and smaller, ray-like bones radiate from this base (Fig. 8.1) (Zhu and Yu, 2009; Zhu et al., 2009). The tail fin is diphycercal, being remarkably similar to that seen in coelacanths, and must have provided powerful thrust. Finally, *Guiyu* was covered in a heavy jacket of interlocking scales (Zhu et al., 2009), suggesting that, like the bichir, it had some sort of suction-based mechanism for breathing air.

Looking superficially like moray eels, an odd group of fanged sarcopterygians generally grouped together as onychodontiforms populated areas of Asia and Australia. One of the best known of these animals is *Onychodus* from the ancient reef system where we have already searched for placoderms in what is now Western Australia (375 Ma; Card 92) (Fig. 8.6) (Long, 2001). The head of *Onychodus* is long, and it possessed a gaping maw full of pointy teeth. Curiously, a prong of elongate and wickedly recurved teeth are perched atop the cranial end of the dentary bone in these fishes. Based on articulation of these elements in exquisitely preserved three-dimensional specimens, it appears that *Onychodus* and other onychodontiforms could actively rotate this tooth whorl toward prey: much like the front shovel of a bulldozer, these fishlike sarcopterygians could impale and then rotate hapless prey back into their waiting mouths (Long, 2001, 2006)! In fact, *Onychodus* and its relatives reduced the bones in their upper palate to make room for these impressive fangs once they were rotated inside the mouth (Long, 2001; Benton, 2005). As with coelacanths and *Guiyu*, an intracranial joint was present in these animals.

8.6. The primitive fishlike sarcopterygian *Onychodus* from western Australia some 375 Ma (Card 92). Note the tooth whorl on the end of the dentary bone that was capable of rotating toward and impaling prey. The pectoral and pelvic fins are fleshy, but not as robust as in sarcopterygians closer to the common ancestor of tetrapods. *Onychodus* skull and body after Long (2001, 2006).

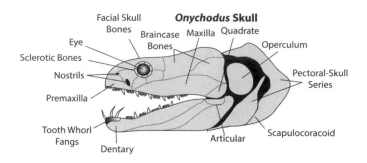

Onychodus Skull

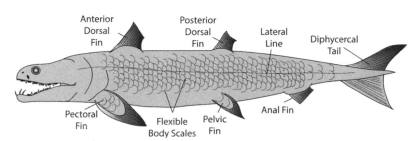

Onychodus Full Body Restoration

The body form of onychodontiforms was similar in overall appearance to *Guiyu*, but was more slender and lacked fin spines (Fig. 8.6). The fins, while fleshy, were relatively small, although the pectoral fins are robust. One gets the impression that onychodontiforms were sit-and-wait predators, striking out with their muscular bodies, fins, and mobile fangs at unsuspecting fishes among the Devonian reefs (Long, 2006). Like a powerful truck with front-wheel drive, the heavy pectoral fins would act to provide traction and thrust to these large predators, flinging them at their prey much like an SUV lurching suddenly out of a driveway onto a busy road.

Coelacanths make their debut in the Devonian seas as well, appearing in many aquatic environments across the world. Approximately 375 Ma (Card 92) to 250 Ma (Card 95), a variety of coelacanths evolved diverse forms that inhabited marine and freshwater environments (Lund and Lund, 1984; Benton, 2005; Friedman and Coates, 2006). Some of these fossil coelacanths were very similar in overall appearance to their modern descendants, whereas others had incredibly deep bodies or were overly elongate. The fossil record shows that these animals, once thought to be a textbook case of an unchanging and somewhat uninteresting branch of the vertebrate family tree, were primitively quite diverse in their body forms and habits (Friedman and Coates, 2006) (Fig. 8.7).

Lungfishes and the closely related porolepiforms also radiated into many diverse forms over the same period of time (375–250 Ma; Cards 92–95) (Clack, 2002). Intriguingly, the intracranial joint was present in porolepiforms and perhaps the earliest lungfish, but was ultimately fused in later species of the lungfish lineage, probably to assist them with crushing shelled invertebrates (Janvier, 1996). One of the best-known fossil

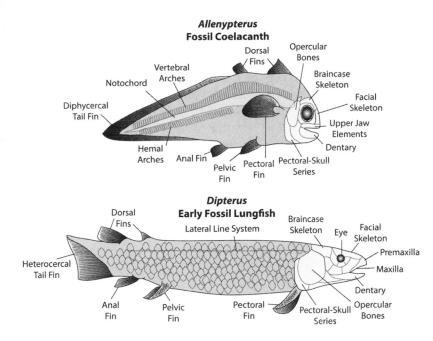

Allenypterus
Fossil Coelacanth

Notochord · Vertebral Arches · Dorsal Fins · Opercular Bones · Braincase Skeleton · Facial Skeleton · Upper Jaw Elements · Dentary · Diphycercal Tail Fin · Hemal Arches · Anal Fin · Pelvic Fin · Pectoral Fin · Pectoral-Skull Series

Dipterus
Early Fossil Lungfish

Dorsal Fins · Lateral Line System · Braincase Skeleton · Eye · Facial Skeleton · Premaxilla · Maxilla · Dentary · Opercular Bones · Heterocercal Tail Fin · Anal Fin · Pelvic Fin · Pectoral Fin · Pectoral-Skull Series

8.7. A fossil coelacanth and lungfish. *Allenypterus* is an odd fossil coelacanth from the Bear Gulch Formation in Montana from 318 Ma (Card 93) that resembles teleost actinopterygians in many ways. Its deep body and cranially placed pectoral and pelvic fins suggest it was capable of turning and maneuvering in complex environments such as modern reefs. *Dipterus* is one of the best-known Devonian lungfishes. Unlike modern lungfishes, *Dipterus* possessed many of the skull bones we are familiar with, had a tough jacket of scales, and retained a heterocercal tail. However, like modern lungfishes, this animal was probably capable of burrowing into the soil and aestivating for extended periods, based on fossil lungfish burrow casts. *Allenypterus* diagram after Lund and Lund (1984); *Dipterus* after Janvier (1996) and Benton (2005).

lungfish, *Dipterus*, found in Devonian rocks from Scotland, already has the crushing palatal teeth seen in its modern descendants, but there are some differences in the chassis (Fig. 8.7) (Carroll, 1988; Pough, Janis, and Heiser, 2002; Benton, 2005). The skull is more elongate, the tail is heterocercal, and the body is lengthy and covered in a jacket of scales somewhat reminiscent of the bichir fish discussed previously (Carroll, 1988). Perhaps most amazing of all is that we have fossil lungfish burrows dating back to the Devonian and throughout most of the vertebrate fossil record (Hasiotis et al., 2004). This means that we can definitively say that at least some portion of the lungfish lineage have been aestivating for hundreds of millions of years when the rivers ran dry. It boggles the imagination to know that a behavior we observe in modern lungfishes existed long before anything even resembling a frog was hopping about the land.

Approximately 375 Ma (Card 92), a group of sarcopterygians appears in the fossil record that would eventually give rise to the tetrapods. Again, there is controversy surrounding the relationships of these vertebrates to one another, and regarding whether certain trait states also appear in animals like lungfishes and porolepiforms. For simplicity, we will refer to this group of sarcopterygians as the tetrapodomorphs. These animals were relatively large predators (from approximately 30 cm long to some giant forms over 6 meters in length) and show a diversity of body forms. Several tetrapodomorphs are known: some are fishlike in overall appearance, but others are closer to the anatomy of what we encounter in a limbed tetrapod. Here, we will focus on one of the fishlike tetrapodomorphs in detail. We do this to streamline our discussion and to highlight the undergirding of what would become the basis of the tetrapod chassis. Transitional tetrapodomorphs that possess a combination of fish- and tetrapod-like traits will be discussed in more detail in chapter 10.

8.8. The well-known tetrapodomorph, *Eusthenopteron*. Although this animal has a number of features that place it close to the common ancestor of tetrapods, it is still a 'fish' for all intents and purposes. AM marks the region of the adductor mandibulae muscle, hidden beneath the squamosal in an adductor fossa. The quadrate-articular jaw joint is circled. In palate view, notice the large, fang-like teeth on the palatal bones. Also, note the presence of the choanae just behind the premaxilla and next to the vomer bones. *Eusthenopteron* lacked a vomeronasal organ, but in tetrapods with this trait, it is typically located above the vomers. There is an intracranial joint in the skull as well as large pterygoid bones that housed the deep jaw-closing pterygoid muscles. Each vertebra is made up of two centra (an intercentrum and pleurocentrum, not distinguished in the diagram), a neural arch, and, in some cases, a small rib. In the tail, hemal arches to protect delicate blood vessels are present. *Eusthenopteron* anatomy modified from Carroll (1988), Clack (2002), and Benton (2005).

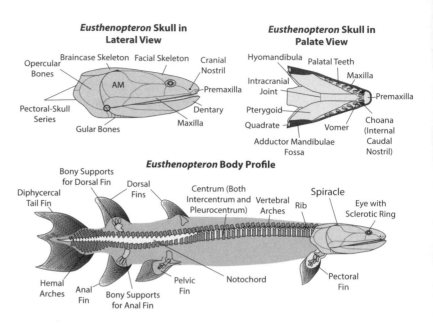

Eusthenopteron **Skull in Lateral View**

Braincase Skeleton — Facial Skeleton — Cranial Nostril — Opercular Bones — AM — Premaxilla — Pectoral-Skull Series — Dentary — Gular Bones — Maxilla

Eusthenopteron **Skull in Palate View**

Hyomandibula — Palatal Teeth — Intracranial Joint — Maxilla — Pterygoid — Premaxilla — Quadrate — Vomer — Choana (Internal Caudal Nostril) — Adductor Mandibulae Fossa

Eusthenopteron **Body Profile**

Bony Supports for Dorsal Fin — Dorsal Fins — Centrum (Both Intercentrum and Pleurocentrum) — Vertebral Arches — Rib — Spiracle — Diphycercal Tail Fin — Eye with Sclerotic Ring — Hemal Arches — Anal Fin — Bony Supports for Anal Fin — Pelvic Fin — Notochord — Pectoral Fin

The fishlike tetrapodomorph in question is *Eusthenopteron*, a good-sized predator nearly a meter long from rocks 375 Ma (Card 92) in Canada (Fig. 8.8) (Carroll, 1988; Pough, Janis, and Heiser, 2002; Benton, 2005). Superficially, it looks a great deal like many of the fishlike sarcopterygians we have encountered, but its underlying chassis is a different matter. At first blush, the skull of *Eusthenopteron* is nothing remarkable, looking somewhat reminiscent of a bowfin or a pike. But looks are deceiving. In lateral view, one can see that palatopterygoid bones that make up the temple of *Eusthenopteron* have now fused into a large squamosal bone (Carroll, 1988; Benton, 2005) (in Fig. 8.8, this is the area labeled AM). As we discussed previously, this region of the skull houses the adductor mandibulae and pterygoid muscles that close the jaws. The large size of the squamosal bone in *Eusthenopteron* is therefore suggestive of fairly powerful jaw-closing muscles. In fact, a large, hollow chamber beneath the squamosal region of the skull (called the subtemporal fossa) is further evidence of a large adductor mandibulae group of muscles in this location (Carroll, 1988). The presence of a squamosal bone itself, rather than the palatopterygoid bones, is telling: this is the condition retained in tetrapods, and it speaks to the close relationship of *Eusthenopteron* with our common ancestors.

A prominent intracranial joint is present in this animal and, as with coelacanths, this suggests that the snout of this animal flexed downward powerfully when biting prey (Carroll, 1988; Benton, 2005). Speaking of biting, there are also plenty of small, pointed teeth along the premaxilla, maxilla, and dentary bones, and larger, fang-like teeth are interspersed along the palate and inside of the lower jaw (Fig. 8.8) (Carroll, 1988). On the surface of the skull, one can see a pair of cranial nostrils, like in many fishes we have encountered, but no trace of the caudal nostrils.

This is because the caudal nostrils have shifted. Take the skull of *Eusthenopteron*, flip it over so you are looking at the roof of the mouth (palate), and you will notice two internal nostrils opening directly into the mouth just behind the premaxilla (Fig. 8.8). These internal nostrils are called choanae, and they are the modified caudal nostrils we have encountered previously in other fishes. A hollow passage connects the external to the interior nostrils such that air or water sucked through the former would pass into the mouth through the latter. It would be tempting to assume that *Eusthenopteron* was using its nostrils like a snorkel, taking in air through its nose as we might do if our mouths were submerged below the waterline. However, it does not appear that *Eusthenopteron* used its nose in this way because its skull lacks any evidence of two important traits: a tear duct or a vomeronasal organ (Carroll, 1988; Janvier, 1996; Benton, 2005).

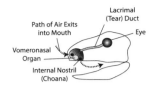

8.9. Diagram of the relative position and interconnectedness of the lacrimal (tear) duct and associated eye and nose structures.

Crying is something we do well from infancy. During all the times you've cried, you may have noticed that you also get the bonus of a runny nose. It is actually so commonplace that few of us stop to think about why the weeping of tears and the loosening of nasal mucus should go hand in hand. It turns out that your tears, which act to lubricate and moisturize your eye, have to drain somewhere. That somewhere is a pore located at the inside corner of your eye which is the opening for the lacrimal duct (lacrimal means 'tear') (Moore and Dalley, 1999). The lacrimal duct drains tears into the nose, and there it serves to keep the areas for odor detection (the olfactory cells embedded in a thin membrane) moist (Janvier, 1996). All that breathing in and out through our noses would dry out our odor detectors were it not for the constant dripping of the tears from the lacrimal duct. Crying simply opens the flood gates, and more tears than usual flow down the lacrimal duct, causing excessive moisture inside the nose. All tetrapods have a lacrimal duct, which typically shows itself as a pore or opening through a lacrimal bone at the corner of their eye (Fig. 8.9) (Pough, Janis, and Heiser, 2002). No such lacrimal pore is present in *Eusthenopteron*, and so we can be fairly certain that its nose was not used as a conduit for breathing. Such a function would dry out the odor-detecting cells, and that in turn would require the presence of a lacrimal duct to hydrate them.

There is also no indication of a vomeronasal organ in *Eusthenopteron*. The vomeronasal organ is known by various names, the most familiar of which is the Jacobsen's organ. It is located in a special hollow just above the mouth in many tetrapods, where it detects species-specific airborne chemicals called pheromones (Fig. 8.9) (Bertmar, 1981). Pheromones are odorless signals but are detected by the vomeronasal organ in much the same way that odors are detected by olfactory cells in the nose (Liem et al., 2001; Pough, Janis, and Heiser, 2002; Kardong, 2012). Many pheromones trigger reproductive responses between members of the opposite sex in a given species. Not all tetrapods have a vomeronasal organ (including humans), but given its distribution it was probably present in the common ancestor of all major living tetrapod groups (Bertmar, 1981).

The lack of any space for a vomeronasal organ in *Eusthenopteron*, an organ adapted to detecting airborne pheromones, further suggests that this tetrapodomorph did not use its nose to breathe in air.

The idea that *Eusthenopteron* spent a majority of its time in water is bolstered by the presence of lateral line canals in the skull and along the body, the retention of the gill-aerating gular and opercular bones, and numerous gill rakers along the pharyngeal arches (Janvier, 1996; Benton, 2005). A fishlike spiracle opening is also present as a notch behind the braincase, and this may have either contained a small gill filament or allowed for air breathing as we described previously for the bichir in chapter 7. Certainly, this animal must have had occasion to breathe air, and might have had a heart-lung system like that of lungfish. However, it is clear that *Eusthenopteron* was not a land-dwelling sarcopterygian (Clack, 2002).

Unlike in many of the fishlike sarcopterygians we have previously encountered, the vertebral column of *Eusthenopteron* is more robust and the notochord is somewhat reduced. U-shaped vertebral bodies (centra) surround the notochord ventrally, and each vertebral segment is actually composed of two interlinked pieces (Fig. 8.8) (Carroll, 1988; Janvier, 1996; Clack, 2002). Above the vertebral bodies sit the neural arches and spines that protect the spinal cord, and in the tail are the now-familiar downward-pointing hemal arches that protect delicate blood vessels. Two dorsal fins and an anal fin are present, along with a well-developed diphycercal tail fin. As with other fishes we have discussed, the vertebral skeleton of *Eusthenopteron* was loosely articulated and worked with the notochord to achieve the flexibility necessary to throw the body laterally against the water (Carroll, 1988; Janvier, 1996; Clack, 2002).

The pectoral girdle remains anchored caudally to the skull, but is very robust. The clavicle and the cleithrum are large elements that dwarf the nearly hidden, tiny scapulocoracoid (Carroll, 1988; Clack, 2002; Benton, 2005). The pectoral fin is short but stout in *Eusthenopteron*. The humerus is rectangular and sports a large crest that presumably offered attachment sites for fin-moving muscles (Carroll, 1988; Benton, 2005; Boisvert, 2009). Crests and other such landmarks on the skeleton are the result of the bone reinforcing itself where a muscle or tendon is pulling. The size and shape of these landmarks can be indicators of the relative power and direction of pull for a particular muscle or group of muscles. The crest on the humerus of *Eusthenopteron* is an ancient landmark called the deltopectoral crest. As the name suggests, this landmark is produced by forces from two large blocks of muscles, the pectoralis (chest) group and the deltoid (shoulder) group.

To understand the significance of the deltopectoral crest and its associated muscle groups, we need to introduce two more directional terms. Raising your forelimb away from your body is called abduction, whereas tucking your arm against your body is called adduction (Fig. 8.10). You can remember the difference between these two motions by thinking of *absent from the body* for *ab*duction and *adding to the body* for *ad*duction.

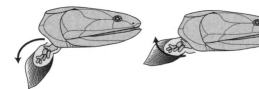

Abduction = Fin Moves Away from Body **Adduction= Fin Moves against Body** **Longitudinal Rotation of Fin**

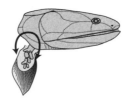

8.10. Functional anatomy of the pectoral fin in *Eusthenopteron*. The shapes of the scapulocoracoid and humerus allowed tetrapodomorphs like *Eusthenopteron* to abduct, adduct, and longitudinally rotate the pectoral fin. These basic movements carried over into the common ancestor of the tetrapods.

In a typical 'fish,' fin movements are almost solely abduction and adduction. In tetrapods, one of the major functions of the pectoralis muscles is adduction of the forelimb, whereas deltoids are involved in abduction. Most 'fish' lack a deltopectoral crest, so the presence of this landmark on the humerus of *Eusthenopteron* suggests that it could powerfully abduct and adduct its pectoral fin (Benton, 2005). But *Eusthenopteron* could do more than just abduct and adduct its pectoral fins. The head of the humerus is the joint surface that articulates with the glenoid (shoulder socket) formed by the scapulocoracoid. In *Eusthenopteron*, the glenoid is a shallow socket whereas the humeral head is wide and convex. This particular arrangement, elaborated on in modern tetrapods, allows a combination of abduction and adduction as well as craniocaudal rowing movements. These features of the humerus and scapulocoracoid in *Eusthenopteron* suggest its body could be supported, moved, and stabilized on the sediment.

Two other pectoral bones become distinct enough in *Eusthenopteron* to warrant naming them: the radius and the ulna. These bones connect to the distal end of the humerus, forming the elbow joint. The articular surfaces of the radius, ulna, and distal end of the humerus were relatively flat and underdeveloped in *Eusthenopteron*, suggesting that the elbow was stiff and had a limited range of movement (Benton, 2005). A few more blocky bones branch from the radius and ulna, somewhat equivalent to our wrists, but no fingers were present. Instead, the pectoral fin was fringed with thin fin rays (Fig. 8.11).

The pelvic girdle is much smaller than the pectoral girdle in *Eusthenopteron*, resembling those of other fishes we have encountered. Unlike our own legs, the pelvic fins of *Eusthenopteron* would not have provided much thrust, particularly because, as in a typical fish, the pelvic girdle was embedded in muscle and not connected to the vertebral column (Carroll, 1988; Clack, 2002; Benton, 2005). Therefore, the pelvic fin could not transfer forward momentum through the body. The pelvic fin does, however, have distinctive bones. Proximally, we find a small femur connected to the pelvic girdle through the acetabulum. The acetabulum is shallow and the femoral head wide, again suggestive of movements such as abduction, adduction, and craniocaudal rowing (Clack, 2002; Benton, 2005). The two bones that branch distal to the femur form the knee joint and are identified as the tibia and fibula (Fig. 8.11). A number of other

8.11. The bones of the pectoral and pelvic girdles and fins in fishlike tetrapodomorphs, based on *Eusthenopteron*.

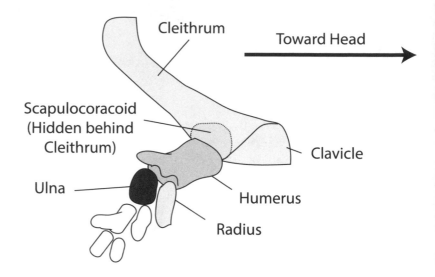

Tetrapodomorph Pectoral Appendage Anatomy

Cleithrum

Toward Head

Scapulocoracoid (Hidden behind Cleithrum)

Clavicle

Ulna

Humerus

Radius

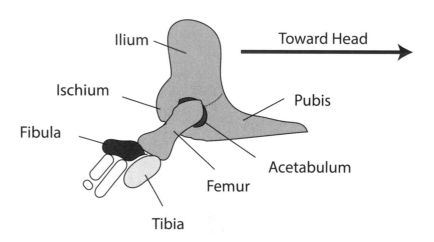

Tetrapodomorph Pelvic Appendage Anatomy

Ilium

Toward Head

Ischium

Pubis

Fibula

Acetabulum

Femur

Tibia

bones (somewhat reminiscent of ankle bones) branch more distally in the pelvic fin, but again there are no digits (no toes in this case) and the pelvic fin is fringed with fin rays.

Although the larger appendicular bones of *Eusthenopteron* seem to suggest they pushed themselves across land, this does not appear likely. As we discussed in chapter 4, bones are living tissues that respond to the major stresses placed upon them by changing their shape and thickness.

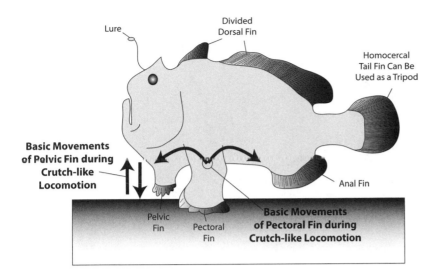

Lure

Divided
Dorsal Fin

Homocercal
Tail Fin Can Be
Used as a Tripod

**Basic Movements
of Pelvic Fin during
Crutch-like
Locomotion**

Anal Fin

Pelvic
Fin

Pectoral
Fin

**Basic Movements
of Pectoral Fin during
Crutch-like Locomotion**

8.12. Diagram of a frogfish, a teleost capable of walking on the seafloor with its fins. In this figure, the frogfish is supporting its body on the pectoral fins while the pelvic fins have been lifted off the seafloor. The movement is similar to an injured person walking with crutches, where the pectoral fins are the crutches and the pelvic fins are the legs. Figure based on Pietsch and Grobecker (1987).

If we were to take a cross-section through the middle of one of your long bones, such as your humerus or femur, what you tend to see is dense, compact bone. This makes sense: you want strong, sturdy bone to hold up the body's weight. It is only toward the middle of the bone (around the marrow cavity) or underneath the joint cartilage that we would find more loosely consolidated, spongy bone. Near the marrow cavity, the stresses are minimal, and so the bone does not have to be so compact to resist the forces placed on it. Much like the shock absorbers in a car chassis, the spongy bone near the joint surfaces provides a little 'give' to absorb the shocks of walking and running. Having compact bone directly under the joint cartilage would be equivalent to having the tires of a car rigidly attached to the chassis: you would feel every bump in the road and damage the car. You can imagine what would occur to your joints! Studies of the microscopic structure of *Eusthenopteron* fin bones show they are not thickly fortified like our limb bones (Laurin et al., 2007). Instead, a cross-section through the middle of the fin bones reveals only a thin rind of compact bone encasing loose, spongy bone (Laurin et al., 2007). This shows that pectoral and pelvic fin bones of *Eusthenopteron* were not routinely exposed to the crushing pressures of gravity. In fact, it is difficult to imagine *Eusthenopteron* being able to adequately support its body out of water for any significant length of time. Therefore, most tetrapodomorph sarcopterygians must have been doing something else with those muscular fins.

What does one do with robust pectoral fins if not walk on land? One example from the living world is particularly illuminating. Actinopterygian fishes called frogfish, a type of anglerfish, have limb-like appendages but never leave the water. In fact, their pectoral and pelvic fins are fleshy and bear digit-like projections (Pietsch and Grobecker, 1987). Some species hide out in seaweed, which they grasp with their hand- and foot-like appendages (Pietsch and Grobecker, 1987). Other frogfish species walk about underwater using the fleshy fins and the back corner of the tail as

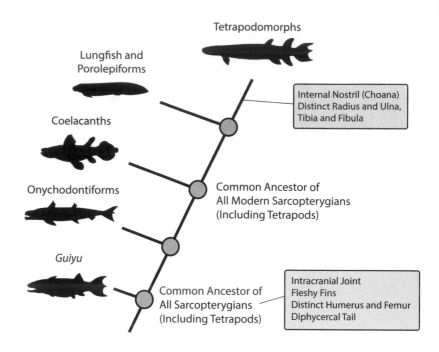

8.13. Pedigree of relationships among the fishlike sarcopterygians. Note the position of the two living groups of fishlike sarcopterygians, coelacanths and lungfishes: sandwiched between some primitive forms and the tetrapodomorphs. Therefore, coelacanths and lungfish cannot provide us with a direct model of an animal transitional to tetrapods. We must rely on the fossil record to decode the sequence of events that led to the tetrapod chassis.

Tetrapodomorphs

Lungfish and Porolepiforms

Internal Nostril (Choana) Distinct Radius and Ulna, Tibia and Fibula

Coelacanths

Onychodontiforms

Common Ancestor of All Modern Sarcopterygians (Including Tetrapods)

Guiyu

Common Ancestor of All Sarcopterygians (Including Tetrapods)

Intracranial Joint
Fleshy Fins
Distinct Humerus and Femur
Diphycercal Tail

stabilizing points (Pietsch and Grobecker, 1987) (Fig. 8.12). In some cases, a frogfish will even walk like a salamander by placing opposite 'limbs' forward in a sequence while bending the body laterally. For example, the right pectoral fin and left pelvic fin are swung forward as the body bends sideways, followed by the same sequence on the opposite side (Pietsch and Grobecker, 1987) (Fig. 8.12). All of this occurs underwater, and usually in the context of staying unnoticed until the frogfish can strike unsuspecting prey.

Another clue about the typical habits of *Eusthenopteron* comes from its habitat: this animal is commonly found in sediments deposited near the mouths of rivers along coastal regions (Clack, 2002; Benton, 2005). Living as *Eusthenopteron* did in these areas, its pectoral fins may have aided this sarcopterygian in supporting itself on the muddy bottom against the current. The whole chassis and body plan of *Eusthenopteron* remain fishlike, and has reminded some paleontologists of fully aquatic ambush predators such as pike (Benton, 2005). Large, fleshy pectoral fins may have been advantageous for 'lurking,' waiting on the sediment behind some weeds until unsuspecting prey swam by. Then, with a flick of its powerful tail, and perhaps a 'launch' from the sediment by its fleshy pectoral fins, *Eusthenopteron* would do lunch. Whatever *Eusthenopteron* was ultimately doing with its fins, it was certainly not spending an inordinate amount of time out of water.

Figure 8.13 provides an overview of the evolutionary pedigree of the odd and intriguing fishlike sarcopterygians. However, overcrowding often leads to exodus. As you are probably beginning to appreciate, by the end of the Devonian, the coastal waters, estuaries, rivers, ponds, and lakes were all becoming crowded with an amazing variety of jawless vertebrates, placoderms, chondrichthyans, and osteichthyans. With many animals vying for ever-dwindling resources, there would be selective pressure to exploit new environments. Much like weary city dwellers who, fed up with the crush of people, escape to the country, so one line of shallow-water sarcopterygians began to exploit food and resources from the land. It is to these animals and the adaptations of their chassis that our story turns next.

The Vertebrate Chassis Moves to Land

4

When something left the ocean, to crawl high above the foam.

RUSH, *High Water*

9.1. Comparing the skull of a fishlike sarcopterygian (*Eusthenopteron*) to that of a generalized tetrapod. Note that there are fewer bones in the tetrapod skull than in fishlike vertebrates. The regions of the cheek (J, the jugal [cheek] bone) and squamosal (AM) are expanded along with the associated adductor mandibulae muscle in tetrapods, and the opercular, gill, and pectoral-skull bones are lost. The deep pterygoid bones are also expanded in tetrapods to house larger pterygoid muscles, which assist the adductor mandibulae in powerfully closing the jaws. Prominent occipital bones project underneath the exit for the spinal cord (foramen magnum) and connect with the first vertebra to allow independent movements of the skull on the neck. In tetrapods, the lower jaw now has a projection into which jaw-lowering muscles (the depressor mandibulae) insert. Unlike those of fishlike sarcopterygians, the skulls of tetrapods have a longer facial region and a more compact braincase region. *Eusthenopteron* anatomy modified from Carroll (1988), Clack (2002), and Benton (2005). Generalized tetrapod skull anatomy derived, in part, from Liem et al. (2001).

Eusthenopteron **Skull (Fishlike Sarcopterygian)**

Cranial Nostril; Facial Skeleton; Braincase Skeleton; Opercular Bones; Premaxilla; AM; Pectoral-Skull Series; Facial Region; Braincase Region; Dentary; Maxilla; Postdentary Bones; Gular Bones; Quadrate-Articular Jaw Joint

Generalized Tetrapod Skull (Based on Salamanders and Lizards)

Facial Skeleton; Parietal; Opening to Inner Ear; Eye; Parietal Opening; Lacrimal; Quadrate; Anterior Nostril; Occipital Bones; Premaxilla; AM; J; Attachment for Jaw-Opening Muscles; Pterygoids; Maxilla; Dentary; Postdentary Bones; Articular; Facial Region; Braincase Region

The Tetrapod Chassis: A Primer

IN THE MOVIE *COMMANDO*, ARNOLD SCHWARZENEGGER'S CHARACTER holds a villain over a cliff and says, "Listen, loyalty is very touching. But it is not the most important thing in your life right now. But what *is* important is gravity!" Such a quote could equally well apply to our common ancestor: without the buoyancy of water, gravity was one of the most 'pressing' issues facing the early tetrapods. If we were being strictly chronological, then the next order of business would be an examination of these early transitional tetrapods and their relatives. Until recently, that would have been very difficult to do, because all that was known were animals like *Eusthenopteron* and primitive but full-blown tetrapods. However, the last few decades have seen an explosion of new finds of early tetrapods and transitional tetrapodomorphs that have given us some incredible insights into the origins of the tetrapod chassis. In addition, new research on previously known specimens has revealed some surprising results.

But I have found for both myself and my students that following in the footsteps of the mid-twentieth-century paleontologists is more instructive and helpful than simply diving into the new transitional forms. This is because most of us are intuitively more accustomed to thinking about tetrapod anatomy: most people have a better innate grip on how a skeleton like their own works than that of a transitional tetrapod. Knowing typical 'fish' anatomy and then comparing that to what has changed in a modern tetrapod is a useful exercise. This is also the order of things when my students move on from their sharks to a salamander in comparative anatomy.

Therefore, as in previous chapters, if we want to examine the fossil record for transitional and early tetrapods, we need a search image based on what we know about modern tetrapods compared with fishlike vertebrates. We want to come up with a hypothesis of what the first tetrapod chassis would be predicted to look like based on the physical challenges of living on land as well as anatomical information from modern tetrapods, just as earlier paleontologists had done. The tetrapod family tree consists of two major branches: the Amphibia and the Amniota. The Amphibia today comprises tetrapods mostly still tied to the water for reproduction, such as frogs and salamanders. The Amniota are those tetrapods that are fully adapted to a terrestrial existence, such as reptiles, birds, and mammals. As we will see, each of these groups has its own specialized adaptations, but the undergirding of the chassis is remarkably similar. Living salamanders, from the Amphibia, provide us with a good counterpoint to the anatomy we have come to know in fishlike vertebrates. These amphibians have retained a number of primitive trait states, and can therefore serve as a reference to help us predict what the first tetrapods may have been like. I emphasize here that this is a generalized comparison. As detailed later, salamanders are not transitional tetrapods, and their anatomy will receive a proper treatment within the context of their family trees.

This chapter ultimately serves as a primer to acquaint us with the challenges faced by the earliest tetrapods venturing onto land. We will explore the effect that the physical differences between water and air as fluids have had on modern tetrapod anatomy. We also want to (briefly) consider why it would be that the vertebrates, so supremely adapted to an aquatic existence, would make the switch to land in the first place. This chapter will ultimately prepare us for appreciating the evolution and adaptations of the various tetrapod groups that take up the rest of this book.

Gravity and the Skeleton

It bears repeating: water is 800 times denser than air (Vogel, 2003). This means that the force of buoyancy is much stronger than gravity on animals that live among the waves. In contrast, air is thin and wispy, and cannot buoy up a vertebrate chassis. Therefore, on land gravity is a much more powerful force, and it is no small consequence that its terrestrial intensity is what kills many a beached dolphin or whale. Resisting gravity and moving under its force led to selection of fewer, sturdier bones and strengthening of the skeleton.

Let's start with the head. In salamanders, there are fewer bones than in fishy sarcopterygians, and the skull is less flexible and more robust (Fig. 9.1). In general, the premaxilla, maxilla, and other upper jaw elements are fused to the rest of the skull, making it a more solid unit. In our fishy friends, the face (the region between the eyes and the nose) was short, but in salamanders the face is longer while the braincase becomes shortened. In addition, most of the mobility within the tetrapod skull is, at least initially, restricted to the opening of the lower jaw. Now, instead of the complex interactions of skull and jaw bones required for opening the mouth in fishes, a hinge-like joint between the quadrate and articular allows the lower jaw to drop with assistance from gravity (Fig. 9.1) (Radinsky, 1987; Liem et al., 2001; Pough, Janis, and Heiser, 2002).

These changes in the skull are related not just to gravity but also to feeding out of water. As we have seen for many fishlike vertebrates, some form of suction is used to draw prey into the mouth. This strategy evaporates into thin air because opening or protruding the mouth quickly does not create enough of a vacuum to suck in lunch. Instead, prey must now be caught directly with the mouth, and here's where longer snouts come in. As we saw for shears and scissors, longer blades and shorter handles result in quick opening and a wide arc of movement. So, too, the longer jaws and shorter braincase of tetrapods like salamanders allow them to more quickly snap up prey in an airy environment. The fewer, fused bones in the skull further endow tetrapods with the ability to subdue struggling prey without breaking their heads in the process. The squamosal bone expands in concert with the larger adductor mandibulae muscles, a more prominent cheek bone (the jugal) develops, and the pterygoid bones enlarge with their associated pterygoid muscles (Fig. 9.1). The role of these bones and muscles in quickly clamping the jaws shut in air seems to have selected for their expansion.

Changes behind the skull took place in tetrapods as well. First, there was the loss of the gular, opercular, and other gill-bearing bones as respiration switched to rely on the lungs (Clack, 2002; Pough, Janis, and Heiser, 2002). As we will see, some tetrapods, such as certain salamander species, continue to develop gills as larvae or as adults, but, unlike in fishes, these respiratory filaments tend to be mounted to external arches. Second, in the dense fluid of water, having the head seamlessly joined to the body was valuable because it improved the swimming hydrodynamics of the animal. In air, such streamlining doesn't matter anymore. In tetrapods, we see the loss of the pectoral-skull series and the evolution of a distinct neck (Clack, 2002). A neck is useful for a great many things, two of which are thrusting the mouth at prey and being able to swing the head around to look at the surroundings while standing still. Occipital bones that project underneath the opening for the exiting spinal cord (the foramen magnum) form a ball that connects with a socket on the first vertebra, which provides additional mobility for the skull independent of the body or neck (Fig. 9.1) (Liem et al., 2001; Pough, Janis, and Heiser, 2002; Kardong, 2012).

In gravity, the vertebral column must take on a new role supporting the body. Imagine a chain stretched across an open space and anchored to a support on either end. Now, imagine hanging a bundle of weights from the middle of the chain. You intuitively know that this setup will give you a sagging chain (Fig. 9.2). Replace the chain with the vertebral column, the supports with the appendages, and the weight with the body organs, and you end up with a sagging tetrapod dragging its gut on the ground (Fig. 9.2). Now imagine taking the chain and adding special connectors between each link that still allow side to side motions but that resist downward movements (Fig. 9.2). To further pull the weight off the ground, imagine attaching loops to the top of each chain link through which you thread a sturdy wire or rope that can bend and maintain the chain into an inverted U-shape (Fig. 9.2). To improve the resistance and weight-bearing capacity of the chain even further, we could extend curved wires or struts out from the sides of each link, and string additional wires or ropes perpendicular to these. Essentially, we would have a basket or truss that would further strengthen the chain while surrounding and supporting the weight suspended beneath it (Fig. 9.2).

In tetrapods, new joints called zygapophyses developed between the vertebrae that, like connectors in our imaginary chain, allow side to side movements but resist downward bending (Fig. 9.3) (Radinsky, 1987; Liem et al., 2001; Pough, Janis, and Heiser, 2002; Kardong, 2012). A look at the vertebrae of salamanders shows these joints to good effect. The neural arches of tetrapod vertebrae tend to fuse with the vertebral bodies (the

9.2. The vertebral column as a chain. The vertebral column of a fishlike vertebrate is like a chain link with no support. Just like a chain strung across two pillars and supporting a weight in gravity will sag, so, too, would the vertebral column in a tetrapod without certain anatomical adaptations. To brace a chain against the weight it is bearing without sagging, the chain must be held up by a suspension cable, and suspension struts and trusses can be added. In tetrapods, the evolution of zygapophyses (intervertebral joints) that resist gravity while allowing other movements, the expansion of epaxial muscles and ligaments along the vertebral column, and the development of ribs and their associated musculature and ligaments act similarly to hold the body off the ground.

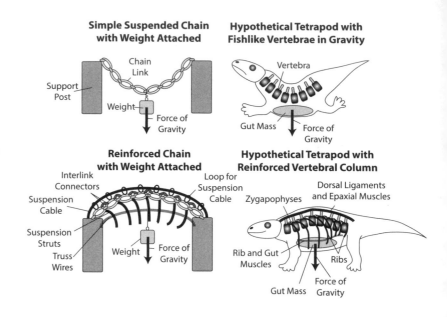

centra), and these act like the loops on our hypothetical chain. The centra themselves have become solid, single elements, unlike the two pieces we saw in animals like *Eusthenopteron*. Ligaments and strong epaxial muscles, the wire or rope in our thought experiment, span the neural arches and place tension on them, collectively bending the vertebral column into an inverted U that further resists gravity and suspends the organs. Finally, we see the development of strong ribs, which, spanned by muscles and ligaments, help hold up the organs and further improve resistance to gravity, much as our truss did in our thought experiment (Fig. 9.2). Salamanders and modern amphibians lack well-developed ribs for reasons we will explore in later chapters.

Without the density of water and the problems associated with roll and pitch, the need for fins has diminished. In terrestrial salamanders, the now-familiar dorsal, anal, and tail fins have all disappeared (Clack, 2002). We saw that for fishlike vertebrates the notochord was commonly prominent. In salamanders, the vertebral column is now well developed, and the remnants of the notochord form pulpy discs between the vertebrae. In fact, the vertebral column itself is now divided into distinct sections (Fig. 9.4). With the head separated from the pectoral girdle, we now have a neck, and the vertebrae associated with the neck are called the cervical series. The trunk vertebrae that bear the ribs are referred to as the dorsal series. There are now vertebrae associated with the pelvic girdle called the sacrum or sacral series, and the tail vertebrae are known as the caudal series. The lateral bending of the vertebral column we witnessed in fishes is still conserved in salamanders, but because air is not very dense, simply throwing the body against it will get you nowhere. Therefore, salamanders couple the fishlike lateral bending of their body with movements of the appendages, and it is this combined movement that allows them to put their best feet forward (Kardong, 2012).

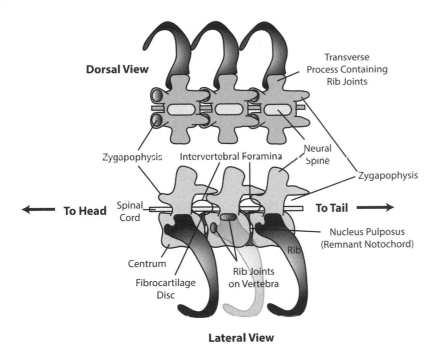

Dorsal View

Transverse Process Containing Rib Joints

Zygapophysis   Intervertebral Foramina   Neural Spine

Zygapophysis

← **To Head**   Spinal Cord   **To Tail** →

Nucleus Pulposus (Remnant Notochord)

Rib

Centrum

Fibrocartilage Disc

Rib Joints on Vertebra

Lateral View

9.3. Diagram of generalized tetrapod vertebrae. Note that the body of the vertebra (centrum) is large in tetrapods and surrounds the notochord completely. In many tetrapods, the notochord is reduced to the pulpy discs (nucleus pulposus) between the vertebrae. Note also the zygapophyses that link the vertebrae together, prevent the vertebral column from sagging, and allow lateral movements. Tough, fibrous cartilage (fibrocartilage) further helps link the vertebrae to one another. Spaces between the vertebrae called intervertebral foramina allow nerve tracts to pass from the spinal cord out to the body and appendages. Ribs are prominent in most tetrapods, but not (as we will see) in modern amphibians.

Moving and supporting the body out of water required further development of the appendages and their girdles. As we see in the pectoral girdles of modern salamanders, the scapulocoracoid is large, now taking the brunt of the forces transferred through the humerus. Bones that previously linked the pectoral girdle to the skull, such as the clavicle and cleithrum, become smaller, either fusing to the scapula or acting as pivots about which the forelimb can rotate (Clack, 2002). The humerus and femur project laterally from the body in salamanders, much like fish fins. This orientation creates twisting and tensile forces across the chest, forces you feel when you do pushups or bench presses. Because the pectoral girdle is now decoupled from the head, there was selection for the evolution of elements to resist these new forces of locomotion. In the tetrapod chassis, we see the evolution of the sternum, a collection of bones or cartilage that resist and redistribute the powerful twisting and stretching forces of forelimb movement. In tetrapods such as salamanders, the coracoid portion of the scapulocoracoid now links into a joint on the sternal bones, allowing further movement and relieving some of the stresses from walking or standing (Fig. 9.4). In many tetrapods, the sternum also receives flexible, cartilaginous portions of the ribs, which also help to carry and transmit the forces of movement to the body core.

As we saw for *Eusthenopteron*, the closest relatives to the early tetrapods were front-wheel-drive organisms: their pectoral fins were larger than their pelvic fins and did the lion's share of body support and movement. However, the hind limbs were ultimately in a better anatomical position to provide the forward thrust that could no longer be achieved by simply wagging the tail. In fishlike sarcopterygians, the pelvic girdle is small, is embedded in the body wall, and does not make a direct connection with

9.4. Schematic of generalized tetrapod showing vertebral sections and girdles. Unlike in fishlike sarcopterygians, a distinct neck and associated vertebrae (cervical vertebrae) are present in tetrapods, and the axial fins (dorsal, anal, tail) are reduced or absent. Hemal arches are still present in the tail as they were in many fishlike vertebrates we have already encountered. The pectoral girdle is composed of a large (and typically fused) scapula and coracoid, and smaller cleithrum and clavicle. A typically cartilaginous sternum is present that resists the twisting forces generated by the movements of the forelimbs, and commonly an episternum projects forward toward the head. The coracoid typically fits into a sliding joint on the sternum, which improves the range of motion in the forelimb. Cartilaginous sternal ribs help connect the bony ribs to the sternum, completing the truss-like structure of the ribcage. In the pelvic girdle, the ilium is expanded and connects directly to the sacral vertebrae through special sacral ribs to transfer the force of the hind limb directly down the body axis.

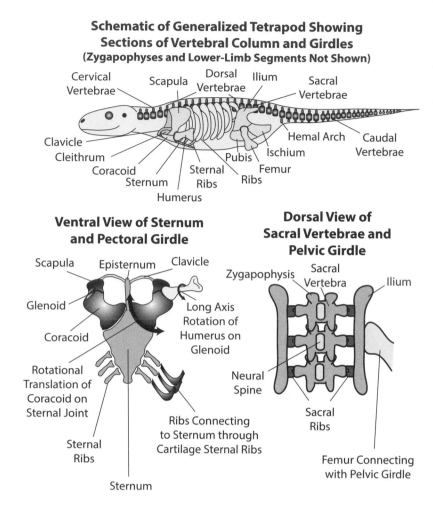

Schematic of Generalized Tetrapod Showing Sections of Vertebral Column and Girdles
(Zygapophyses and Lower-Limb Segments Not Shown)

Cervical Vertebrae — Scapula — Dorsal Vertebrae — Ilium — Sacral Vertebrae — Hemal Arch — Caudal Vertebrae — Clavicle — Cleithrum — Coracoid — Sternum — Sternal Ribs — Ribs — Femur — Ischium — Pubis — Humerus

Ventral View of Sternum and Pectoral Girdle

Scapula — Episternum — Clavicle — Glenoid — Coracoid — Long Axis Rotation of Humerus on Glenoid — Rotational Translation of Coracoid on Sternal Joint — Sternal Ribs — Ribs Connecting to Sternum through Cartilage Sternal Ribs — Sternum

Dorsal View of Sacral Vertebrae and Pelvic Girdle

Zygapophysis — Sacral Vertebra — Ilium — Neural Spine — Sacral Ribs — Femur Connecting with Pelvic Girdle

the rest of the skeleton. Such an arrangement prevents thrust generated by the hind limb from being transferred into forward momentum because there is no direct connection between the hind limb and body. This is somewhat similar to the odd situation with airplane wheels: they contribute nothing to thrust when the plane takes off (Bloomfield, 2006). For example, if you placed a plane on a treadmill running backward, the plane could still easily take off with little resistance. This is possible because, whereas the wheels support the aircraft when it is stationary and allow it to roll, they are not connected to the major thrust generator: the engine. It is the engine and propeller that provide the forward momentum and thrust that get the plane off the ground. On the other hand, in a car, the wheels are connected directly into the engine through the chassis, and so apply force to the ground, which pushes the car forward (Newton, 1999; Bloomfield, 2006). In tetrapods like our salamander, the pelvic girdle greatly expands, and its upper portion, the ilium, attaches directly to the vertebral column (Fig. 9.4). In turn, the portion of the vertebral column that receives the ilium becomes more robust, and commonly the vertebrae fuse together into the more cohesive unit called the sacrum. As with the wheels of a car, the thrust of the legs is transferred directly into the chassis (the vertebral column) to push these tetrapods forward (Fig. 9.4) (Liem et al., 2001; Pough, Janis, and Heiser, 2002; Kardong, 2012).

We have already described the major bones in the pectoral and pelvic appendages in fishlike sarcopterygians, and these elements were retained and enlarged in tetrapods. The humerus and femur typically have an hourglass shape, their proximal and distal ends expanded to increase movement and weight support at the joint surfaces of the elbow and knee. In salamanders, adduction, abduction, and longitudinal rotation continue to be the primary modes of movement at the shoulder and hip (Fig. 9.5) (Reilly et al., 2006). However, hinge-like and rotational movements are possible at the elbow and knee so that forearm (radius and ulna) and shin (tibia and fibula) can be oriented vertically, pushing the salamander up off the ground and forward as it walks (Radinsky, 1987). So far as is known, the knees of all tetrapods have cartilaginous discs called menisci that provide extra cushioning and that deform to cup and cradle the ends of the femur as the knee bends and rotates (Haines, 1942; Liem et al., 2001). In addition, there has been the development of two ligaments arising from the tibia that cross each other, called the cruciate ligaments (Haines, 1942; Liem et al., 2001). These ligaments act like two taut rubber bands that allow some twisting at the knee to occur but tighten as they are twisted to prevent overrotation (Haines, 1942; Moore and Dalley, 1999). Hands and feet with jointed fingers and toes have replaced the fringe of delicate rays seen in fishy sarcopterygians. In addition, the development of wrists and ankles that allow certain motions but inhibit others further enhances the transfer of thrust from the forelimbs and hind limbs into the body core (Fig. 9.5) (Radinsky, 1987; Liem et al., 2001). In particular, the hands and feet are said to be pronated: the palm and sole are planted on the ground with the fingers and toes facing cranially. This posture ensures that the hands and feet push the tetrapod forward rather than sideways.

All of these bones and joints allow the forelimb and hind limb to act collectively like a crank (Radinsky, 1987; Carroll, 1988; Liem et al., 2001). To plant the hand or foot on the ground, the humerus or femur is rowed cranially (protracted) and adducted, which lowers the vertical forearm or shin toward the ground. With the hand or foot now supporting the body, the humerus or femur is then rowed caudally (retracted). This action swings the vertical forearm or shin rearward, which pushes the body forward over the supporting hand or foot. When the end of a stride is reached, the humerus or femur is abducted, elevating the forearm or shin and lifting the hand or foot from the ground. Finally, the humerus or femur is protracted, bringing the forearm, shin, hand, and foot into position to support the body, and the whole process repeats. So that the body is continuously supported and propelled forward, limbs on diagonally opposite sides move in tandem (Reilly et al., 2006). In other words, while the left forelimb and the right hind limb are planted, the right forelimb and left hind limb are off the ground. The stride length is further increased by lateral flexion of the vertebral column (Schaeffer,

9.5. Generalized anatomy of tetrapod forelimbs and hind limbs. As with fishlike sarcopterygians such as *Eusthenopteron,* the forelimb of tetrapods typically has an upper humerus followed distally by the radius and ulna. A series of multifaceted wrist bones follows the radius and ulna, and the forelimb terminates in a number of digits (shown as five here, but there is much variation among tetrapods). In tetrapods with five digits, the radius always aligns with digit 1 (what you would call your thumb), and the ulna always aligns with digit 4 (what you call your ring finger). The hind limb repeats a similar pattern, with a single femur proximally, a tibia and fibula distally, a series of interlinking ankle bones, and, finally, digits. As with the forelimb, the tibia aligns with digit 1 (the big toe) and the fibula aligns with digit 4 (one of the smaller toes). The reasons behind the alignment of particular digits with particular bones will be explained in chapter 10.

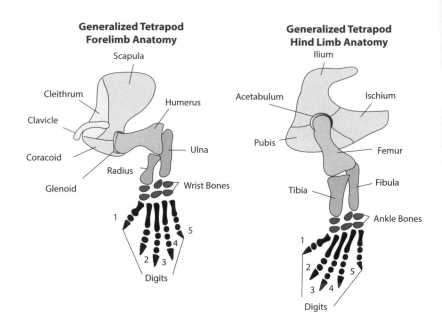

1941; Rewcastle, 1983). This basic pattern, still present in modern salamanders, was present in the common ancestor of all modern tetrapods and forms the basis from which other, more derived tetrapod locomotor movements have evolved (Reilly et al., 2006).

Many of the adaptations of tetrapods for living in air are not preserved in their skeletons, and thus not immediately apparent in fossils. Yet, based on a long tradition of comparative anatomical study, we know that certain soft tissues and systems tend to appear in tetrapods. As is common in paleontology, we are drawing inferences based on correlation. In other words, all living tetrapods, which are built around a characteristic skeletal chassis, always possess a particular suite of soft tissue features. Although the timing and appearance of various soft tissue features are debated, the simplest explanation in the absence of competing data is that the soft tissue characteristics described below were present in most fossil tetrapods. The exceptions and changes to these features will be noted as we discuss the various groups of tetrapods throughout this book.

Air, Sound, and Vision

The air density problem brought with it challenges that the early tetrapods had to overcome. One was a question of hearing sounds. In the dense medium of water, the molecules are squished so close together that sound waves can travel fast, far, and wide across large distances (Vogel, 2003; Bloomfield, 2006). In air, with the molecules spaced farther apart, sounds travel more slowly, fade more quickly from their source, and go nowhere near as far (Vogel, 2003; Bloomfield, 2006). Moreover, on land the density of an early tetrapod was similar to that of water, not air. Even if airborne sounds reached these animals, they couldn't pass through their skulls or up a lateral line canal: instead, the sounds would simply bounce off the creature, much as a rubber ball will bounce off concrete.

This doesn't mean that tetrapods can't sense vibrations in air. If you've ever been to a loud rock concert, stood close to a passing train, or watched a fireworks display, you know that certain low-frequency sounds can pass as vibrations through your body and head. Even without a means to pick up high-frequency airborne vibrations, the earliest tetrapods would certainly have been capable of detecting low-frequency thumps and rumbles through their skulls, bodies, and limbs. Modern snakes, for example, have no external ears but are still capable of detecting low-frequency vibrations transmitted through the ground into their bodies and skulls (Liem et al., 2001).

The direction from which a sound is coming in water is deciphered by differences in the timing of vibrations hitting the lateral line system and skull (Liem et al., 2001; Pough, Janis, and Heiser, 2002). The trouble with low-frequency sounds in air is that it is very difficult to tell where they are coming from. The rumbling permeates the skull and body, and this makes it nearly impossible to identify its direction. Makers of various surround sound systems have taken advantage of this: you can hide a subwoofer anywhere in a room or car and no one will know the difference. Because the bass speaker is the biggest and heaviest part of the system, you can now take the separate and lighter high-frequency speakers and mount them where the directions of those sounds will be distinctly detected, such as on either side of the TV or in the door panels of a car.

So, in air, high-frequency sounds are more directional, and better allow you to detect whether prey or predator is coming toward you. But to hear high-frequency sounds requires a way to amplify and direct sounds into the inner ear. In modern microphones, the high-frequency vibrations of a singer's voice are picked up by a thin aluminum membrane. In turn, the vibrations of this thin metal are amplified and converted into electrical signals (Bloomfield, 2006). Salamanders lack an external ear, but frogs, their close relatives, have an ear drum, a membrane stretched taut over the opening to the inner ear, which amplifies high-frequency sounds bouncing against it (Fig. 9.6) (Liem et al., 2001; Carroll, 2009). Like the large orchestral drums called tympani, the anatomical name for the tetrapod ear drum is the tympanum. But the vibration of the tympanum alone would not be able to transmit high-frequency sounds into the dense fluid of the inner ear. A way must be found to convert the high-frequency vibrations of the tympanum into pulses strong enough to overcome the viscous fluid of the inner ear.

Frogs and most modern tetrapods have solved this problem by adapting the old hyomandibula into a sound transducer. As you may recall, the hyomandibula was an important element for suspending and bracing the jaws of various fishlike vertebrates. In tetrapods, the upper jaws (premaxilla, maxilla, and other bones that lie over the old palatoquadrate cartilage) are fused with the skull, and so the hyomandibula is no longer required as a jaw suspensor. It turns out that the hyomandibula is already in an ideal place to transmit sounds into the inner ear because it is naturally braced against the otic region of the braincase. In frogs, the end that

9.6. Diagram of differences between the inner ear and jaws of fishlike vertebrates and tetrapods. In many fishes, the hyoid arch, particularly the hyomandibula, suspends the jaws underneath the braincase. In tetrapods, the upper jaw is typically fused to the skull, freeing the hyomandibula from its original role. Still braced against the otic capsule, the hyomandibula is now in an ideal position to pass vibrations into the inner ear. Its free end is usually anchored to a tympanic membrane that amplifies high-frequency vibrations from the air. Inside the tetrapod skull, the hyomandibula (now called the stapes) vibrates against a gel-like fluid-filled chamber that passes vibrations into the portion of the inner ear capable of detecting sounds. This region is located adjacent to regions of the inner ear still involved with balance and orientation. Inner ear diagram based on figures from Liem et al. (2001).

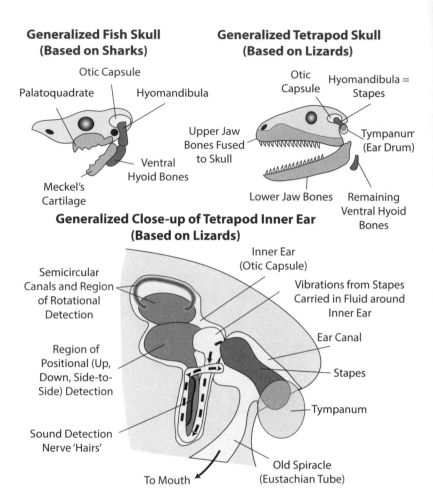

used to suspend the jaws in their fishlike ancestors is now free and lies against the tympanum. As the tympanum vibrates, the hyomandibula is moved back and forth against the inner ear, transmitting strong pulses that can be detected as sound by the brain (Fig. 9.6) (Liem et al., 2001; Pough, Janis, and Heiser, 2002; Carroll, 2009).

The detailed anatomy of the inner ear goes beyond the scope of this book. In simple terms, vibrations from the hyomandibula pass into a gel-like fluid-filled sac around the inner ear. These vibrations are then passed into a region of the inner ear that detects sounds, causing hairlike nerve cells within to vibrate at various frequencies. These vibrations are subsequently converted into electrical impulses that can be deciphered by the brain as sounds. Differences in the timing of when high-frequency sounds reach the right and left tympanic membranes allow tetrapods like frogs to detect the direction of a given airborne noise (Liem et al., 2001). In its modified role as a sound transducer, the hyomandibula is renamed the stapes in tetrapods. The old spiracle comes aboard here as well. Lying as it does between the jaws and hyoid arch, it is in an ideal place to act as a valve to equalize air pressure on either side of the tympanum (we have already discussed this in chapter 5) (Hildebrand and Goslow, 2001;

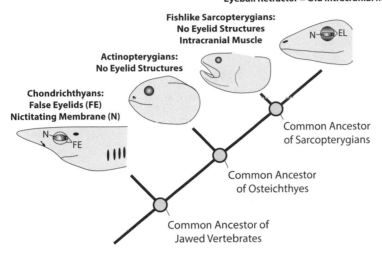

Tetrapods:
Two Outer, Movable Eyelids (EL)
Nictitating Membrane (N)
Eyeball Retractor = Old Intracranial Muscle

Fishlike Sarcopterygians:
No Eyelid Structures
Intracranial Muscle

Actinopterygians:
No Eyelid Structures

Chondrichthyans:
False Eyelids (FE)
Nictitating Membrane (N)

Common Ancestor
of Sarcopterygians

Common Ancestor
of Osteichthyes

Common Ancestor of
Jawed Vertebrates

9.7. Distribution of eyelid features in vertebrates. True, muscle-powered eyelids and a nictitating membrane appear together only in tetrapods, where they clean, protect, and remoisten the eyes. Actinopterygians and fishlike sarcopterygians lack eyelids or a nictitating membrane. Although chondrichthyans like sharks have a nictitating membrane, it does not develop from the same anatomical origins as the one in tetrapods. Furthermore, the 'false' eyelids of these vertebrates are not muscular or mobile. Thus, only tetrapods have true, moving eyelids and a nictitating membrane, and these features go hand in hand with living in an airy environment.

Liem et al., 2001). How the spiracle came to possess this role after serving faithfully as a small gill opening in many fishlike vertebrates will be discussed in the next chapter.

Seeing in air is also different from seeing in water. On the plus side, because air is much less dense than water, you can, as the Who once sang, see for miles and miles. But there are a number of problems associated with seeing in air, not the least of which is having your eye dry out. We have previously alluded to the tear (lacrimal) duct and its role in moisturizing the olfactory membranes in the nose. Tear glands have evolved in tetrapods to moisten the eye and to wash away bacteria, fungi, and debris. Being able to blink is essential to remoisten the eye and, as windshield wipers do on a car, clear away particles on the eye's surface. All tetrapods have evolved some sort of eyelids, and have adapted portions of some of the eye-moving muscles to help them blink (Fig. 9.7) (Young, 2008). As we discussed in chapter 8, there was a basicranial muscle present in the common ancestor of sarcopterygians that flexed the snout during feeding. In tetrapods, however, the intracranial joint has fused, and the basicranial muscle had to go somewhere. In salamanders and many other tetrapods the basicranial muscle has been retained as a retractor that pulls the eyeball into the eye socket (Levine, 2004). We can be fairly confident that the eye retractor and basicranial muscle are one and the same because they are both supplied by the same specialized cranial nerve (Liem et al., 2001). When the eye is depressed into the socket, the eyelids blink over the eye, moisturizing and cleaning its surface. Many tetrapods have also evolved a third eyelid, a clear membrane that slides sideways across the eye, called the nictitating membrane (Fig. 9.7). This third eyelid is also powered by derivates of the eye-moving muscles.

The other major problem with seeing in air has to do with refraction. When light passes from one medium into another, differences in density

9.8. Comparing 'fish' eyes to tetrapod eyes. In each diagram, the eye in question is looking at the image of a cat. Light rays reflect off the cat image and pass through the cornea, pupil, and lens before striking the light-detecting retina layer. The choroid is a layer that supplies blood to the retina. Because 'fish' eyes are immersed in water, most of the refraction and focusing happens in a spherical lens whose microscopic layers bend light toward a focus point on the retina. 'Fish' eye lenses can be physically moved back and forth to improve focus. In tetrapods, the difference in density between air and eye is great, and much of the primary focus and refraction of light occurs in the outer rind of the eye called the cornea. The lens also focuses light on the retina, but it is thin and flexible, and focusing happens by its distortion. General properties of optical physics cause the focused image to be inverted on the retina, and the optic lobes of the brain re-invert the image after the fact. Eye diagrams based on Liem et al. (2001) and Wilt and Hake (2003).

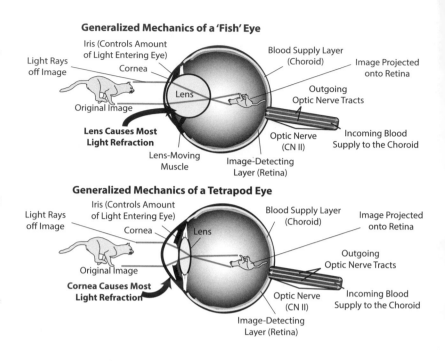

Generalized Mechanics of a 'Fish' Eye

Iris (Controls Amount of Light Entering Eye)
Light Rays off Image
Cornea
Blood Supply Layer (Choroid)
Image Projected onto Retina
Lens
Original Image
Outgoing Optic Nerve Tracts
Lens Causes Most Light Refraction
Optic Nerve (CN II)
Incoming Blood Supply to the Choroid
Lens-Moving Muscle
Image-Detecting Layer (Retina)

Generalized Mechanics of a Tetrapod Eye

Iris (Controls Amount of Light Entering Eye)
Light Rays off Image
Cornea
Lens
Blood Supply Layer (Choroid)
Image Projected onto Retina
Original Image
Outgoing Optic Nerve Tracts
Cornea Causes Most Light Refraction
Optic Nerve (CN II)
Incoming Blood Supply to the Choroid
Image-Detecting Layer (Retina)

change the direction and angle of the light. For example, as light goes from air (less dense) to water (denser), it bends. If you have ever tried to retrieve something from the bottom of a pool, you have probably noticed that where the item appears to be when you look at it through the water is not where it is actually located. The bending of the light rays (their refraction) distorts where the object appears to be underneath the water. For our fishy friends, the density of water and the eye are nearly the same, and so very little bending of light occurs when it enters the eye (Fig. 9.8) (Liem et al., 2001). In contrast, the difference in air and eye density is much greater, and so light waves are much more greatly refracted as they enter the eyes of a tetrapod (Fig. 9.8) (Liem et al., 2001). Nowhere is the effect of this difference between seeing in water and seeing in air more apparent than in lens shape. In fishlike vertebrates, the lens is large and spherical, and is physically moved back and forth to focus light on the light receptor cells at the back of the eye called the retina (Fig. 9.8) (Liem et al., 2001; Kardong, 2012). Because there is little difference in density between the cornea and water, there is very little change in light direction at this boundary in fishlike vertebrates. Therefore, most focusing in fish eyes happens through physical movement of the lens. In contrast, the lens of most tetrapods tends to be a thin disc that can be tightened or relaxed to fine-tune the focus of a particular image (Liem et al., 2001; Pough, Janis, and Heiser, 2002; Kardong, 2012). However, much of the proper focus of an image happens at the cornea, where most of the light bending occurs (Fig. 9.8) (Liem et al., 2001). This is why conditions that affect cornea shape in humans, such as astigmatism, can be corrected only externally using the appropriate contacts or eyeglasses. We will see that modern amphibian eyes contain a mixture of tetrapod- and fishlike features.

Differences in air and water density bring yet more problems related to respiration and dehydration. Gills simply can't work in air because there is not enough density to hold their delicate membranes apart. Gills in air collapse into a wet lump that has too little surface area available for gas exchange. We have previously discussed lungs, organs better equipped to extract oxygen from air (see chapter 8). Along with a greater or exclusive reliance on lungs for respiration, we see the reduction and disappearance of the gular bones, parts of the hyoid apparatus, and pharyngeal arches devoted to holding up and moving the gills. We also see the enlargement and retraction of the choanae (internal nostrils) on the palate so that air inhaled through the external nostrils is transferred to the mouth (Carroll, 1988; Liem et al., 2001; Pough, Janis, and Heiser, 2002). The ribs themselves enhance breathing. When the ribs are collectively rocked caudally, the chest cavity is constricted, and they force air (exhalation) out of the lungs. When the ribs are rocked cranially, the chest cavity increases, sucking air into the lungs (inhalation). The motion of the ribs is best described through analogy with a bucket handle (Fig. 9.9). On a paint bucket, there is typically a U-shaped handle that you lift up to carry your color of choice from one room to another. When you lift up that handle, it rotates up and away from the bucket's lid, increasing the space in between. Now, imagine all the rib pairs as U-shaped bucket handles (Fig. 9.9). As they are rocked cranially, they rotate up and away from the body cavity much like the bucket handle rotates up and away from the bucket lid. If you imagine the space between the bucket lid and handle as the space inside the ribcage, you can appreciate that when the ribs are rocked forward, the chest cavity's space increases. The expanding chest in turn sucks air into the body through the nose, mouth, and trachea, and it is channeled into the lungs. When you drop the handle on your paint bucket, the space between it and the bucket lid diminishes. By the same principle, exhaling air from the lungs is caused by collectively rocking the ribs caudally, which decreases the size of the chest cavity and forces air out.

In tetrapods, a variety of muscles are associated with the ribcage and abdomen. A series of external and internal rib muscles controls the movements of the ribs in most vertebrates to enhance inhalation and exhalation. In addition, oblique flank muscles and a pair of transverse and longitudinal abdominal muscles collectively act to constrict the abdomen and force air out of the lungs (Kardong, 2012).

Eating also becomes a problem. As we have seen, for many fishlike vertebrates, some form of suction is used to draw prey into the mouth, and water itself is used to lubricate and push prey down the gullet. Tetrapods cannot use suction nor rely on water to lubricate food and move food down the pharynx. It should be no surprise, then, that in salamanders and other tetrapods salivary glands are present that moisten and lubricate food, making it easier to swallow (Liem et al., 2001; Pough, Janis, and Heiser, 2002; Kardong, 2012). But saliva by itself would not be enough to overcome the problems of manipulating and pushing food down the pharynx, let alone capturing prey to begin with.

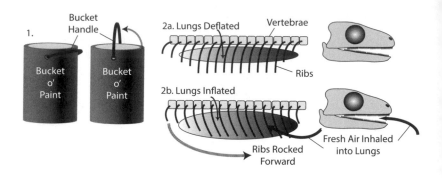

9.9. How the ribs of tetrapods work in aiding respiration using a 'bucket-handle' motion: (1) just as lifting the U-shaped handle on a bucket of paint increases the space between it and the lid, (2) collectively rocking the ribs forward increases the space of the chest and body cavity, allowing the lungs to inflate. This type of rib motion is best exemplified in amniote vertebrates, discussed in chapter 12.

Enter the fleshy, muscular tongue of tetrapods. All vertebrates have some kind of tongue, and most fishlike vertebrates have a relatively stiff and immobile tongue stuck to the bottom of the mouth (Fig. 9.10). Sometimes, the tongues of various fishlike vertebrates even sport teeth! In contrast, the tetrapod tongue is strong and mobile, and capable of manipulating food in the mouth and pushing it into the pharynx. In some salamanders and lizards the tongue can even be projected at distant prey, and many salamanders capture prey by 'slapping' it with their moist and sticky tongues (Reilly and Lauder, 1989; Deban et al., 2007). But where did this tongue come from? Observations of tetrapod embryo development and the tracing of nerve tracts have revealed that the tongue is composed of the same muscles as those that open the jaw in fishy vertebrates. Many fish have a muscle called the coracomandibularis that opens the lower jaw like a chain opening an inverted drawbridge. Such an arrangement is necessary to open the jaw against water pressure, but in air and gravity, this muscle is not mechanically effective. Instead, simply relaxing the jaw-closing muscles (adductor mandibulae and pterygoids) allows the jaw to drop (just ask Jacob Marley). Although redundant as a jaw-opening muscle, this muscle, running as it does between the lower jaw bones, could take on a new role as the fleshy tongue (Iwasaki, 2002). This new tongue encapsulates the old, stiff fish tongue, and can be used to capture prey, to manipulate prey once in the mouth, and to push prey, coated with a liberal amount of saliva, into the pharynx (Fig. 9.10) (Iwasaki, 2002). Essentially, the fleshy tongue takes the place of water as a tool for capturing and swallowing prey.

This should not give the impression that jaw opening is only passive: active jaw opening is required to control when and how widely the jaws open. But if the original jaw-opening muscle became the tongue, what muscles open the jaws? It turns out that muscles that used to aid in jaw suspension are now in an ideal place to rotate the jaw open. These muscles are associated with the hyomandibula in fishes, but because this bone is now employed for transmitting sounds into the inner ear in tetrapods, the muscles now attach directly to the lower jaw behind the jaw joint (the quadrate and articular, as you will recall). If you imagine the jaws as a pair of scissors, the pivot about which the blades rotate would be the quadrate and articular jaw joint, and the handles behind the pivot

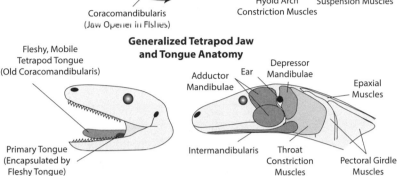

Generalized Fishlike Vertebrate Jaw and Tongue Anatomy

Accessory Jaw Adductors
Spiracle
Epaxial Muscles
Pectoral Girdle Muscles
Adductor Mandibulae
Gill Muscles
Primary Tongue (Nonmobile)
Intermandibularis
Hyomandibula Jaw Suspension Muscles
Hyoid Arch Constriction Muscles
Coracomandibularis (Jaw Opener in Fishes)

Generalized Tetrapod Jaw and Tongue Anatomy

Fleshy, Mobile Tetrapod Tongue (Old Coracomandibularis)
Depressor Mandibulae
Adductor Mandibulae
Ear
Epaxial Muscles
Primary Tongue (Encapsulated by Fleshy Tongue)
Intermandibularis
Throat Constriction Muscles
Pectoral Girdle Muscles

9.10. Homology and development of jaw and tongue muscles in tetrapods. In tetrapods, muscles once used for jaw opening and gill movements in fishes have been modified into the fleshy tongue and a new jaw-opening muscle, the depressor mandibulae. Based on diagrams in Romer (1962) and Liem et al. (2001).

would represent attachment sites for jaw-closing and -opening muscles. When you open a pair of scissors, you pull up on the lower blade's handle, which causes this blade to rotate downward. In similar fashion, the old jaw suspension muscles pull up on the portion of the lower jaw behind the jaw joint, causing the lower jaw to pivot downward (Fig. 9.8) (Romer, 1962; Liem et al., 2001). These old jaw suspension muscles get a different name in tetrapods to reflect their new function: depressor mandibulae (Romer, 1962; Liem et al., 2001) (Fig. 9.10).

Odors are detected in air using the nostrils and sometimes the tongue. Breathing in through the nose is now possible, and a moisturized layer of cells lining the nasal cavities contains the olfactory nerve endings that lead back to the brain through the first cranial nerve (Liem et al., 2001; Pough, Janis, and Heiser, 2002; Kardong, 2012). Airborne odorant particles are inhaled and stick to the wet mat of olfactory tissues. Like a lock and key, different odorant particles can fit only with certain nerve endings. When they dock with their odor receptors, chemical signals travel up the first cranial nerve to the brain and register a particular odor (Liem et al., 2001; Pough, Janis, and Heiser, 2002; Kardong, 2012). As you will recall from chapter 8, the nasal tissues and olfactory membranes are kept moist by tears that slowly trickle through the lacrimal (tear) ducts at the corners of the eyes.

Physiology

We now move from the mouth to the heart. The circulatory system of tetrapods had to change as well to cope with gravity. The heart of a typical fishlike vertebrate is small and located just behind the pharynx because it is in the best place to pump blood to the gills. But under gravity's influence, the heart would not be able to pump efficiently, blood would pool in the appendages and body cavity, and not enough blood would get to the head and brain. More to the point, the pharynx of tetrapods became smaller and less directly involved with respiration with the reduction or loss of gills. The position of the heart in tetrapods is farther away from the head because they now have a neck. Residing as it does in the chest, the heart must now pump blood a greater distance to reach the brain (Liem et al., 2001; Pough, Janis, and Heiser, 2002; Kardong, 2012). But directing high-pressure blood to the lungs would be a disaster: the tiny capillaries in the lungs would rupture, filling them with blood. As we have seen in lungfish, the heart in the ancestors of the tetrapodomorphs was probably already a double-pump, dividing blood into a lung (pulmonary) circuit and body (systemic) circuit. In tetrapods such as salamanders the right side of the heart tends to have less powerful muscles than the left side. Because oxygen-poor blood going to the lungs enters the heart on the right side, the less powerful muscles on this side gently pump the blood under low pressure to the lungs. In contrast, oxygen-rich blood returning to the heart from the lungs enters the more muscular left side, where it can be accelerated at high pressure to reach its various distant destinations (Liem et al., 2001; Pough, Janis, and Heiser, 2002; Kardong, 2012).

Temperature was yet another hurdle for the earliest tetrapods. The density of water and its physical properties (such as the polarity of water molecules) are such that it holds onto heat, what scientists might call heat inertia (Bloomfield, 2006). It takes water a long time to heat up or cool down, and so organisms living in this medium are insulated from rapid environmental changes that could be damaging or fatal. In contrast, the low density of air means that temperatures can change much more quickly on land, and air is nowhere near as insulating. Two examples illustrate these differences. If you live in regions where you get cold winters, you have probably noticed that rivers and lakes do not freeze quickly and in fact can be seen to 'steam' in the cold air. In this example, the air temperature has dropped much more quickly than the water temperature, and so the water is much warmer compared to the surrounding air, at least for a time. As another example, I have participated in paleontological field work in areas of the United States and Africa that are desert environments with very little moisture in the air. When I visit these areas, I am always struck by how quickly the air temperature can become blisteringly hot during the day, and then, just as quickly, come crashing down to an uncomfortably cold temperature at night. By contrast, having spent most of my life Illinois, I can attest to the very humid summers characteristic of the Midwest. The difference in temperature between night and day is nowhere near as extreme as in a desert, and nights can remain quite muggy and hot because the excess moisture in the air is a better insulator, holding onto all the heat it stored up during the day. There would have been strong selective pressure on the earliest tetrapod populations to evolve means of dealing with rapidly changing environmental conditions.

Most vertebrates, including salamanders, are 'cold-blooded' animals, which modify and control their internal body temperature using external sources of heat or cold. 'Cold-blooded' is a poor term because many such animals can actually keep their blood and bodies quite warm. Instead, it is more appropriate to use the term 'ectothermic' to identify fishlike vertebrates and tetrapods that mainly regulate their body temperature externally (Liem et al., 2001; Pough, Janis, and Heiser, 2002). Due to the heat-holding properties of water, a fishlike vertebrate has an easier time maintaining an appropriate body temperature: find an ideal environment and stay there as long as you can. In air, the rapidly changing temperatures during the day make it difficult or impossible for ectothermic tetrapods to follow a fishlike strategy. Instead, tetrapods such as lizards seek out places to sun themselves when temperatures are cooler, and seek refuge from the heat of the day by finding shady areas, burrowing into the soil, or even immersing themselves in water (Liem et al., 2001; Pough, Janis, and Heiser, 2002).

Our old friend the pineal organ plays a crucial role in temperature regulation in ectothermic tetrapods. You will recall that the pineal organ develops a parietal eye that peeks out from an opening in the skull (the parietal bone) to detect light levels and conveys this information to the brain to regulate cycles of activity. For ectothermic tetrapods, the differences in light levels detected by the parietal eye and conveyed to the pineal organ help these animals determine when to seek a sunny rock or when to seek shelter from the sun (Liem et al., 2001). Although we haven't drawn attention to this with the fishlike vertebrates, most ectothermic vertebrate skulls have a small parietal opening through which the parietal eye peeks (Fig. 9.1). As we will see, the skulls of early tetrapods and many modern ectothermic tetrapods possess some form of the parietal opening (Carroll, 1988; Liem et al., 2001).

Finally, we must briefly touch on development. Most salamanders retain a primitive, water-based form of development (although some salamanders do lay eggs in moist places on land), and so we focus on the basic aspects of their development as a template for what may have been occurring in early tetrapods. Salamander embryos typically develop within small, jelly-coated eggs that float in mats or are attached to vegetation within shallow pools of water, much like many freshwater fishes (Gilbert and Raunio, 1997; Gilbert, 2010). During development inside the egg, salamanders feed off yolk proteins that were placed there by the mother when she produced the eggs in her ovaries and reproductive tract. Most salamanders and many amphibians have an aquatic larval stage. Essentially, after a typically brief developmental period inside the egg, the salamander embryo hatches out as an aquatic, carnivorous larva with rudimentary limbs, external gills, functional lungs, and a broad tail fin (Gilbert and Raunio, 1997; Gilbert, 2010). Salamander larvae typically feed like fish, using rapid mouth opening to suck in potential prey. Some may lie in wait beneath the waves, ambushing small animals at the water's surface. The lungs, gills, and heart work together as discussed for lung-fishes. In many salamander species, metamorphosis eventually occurs, during which time the external gills and tail fin are lost, lungs become the primary organ of respiration, the limbs become more prominent, and the skull is remodeled to better grab prey in air (Gilbert and Raunio, 1997; Gilbert, 2010).

Plants and Changing Continents

There is a saying coined by James Carville: "It's the economy, stupid." The economic environment in any country has a huge impact on the products, jobs, and services offered. Currently in the United States, uncertainties surrounding oil supply and increasing oil prices have opened opportunities for new car designs. There are now several hybrid cars that use a combination of an electric motor and a gas engine. These seemingly 'odd' cars would not have been viable in U.S. markets a decade or more ago because there was no economic incentive to buy them. However, in an economy where citizens are feeling the crunch of dwindling budgets, hybrid cars make more sense (and more 'cents') in that they will save you money. The reality is that economic factors matter from a human survival perspective: access to money is the environmental factor that selects for what people will purchase and what they will not, and this in turn affects the 'survival' of particular products and industries.

In a similar way, changing natural environments have greatly influenced the selection and success of various vertebrate chassis forms. A niche is a specialized 'job' a particular animal has in an environment. Like the job market, as niches begin to fill up, the pressure to survive becomes more and more competitive. The opening of new niches has often sparked the evolution of new species. So, what changed during the Devonian, the great 'Age of Fishes,' that touched off the evolution of the tetrapod common ancestor? Simply put: the land was finally a viable place for vertebrate animals to exploit.

The invasion of land by vertebrates was preceded by plants and invertebrates. When the first vertebrates were swimming the seas (approximately 540 Ma; Card 89), the land was barren and quiet, but by 440 Ma (Card 91) some low-growing plants were gaining a foothold in moist areas around the continental margins (DiMichelle and Hook, 1992). The evolution of plants from aquatic to terrestrial organisms is itself a fascinating and remarkable story that goes far beyond the scope of this book. Suffice it to say that the colonization of the land by plants provided some niches for several invertebrate animal groups to exploit, and any sounds produced by these creatures finally broke the nearly 4 billion years of silence that preceded them. By the Devonian (or perhaps earlier), the first vascular plants that could support themselves against gravity had appeared, and this opened additional niches for various invertebrates (DiMichelle and Hook, 1992). Throughout the Devonian, land plants became more numerous and larger, so that by the end of the period some 360 Ma (Card 93), there were trees related to the common ancestor of ferns and conifers that stood over 18 meters in height (Stein et al., 2007).

Land plants had other effects besides giving invertebrates places to live and things to eat. By the Devonian, several land plant species were deciduous, meaning they regularly shed their leaves. The increasing influx of leaf litter would have deepened and enriched the soils on the continents, providing a further source of nutrients that were washed into the estuaries and shallow seas on the continental shelves (DiMichelle and Hook, 1992; Stein et al., 2007). Additionally, the growing number of land plants seems to have accelerated the accumulation of oxygen in the atmosphere, so that by the end of the Devonian levels of this precious gas approximated those present now. The ozone layer, which protects land animals from excess ultraviolet radiation, was also approaching modern levels, and carbon dioxide levels were dropping, being taken up instead by plants to create sugars. By the middle of the Devonian, there was more free oxygen available for respiration in air than in water (DiMichelle and Hook, 1992).

Continental movement also contributed to a changing landscape. At the beginning of the Devonian (~416 Ma, Card 91), there were roughly two large landmasses: Euramerica to the north and Gondwana to the south (Blakey, 2010). Over the Devonian period, the two landmasses drifted closer together. Generally speaking, Euramerica was situated over the equator and did not migrate very far during this time, whereas most of Gondwana started in the Southern Hemisphere but drifted northward (Blakey, 2010). Because Euramerica was equatorial, perhaps it is not surprising that fossils of the earliest tetrapods and most tetrapodomorphs are to be found in rocks from North America, Greenland, Iceland, and Europe today. By the end of the Devonian, Euramerica and Gondwana were starting to amalgamate into what would eventually become the supercontinent Pangea – the final assembly of this supercontinent would have to wait until the end of the Carboniferous period, about 299 Ma (Card 94). However, during the Devonian, many parts of the continents were submerged, forming shallow seas where many of the diverse vertebrates we have already encountered flourished (Benton, 2005; Blakey, 2010).

Leading up to the Devonian, the northern continental mass of Euramerica (present-day North America) had experienced collisions with smaller landmasses (present-day Iceland, Greenland, and Europe), and these events formed a long mountain chain that stretched from the southwest to the northeast. Part of this old mountain range still exists on the East Coast of North America today as the Appalachian Mountains (Redfern, 2003; Blakey, 2010). Throughout the Devonian, this mountain chain rose and was the source of various rivers that carried sediments from these highlands west to wide, tropical floodplains and then down into deltas along miles and miles of shallow ocean waters. This influx of sediments provided nutrients for the newly evolving land plants and the invertebrate denizens inhabiting the tropical soils (DiMichelle and Hook, 1992).

This overly brief summary of major events in plant and animal evolution, as well as continental movements, can in no way do justice to this important time in earth's history. For our purposes, it is enough to say the time was ripe for the larger vertebrate animals to begin exploiting land resources. The Devonian landscape offered new sources of food, more oxygen for respiration and protection from ultraviolet rays, and plenty of warm and shallow waters from which the common ancestors of the tetrapods could take their first steps.

Next Stop: Early Tetrapods and Amphibian Diversity

Based on what we know of fishlike sarcopterygian anatomy and the generalized anatomical details we have explored in salamanders and some other tetrapods, we can make the following predictions about the earliest tetrapods. The skull of early tetrapods should show less mobility than that typically found in fishlike sarcopterygians, and the intracranial joint should be fused. The premaxilla, maxilla, and other upper jaw bones should be fused with the skull, and the major jaw-opening movements should occur between the quadrate and articular bones that allow the lower jaw to drop. The snout should be longer than the braincase, and the squamosal and pterygoid bones should be expanded compared with fishlike sarcopterygians to house larger jaw-closing muscles We should also see the reduction or loss of the opercular, gular, and other gill-bearing bones, and the detachment of the pectoral girdle from the back of the skull. A parietal opening in the skull for the pineal organ should be present as well, helping an early tetrapod to sense when to warm and cool itself in an airy environment.

The vertebral column should be more robust than that of fishlike sarcopterygians and divided into cervical, dorsal, sacral, and caudal series. Zygapophyses should span each vertebra to prevent the body from sagging, neural arches and spines should be prominent and fused to the vertebral bodies (centra), and ribs should articulate with the dorsal vertebrae. Body axis fins should be reduced or absent. The pectoral girdle should have a large scapulocoracoid and reduced clavicle and cleithrum, whereas the pelvic girdle will articulate with the sacral vertebrae. Fewer, sturdy long bones should comprise the forelimb and hind limb, and jointed digits (fingers and toes) are predicted to be present. One would predict some ability to hear and see in air, the presence of a fleshy tongue, and at least the beginnings of the physiological adaptations we have discussed for the circulatory and respiratory systems. In addition, it would not be surprising to find evidence of an aquatic larval stage of development.

As we have briefly discussed, the Devonian was an important time in earth history. Significant land plant communities became firmly established, and land-living invertebrates came along to exploit these resources. The time was ripe for vertebrates to do the same. We now turn to the fossil record at the end of the Devonian to check our predicted tetrapod against reality.

10.1. Diagram of the crocodile-like transitional tetrapodomorph, *Tiktaalik* (after Daeschler, Shubin, and Jenkins [2006], and Shubin, Daeschler, and Jenkins [2006]). The skeleton of *Tiktaalik* shows a number of tetrapod-like features, including a distinct neck and well-developed forelimbs. The forelimbs of *Tiktaalik* show a tetrapod-like series of bones, including fingerlike bones, encased in a jacket of scales and fin rays. The large spiracles atop the flattened skull suggest this animal may have utilized them for breathing air while submerged in wait for prey. Note the large area (AM) for the adductor mandibulae and other jaw-closing muscles. The quadrate-articular jaw joint is circled.

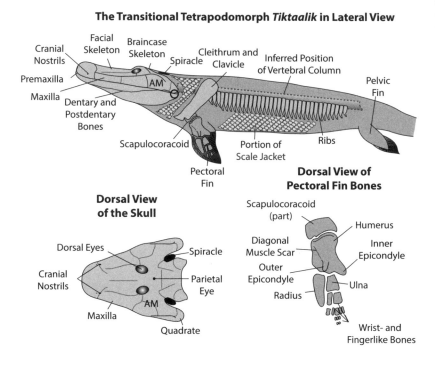

The Transitional Tetrapodomorph *Tiktaalik* in Lateral View

Dorsal View of the Skull

Dorsal View of Pectoral Fin Bones

The Tetrapod Chassis in Transition

10

IF YOU LIVE IN THE UNITED STATES AND USE A STANDARD ENGLISH computer keyboard, chances are excellent that you have what is called the QWERTY design. The design name comes from the arrangement of the first six letters, from left to right, in the first letter-bearing row. The history of the QWERTY keyboard is actually controversial and even contentious in some circles, and it has been used as an example of an inefficient design that has persisted despite the presence of superior alternatives (David, 1985; Liebowitz and Margolis, 1990; Gould, 1991; Hossain and Morgan, 2009). The argument goes that because so many people adopted the QWERTY keyboard and got used to it, there eventually was little demand for other key arrangements. What can be agreed upon is that this standard arrangement of keys was first developed to prevent key-jamming on old typewriters: the familiar arrangement of letters and numbers allowed keys to be struck in a sequence that typically did not lead to them sandwiched on top of one another. However, although computer keyboards no longer suffer from key jams the way old typewriters did, the argument goes that the old, inefficient QWERTY combination has remained only because of familiarity (David, 1985; Gould, 1991). Recent studies, however, have suggested that claims of QWERTY's inefficiency are a bit exaggerated, and that they miss a crucial advantage to the design: one of finger speed (Hossain and Morgan, 2009). It turns out that the QWERTY design, although less efficient than others in terms of hand use symmetry, favors alternating hand sequences and minimizes same-finger typing (Hossain and Morgan, 2009). What is peculiar about all this is that the designers of QWERTY were focused on preventing key jamming. It turns out that the gain in typing speed due to the arrangement of the keys was probably a fortuitous accident rather than an intentional design feature.

QWERTY design serves as a good analogy for tetrapod chassis evolution. The common ancestor of the tetrapods had inherited a body plan adapted to living in shallow water and ambushing prey. In our *Eusthenopteron* example in chapter 8, we have a fishlike sarcopterygian adapted to hold itself against the water current using sturdy pectoral fins and a more robust skeleton. Unlike the delicate fin skeleton of actinopterygians, this chassis provided a foundation on which further refinements could be added to allow more and more effective travel and support of the body in air. Technically, we call this sort of unintentional adaptation to a new environment an exaptation: an anatomical arrangement that evolved under one set of conditions but, by luck, is also useful under different circumstances (Gould and Vrba, 1982). Like the serendipitous increase in typing speed that resulted from other design considerations with the QWERTY keyboard, so the robust skeleton and modified anatomy of fishlike sarcopterygians set them up best for colonizing the land. In the previous chapter, we have predicted what we should find in the earliest tetrapods. It is now time to put these predictions to the test.

Transitional Tetrapodomorphs

Pedigree Tetrapodomorphs That Form Close Outgroups to True Tetrapods

Date of First Appearance ~385 Ma (Card 92)

Specialties of Skeletal Chassis Combinations of Fishlike and Tetrapod-like Features

Eco-niche Fishy Predators in Estuaries and Shallow, Inland River Channels

Work over the past 30 years by a dedicated group of paleontologists has unearthed amazingly well preserved fossils of vertebrates that blur the line between fishlike and tetrapod-like anatomy (Clack, 2002, 2009a). Most of these fossils have been found in rocks from the old Euramerican continental mass discussed in the previous chapters (i.e., North America and Eurasia). These transitional tetrapodomorphs, as we will call them here, allow us to test hypotheses about the earliest steps toward becoming a true tetrapod. In recent years, one particular fossil above all others illuminates the evolution of a vertebrate chassis 'caught in the act' of transformation from a fishlike frame to one of a tetrapod. That fossil is *Tiktaalik* (Plate 9).

This remarkable tetrapodomorph from Canada is known from several well-preserved specimens approximately 1–2 meters in length (Fig. 10.1) (Daeschler, Shubin, and Jenkins, 2006; Shubin, Daeschler, and Jenkins, 2006; Downs et al., 2008). Overall, *Tiktaalik* is a relatively dorsoventrally flattened animal and looks remarkably crocodylian in its general body outline. A large, flat head gives way to a broad and flat body encased in a jacket of overlapping scales. The pectoral girdle and fins of *Tiktaalik* are robust (Shubin, Daeschler, and Jenkins, 2006), and recently described pelvic material shows that a robust pelvis was present (Shubin, Daeschler, and Jenkins, 2014), but unfortunately nothing is known so far about the tail as this region of the animal is not preserved in any of the described fossils (Daeschler, Shubin, and Jenkins, 2006).

The skull is robust, the eyes are perched dorsally, and the intracranial joint is, as predicted for ancestral tetrapods, fused (Fig. 10.1) (Downs et al., 2008). What is fascinating about *Tiktaalik* is just how crocodile-like its skull is. The distance between the eyes and snout has increased compared to that of animals like *Eusthenopteron*, so that the skull of *Tiktaalik* has an almost 'reptilian' look to it. Like a crocodile or alligator, the snout is triangular when viewed from above and broadens toward the back of the skull (Fig. 10.1) (Daeschler, Shubin, and Jenkins, 2006; Downs et al., 2008). The squamosal and jugal regions of the skull are greatly expanded and probably housed large adductor mandibulae and pterygoid muscles for clamping the jaws shut on prey. Most curious of all is the pitted and dimpled texture of its skull. Such a skull texture is precisely what is observed in modern crocodylians and other tetrapods where the skin is tightly affixed to the head (Witzmann et al., 2010). By analogy, a bumpy, tough skin probably adorned the head of *Tiktaalik* in life. The premaxilla, maxilla, and dentary bones were studded with sharp, cone-shaped teeth much like those of modern crocodylians, and the lower jaw itself was long but broad (Daeschler, Shubin, and Jenkins, 2006; Downs et al., 2008).

The cone-shaped teeth in crocodylians are ideal for both puncturing and crushing prey, and by analogy the teeth of *Tiktaalik* suggest a similar function. As we predicted for the earliest tetrapods, this animal had a neck: the head is completely decoupled from the body, and the operculum and most (but not all) of the pectoral-skull series of bones have vanished. Scarring on the back of the skull just in front of the neck (Downs et al., 2008) suggests large and powerful epaxial muscles were present that could swing the head laterally.

That *Tiktaalik* spent a majority of its life in the water is indicated by the retention of the gular and hyoid-derived bones that supported gills (Daeschler, Shubin, and Jenkins, 2006). In fact, some of these bones have distinct channels embossed into their surfaces, a situation observed in all fishes with operational gills. These channels house the respiratory vessels that supply the gills, and suggest similar anatomy was present in *Tiktaalik*. A large spiracular notch was present caudal to the eyes into which a now smaller hyomandibula crossed (Fig. 10.1). Additionally, the hyomandibula of *Tiktaalik* was no longer a vertical strut as it is in many fishes but was angled somewhat horizontally (Daeschler, Shubin, and Jenkins, 2006; Downs et al., 2008). With the absence of opercular bones, the once stationary end of this bone was now 'free' to move, but probably did not transmit sound into the inner ear. Instead, curiously, the hyomandibula appears to have articulated with the remaining gular and gill bones (Daeschler, Shubin, and Jenkins, 2006; Downs et al., 2008) and may have played some role in respiration. If you recall from chapter 7, bichir fish breathe through the spiracle, and it is possible that the larger and more dorsal position of the spiracle, in combination with movements of the gular and gill bones leveraged by the hyomandibula, played a similar role. In fact, some paleontologists have gone so far as to suggest that animals like *Tiktaalik* may have been breathing through their spiracles (Brazeau and Ahlberg, 2006; Clack, 2007).

What is known of the vertebral column and ribs is intriguing. Although a number of relatively complete specimens are known, not one *Tiktaalik* fossil shows any direct traces of a vertebral column (Daeschler, Shubin, and Jenkins, 2006). This suggests that the vertebral column was cartilaginous and that the notochord was prominent. So, this was an animal that had a lot of lateral mobility but probably limited ability to traverse dry land for extended periods of time. That said, nearly 45 sets of thick, well-ossified ribs are present (Fig. 10.1). Because ribs articulate with vertebrae, an equal number of cartilaginous vertebrae must have been present in *Tiktaalik* (and obviously more given that not all vertebrae bear ribs) (Daeschler, Shubin, and Jenkins, 2006). Many of the ribs overlap one another and appear to have been capable of sliding over one another. Some of the ribs even possess expanded areas reminiscent of similar features on bird ribs that provide lever-like attachment sites for ribcage-expanding muscles (Tickle et al., 2007). These large ribs, in combination with the jacket of overlapping scales, suggest *Tiktaalik* may have possessed a respiratory pattern similar to that of the bichir. In particular, the ribs may have acted, as they do in the bichir, to resist the collapsing scales as *Tiktaalik* exhaled stale air.

The pectoral girdle and fin of *Tiktaalik* are almost precisely what you would predict in an animal transitioning from water to land (Fig. 10.1) (Shubin, Daeschler, and Jenkins, 2006). The excellent preservation and numerous specimens of *Tiktaalik* have allowed certain pectoral fins to be prepared out of the rock so that the individual bones can be articulated together. The cleithrum, clavicle, and other remnants of the skull-pectoral series are still prominent but reduced. The scapulocoracoid has become enlarged and more robust. The head of the humerus is more convex than in *Eusthenopteron*, and the glenoid more concave, so that something resembling a primitive ball-and-socket joint has developed (Shubin, Daeschler, and Jenkins, 2006). Articulation of the humerus with the glenoid shows that some longitudinal rotation, abduction and adduction, and protraction and retraction were possible. But the leading edge of the glenoid has a distinct knob that fits snugly with a divot on the humerus when it has been rotated, abducted, and protracted, preventing movement and providing a very stable platform against which the body could be supported (Shubin, Daeschler, and Jenkins, 2006). At the elbow, the humerus expands into two distinct, convex joint surfaces that accepted the cup-shaped ends of a bladelike radius and stout ulna. These joint shapes appear to allow independent rotational and piston-like movements of the radius and ulna relative to one another that collectively would allow the 'forearm' to assume a more vertical orientation beneath the humerus. Numerous wrist-like bones with well-defined joint surfaces are present in *Tiktaalik*, fingerlike bones extend from these, and the entire pectoral fin is covered in fin rays (Shubin, Daeschler, and Jenkins, 2006). The well-developed joint surfaces between the wrist- and fingerlike bones were arranged in three horizontal rows that, like the spaces between planks on a wooden bridge, must have enabled the end of the fin to flex and extend (Fig. 10.2) (Shubin, Daeschler, and Jenkins 2006).

Collectively, the pectoral fin joints in *Tiktaalik* would have allowed the animal to lift and prop its body, head, and neck out of the water or even just off the bottom, much like doing a pushup. The specialized part-counterpart knob and socket joints between the glenoid and humerus further enhanced this movement, bracing and stabilizing the animal in a propped-up position (Fig. 10.2). At first, it appeared that *Tiktaalik*, like *Eusthenopteron*, was an animal in front-wheel drive. What was known of the pelvic fin suggested that it was much smaller than the pectoral fin and that it was not attached to the vertebral column (Daeschler, Shubin, and Jenkins, 2006). However, the pelvic girdle of *Tiktaalik* is now known to be more robust than in tetrapodomorphs such as *Eusthenopteron*, and may have allowed more forward propulsion than previously suspected (Shubin, Daeschler, and Jenkins, 2014).

Where we find *Tiktaalik* fossils fills in more of the story. *Tiktaalik* is found in freshwater sediments indicative of a shallow, braided river system stretching through vegetation-choked tropical environments (Daeschler,

**Tiktaalik Pectoral Fin
in Dorsal View**

Scapulocoracoid

Part-Counterpart
Articulating
Surfaces
between
Scapulocoracoid
and Humerus

Humerus

Ulna

Radius

Slat-like Joints
That Allow Fin
to Flex and Support
Body against
River Bottom

**Tiktaalik Pectoral Fin Supporting Body
against River Bottom (Cranial View)**

Inferred
Position of
Notochord
and Vertebrae

Ribs

Cleithrum and
Clavicle

Humerus

Ulna

Wrist- and
Fingerlike
Bones

Scapulocoracoid

Radius

River Bottom

10.2. Detail of the workings of the forelimb of *Tiktaalik* (based on Shubin, Daeschler, and Jenkins [2006]). The forelimb of this transitional tetrapodomorph could apparently support the body against muddy river bottoms. Note the part-counterpart joint of the shoulder girdle that would allow the forelimb to be braced in a support position.

Shubin, and Jenkins, 2006). You can find such river systems today across the vast floodplains of Africa or meandering through the rainforests of South America. Against this environmental backdrop, the chassis of *Tiktaalik* exudes crocodylian overtones. In shallow freshwater along vegetation-rich shores, there are plenty of places for a large, flat, log-looking animal to remain inconspicuous. One can imagine *Tiktaalik* floating motionless, only its eyes and spiracles poking above the water's surface, waiting for an opportunity to strike at careless arthropods that came too close to the shore. Once in striking range, *Tiktaalik* could quickly prop itself up on its forelimbs and strike out with its heavy, flat head, catching and crushing the prey in its strong jaws. That a fishlike animal could and would lunge from the safety of the water is not science fiction: certain catfish have been documented stalking pigeons from the water's edge and 'pouncing' on them for a tasty meal (Cucherousset et al., 2012)!

But the sedimentology tells us more than just a story about a potential predatory strike. Most importantly, the shallow snaking rivers that *Tiktaalik* swam in provided opportunities for limb and neck use (Daeschler, Shubin, and Jenkins, 2006). In shallow, warm waters, oxygen levels are volatile and often drop to levels that cannot sustain larger vertebrates. It is here that we see the pieces of the tetrapod chassis coming together. In *Tiktaalik*, we have an animal that can prop its body up on sturdy appendages, allowing its head to breach the water's surface to breathe in life-giving oxygen. With its head elevated above the waves, *Tiktaalik* could use its powerful neck to scan the horizon for potential arthropod prey or for other routes to fresher water. Regardless of the precise behavioral repertoire of this transitional tetrapodomorph, one thing is clear: the sturdy limbs that could support the body on land evolved in the water (Daeschler, Shubin, and Jenkins, 2006; Shubin, Daeschler, and Jenkins, 2006; Clack, 2009a, 2009b). *Tiktaalik* shows us that powerful appendages were an adaptation in the common ancestors of the tetrapods for navigating a shallow, vegetation-choked environment. Like the QWERTY keyboard, this adaptation exapted the descendants of animals like *Tiktaalik* for navigating the land.

Basal Tetrapods

Pedigree Animals Close to the Common Ancestor of All Other Tetrapod Descendants

Date of First Appearance 374–359 Ma (Cards 92–93)

Specialties of Skeletal Chassis First True Limbs with Digits

Eco-niche Amphibious Predators of Shallow, Inland River Channels

The transition from horse-drawn carriage to automobile was a gradual one, with fits and starts along the way. The earliest vehicles that could be described as 'cars' were essentially horseless carriages with a retrofitted motor and steering wheel. Even the wheels themselves were still large, hard, and filled with spokes (Setright, 2003). At some point, enough of the components of the modern automobile were in place that the horseless carriages could now be called 'cars' in the truest sense of the word. Few people would argue, for example, that the first mass-produced automobile, the Ford Model T, was not a car (Setright, 2003). The debate about the exact moment of transition between a car-like carriage and a car itself will probably be eternal because it is dependent on human definitions, categories, and what importance you lend to those distinctions.

The same goes for vertebrate paleontology. It is often difficult, as my colleague Adam Yates once remarked during our research on South African dinosaurs, to place a "golden spike" demarcating the beginning of a group of vertebrates. Evolution is "smeary" (another of Adam's sayings) in that, given descent with modification, populations change in a mosaic fashion. That is, you get a peppered accumulation of new or modified traits intermixed with primitive traits over large spans of time. Deciding exactly when one category of animal changes officially into a new category of animal is bound to be fraught with difficulties because nature is seamless. Categories are something we humans invented to make our lives easier: they help tremendously in organizing information and discussing the science of vertebrate paleontology, but transitional forms always pose problems.

Although debate will undoubtedly continue about where the transition from a fishlike chassis to a tetrapod chassis officially occurred, most paleontologists agree that the possession of limbs with digits (fingers and toes) is an undisputable trait state of tetrapods and tetrapods alone (Benton, 2005; De Iuliis and Pulerà, 2011). Enter the basal tetrapods, which still had some fishy elements about them, but are distinguished from animals like *Tiktaalik* by their digit-bearing limbs. Here, as Adam Yates would say, is where we put our golden spike labeled, "Tetrapods."

Perhaps the most astounding thing about the fin-to-limb transition is the way embryology informs us about how this occurred. Specialized body-patterning *HOX* genes are responsible for fin and limb development and form. *HOX* genes are like master switches: when expressed, their proteins activate multiple series of other genes that in turn produce various anatomical forms. Thus, the *HOX* genes that activate and control the formation of a fish fin are the same as the ones that pattern a tetrapod limb. The difference is in where and when they are expressed.

Sarcopterygian 'Fish' Fin

Fin Rays
Distal Elements
Ulna
Radius
Humerus

Straight Major Axis with Branching

Unchanged Gene Expression Regions

Tetrapod Limb

Digits
Wrist Bones
Ulna
Radius
Humerus

Major Axis Flips at Wrist to Form Digits

Gene Expression Flips at Wrist

■ Major Axis Gene Expression   ▨ Branching Gene Expression

10.3. Diagrammatic comparison of the developing appendage in a sarcopterygian 'fish' fore-fin and a generalized tetrapod forelimb. The same genes (called *HOX* genes) are responsible for the development and pattern of the bones in the 'fish' fin and tetrapod limb. However, a simple flip of the pattern of *HOX* gene expression at the wrist in tetrapods changes the direction of the branching axis so that digits form at the end of the limb. Another gene, *Sonic hedgehog* (Shh), produces a gradient of Shh proteins that help determine the identities and number of digits. See text for more details. Based on Gilbert (2010).

In fishlike sarcopterygians, a single upper bone (humerus/femur) forms first, followed by a branching into two bones (radius and ulna/tibia and fibula), followed by multiple branching distal elements (distal fin elements). It turns out that the branching pattern always occurs along the axis of the ulna or fibula: after the fin elements branch into a radius and ulna (or tibia and fibula), only the ulna (fibula) side of the appendage gives rise to the next set of branching elements (Gilbert, 2010) (Fig. 10.3). No significant successive branching elements emanate from the radius (tibia) side. These branching events are correlated with the expression of specific *HOX* genes in a sequence down the developing appendage in the embryo: there is a major-axis *HOX* gene expressed on the ulna/fibula side, and a branching *HOX* gene expressed on the radius/tibia side (Fig. 10.3). But something marvelous and unusual occurs in tetrapod embryos: the branching axis (and *HOX* gene expression) flips at the wrist or ankle so that branching elements now radiate distally away from the appendage (Fig. 10.8). Amazingly, fingers and toes result from a simple reorientation of the branching axis. Put another way, getting a fin or a limb with digits requires the same sequence of *HOX* genes, but in tetrapods the region of *HOX* gene expression flips at the wrist, resulting in outwardly radiating branches that become the digits (Gilbert, 2010) (Fig. 10.3). Thus, when we find fossils that have distinctive fingers and toes, we know we have vertebrates in which this *HOX* gene flip has occurred.

Among the many fossil examples now available to document the earliest tetrapods, perhaps the best examples are *Acanthostega* and *Ichthyostega*. Let's start with *Acanthostega*. This tetrapod is known from numerous well-preserved fossils excavated in Greenland (and thus we still have a linkage between the evolution of early tetrapods and the Euramerican

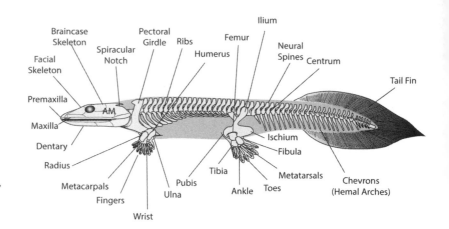

10.4. Skeletal diagram of the early tetrapod, *Acanthostega* (based on Clack [2002]). Probably one of the best-known early tetrapods, the skeleton of *Acanthostega* shows this was an animal mostly at home in the water. In particular, lateral line canals on the skull, slender ribs, and the broad caudal fin all point toward an actively swimming predator. The hand and foot sported eight and seven digits, respectively, and *Acanthostega* has been significant in understanding the origin and development of the hand and foot. As with *Tiktaalik*, large spiracular openings on top of the skull suggest *Acanthostega* may have been able to breathe while submerged and waiting for prey.

continent) (Clack, 2002). Overall, *Acanthostega* was approximately 60 cm in length (Fig. 10.4). The exquisitely preserved skulls of *Acanthostega* specimens provide an extraordinary amount of anatomical data. Like *Tiktaalik*, the skull of *Acanthostega* was flat, broad, and sculpted by pits and grooves, which suggest that the skin of this animal was tightly integrated with the head. The snout was long, and the braincase region was relatively short but broad (Clack, 2002). The expanded squamosal and jugal regions of the skull leave no doubt that *Acanthostega* possessed large adductor mandibulae and pterygoid muscles. Moreover, a large subtemporal fossa provided a hollow space beneath these bones with ample room for such muscles (Clack, 2002). Coupled with a long snout, the size and position of the jaw-closing muscles would have allowed *Acanthostega* to quickly open and snap its jaws shut (the scissors analogy again), driving its small but plentiful teeth deep into its prey. The eyes were, as in *Tiktaalik*, dorsally placed and supported by sclerotic rings. Just caudal to the eyes and centered above the braincase region again was a pineal opening for the parietal eye (Clack, 2002). As with lungfish and the tetrapodomorphs, the external nostrils were located low on the skull just above the mouth, and prominent, larger choanae were present on the palate near the front of the snout. The lower jaw was long and studded with small, sharp teeth.

That *Acanthostega* spent a majority of its time in the water is of little doubt (Clack, 2002, 2009a). Prominent lateral line canals were present on the skull and jaw, a large spiracular notch was present, and a series of gill-bearing bones persisted behind the jaws. Intriguingly, the lateral line canals are concentrated along the lower half of the skull and are conspicuously absent along its dorsal side (Clack, 2002). This would make functional sense if the head of *Acanthostega* was partly submerged with the eyes and spiracles protruding above the surface of the water. The hyomandibula is even smaller than in animals like *Tiktaalik*, but still apparently served a gill- and spiracle-supporting role rather than that of an auditory conductor (Clack, 2002). The gill-supporting bones continue to bear slots that would have accommodated blood vessels that supplied these respiratory structures. Skipping for a moment to the shoulder girdle, a distinctive slot also runs along the leading edge of the cleithrum bone (Fig. 10.5). Such a slot is present in the African lungfish, and in this animal it acts as a conduit to direct water from the gills out of the gill chamber. The presence of this slot on the shoulder demonstrates that functional and prominent gills were present in *Acanthostega* (Clack, 2002). Overall, *Acanthostega* clearly had a head that was submerged most of the time, relying on the lateral line system to detect water-borne vibrations and using gills as the predominant respiratory organ.

The vertebral column is known in detail from a number of specimens (Fig. 10.4). The shoulder girdle is separate from the skull, so *Acanthostega* had a distinct neck. As we discussed and predicted for the ancestor of tetrapods, there are cervical, dorsal, sacral, and caudal sections to the vertebral column. However, the variation in vertebral shape along the spine is minimal, and the zygapophyses connecting these bones to one another are weakly developed (Clack, 2002). Even the vertebral bodies (centra) are unfused, remaining as paired elements, and are not very robust. Distinct cervical ribs are present, and based on comparative anatomical data with other tetrapods, these elements probably provided anchoring spots for muscles that controlled head movements. The ribs along the dorsal vertebrae are rather small and not all that much larger than the cervical ribs (Clack, 2002). Finally, the tail is narrow but tall, and sports extensive fin rays along its axis. In modern salamanders or alligators with similar tail dimensions, this part of the vertebral column has become an effective sculling organ. By analogy, it seems most likely that *Acanthostega* could have used its tail in a similar way, perhaps even venturing into deeper water. Overall, the vertebral column of *Acanthostega* is reminiscent of fishlike sarcopterygians such as *Eusthenopteron*, and this further bolsters a mostly aquatic existence for this animal. The reduced zygapophyses, vertebral development, short ribs, and tall tail strongly suggest that *Acanthostega* waggled its body laterally like many of the fishlike vertebrates we have previously encountered (Clack, 2002).

Returning to the bones in the forelimb of *Acanthostega*: the shoulder girdle, like that of *Tiktaalik*, still retains a sizable cleithrum and clavicle, but the scapulocoracoid is a more robust element (Fig. 10.5). The glenoid (shoulder socket) is shallow and faces almost directly laterally, and this would have oriented the forelimb laterally away from the body (Clack, 2009b). The humerus is robust and L-shaped in profile, and the proximal end is broad and hemispherical, a shape that would have allowed crank-like movements of abduction, adduction, protraction, retraction, and possibly limited longitudinal rotation. Overall, the forelimb as a whole would work well as a paddle (Clack, 2002, 2009b).

The distal end of the humerus flares substantially and is divided into two joint surfaces for the radius and ulna, flanked by bony projections called epicondyles (Fig. 10.5) (Clack, 2002). In tetrapods, the outer epicondyle serves as the anchoring point for muscles that extend and abduct the wrist and hand. By contrast, muscles that flex the wrist and hand originate from the inner epicondyle (Liem et al., 2001; Pough, Janis, and

10.5. Detail of the shoulder girdle and forelimb of *Acanthostega* (based on Clack [2002]). The shoulder girdle shows the 'gill groove' along the scapulocoracoid, which indicates the presence of well-developed gills in this early tetrapod. The forelimb includes a robust humerus with prominent muscle attachment sites: the inner and outer epicondyles.

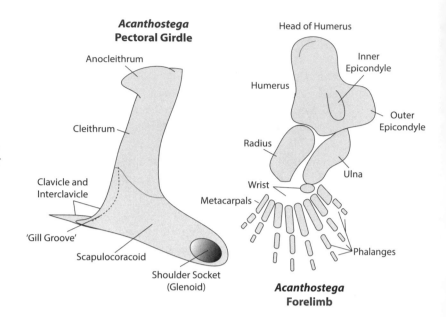

Heiser, 2002). Given that the inner epicondyle is huge compared with the much smaller, outer epicondyle in *Acanthostega*, we can infer that wrist and finger flexion were well developed. Although it would be tempting at this point to infer that such wrist and finger flexion would provide the hand with appropriate leverage to push and move the animal forward in air, the forearm and hand provide unexpected insights into the lifestyle of *Acanthostega*.

The joint surfaces of the humerus, radius, and ulna are collectively shallow and relatively flat, a condition we see today in the flippers of porpoises and other seagoing vertebrates where the elbow is relatively inflexible (Clack, 2002). Moreover, the radius and ulna are relatively flat bones, shapes that are again the norm in the flippers of marine tetrapods, where rotation and independent movements of the radius and ulna are either absent or poorly developed (Clack, 2002). Given these shapes, it is simplest to infer that in *Acanthostega* limited, hinge-like movements characterized the elbow, just like those in the flippers of many aquatic tetrapods. Things become more curious as we begin to study the hand of *Acanthostega*: instead of the five fingers we are accustomed to having, this early tetrapod possessed eight! Given the predominantly aquatic habits of *Acanthostega*, it is easy to imagine such a broad hand acting as the functional end of a paddle. In fact, given the probable rotary movements of the humerus, the relatively stiff elbow, and the wide hand of *Acanthostega*, an overall crank-like motion is inferred that would make this appendage an effective oar (Clack, 2002).

The pelvic girdle of *Acanthostega* is less robust than its pectoral counterpart (Fig. 10.4). However, it is still larger and more robust than anything found in fishlike sarcopterygians or transitional forms like *Tiktaalik*. As with other sarcopterygians, the pelvic girdle of *Acanthostega* is composed of an upper, skinny ilium bone and two lower bones, the pubis and ischium. In tetrapods, the ilium connects directly to the vertebral column through one or more ribs off the sacrum. But in *Acanthostega*, neither the ilium nor the ribs of the sacrum are very well developed. Given the presence of one slightly larger sacral rib and some pitting and scarring on the ilium, it is likely that the pelvic girdle was tied into the vertebral column through ligaments (Clack, 2002). This feature alone suggests that whereas some support of the body on land may have been possible for brief periods of time, *Acanthostega* was more at home in the water. The ischium is short but stout and angled down and back, whereas the pubis is both long and broad. In fact, viewed from the below, the two pubic bones form a broad, sled-like platform. All three pelvic bones form the acetabulum, which points almost directly outward, orienting the hind limb laterally from the pelvis.

The sizes and shapes of the pelvic bones are significant in relation to the types of muscles that would have been anchored to them. In tetrapods, the ilium is the anchor for the gluteal muscles, which help with abduction and rotation of the femur, and some of the muscles that extend the knee originate here as well (Romer, 1962; Liem et al., 2001). The ischium and pubis are the anchoring points of the femoral adductors (Romer, 1962; Liem et al., 2001). Given the small size of the ilium and the large size of the ischium and (especially) the pubis, we infer that muscles of femoral adduction were better developed compared with those of rotation in *Acanthostega*. If you think about a scissors kick in swimming, you begin to appreciate why large femoral adductors would be beneficial to a mostly aquatic tetrapod like *Acanthostega*.

The femur is a long, thin bone with its distal end a bit more expanded than its proximal end (Fig. 10.4). A long, roughened ridge crosses much of the inside of the femoral shaft. In most tetrapods, including humans, a ridge, scar, or (in our case) roughened line demarcates the attachment point of the adductors on the femur (Romer, 1962; Liem et al., 2001). We infer that this ridge on the femur of *Acanthostega* is associated with such adductor muscles, and its presence, along with the expanded pubis, suggests pulling the femur toward the body wall was an important movement in this tetrapod (Clack, 2002). As with the radius and ulna, the tibia and fibula are stout, flat bones that would not have allowed much independent rotation but whose flattened profile would have emphasized an overall paddle-like shape. As with the hand, multiple toes (seven) adorn the foot, and given the mostly aquatic lifestyle indicated by the skeleton as a whole, such a broad foot would again serve as a good flipper paddle (Fig. 10.4) (Clack, 2002).

A more robust, more land-worthy chassis is present in another early tetrapod, *Ichthyostega* (Fig. 10.6). This early tetrapod is not much larger than *Acanthostega* in overall body plan (approximately 1.3 meters long), but there are some telling differences between the chassis structure of these two animals. Like *Acanthostega*, *Ichthyostega* has a large, flattened skull with sculptured pits and expanded spiracular notches. Lateral line canals are present, but these are not as superficial as in *Acanthostega* and

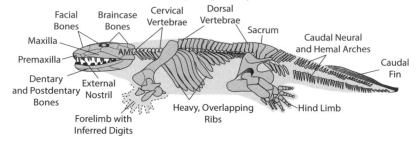

The Early Tetrapod *Ichthyostega* Body Profile in Side View

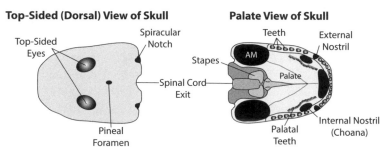

Top-Sided (Dorsal) View of Skull

Palate View of Skull

are 'buried' beneath the pockmarked skull bones. The teeth of *Ichthyostega* are much larger than those of *Acanthostega* and look like sharply pointed bony daggers: this animal would have flashed a terrifying smile. Intriguingly, there is no evidence of any gill-bearing skeletal elements in *Ichthyostega* (Fig. 10.6). If this absence is indeed biological and not the result of poor preservation, it suggests that *Ichthyostega* relied much more on lung-based respiration than did *Acanthostega*, a hypothesis tentatively supported, as we will soon see, by the shape of its ribs. The otic region of *Ichthyostega* contains a bulky but somewhat movable stapes that might have served to transmit sounds directly to inner ear (Clack et al., 2003). However, it seems most likely whatever sounds were detected came from the water rather than the air (Clack et al., 2003).

The vertebral column is less uniform and more robust than that of *Acanthostega*. In fact, along the vertebral column the neural spines of numerous vertebrae vary in their size and shape (Fig. 10.6). We see a similar pattern of neural spine variation in land-living mammals where the vertebral column can flex and extend in a vertical (dorsoventral) plane. Such movements in mammals increase the length of their stride (more on this in future chapters). This similarity in neural spine variation might indicate that *Ichthyostega* had more subtle control over the flexion and extension of its back. There has even been the suggestion that, like a strange inchworm, *Ichthyostega* could draw its pelvis toward its pectoral girdle, moving it across land from one water source to another (Ahlberg, Clack, and Blom, 2005). However, it remains to be seen if these movements were indeed possible. More to the point, if the shapes of the ribs are significant, it looks as if *Ichthyostega* would have had a relatively stiffened body core. The ribs are thick, massive, and overlap each other (Fig. 10.6). Unlike in *Acanthostega*, there are processes on the ribs that could have served as attachment sites for breathing muscles (although this is disputed) but that definitely would have limited lateral and vertical movements of the body core. Given the absence of gill-bearing elements in *Ichthyostega*, having such well-developed ribs would be advantageous for supporting the body and sucking air into the lungs. If we consider our hypothesis in chapter 9 for what we would find in the vertebral column and ribs of a tetrapod, *Ichthyostega* is certainly showing early trends toward strengthening these regions against the force of gravity. However, even this animal still possesses a two-part centrum (composed of an intercentrum and pleurocentrum for each vertebral segment). Moving caudally, the tail of *Ichthyostega* is narrower than that of *Acanthostega*, and the fin rays are smaller. This tail shape suggests that *Ichthyostega* could not waggle its tail fin as effectively as *Acanthostega* for swimming.

The pectoral and pelvic girdles and limbs are similar in many ways to those of *Acanthostega*, and so we will only briefly highlight what is different about these elements in *Ichthyostega*. In general, the girdles and limb elements are more robust than in *Acanthostega* (Fig. 10.6) (Pierce, Clack, and Hutchinson, 2012). The humerus is a bit larger than that of *Acanthostega*, but otherwise has a similar L-like shape, retains dorsal and ventral shaft ridges, and sports a large, outer epicondyle. As we have discussed for *Acanthostega*, the ridges were likely associated with abductors and adductors that propped up the body, and such a prominent epicondyle probably served to anchor large wrist and finger extending muscles. The ulna of *Ichthyostega* has developed a prominent olecranon process. This process, which you identify as your 'elbow,' is an area for the attachment of the triceps muscle. The triceps in all quadrupedal tetrapods is an antigravity muscle: it extends the forearm against the ground, holding the body of the tetrapod up and off the mud. The olecranon process acts like a lever handle for the triceps to pull against, and so the size and shape of this feature indicates in a general way how crucial forearm extension is in various tetrapods (Fig. 10.6). Unfortunately, as of this writing, we know nothing about the shape of the hand or how many digits it possessed. Recent study of three-dimensional CT scans of an *Ichthyostega* specimen suggests that some elbow movement was possible (Pierce, Clack, and Hutchinson, 2012).

The ilium of the pelvic girdle is still somewhat Y-shaped in *Ichthyostega* as it is in *Acanthostega*, but less so, with a broader surface for limb muscle attachments (Fig. 10.6). There appears to have been space for a prominent sacral rib, which has not been found, to attach the pelvis directly into the vertebral column. The pubis and ischium are absolutely huge compared to *Acanthostega*, and this suggests that *Ichthyostega* had very powerful hind limb adductors. The femur has a prominent adductor ridge, and like in *Acanthostega* this would have been the attachment point for muscles that caused a scissors-kick hind limb stroke. The tibia and fibula are large and flat bones, and the knee joint consists of relatively flattened surfaces that would have limited the extension and flexion of the knee. Several block-like ankle bones are present, and seven toes adorn the foot. The first three toes are essentially duplicates of the big toe, and the remaining four toes have the typical digit identities.

Overall, the chassis of *Ichthyostega* suggests that while this animal probably did not venture far from the water, it was much more capable of traversing land than *Acanthostega*. In particular, its robust ribs and limbs suggest that *Ichthyostega* perhaps had regular sojourns across land. It seems like an animal capable of using its lungs as a primary respiratory organ, and its large ribcage and sturdy limbs would have kept its body weight from crushing these organs while on land. Some paleontologists have envisioned the locomotion of *Ichthyostega* as somewhat akin to the vertically undulating movements of a seal (Ahlberg et al., 2005). One can imagine *Ichthyostega* using its robust forelimbs to pull itself forward and to prop up its torso while its vertically flexing vertebral column could draw the hind limbs into a mechanically advantageous position to scoot the body forward. We also know from recent three-dimensional modeling efforts of the *Ichthyostega* skeleton that whereas the limbs could move and aid propulsion on land, they could not have worked to turn the hands and feet forward as in modern tetrapods (Pierce, Clack, and Hutchinson, 2012). Whether or not this locomotor scenario turns out to be correct, it is clear that *Ichthyostega* was a much more terrestrial animal than was *Acanthostega*.

What is fascinating about the anatomical differences between *Acanthostega* and *Ichthyostega* is that they are consistent with the sedimentary rocks in which these tetrapods are found. *Acanthostega* fossils are found well preserved and usually well articulated in sandstones and siltstones indicative of flowing river systems, and there are many fish scales and fish carcasses that support this environmental interpretation (Clack, 2002). How can we be confident that these fossils were not simply washed in from somewhere else and that *Acanthostega* actually lived in these environments? The answer is actually quite simple: skeletons that travel far and tumble down rivers usually do not remain articulated to arrive at their destination in one piece. Moreover, the waggling body, presence of gills, prominent fins, and paddle-like limbs all make sense in the context of a river-living animal (Clack, 2009a, 2009b). In contrast, fossils of *Ichthyostega* tend to be disarticulated and less well preserved, suggesting they were transported greater distances from where they lived. These fossils are found in sedimentary rocks deposited as sand lenses (that is, clusters of sandstone sandwiched in between other rocks) inserted into other sandstones with no tetrapods but many fossil fishes (Clack, 2002). These data suggest that *Ichthyostega* was an animal living in upland areas away from where it was deposited, and the sand lenses indicate flooding events. In other words, *Ichthyostega* may have been living and dying in drier regions, and their carcasses were later swept away by flood waters to be bounced and disarticulated down river channels into low-lying ponds, lakes, or inland seas.

After the Devonian, there was a major change in the environment and climate, a time of the great coal forests known as the Carboniferous (354–299 Ma; Cards 93–94). During this time, the continents were beginning to merge into the Supercontinent Pangea (DiMichelle and Hook, 1992; Redfern, 2003; Clack, 2007). A huge swath of land in this swarming mass of continents sat right along the equator. This region of the world is now broken up into North America, Europe, and Asia, and in all these places we find an important mineral resource: coal. The coal is evidence of the huge stretch of tropical and subtropical forests that blanketed these areas of the world during the Carboniferous. In this belt of hot and humid weather, the first major rainforests thrived (DiMichelle and Hook, 1992).

At a glance, these places would be reminiscent of regions of the Amazon or Congo today, but a closer inspection would reveal strange plants and a somewhat different environment. During the Carboniferous, there were none of the flowering plant species that make up the majority of tropical flora today. Instead, there were humongous club moss 'trees' that stood over 30 meters tall, horsetails nearly 15 meters high, and a lower understory of ferns and fern trees (Carroll, 1988, 2009; DiMichelle and Hook, 1992). When modern trees are cut down, we expect to see tree rings that mark the years and seasons of a plant's growth. In contrast, cross-sections of many Carboniferous plants show no tree rings at all (DiMichelle and Hook, 1992). This means that the warm, tropical conditions were so ideal for plant growth that they literally just kept on growing, not having to slow for cold seasons or times of dryness. As plants within these forests died, their leafy remains fell into fetid swamp waters made anoxic by the massive accumulation of so much dead plant material. Unlike in modern rainforests, many of these dead plants were never completely broken down (the modern fungi that do this job were yet to evolve), but were instead 'preserved' in the oxygen-poor swamp waters into which they fell (DiMichelle and Hook, 1992). Over time, the plant remains were first converted to peat, and after many millions of years of burial, the peat was converted into coal. It is strange to think about, but given that so much of the industrialized world runs on coal, much of today's power comes from the release of the Carboniferous sun's energy tied up in these ancient plants.

There was another environmental consequence of these massive coal forests. Because so much carbon was tied up in living and dead plant material at the time, and because there were so many living plants pumping oxygen into the atmosphere, carbon dioxide levels dropped while oxygen levels rose. In fact, oxygen levels reached nearly 35% of the Carboniferous atmospheric composition (compared to the modern level of 21%) (DiMichelle and Hook, 1992). This higher level of oxygen had two important effects for the Carboniferous environment. First, it allowed many insects and other arthropods to become larger. Terrestrial arthropods breathe through pores on the sides of their bodies that transfer oxygen to a network of tubules situated directly around the major body organs. Imagine if instead of breathing through your nose and mouth to saturate your blood with oxygen, pores along your body allowed oxygen into the bronchi of your lungs, which then transferred this gas directly to your organs. This method of breathing works best at small body size because organ surface area is great compared with volume. Atmospheric oxygen content will therefore limit terrestrial arthropod body size, which partly explains why we don't see rhino-sized rhinoceros beetles. However, during the Carboniferous with atmospheric levels of oxygen approaching 35%, there were dragonflies with wingspans nearly 60 cm across, and in the undergrowth crawled millipedes nearly 3 meters long (DiMichelle and Hook, 1992)!

The second consequence of sequestering vast amounts of carbon and pumping out large quantities of oxygen was the increased frequency of forest fires. Firefighters, fire scientists, and any campers worth their salt all know about the fire triangle, a simple diagram showing the three ingredients for starting a fire: fuel, oxygen, and heat. The Carboniferous rainforests were a fire triangle writ globally waiting to be completed: their biomass was the fuel, atmospheric oxygen was high, and heat could conveniently arrive from a lightning strike. Not surprisingly, the presence of fusain, the technical name for fossilized charcoal, becomes common at the opening of the Carboniferous (DiMichelle and Hook, 1992; Benton, 2005). Just as modern forest fires have a profound effect on their animal and plant communities today, so too the frequent burnings in the Carboniferous must have played an important role in vertebrate evolution. It is under these environmental conditions that the early tetrapods began to evolve and diversify. An abundance of arthropod prey, an increase in atmospheric oxygen, and warm, humid weather must have been part of a series of interrelated factors that spurred on this evolutionary jump.

Stem Tetrapods

Pedigree Animals Close to the Common Ancestor of Amphibians and Amniotes

Date of First Appearance ~350 Ma (Card 93)

Specialties of Skeletal Chassis Tetrapods with Five or Fewer Toes That Are Not Amphibians or Amniotes

Eco-niche Small to Medium-Sized Terrestrial to Aquatic Insectivores and Carnivores

We now consider a group of animals often referred to as 'stem tetrapods' because they represent the populations among which the common ancestors of the modern amphibians and amniotes arose or 'stemmed.' These animals were neither amphibians nor amniotes, but instead were members of the original populations to give rise to both of these later tetrapod groups. Some of the stem tetrapods returned to watery haunts, whereas others were more terrestrial in their habits (Clack, 2002; Benton, 2005; Carroll, 2009). Although the aquatic stem tetrapods are intriguing animals in their own right, for our purposes we will focus on one of those animals that forged its way further toward land: *Pederpes*.

Pederpes is known from a single, well-preserved fossil discovered from the earliest portion of Carboniferous time (~350 Ma; Card 93) (Clack and Finney, 2005) (Fig. 10.7). This animal is important to our story both because it is the earliest known tetrapod from the Carboniferous and because its skeletal chassis is a blend of basal and derived tetrapod elements. Discovered in Scotland, *Pederpes* was almost 60 cm in length from its head to its pelvis (most of the tail is missing) (Fig. 10.7). Unlike

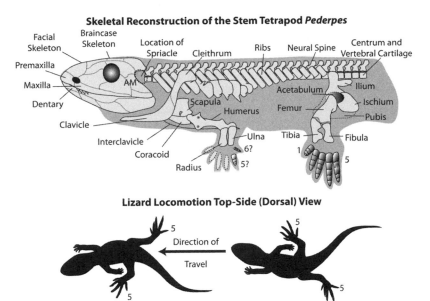

Skeletal Reconstruction of the Stem Tetrapod *Pederpes*

Facial Skeleton
Braincase Skeleton
Location of Spiracle
Cleithrum
Ribs
Neural Spine
Centrum and Vertebral Cartilage
Premaxilla
Maxilla
AM
Dentary
Acetabulum
Ilium
Scapula
Ischium
Humerus
Femur
Pubis
Clavicle
Interclavicle
Ulna
Tibia
Fibula
Coracoid
6?
1
Radius
5?
5

Lizard Locomotion Top-Side (Dorsal) View

5
5
Direction of Travel
5
5

10.7. Skeletal reconstruction of the stem tetrapod *Pederpes* (based on Clack and Finney [2005]) from the Early Carboniferous. Note that the skull of *Pederpes* is tall and more like the predicted shape for a tetrapod head than in animals like *Acanthostega* and *Ichthyostega*. The hind foot of *Pederpes* is significant because it has begun to resemble the asymmetrical digit pattern present in modern sprawling tetrapods, a pattern known to be associated with a more pronated foot. The inset shows a silhouette of a walking lizard in dorsal view. Notice how the asymmetrical shape of the digits on the foot plays a role in helping the foot to push the lizard forward even when the foot becomes directed sideways at the end of each step.

in *Acanthostega*, the skull of *Pederpes* is tall dorsoventrally, giving the head a hatchet-like appearance in side view (Clack and Finney, 2005). The snout is long, the braincase region is short, with tall squamosal and jugal bones, and the orbits are large and laterally facing. As you will recall from chapter 9, these are the hallmarks we predicted for tetrapods: the long snout would act as scissors blades for quickly snapping up prey, whereas the expanded squamosal and cheek region must have housed powerful adductor mandibulae and pterygoid muscles. The premaxilla, maxilla, and dentary bones were lined with large, recurved teeth as in many tetrapods, but the roof of the mouth still contained many smaller teeth and fang-like projections, as in many fishlike sarcopterygians. As with transitional and basal tetrapods, the skull of *Pederpes* is pitted and sculpted, and some lateral line canals are present within the skull bones. The hyomandibula (now called the stapes) is beautifully preserved in *Pederpes*. This bone was likely not involved in detecting airborne sounds because, like in *Acanthostega*, it is intimately connected with the spiracle and other skull bones, and probably continued to play a role in respiration and/or jaw support (Clack and Finney, 2005).

The vertebral column is reminiscent of *Acanthostega*, but the zygapophyses are better developed (Clack and Finney, 2005) (Fig. 10.7). The neural spines and vertebral bodies are not fused together, and irregular pitting and scarring on the vertebral elements indicates that a fair amount of cartilage was present. As with *Acanthostega*, the vertebral column is not very differentiated, but the ribs of *Pederpes* are robust and expanded. This may indicate that this animal was using its ribcage to actively inhale air, and it certainly points to an animal that could support its body out of water (Clack and Finney, 2005).

The limbs and girdles of *Pederpes* are much closer to the condition predicted for tetrapods in chapter 9. In the forelimb, the scapula has become a large, robust element, although the clavicle and cleithrum are still large and form most of the shoulder 'blade.' Curiously, the coracoid bone does not appear to have ossified in *Pederpes*, but instead formed a cartilaginous link between the shoulder and the chest (Clack and Finney, 2005). The humerus resembles that of *Acanthostega* in being L-shaped and having prominent muscle scars, but the scars are more distinct, and a deltopectoral crest is present as in modern tetrapods. The implication of this change is that the limbs were being used in a greater range of activities that required specialized subdivisions of the muscles during locomotion rather than primarily as paddles or struts for elevating the body off a river bottom (Clack and Finney, 2005; Clack, 2009b). As with *Acanthostega*, at the elbow the joints of the humerus, radius, and ulna were rather flat and underdeveloped, suggesting that extensive flexion at the elbow was probably limited. Only two digits of the hand are preserved in *Pederpes*; it is not entirely clear where these bones belong, and so it is unknown if *Pederpes* actually had more than five fingers (Clack and Finney, 2005).

The pelvic girdle possesses a Y-shaped ilium whose scarring suggests an attachment to the sacrum through one or more sacral ribs, although none are known for *Pederpes* (Clack and Finney, 2005) (Fig. 10.7). A small but broad ischium is present, but there is no trace of the pubis. Either the pubis was disarticulated and transported away from the carcass before its preservation, or, as in some modern tetrapods, the pubis may have been entirely cartilaginous.

The femur of *Pederpes* is a more robust element than that observed in *Acanthostega*, and has broad proximal and distal ends (Clack and Finney, 2005) (Fig. 10.7). Presumably, these expanded joint surfaces would allow the femur to assume a number of postures and to be rotated through a large range of motion. As with the humerus, the muscle insertion scars on the femur are more distinct and less continuous. These features again suggest that the division of labor among the limb muscles was becoming segregated by function, a feature typical of modern tetrapods. At the distal end of the femur, we see for the first time the wrapping of joint surfaces onto the ventral surface of the knee (Clack and Finney, 2005). This change in joint orientation is significant in that it would allow the knee to bend in such a way that the tibia and fibula would be directed more vertically instead of mostly outward as in basal tetrapods. This condition is similar to that observed in tetrapods like salamanders, and such a posture helps to direct the foot cranially. The tibia and fibula are both shorter than the femur, and relatively robust bones as well. No ankle bones are preserved in the only specimen of *Pederpes*, but given its well-articulated condition, this suggests that these elements remained cartilaginous into adulthood.

The hind foot is remarkable in that it is composed of five sturdy toes that are arranged in an asymmetrical pattern (Clack and Finney, 2005). This is significant in two respects. First, *Pederpes* may well be the first example of the five-toes-or-fewer condition observed in the radiations of all modern tetrapods (Fig. 10.7). Given the excellent preservation of the specimen, it is unlikely that additional digits were present but not preserved. Second, the asymmetrical form of the hind foot is reminiscent of that observed today in many tetrapods, especially those that walk with a sprawled gait. In lizards, for example, the toes become longer from digit 1 out to digit 4, and then digit 5 becomes a bit shorter again. When lizards walk, their hind feet rotate laterally, and the asymmetrical form of the hind foot helps them press effectively against the ground to push them forward (Schaeffer, 1941; Rewcastle, 1983) (Fig. 10.7). The five-toed, asymmetrical hind foot of *Pederpes* strongly suggests that it was used in a similar way, and this would further support the hypothesis that this animal was more terrestrial in its habits than *Acanthostega*.

What is perhaps most intriguing about *Pederpes* is that it is a rare example of a tetrapod becoming more terrestrial. In fact, during the early Carboniferous (~350–318 Ma; Card 93), several radiations of stem tetrapods appeared that mostly retained aquatic habits and life histories (Clack, 2002). Some of these stem tetrapods took on crocodile-like body forms, as with *Tiktaalik* before them, with short limbs and flat heads, whereas others greatly reduced their appendages and swam in shallow waters like amphibious snakes (Clack, 2002; Benton, 2005). During this time, only the close North American relative of *Pederpes*, an animal named *Whatcheeria*, is known to have followed a more terrestrial pattern of development (Lombard and Bolt, 1995). Although we will not detail the anatomy of these forms here, this diversity of mostly aquatically adapted stem tetrapods shows again that the evolution of the tetrapod chassis was not directly linked to moving away from the water.

The Tetrapod Chassis: Where We've Been, Where We're Going

All current fossil evidence strongly suggests that the tetrapod chassis developed in the water and only later was adapted to a landlubbing existence (see Fig. 10.8 for relationships and basic evolutionary pattern). Transitional tetrapodomorphs such as *Tiktaalik* show that, early on, there was selective pressure to develop a chassis with a flattened profile, especially along the skull, and to lose the dorsal and anal fins. The forelimbs were modified into structures that could hold the body against strong water currents or prop it above the waves to glimpse potential arthropod prey on river banks or delta muds. The roles of the hyomandibula and spiracle appear to have been modified to perform respiratory functions, perhaps helping to draw in air through the top of the head. In basal tetrapods such as *Acanthostega* and *Ichthyostega*, the vertebral column and ribcage become more prominent, and the limb elements become more robust and ended in distinct digits. Such limbs were likely useful as paddles, especially in *Acanthostega*, but must have been essential in supporting the body against gravity during forays out of the water. Finally, stem tetrapods like *Pederpes* show us that as the Carboniferous opened, most of these animals were still tied to water, but some were developing features in their limb skeletons, for example, that may have given them an advantage in moving about on land when necessary.

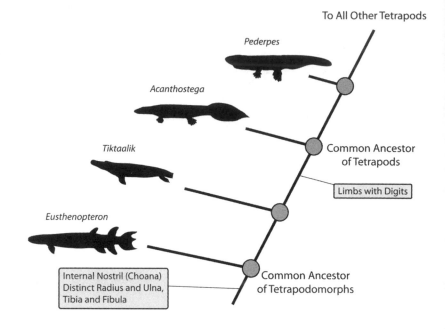

10.8. Pedigree of relationships among the transitional tetrapodomorphs and early tetrapods. See text for details.

To All Other Tetrapods

Pederpes

Acanthostega

Tiktaalik

Common Ancestor
of Tetrapods

Limbs with Digits

Eusthenopteron

Internal Nostril (Choana)
Distinct Radius and Ulna,
Tibia and Fibula

Common Ancestor
of Tetrapodomorphs

Not everything we predicted in chapter 9 regarding what should be in the earliest tetrapods is supported by the fossils. In particular, the inner ear and hyomandibula still functioned more or less like those of fish, lateral line canals were still present, and in some forms the vertebral column and ribcage remained underdeveloped. This transitional series of tetrapod chassis shows that the anatomical developments we take for granted in the two major groups of modern tetrapods (Amphibia and Amniota) were acquired piecemeal over many generations.

It is now time to retire our ship, the *Devonian Destiny,* and climb ashore. Soon after the initial radiation of the stem tetrapods, some 330 Ma (Card 93), a momentous split in the tetrapod family tree occurred. It is at about this time that the common ancestors of the modern amphibians (Amphibia) and amniotes (Reptilia and Synapsida) appeared. The Amphibia would go on to become specialized tetrapods that could breathe through their moist skins and pass through a fishlike larval stage before changing (metamorphosing) into a limbed adult. The Amniota would take a different tack, evolving the ability to lay shelled eggs on land containing embryos that could develop their own life-support membranes. By the time the Carboniferous came to a close, the two lines of tetrapods were distinct and separate from one another, and would go on to populate the terrestrial world in their own unique ways. Initially, it was the Amphibia that were flush with success as terrestrial vertebrates. In chapter 11, we hop back to the ponds, rivers, lakes, and streams where the Amphibia got their start.

11.1. Common characteristics of lissamphibians.

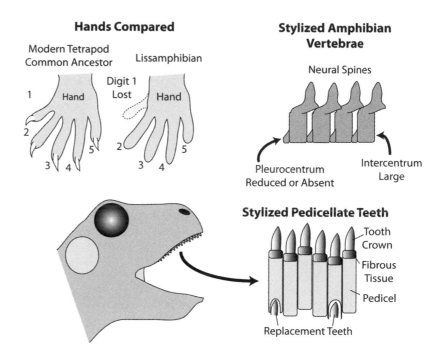

Hands Compared

Modern Tetrapod Common Ancestor

Lissamphibian

Digit 1 Lost

1
Hand
2
5
3   4

Hand
2   5
3   4

Stylized Amphibian Vertebrae

Neural Spines

Pleurocentrum Reduced or Absent

Intercentrum Large

Stylized Pedicellate Teeth

Tooth Crown
Fibrous Tissue
Pedicel

Replacement Teeth

The Amphibian Chassis

11

ONE OF THE KEY COMPONENTS OF ANY VEHICLE IS ITS HORN. Whether you are cut off in traffic or alerting other drivers to your intentions, a press of the steering column produces a loud beep or honk. When the horn button is pressed, it causes an electromagnet to draw a cone back and forth in rapid succession, producing vibrations that generate the distinctive sound most of us are familiar with on the highway (Newton, 1999; Bloomfield, 2006). But horns are effective only because we can hear them: if they were in a register too low or too high to detect, we would never notice. By the same token, on any given spring night the air comes alive with a variety of animal sounds. Frogs and toads form some of the key members of this evening orchestra, and the energy they expend in this chorus of peeps, croaks, and rumbles pays off because it can be heard by other members of the same species. The ancestors of frogs and toads, as we will see, were among the first terrestrial vertebrates to hear high-frequency vibrations in air. Given that they appear early during the age of coal forests, those ancient landscapes were probably the first whose air reverberated with both insect wings and vertebrate songs.

The life cycles of modern amphibians have always inspired curiosity and fueled our collective imaginations about how tetrapods ventured landward while still tethered firmly to the water. Here are animals, such as frogs, which start life as aquatic, gill-breathing tadpoles, only to develop strong legs and lungs that enable them to hop away from their ancestral ponds. Thus, it is not surprising that early paleontologists and the general public alike have imagined amphibians as a way-station on the path to 'true' landlubbing tetrapods like lizards. However, this scenario gives the false impression that modern amphibians are living intermediates between fishlike vertebrates and so-called reptiles. Instead, modern amphibians are very derived and specialized animals whose anatomy could not simply be a step away from a lizard. In fact, looked at from the perspective of the vertebrate family tree, modern amphibians share only a distant common ancestor with the great reptile and mammal lines. Thus, whereas modern amphibians can enlighten us about certain key innovations of the vertebrate chassis that appeared during the rise of the coal forests, to paraphrase the old Frank Sinatra song, amphibians did it their way.

The amphibian portion of the vertebrate family tree has been difficult to unravel for a variety of interrelated reasons that go well beyond the scope of this book, and the origins of modern amphibians are contentious at best among the paleontologists who study these issues. Some paleontologists view the modern amphibians as descendants of a single common ancestor nested among fossil animals called the temnospondyls (a name essentially meaning 'cut vertebrae,' referring to the presence of a two-piece vertebral body) (Warren and Yates, 2000; Ruta et al., 2007). Others argue that modern amphibians arose from different ancestors among the temnospondyls and another tetrapod group dubbed the lepospondyls, which we will return to in the next chapter (e.g., Carroll, 2009). This vigorous debate can make the evolutionary history of the amphibian chassis difficult to appreciate for the nonspecialist. For these reasons, and because our focus is on the major changes in the skeletal chassis of various vertebrate groups, I simplify what is a much more complex evolutionary history. For our purposes, we will assume that all modern amphibians are descendants from a single common ancestor among the temnospondyls. In turn, we will use the term 'Amphibia' to mean the temnospondyls plus modern amphibians, a group called the Lissamphibia. As in previous chapters, if we want to examine the fossil record and understand the origins of amphibians, we need a search image based on the chassis of the living forms.

Amphibia: Tetrapods Living a Double Life

Pedigree Tetrapods with Four Fingers Capable of Metamorphosis

Date of First Appearance 340 Ma (Card 93)

Specialties of the Skeletal Chassis Four Digits on the Hand and Short Ribs

Eco-niche Terrestrial or Semiaquatic to Aquatic Predators and Omnivores

Lissamphibia, the modern amphibian group, is a fairly diverse collection of seemingly unrelated creatures: some hop, some walk, some swim, and some are legless (Plates 10–12). Yet, much like comparing what you find lifting the hood on a compact, sedan, and SUV designed by the same manufacturer, evidence of the common chassis of lissamphibians can be revealed from direct observations of their skeletons and soft anatomy. Skeletally, an easily definable characteristic of lissamphibians is observed in their hands. In tetrapods, the digits are numbered from 1 to 5. The primitive members of the modern tetrapod lineages (excluding early tetrapods such as *Acanthostega*) typically had five digits on their hands and feet, with digit 1 being the thumb or big toe and digit 5 being the 'pinky' finger or toe. In lissamphibians, we see the loss of digit 1 in the hand so that only digits 2–5 remain (Fig. 11.1). The functional explanation for this remains unclear, but its presence in all lissamphibians indicates it was present in their common ancestor (Carroll, 1988, 2009; Benton, 2005). Another characteristic of modern lissamphibians is that the vertebral bodies (centra) have come to be dominated by the portion called the intercentrum, whereas the remaining portion, the pleurocentrum, has either fused with the former bone or disappeared entirely (Fig. 11.1) (Benton, 2005; Carroll, 2009). Functionally, this appears to be a way to strengthen the vertebral column against gravity by eliminating additional movements within each vertebra. Lissamphibians are also characterized by their pedicellate teeth. In such teeth, the crown of the tooth is separated from the root by fibrous tissue (Fig. 11.1) (Pough, Janis, and Heiser, 2002; Benton, 2005). The root of the tooth itself is attached to the jaw bones superficially by a type of bony cement. When lissamphibian teeth break or fall out along the fibrous tissue line, they can be replaced through the growth

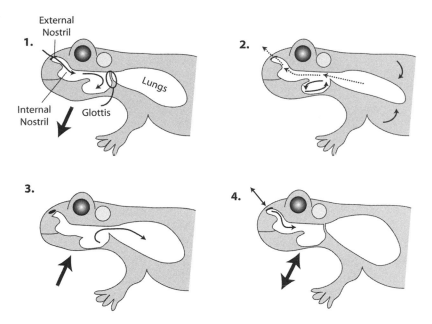

11.2. Stylized cycle of breathing in a lissamphibian such as a frog. (1) At the start of the cycle, fresh air is inhaled through the nose into the mouth, with suction being applied by dropping the throat (large arrow). (2) Next, the fresh air is held in the mouth while stale air from the lungs (released with the opening of the glottis) escapes through the nose. (3) Fresh air held in the mouth is then swallowed into the lungs. (4) Finally, the glottis closes and there is a period of apnea while the oxygen within the lungs is extracted. Before the next cycle, there is a period of buccal pumping (large double-headed arrow) to refresh the air in the mouth. Diagrams after Liem et al. (2001).

of a new tooth within the pedicel (Fig. 11.1) (Liem et al., 2001; Carroll, 2009). It is not clear what functional value, if any, pedicellate teeth have over the socketed teeth we will later see in other tetrapods.

Two other features that go together in lissamphibians are reduced ribs and the possession of a smooth, glandular skin that either sports tiny scales or lacks scales altogether (Pough, Janis, and Heiser, 2002; Benton, 2005). The reduction in rib size is significant in relation to respiration. Many tetrapods employ movements of the ribs to expand and constrict the chest cavity to ventilate their lungs. Breathing in lissamphibians does not rely on rib movements, but instead is accomplished by swallowing air. The glottis is the muscular valve that opens and closes the trachea, the so-called wind pipe that transports air to and from the lungs. A lissamphibian like a frog sucks fresh air into its closed mouth through its external nostrils, while simultaneously its glottis opens and releases the stale air in its lungs (Fig. 11.2). Because the stale air is warmer than the incoming cooler fresh air, it rises into the choanae and is pushed out the external nostrils, whereas the fresh air can now flow into the trachea and to the lungs. The glottis then closes, holding the fresh air in the lungs for a period of time to allow enough oxygen to be dissolved in the blood. Then the pattern repeats (Liem et al., 2001) (Fig. 11.2). This breathing repertoire relies on pumping actions of the hyoid bone. As discussed in chapter 9, oblique, transverse, and longitudinal abdominal muscles are present in lissamphibians, which further enhances exhalation. Given their reduced or absent ribs, there are no rib-associated muscles. Overall, whereas the breathing mechanism in lissamphibians would be terribly inefficient in vertebrates such as humans, it is successful for lissamphibians given their small body size and low metabolic rate.

11.3. The generalized mechanics of the lissamphibian eye. As with all tetrapods, the cornea causes most of the refraction of the light into the eye, but the lens is more fishlike and can be drawn forward with a special protractor muscle to focus on objects near the animal.

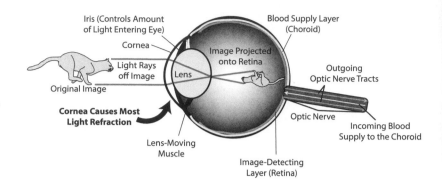

Iris (Controls Amount of Light Entering Eye)

Cornea

Light Rays off Image

Lens

Original Image

Cornea Causes Most Light Refraction

Image Projected onto Retina

Blood Supply Layer (Choroid)

Outgoing Optic Nerve Tracts

Optic Nerve

Incoming Blood Supply to the Choroid

Lens-Moving Muscle

Image-Detecting Layer (Retina)

Returning to the smooth skin of lissamphibians: one of the major roles of this organ is respiration (Pough et al., 1998). In fact, in some salamanders, lungs are nonexistent and the skin is *the* respiratory organ. Therefore, large, mobile ribs for ventilating the lungs are not necessary given the importance of the skin for oxygen extraction and carbon dioxide excretion. A vast supply of blood vessels invests lissamphibian skin, and oxygen picked up from the surrounding environment is passed on to the right side of the heart, where it mixes with oxygen-poor blood returning from the brain, head, and body (Pough et al., 1998). Due to their relatively small size and low metabolic rate, this mixing of oxygen-rich and oxygen-poor blood does not seem to affect lissamphibians' ability to thrive. Many are, in fact, fairly active animals capable of rapid locomotion and, in the case of frogs and toads, hopping and jumping. The smooth skin serves numerous functions in addition to respiration, including mucus production to protect the skin from drying out and the generation of noxious substances to deter predators (Pough et al., 1998; Pough, Janis, and Heiser, 2002; Liem et al., 2001). It should be noted that some lissamphibian species develop small dermal scales and bones called scutes within their skin (Pough et al., 1998). However, this does not appear to interfere with their skin's gas-exchanging abilities.

Vision is fairly well developed in lissamphibians. The eyes of these animals remain in a dorsally facing position on their typically flat and broad skulls, and many lissamphibians see across a range of colors. The eyes of lissamphibians are somewhat of a mosaic of tetrapod- and fish-like characteristics (Fig. 11.3). As with tetrapods, most of the refraction and initial focus occurs at the cornea. However, the lenses are relatively large and accommodate close images somewhat as in a fish, in that they use a protractor muscle within the eye to pull the lens toward the cornea (Pough et al., 1998; Liem et al., 2001). One other intriguing feature all lissamphibians share is an eye-elevating muscle that causes the eyes to bulge out when the muscle is contracted, which probably serves to increase the size of the oral cavity when feeding or breathing (Pough et al., 1998; Levine, 2004). It works with the old eye-retractor muscle discussed in previous chapters during feeding, and we will return to this topic shortly.

Most lissamphibians are also characterized by their dual lives. Many species begin life as an aquatic larva with gills and other fishlike features,

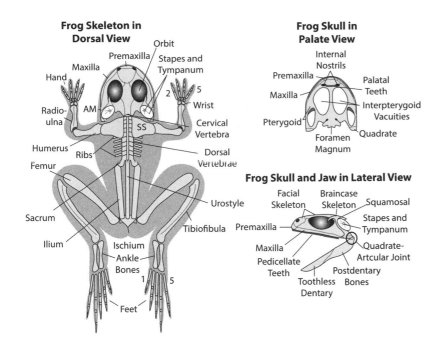

Frog Skeleton in Dorsal View

Orbit
Premaxilla
Maxilla
Hand
Stapes and Tympanum
Radio-ulna
AM
2
5
Wrist
SS
Cervical Vertebra
Humerus
Ribs
Femur
Dorsal Vertebrae
Urostyle
Sacrum
Ilium
Ischium
Ankle Bones
1
5
Feet

Frog Skull in Palate View

Internal Nostrils
Premaxilla
Maxilla
Palatal Teeth
Interpterygoid Vacuities
Pterygoid
Foramen Magnum
Quadrate

Frog Skull and Jaw in Lateral View

Facial Skeleton
Braincase Skeleton
Squamosal
Premaxilla
Stapes and Tympanum
Maxilla
Quadrate-Artcular Joint
Pedicellate Teeth
Postdentary Bones
Toothless Dentary

11.4. Generalized schematic of a frog skeleton and skull detail, based on the skeleton of a bullfrog (*Lithobates catesbeiana*). AM, adductor mandibulae muscle group region; SS, suprascapular cartilage.

only to undergo a process of metamorphosis where more familiar tetrapod features arise in relatively rapid succession, replacing or augmenting the original larval traits. Many frogs and toads show this classic mode of development, but other lissamphibians may retain larval features into adulthood, while still others will develop directly by hatching out of their eggs as miniature versions of the adult (Pough et al., 1998; Pough, Janis, and Heiser, 2002).

Perhaps the most familiar and diverse lissamphibians belong to the frog and toad lineage (hereafter referred to simply as frogs), and it is instructive to examine the chassis of these animals as a starting point (Plate 11). Compared to what we've already encountered with the early tetrapods and the transitional tetrapodomorphs, the skeletons of frogs appear extremely simplified (Fig. 11.4). Like early tetrapods, the skulls of most frogs are broad and flattened dorsoventrally. One of the features that you tend to notice almost immediately is the large eye sockets that gape into the skull, forming large bay-like openings that perforate through to the roof of the mouth (Liem et al., 2001; De Iuliis and Pulerà, 2011). In early tetrapods, the eyes tended to face sideways. In contrast, the eyes of frogs are situated much closer together – so much so that the images collected by each eye overlap those of the other to a significant extent (Fig. 11.5). Thus, most frogs have a good degree of binocular vision (Pough et al., 1998; Liem et al., 2001).

There is now a new generation of 3-D movies, but the illusion of depth is still accomplished the same way: by showing two overlapping sets of images filmed at slightly different angles. When these are viewed through polarizing 3-D glasses that separate the two images into single left and right pictures, your brain is 'tricked' into fusing these images into

11.5. Frog eating mechanics. Upper left: the overlap of the visual fields in frogs allows them to perceive depth to capture prey. Upper right: the old basicranial muscle, now an eyeball retractor, assists in swallowing prey (after Levine [2004]). Lower sequence: typical tongue projection in a frog from: (1) preparation, (2) projection, and (3) capture (after Pough, Janis, and Heiser [2002, 2013]).

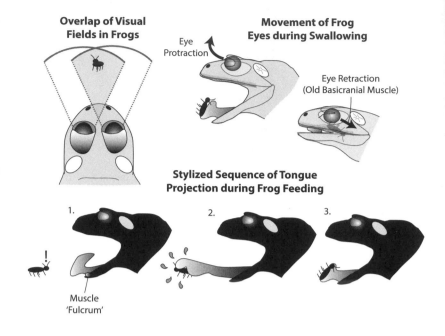

a single, three-dimensional projection. The basic rule with binocular or 3-D vision is that depth is judged based on the difference in the angle between two images. Objects that are far away will tend to look about the same in each eye, whereas objects that are closer to the eyes will show greater differences in their angle. You can demonstrate this to yourself by centering an object within a few centimeters of your eyes and then winking one eye and then the other shut back and forth in rapid succession. The object will appear to bounce left and right compared to the background. Your brain uses this information to compute how close or how far away something is from you. As such, binocular vision is beneficial to frogs because it helps them track and accurately target insects with their sticky tongues.

The large eye openings that pass into the palate are the interpterygoid vacuities, so called because they are situated between the pterygoid bones that anchor the deep pterygoid jaw-closing muscles (Fig. 11.4). Frogs blink by retracting their eyeballs into these openings through the eye sockets, using the old basicranial muscle (now the eye retractor) we first encountered in coelacanths (Levine, 2004). Intriguingly, eye retraction and prey swallowing are coordinated in frogs. Cineradiography (X-ray movies) of leopard frogs swallowing crickets show that the eye retractor pulls the eyeballs well into the oral cavity, pushing the eyes against the palate, which in turn pushes on the prey (Fig. 11.5) (Levine, 2004). In frogs where the nerves supplying the eye retractor muscles were severed, swallowing prey was more difficult and required almost 75% more swallowing movements to push their prey down their esophagus (Levine, 2004). The eye-elevating muscle unique to lissamphibians works with the eye retractor to enhance swallowing (Levine, 2004).

In general, the skulls of frogs are lightly constructed and airy because many of the skull and jaw bones have become reduced, lost, or fused. However, most bones now familiar to us are present. For example, the premaxilla and maxilla continue to be the tooth-bearing bones of the upper jaw. The palate also continues to house several small denticles, and choanae are present on either side of these palatal teeth. However, many frogs lack teeth on the dentary bone, and for a very good functional reason (Pough et al., 1998; Liem et al., 2001; De Iuliis and Pulerà, 2011). During the holidays when you are packing boxes for shipping gifts, you go through many rolls of packing tape. Usually, the clear tape comes on rolls from which it is pulled across a serrated metal strip—a quick jerk downward slices the tape off. Replace the tape with a frog's projectile tongue, and the serrated metal with teeth on the dentary, and it becomes clear why this would be, well, a 'selectively disadvantageous' situation.

In most frogs, the evolution of a projectile tongue serves as the replacement for relying on suction feeding in the adult. The tongue of a frog is shaped somewhat like a mushroom: the front portion of the tongue is broad (the cap) and its base, which is connected to the hyoid bones, is narrow (the stalk) (Fig. 11.5). When a frog focuses its keen binocular eyes on potential prey, a number of fascinating events happen in quick succession. First, the tongue muscle contracts and stiffens, and this has the effect of elevating the cap-like front of the tongue above the base. At the same time, a short muscle at the front of the mouth contracts, stiffens, and forms a fulcrum. Next, continued contraction of tongue muscles and forward movement of the hyoid apparatus pushes the stiffened tongue forward where its 'cap' flips over, out, and downward across the raised muscle fulcrum. This sticky 'cap' slaps and stuns prey and sticks it to the tongue, at which point hyoid muscles retract the tongue 'cap' back into the mouth, pulling the hapless prey into the jaws of defeat (Fig. 11.5) (Pough et al., 1998; Pough, Janis, and Heiser, 2002).

The squamosal bone of frogs is typically T-shaped and does not cover the adductor chamber. Therefore, the adductor mandibulae muscles can now expand out of the skull rather than being tucked inside, allowing them to contract uninhibited over a greater range, increasing the power with which the jaws shut. We will encounter a similar situation in the skulls of amniote vertebrates. One arm of the squamosal 'T' extends forward toward the eye socket as a flange, whereas the rearward-facing arm joins what is termed the 'annular cartilage,' which supports the large tympanic membrane of the ear (De Iuliis and Pulerà, 2011). A thin stapes bone spans the gap between the tympanic membrane and the otic capsule, and as described in chapter 9, high-frequency vibrations impinging on the ear drum are ultimately detected as sounds (Fig. 11.6) (Mason, 2006). Frogs make their high-frequency calls by passing air across vocal cords, causing the thin folds perched above their trachea to vibrate. Both sexes can make sounds, but only male frogs have vocal sacs that amplify their calls (Pough et al., 1998).

11.6. Schematics of hearing in high and low frequencies in modern frogs. For high frequencies, frogs rely on a thin stapes that transmits vibrations from a large tympanic membrane to the inner ear. For low frequencies, frogs utilize a modified levator scapulae muscle, the opercular muscle, to transmit vibrations from their limbs and body into the otic capsule. SS, suprascapular cartilage. After Pough et al. (1998).

Dorsal See-through View of Frog Head Showing Hearing Mechanisms

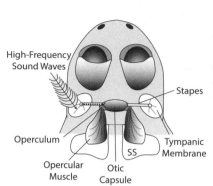

See-through View of Frog Showing Hearing Mechanisms

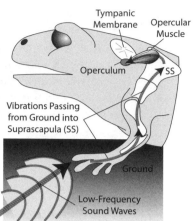

Unique to lissamphibians, a disc-like element called the operculum lies adjacent to where the stapes connects with the otic capsule (Fig. 11.6). The name for this bone is unfortunate: the operculum of lissamphibians is not homologous to the gill covering we saw in various fishlike vertebrates. The lissamphibian operculum connects with the stapes on the otic capsule, and a tensed and somewhat rigid muscle runs from this bone to the shoulder girdle (Fig. 11.6). The operculum muscle itself is actually a derivative of the shoulder-shrugging muscle (the levator scapulae). As low-frequency vibrations pass through the body and up through the forelimbs of frogs, the operculum muscle, much like a taut guitar string, passes these low-frequency vibrations from shoulder girdle to operculum to inner ear (Pough et al., 1998; Mason and Narins, 2002; Mason, 2006; Carroll, 2009). Like two sets of high-end studio microphones, one tuned to high frequencies and one to low, the stapes and operculum of frogs provide these animals with the ability to hear in two registers. The distinct separation in frequency detection observed in frogs probably has its evolutionary roots in the vocal calling these animals do. Because most adult frogs lack a lateral line system, the operculum has become a substitute low-frequency detector.

Before continuing on to the body, let us briefly peer into the braincase of a typical frog. Lissamphibians remain one of the least-studied groups of vertebrates for cognitive research, and as a consequence we know little about what is actually going on inside their heads. However, we can say a few generalized things about brain function and adaptation in these vertebrates. The brain that we find in these animals is similar in many respects to those of teleost actinopterygian fish (see chapter 7), although lissamphibian brains are simplified in comparison (Fig. 11.7) (Butler and Hodos, 2005). As in actinopterygians, the cerebrum is relatively small but still handles input from the olfactory lobes and makes the frog aware of the world via the integration of stimuli through complex somatosensory tracts. We know that frogs are aware of their environments and that they

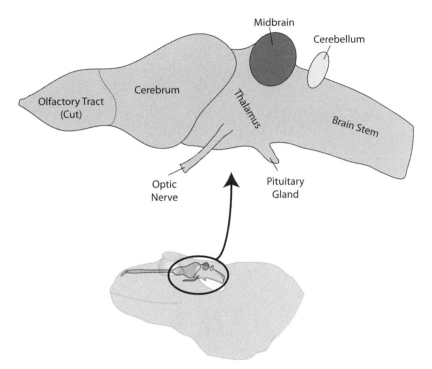

11.7. Schematic of frog brain, shown to scale inside the head of a bullfrog. Notice the similarities to those of actinopterygian fishes (Fig. 7.10). The cerebellum is relatively small as a result of the less complex hopping movements employed by many frogs and the loss of swimming maneuvers. Diagram after Liem et al. (2001).

Midbrain

Cerebellum

Cerebrum

Olfactory Tract
(Cut)

Thalamus

Brain Stem

Optic
Nerve

Pituitary
Gland

flexibly respond to stimuli: experimental removal of the cerebrum in frogs renders them incapable of seeking shelter or feeding themselves (Liem et al., 2001; Butler and Hodos, 2005). The optic tectum of the midbrain is quite large, and there is some evidence to suggest that much of the locomotor and feeding movements of frogs are controlled by this part of the brain (Butler and Hodos, 2005). Due to metamorphosis, adult frogs are more terrestrial and lose their lateral line system and associated motor coordination for swimming. Not surprisingly, the cerebellum of frogs is very small when compared with most fishlike vertebrates, reflecting the more 'simplified' form of hopping or jumping locomotion these animals undertake (Liem et al., 2001; Butler and Hodos, 2005). Overall, more research is needed to understand what these animals are thinking and how they perceive the world.

Jumping ahead to the vertebral column, many frogs have a rather strange backbone that magnifies their leaps (Fig. 11.8). First, there is only one cervical (neck) vertebra, so that whereas frogs have a neck, it is a very short one indeed. Second, the back itself is composed of a small chain of articulated dorsal vertebrae that slope downward from the neck to the pelvis, giving a sitting frog a somewhat swaybacked appearance in side profile. These dorsal vertebrae bend only dorsoventrally and cannot bend laterally. Third, the pelvic and tail vertebrae are most peculiar. There is one sacral bone that articulates with the pelvis such that some movement and rotation can occur between these bones, and the tail vertebrae have fused into a long rod called the urostyle (Figs. 11.4 and 11.8) (Pough et al., 1998; Liem et al., 2001; De Iuliis and Pulerà, 2011). There is a significant functional reason for this interesting vertebral setup. One can think of the

11.8. Mechanics related to frog jumping. At top: (A) the stylized mechanics of a frog jumping with the urostyle acting as a stiffened stick that swings the chain of vertebrae out in front of it during a leap and (B) the whole body motion of the frog. Bottom left: the two types of frog pectoral girdles that allow give at landing. Bottom right: the typical orientation of a frog's hand based on rotary movements at the elbow and sliding mechanisms at the wrist joint. Jumping diagrams based on Jenkins and Shubin (1998); pectoral girdles after Emerson (1983).

Stylized Mechanics of a Frog Jump

Back Curvature
Vertebral 'Chain'
Urostyle and Ilia as a 'Stick'

Two Different Frog Pectoral Girdles Viewed from Beneath (Ventral)

Clavicles
Scapula
Episternum
Clavicle
Epicoracoids
Glenoid
Epicoracoid
Coracoid
Sternum

Orientation of Frog Hand

SS
Humerus
Scapulocoracoid
Radio-ulna
Elbow
Ball Joint
Specialized Wrist Joint
Inward Rotation of Frog Hand

urostyle and sacrum as a stick to which a chain is attached – that chain being comprised of the dorsal vertebrae. When a frog begins to jump, the urostyle and sacrum swing upward, pulling the vertebral 'chain' with them. This movement converts the sagging back into an outstretched, extended structure that adds length to the leap (Fig. 11.8) (Jenkins and Shubin, 1998). Of course, leaping involves more than this.

The ilium bones on either side of the sacrum are incredibly long. Given that the ilium supports the large hip rotators (gluteal muscles) and knee-extending or kicking muscles (quadriceps), actions required for jumping, it should not be surprising that the ilia of frogs are such lengthy structures. In contrast, the pubis and ischium are much reduced. In fact, the pubis is typically made mostly of cartilage. This, too, seems to have functional consequences: when frogs land after jumping and their pelvis contacts the ground, having some 'give' in this region of the anatomy may spare their pelvic bones from damage. The hind limb bones of frogs are incredibly lengthened (Fig. 11.4). The femur is long, the tibia and fibula have been fused into a single tibiofibula bone, the proximal ankle bones have been so elongated that they are essentially another leg segment, and even the five digits of the hind foot have become spindly. All this lengthening greatly increases the jumping distance of the frog (Jenkins and Shubin, 1998).

As the old saying goes, what goes up must come down. The sternum in most frogs apparently plays a significant role in absorbing the stress of landing. The upper portion, called the epicoracoid, is either cartilaginous or bony, but regardless can flex and dissipate the stress of landing (Emerson, 1983). The forelimbs themselves are quite modified compared with early tetrapods (Fig. 11.8). The elbow is a ball-and-socket joint, and the distal end of the humerus consists of a single rounded condyle that accepts the cupped end of the fused radio-ulna (Sigurdsen and Bolt, 2009). Given that frogs land on their forelimbs after a jump, having a single forearm bone provides a simple and strong strut for the frog to land on and pivot over (Fig. 11.8) (Sigurdsen and Bolt, 2009). Moreover, we see the evolution and development of a specialized joint between the radius part of the radio-ulna and the wrist bones on the inside of the hand (Sigurdsen and Bolt, 2009). This joint allows the hand to pivot about the end of the radio-ulna so that it faces medially when the frog is at rest on its forelimbs (Sigurdsen and Bolt, 2009). When the frog goes to jump, the hand is now in a mechanically effective position to push the animal forcefully off the ground, and the hand can be rotated back to this orientation for landing (Fig. 11.8).

Hopping and jumping are derived forms of locomotion in tetrapods, and there have been numerous hypotheses regarding its origin in frogs. The classic hypothesis suggests that the ancestor of frogs developed jumping as a means of getting away from predators by leaping back into the water. Splashing into water would not require precise landings, and, intriguingly, recent studies of jumping and hopping in primitive living frogs have shown that these animals land by belly-flopping (Essner et al., 2010). Such a behavior would be consistent with the water-escape hypothesis. However, a recent competing hypothesis suggests that the odd mechanics of frog anatomy developed initially for stabilizing the vertebral column during locomotion (Reilly and Jorgensen, 2011). In the most primitive living frogs, hopping and burrowing are common, but jumping as we have described tends to occur only in the more derived species. Instead, in primitive frogs the urostyle is commonly used as a muscle anchor rather than a lever for jumping, and the sacroiliac joint typically slides instead of remaining rigid in primitive frogs (Reilly and Jorgensen, 2011). The fossil record provides an interesting test of these hypotheses, as we will see.

We have already discussed salamanders in some general detail in chapter 9. As with frogs, the skulls of salamanders tend to be broad and flat, with more dorsally facing eyes (Fig. 11.9). However, unlike frogs, the eyes of salamanders look laterally and don't have much binocular vision. As with frogs, an eye-retractor muscle is present in salamanders, and they blink in part by pulling their eyes into large orbits, which again pass through to the inside of the palate as interpterygoid vacuities. Teeth are present on the premaxilla, maxilla, and dentary bones, as well as across the surface of the palate as little denticles (Fig. 11.9). As with frogs, the bones of the skull are reduced and fused. Salamanders never develop the tympanic ears of frogs, and their stapes functions in bracing the braincase. Salamanders do not have vocal calling, and therefore a specialized high-frequency detector would be an unnecessary evolutionary development. However, salamanders do possess the operculum system discussed previously for frogs, and so these lissamphibians hear the world as a series of low-frequency rumbles, clicks, and thuds (Smith, 1968; Pough et al., 1998; Liem et al., 2001). We will return later to the feeding adaptations of salamanders, which are best discussed within the context of their development.

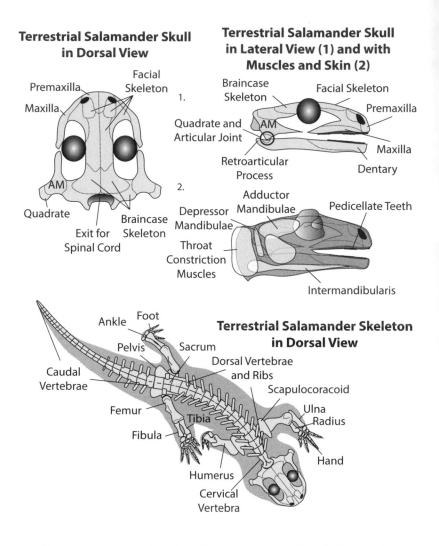

11.9. Salamander skeletal diagrams based on a Tiger salamander (*Ambystoma tigrinum*). Upper left: dorsal view of skull. Upper right: salamander skull in side view showing (1) major aspects of the anatomy and (2) the relationships of the major jaw muscles and skin to the skull. Bottom: skeletal schematic of a salamander in dorsal view. AM indicates the region of the adductor mandibulae muscles. Muscle diagrams after Liem et al. (2001) and De Iuliis and Pulerà (2011).

Terrestrial Salamander Skull in Dorsal View

Facial Skeleton
Premaxilla
Maxilla
AM
Quadrate
Exit for Spinal Cord
Braincase Skeleton

Terrestrial Salamander Skull in Lateral View (1) and with Muscles and Skin (2)

1.
Braincase Skeleton
Facial Skeleton
Premaxilla
Quadrate and Articular Joint
AM
Maxilla
Retroarticular Process
Dentary

2.
Adductor Mandibulae
Depressor Mandibulae
Pedicellate Teeth
Throat Constriction Muscles
Intermandibularis

Terrestrial Salamander Skeleton in Dorsal View

Ankle
Foot
Pelvis
Sacrum
Caudal Vertebrae
Dorsal Vertebrae and Ribs
Scapulocoracoid
Femur
Tibia
Ulna
Radius
Fibula
Hand
Humerus
Cervical Vertebra

However, salamanders do rely more on visual and chemical cues than do frogs. In fact, no experimental evidence has conclusively shown that any species of frogs use chemical signals to attract mates or communicate (Pough et al., 1998). In contrast, many salamander species possess a well-developed vomeronasal organ that can pinpoint pheromones and other chemical cues from members of the opposite sex and from others of the same species, as well as chemosensory information from the environment (Pough et al., 1998). Male salamanders of several species even have specialized glands on the underside of their chins that they rub on females during courtship (Pough et al., 1998). Although difficult to test, perhaps some fossil amphibians possessed similar abilities.

Like frogs, there is only one cervical vertebra in salamanders, but the vertebral column is much longer and has a great degree of lateral flexion. Thus, fishlike undulations of the body combined with limb movements allow salamanders to extend their stride length and cover more ground with each step. The relationship between the coracoids and the sternum is similar to that of frogs, and only limited movements of the scapulocoracoids are possible. As with frogs, there is a single sacral vertebra that articulates with the ilia, followed by a number of caudal vertebrae. The ilium is much more reduced than in frogs, but it is still larger than the pubis and ischium, both of which are mostly cartilaginous (Fig. 11.9) (Liem et al., 2001; De Iuliis and Pulerà, 2011). Locomotion in salamanders was previously described in chapter 9.

Finally, we briefly turn our attention to the odd and fascinating limbless lissamphibians called caecilians (Fig. 11.10). This group of lissamphibians hails from the tropics and is so poorly known and studied that there is not even a common English name for them (Gower and Wilkinson, 2009). Caecilians tend to be small burrowing or swimming animals. Although they retain some of the openings in the skull common to other lissamphibians (for example, the interpterygoid vacuities), the skull's construction is robust and solid (Pough et al., 1998; Gower and Wilkinson, 2009). The sense organs of caecilians are intriguing. The eyes are poorly developed (even absent in some caecilians), and act mostly as light-detecting spots that serve to alert these lissamphibians when it is dusk, a time of when many of these animals become active (Pough et al., 1998; Liem et al., 2001; Mohun et al., 2010). In many the eyes are covered by skin and in some cases by skull bones. Pores on the sides of the skull between the eyes and nostrils house a pair of special sensory tentacles that are used to probe the surroundings for chemical signals (Fig. 11.10). The tentacles are controlled by eye muscles: the eye retractor pulls the tentacles into their pores and the eye elevator moves the tentacles forward (Pough et al., 1998; Liem et al., 2001; Gower and Wilkinson, 2009). Weirder still, some of the tear glands that typically moisten the eyes in tetrapods have been modified to lubricate the tentacles in caecilians. For the ultimate in true amphibian weirdness, in some species of caecilians the eyes are attached to the base of the tentacles such that when the tentacles are protruded, the eyes literally pop out of their sockets (Pough et al., 1998; Liem et al., 2001; Gower and Wilkinson, 2009)!

The body is limbless and snakelike, circular skin folds mark the body segments, and pockets of scales lie within the skin folds. Caecilians move about in several ways. In one form of locomotion, these lissamphibians can throw their bodies laterally like a swimming fish to move across the ground surface or beneath the waves. As we will see later for snakes, caecilians also employ what is known technically as concertina motion: they press portions of their body against the ground or a burrow wall and extend the free parts to move forward (Woltering et al., 2009; Herrel and Measey, 2010). For example, they might press the rear half of their body and tail against the ground while lifting and extending the rest of their body forward. Caecilians have an even stranger trick that snakes cannot match, in which they can move their vertebral column independently of their skin (Herrel and Measey, 2010)! In this case, the caecilian presses its skin against a burrow wall, but its vertebral column moves within the animal. When the skin is released from its anchoring point, it slides and catches up with the skeleton.

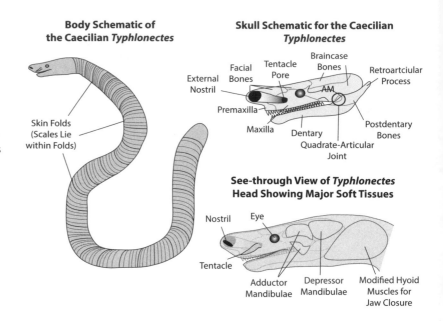

11.10. Schematic of caecilian anatomy based on *Typhlonectes*. Notice the long retroarticular process (upper right) for attachment of the large, modified hyoid muscles (lower right) for improved jaw closure and power. AM, adductor mandibulae muscle group. Skull and muscles after Kleinteich, Haas, and Summers (2008) and the *Typhlonectes* specimen on *DigiMorph* (http://www.digimorph.com/).

Body Schematic of the Caecilian *Typhlonectes*

Skin Folds (Scales Lie within Folds)

Skull Schematic for the Caecilian *Typhlonectes*

External Nostril
Facial Bones
Tentacle Pore
Braincase Bones
Retroartciular Process
AM
Premaxilla
Maxilla
Dentary
Quadrate-Articular Joint
Postdentary Bones

See-through View of *Typhlonectes* Head Showing Major Soft Tissues

Nostril
Eye
Tentacle
Adductor Mandibulae
Depressor Mandibulae
Modified Hyoid Muscles for Jaw Closure

With our survey of living lissamphibians complete, we now have a search image for adult forms in the fossil record. However, given the complex life histories of these vertebrates, it is important for us to briefly review the major aspects of their larval stages and the remarkable transformation into adult anatomy during metamorphosis. This will provide us with an additional set of search images for finding fossil larval forms.

Lissamphibian Reproduction and Metamorphosis

Lissamphibian reproduction and metamorphosis are so varied and complex that only a basic outline can be given here. Fertilization in many frogs is external, with males depositing sperm over eggs. A few frogs give birth to live young (but in this case the embryos develop and hatch inside the female). External fertilization often occurs in ponds or streams, but it can also take place in small pools of water trapped in the leaves of tropical plants. In most salamanders, internal fertilization is common, but it occurs when males deposit a spermatophore (essentially a sperm stalk) that females pick up to fertilize their eggs. Internal fertilization is also common in caecilians, and here males have a penis-like organ that can be protruded from their urogenital opening (cloaca) and placed within that of the female (Gilbert and Raunio, 1997; Liem et al., 2001; Gower and Wilkinson, 2009).

In many lissamphibians, there is only a tiny amount of yolk available for the embryo, and as a consequence few of these animals have anything resembling a yolk sac. Instead, yolk proteins are incorporated into the cells of the rapidly developing embryo, and these are utilized for energy during the first stages of development into a larva (Gilbert and Raunio, 1997; Gilbert, 2010). However, a few frogs have direct-developing young that never pass through a larval stage and emerge from their eggs as tiny

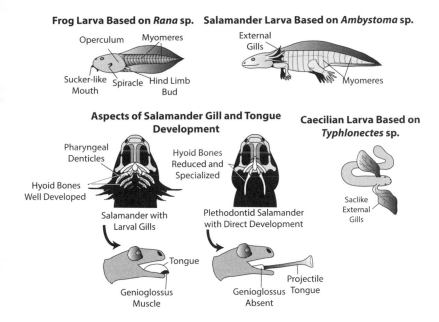

Frog Larva Based on *Rana* sp.

Operculum · Myomeres

Sucker-like Mouth · Spiracle · Hind Limb Bud

Salamander Larva Based on *Ambystoma* sp.

External Gills

Myomeres

Aspects of Salamander Gill and Tongue Development

Pharyngeal Denticles

Hyoid Bones Well Developed

Hyoid Bones Reduced and Specialized

Salamander with Larval Gills

Plethodontid Salamander with Direct Development

Caecilian Larva Based on *Typhlonectes* sp.

Saclike External Gills

Tongue

Genioglossus Muscle

Genioglossus Absent

Projectile Tongue

11.11. Lissamphibian larvae body form and plethodontid salamander tongue development. In salamander larvae that have external gills, the hyoid bones bear denticles. Note that for many salamanders, the gill-bearing larva requires hyoid bones that necessitate the development of a genioglossus muscle, which hinders the tongue from projecting far from the mouth in the adult. In contrast, many plethodontid salamanders, especially those that undergo direct development, lose most of the hyoid bones and genioglossus, and consequently can project their tongue nearly 80% of their body length! Lack of bones and muscles that contribute to buccal pump breathing (hyoids, genioglossus) in plethodontids goes hand in hand with the loss of lungs. Frog and salamander larvae after McDiarmid and Altig (2009). Caecilian larva after Pough, Janis, and Heiser (2002, 2013). Larval salamander hyoids and denticles after Schoch (2009). Plethodontid feeding and tongue projection after Wake and Hanken (1996), Deban and Marks (2002), and Deban et al. (2007).

versions of the adults. This pattern of reproduction foreshadows the development of truly terrestrial eggs (amniotic eggs), like those of lizards, and we will discuss this form of development in chapter 12.

After hatching from their eggs, many lissamphibians spend their larval stages as free-swimming tadpoles. Frog tadpoles are strange creatures with internal gills hidden beneath an operculum (not the sound-detecting organ they develop later but the gill cover similar to those in fish) (Fig. 11.11). Most frog tadpoles are herbivorous or perhaps omnivorous, and commonly have sucker-like mouths with keratinous beaks and teeth. They feed by suctioning or scraping off plant debris, and have long, coiled intestines for breaking down vegetation. There are a few frog larvae that are predatory on small invertebrates or other tadpoles, but this is less common (Gilbert and Raunio, 1997; Wilt and Hake, 2003; Gilbert, 2010). Metamorphosis in frogs is the most dramatic of all lissamphibians, and involves the enlargement of the limbs, the loss of the tail and internal gills, and the simplification of the larval digestive system into one better suited for carnivory. Once metamorphosis begins in frogs, it proceeds rapidly, and for good reason. Metamorphosing frogs that have both tails and hind legs are less efficient jumpers because the tail interferes with locomotion. As a consequence frogs stuck between the stations of larva and adult are the most vulnerable to predation (Gilbert and Raunio, 1997; Gilbert, 2010).

In salamanders, the larvae and juveniles resemble adults in many ways, but typically possess external gills, webbed feet, and a tail fin. Most salamander larvae have up to three pairs of external gills that help aerate their blood, and these structures develop from within the pharyngeal pouches between the slits (Gilbert and Raunio, 1997; Pough et al., 1998). Because there is typically no skeleton that supports these external gills, they are only rarely preserved in fossil amphibian larvae. However, the pharyngeal arches of most larval salamanders sport gill rakers and sometimes denticles, and these features appear to go hand in hand with the presence of external gills (Fig. 11.11). Therefore, finding pharyngeal arches with gill rakers and/or denticles in fossils is usually a good indication that external gills were present (Schoch, 2009; Fröbisch et al., 2010). The circulatory system of a larval salamander is similar to that of lungfishes, where blood can be selectively shunted to gills or lungs depending on the circumstances. Metamorphosis of salamanders into terrestrial creatures results in the permanent shunting of blood to the lungs from the heart as the gill branches atrophy (Gilbert and Raunio, 1997; Pough et al., 1998; Liem et al., 2001; Gilbert, 2010). Some salamander species actually retain their juvenile characteristics into adulthood and reproductive age, a process called neoteny. This tends to happen where water supplies are permanent or perennial, and some salamanders can delay or accelerate the timing of when they metamorphose into terrestrial adults depending on the environment and water availability (Gilbert and Raunio, 1997; Liem et al., 2001; Schoch, 2009).

The process of metamorphosis has had intriguing consequences in a group of salamanders called plethodontids. In adult plethodontids, tongue projection is well developed and is controlled by a highly modified set of hyoid bones (Fig. 11.11) (Wake and Hanken, 1996; Deban and Marks, 2002; Deban et al., 2007). However, given that lissamphibians aerate their lungs using these same bones (as we describe for frogs), something had to be sacrificed. That something in plethodontid salamanders was their lungs! Odd as it seems, a group of terrestrial salamanders has done away with lungs, relying entirely on their skin for respiration to retain tongue projection as a means of securing prey (Wake and Hanken, 1996; Deban and Marks, 2002; Deban et al., 2007). The importance of the hyoid bones in tongue projection has further consequences for these salamanders (Fig. 11.11). In plethodontids that have an aquatic stage, some of the hyoid bones must necessarily function in aerating the gills and sucking in prey. As a consequence, this places constraints on how much the hyoid bones can be modified later during metamorphosis into a tongue-projecting mechanism, and the range of tongue projection in these salamanders is therefore limited (Wake and Hanken, 1996; Deban and Marks, 2002; Deban et al., 2007). In contrast, some plethodontid salamanders have done away with the aquatic larval stage altogether, and their hyoid bones can be put to exclusive use in developing tongue projection. In these salamanders, the tongue can be projected more than half the length of the entire body (Wake and Hanken, 1996; Deban and Marks, 2002; Deban et al., 2007)!

Finally, to end on a truly weird note, we turn to caecilian larvae. Most caecilians develop directly as larvae within the oviducts of their mother. Unlike most lissamphibians, caecilian embryos have well-developed yolk sacs that provide them with some additional energy during development (Wilkinson et al., 2002; Gower and Wilkinson, 2009). However, as the yolk supply dwindles, the larval caecilians hatch out of their egg capsules but remain inside the oviduct. Here, they feed off creamy, milk-like secretions from the oviduct, which they scrape off with specialized mouth parts (Wilkinson et al., 2002). But it gets even weirder: after some species of caecilians are born into the world, they retain their fetal teeth and feed off a lipid-rich skin layer of their mothers. You read that correctly: some caecilian mothers raise their young by allowing them to eat their skin (Kupfer et al., 2006; Kuhel, Reinhard, and Kupfer, 2010; Wilkinson et al., 2013)! In addition, the larval caecilians develop large, branching external gills from their pharynx (Fig. 11.11), and these come in close contact with the vascular walls of the mother's oviduct. It appears that some gas exchange and even some nutrient exchange may occur between the mother and offspring in this way (Wilkinson et al., 2002; Gower and Wilkinson, 2009). By the time most caecilians have hatched, the external gills have diminished or disappeared altogether. Now that we have search images for adult, metamorphic, and larval lissamphibians, we can apply our knowledge to understanding what it is we are finding in the fossil record.

Fossil Amphibians

Fossil amphibians were diverse and extremely varied in their anatomy and eco-niches, and thus we will focus on the major evolutionary changes in the skeletal chassis and associated features among the fossil amphibians leading to the modern forms by selecting a few examples. One of the best known early amphibians is *Balanerpeton*, a salamander-looking creature from Scotland stretching nearly 30 cm long from head to tail (340 Ma; Card 93) (Fig. 11.12) (Milner and Sequeria, 1994). The skull of *Balanerpeton*, like those of salamanders, was flattened with dorsally facing eyes, and the number of bones in the skull was diminished compared with stem tetrapods. Small teeth occupied the premaxilla and maxilla, whereas larger teeth resided on the dentary. Small denticles adorned the palate, and a number of paired fangs jutted out of the roof of the mouth. In some specimens, there is the presence of a sclerotic ring to support the eyes (Fig. 11.12) (Milner and Sequeria, 1994). The function of this sclerotic ring in early amphibians remains unclear. It could have acted, as in fishlike sarcopterygians, to support the eyeball, but it also could have possessed muscles that squeezed the eye, changing its shape to assist in focusing, as in many members of the Reptilia. More astonishing is that some *Balanerpeton* fossils preserve a delicate band of mineralized, fold-like material that curves just above the eye socket, which may be evidence of tetrapod-like eyelids (Milner and Sequeria 1994). Something that may bolster this hypothesis is that the lacrimal bone has now become a significant portion of the orbit. As you will recall, the lacrimal duct is a feature that goes hand in hand with the development of eyelids and tear glands. However, it is not clear from the available *Balanerpeton* specimens if such a duct was present in the lacrimal bone. As we will see in other fossil amphibians, the lacrimal bone becomes a larger portion of the orbit and definitive lacrimal ducts are observed. The squamosal bone has begun to

Balanerpeton in Dorsal View

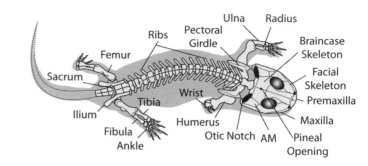

Dendrerpeton in Lateral View

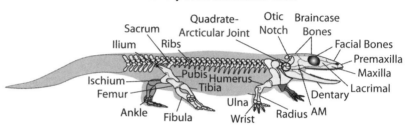

become slightly T-shaped, vaguely resembling what we encountered in frogs, and the backward-facing portion of the T helps form the otic notch at the back of the skull. The vertebral column and ribcage have a lot in common with those of salamanders. A single cervical vertebra is present behind the head, followed by numerous and relatively undifferentiated vertebrae studded with short ribs. However, unlike in salamanders, the vertebral centra remain divided. A single sacral vertebra articulates with the ilia in the pelvis, and numerous caudal vertebrae follow. The pectoral and pelvic girdles are as described for amphibians generally.

Two intriguing aspects of *Balanerpeton* are that numerous specimens of this animal are preserved at different stages of development (Milner and Sequeria, 1994; Fröbisch et al., 2010). The smallest recovered specimen of *Balanerpeton* has a skull length of just about 2 cm long. Most of the skull, some of the girdles, and the proximal elements of the limbs are ossified, but most of the vertebral column and many other skeletal portions are not, suggesting that this individual was in a late larval stage. There is no trace of the hyoids or a pharyngeal skeleton, so it remains unknown as to whether *Balanerpeton* had gilled larvae. As we go from the smallest known specimens of *Balanerpeton* to the largest, the skeleton is ossified in a craniocaudal direction, a trend somewhat similar to that of modern salamanders (Milner and Sequeria, 1994; Fröbisch et al., 2010).

In many of the *Balanerpeton* specimens there are belly scales, sometimes called gastralia, that originate just caudal to the sternum and cover the animal's underside in a V-shaped pattern. In some of the largest individuals, there are also faint patterns and impressions that might indicate the presence of scutes (armor-like dermal bones), as in some modern lissamphibians. However, none of the *Balanerpeton* specimens conclusively has scutes, and this patterning may instead reflect odd preservational conditions or the impressions of folded, desiccating skin. What is perhaps most intriguing is that the presence of gastralia (and maybe scutes) doesn't appear until larger sizes in *Balanerpeton*, which suggests that at most life stages the skin was naked and moist, acting, as in lissamphibians, as a major respiratory organ (Milner and Sequeria, 1994).

Approximately 20 million years after the appearance of animals like *Balanerpeton* (310 Ma; Card 94), more derived species of amphibians began to populate the globe. One such animal was *Dendrerpeton*, another salamander-like creature almost 30 cm in length, hailing from North American and European coal forest deposits (Fig. 11.12) (Milner, 1980; Holmes, Carroll, and Reisz, 1998; Benton, 2005). *Dendrerpeton* differed from more ancestral amphibians like *Balanerpeton* in having a taller skull with expanded squamosal and cheek regions (presumably associated with expanded adductor mandibulae and pterygoid muscles). CT scans of the otic region in a *Dendrerpeton* specimen confirm that its inner ear anatomy and stapes shape most closely resemble those of modern frogs (Robinson, Ahlberg, and Koentges, 2005). We can be confident that eyelids and tear glands were present in *Dendrerpeton* because the lacrimal bone occupies much of the orbit and had distinct lacrimal ducts. The vertebral column was more robust than that of *Balanerpeton*, but each vertebral body was still composed of two pieces (Holmes, Carroll, and Reisz, 1998). As with *Balanerpeton*, the dorsal vertebrae are studded with short ribs, suggesting that *Dendrerpeton* relied on mouth-pump breathing as in frogs and salamanders. In the pectoral girdle, the scapulocoracoid is a larger element than in *Balanerpeton*, whereas the clavicle and cleithrum are somewhat smaller. The ilium is more expanded, but the pubis and ischium remain small. Overall, the girdles, limbs, and feet of *Dendrerpeton* show further trends toward more terrestrial locomotion (Holmes, Carroll, and Reisz, 1998; Benton, 2005).

By approximately 310 million years ago (Card 94), the amphibians had split into two major groups (Benton, 2005). One group consisted of large-bodied predators that were either mostly terrestrial or semiaquatic. The other group was comprised of smaller-bodied amphibians among which the common ancestor of the lissamphibians probably arose. These amphibians are known technically as the dissorophoids, and most were characterized by drastic metamorphosis in which a filter-feeding aquatic larva underwent profound anatomical changes, usually resulting in a terrestrially feeding adult (Fröbisch et al., 2010). A number of larval dissorophoid fossils are known from various species, and it is clear from the presence of gill rakers, hyoid denticles, and occasional fossil impressions of soft tissues that these juveniles had external gills. Among the dissorophoids, the smallest were the amphibamids, and one of these animals, *Doleserpeton* (295 Ma; Card 94), gives us a glimpse at what was likely the beginnings of the ancestral lissamphibian chassis.

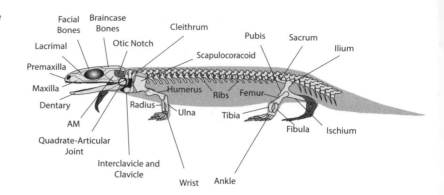

11.13. Skeletal diagram of the amphibamid amphibian, *Dole-serpeton*. Again, the lacrimal bone is prominent and has lacrimal ducts, allowing us to infer that this amphibian had eyelids. AM, adductor man-dibulae muscle group. Skeletal diagram after Sigurdsen and Bolt (2010).

Doleserpeton is very small, measuring only about 7 cm from head to pelvis (Fig. 11.13) (Sigurdsen and Bolt, 2010). The first impression one gets when looking at the skull of *Doleserpeton* is how frog-like it is. The head is relatively short, the orbits are large, and the skull is lightly constructed. The skull bones are relatively unsculpted or ridged, much like those of lissamphibians, in which the skin is not directly attached to the skull. A look at the mouth leaves little doubt as to the relationship of *Doleserpeton* to lissamphibians: it has pedicellate teeth (Sigurdsen and Bolt, 2010). The ribs of *Doleserpeton* are also very short, suggesting that respiration in adults was predominantly mouth-pump breathing. The pectoral girdle of *Doleserpeton* is dominated by a robust and fused scapulocoracoid, and the clavicle and cleithrum are all reduced in size. The pelvic girdle is salamander-like but tiny. The humerus is very frog-like, especially its large, singular ball-like condyle for articulation with the forearm bones at the elbow (Sigurdsen and Bolt, 2010). The radius and ulna remain separate bones, whereas the wrist bones are tiny and rounded, somewhat reminiscent of the condition in frogs, and the hand is again comprised of four digits. The hind limb is essentially a smaller version of that found in amphibians generally, and both forelimbs and hind limbs would again work as modified cranks (Fig. 11.13).

What is especially intriguing about amphibamids like *Doleserpeton* is that they suggest the small body size typical of most lissamphibians probably arose as a consequence of selective pressure to switch from an aquatic to a terrestrial feeder (Schoch, 2002, 2009; Fröbisch et al., 2010). Given that suction feeding does not work well in air, quite drastic changes to the feeding apparatus had to occur, and quickly. This is somewhat equivalent to switching from steam power to internal combustion—nothing in between will work, and you have to overhaul the entire engine so that it will work with the new fuel. So, too, with lissamphibian meta-morphosis: you often switch from an aquatic vegetarian filter feeder to a terrestrial carnivore. Size becomes important here because it is easier to modify something on a small scale than on a large scale. In fact, this explains why so much of the important development and shuffling of anatomical structures occurs in tiny embryos. By the same token, a drastic change of anatomy from a larval to adult stage is best handled quickly at small size (Schoch, 2002, 2009; Fröbisch and Schoch, 2009). True, there are some large modern lissamphibians, the largest being the Chinese giant salamander (*Andrias davidianus*), which reaches over 1.5 meters in length and can weigh close to 64 kg (Pough et al., 1998). But these are rare exceptions, and it seems that the typically small size of modern lissamphibians has occurred, in part, due to the constraints of their drastic metamorphosis (Schoch, 2002, 2003, 2009; Fröbisch and Schoch, 2009; Fröbisch et al., 2010). Of course, after becoming small, the ancestors of lissamphibians now could exploit microenvironments, such as living among damp forest litter, thus expanding their ability to colonize more of the land.

Analyses of DNA trait states consistently suggest that frogs and salamanders share a close common ancestor, followed more distantly by caecilians. Now comes the truly frustrating part: we don't get any fossils that are generally agreed to be lissamphibians until well into the Early Triassic (~250 Ma; Card 95) (Benton, 2005; Anderson, 2008; Sigurdsen and Green, 2011). The earliest fossil recognized as an almost-frog is *Triadobatrachus*, a little animal just over 8 cm long (Fig. 11.14) (Rage and Rocek, 1999). *Triadobatrachus* has many of the signature features we have come to know in frogs, including a flattened head, large orbits, and elongate hind limbs, including the lengthened ankle bones. However, this fossil is incomplete, missing part of the head as well as both hands and feet. In other ways, it less frog-like. The body is longer and more salamander-like, the forearm and leg bones are unfused, and although the ilium becomes narrow and elongated in *Triadobatrachus*, the caudal vertebrae have not fused into a urostyle (Rage and Rocek, 1999). By the Early Jurassic (~185 Ma; Card 96), *Prosalirus*, the earliest recognized frog, appears in North America with a skeletal chassis that conforms in most details to those of modern frogs, including the possession of fused forearm and hind leg bones as well as a urostyle (Fig. 11.14) (Jenkins and Shubin, 1998). There is currently a debate surrounding whether or not *Prosalirus* could jump as modern frogs do because of disagreements about how to interpret urostylic and pelvic anatomy (Reilly and Jorgensen, 2011). For salamanders, a number of Early to Late Jurassic forms are known from several skeletons in varying degrees of preservation (Gao and Shubin, 2001, 2003; Skutschas and Martin, 2011). The skeletal chassis of most known early salamanders is elongate with short limbs, and the skulls of Jurassic salamanders are commonly flattened and broad, somewhat like frogs (Gao and Shubin, 2001, 2003; Skutschas and Martin, 2011). For our purposes, because the skeletal chassis of the earliest known salamanders are so similar to those of their modern brethren, none is detailed further.

A recent discovery from the Early Permian (290 Ma; Card 94) could be a game-changer in terms of redefining the origins of frogs and salamanders. Coming in at just over 10 cm long, an odd amphibamid amphibian called *Gerobatrachus* was unearthed in Texas (Fig. 11.14) (Anderson et al., 2008). This animal has a very wide, flat, frog-like head and an elongate skeletal chassis like that of a salamander, and some of its teeth appear to

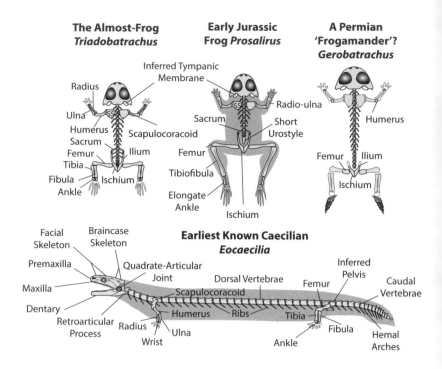

11.14. Early lissamphibians. Note the difference in vertebral and limb proportions between the almost-frog, *Triadobatrachus* (after Rage and Roček [1999]), and the early true frog, *Prosalirus* (after Jenkins and Shubin [1998]). The relationship of the Permian 'frogamander' amphibian, *Gerobatrachus* (after Anderson et al. [2008]), to other lissamphibians remains debated, but it does possess a very lissamphibian-like chassis. The earliest known caecilian, *Eocaecilia* (after Jenkins, Walsh, and Carroll [2007]), had tiny limbs, but the humerus closely resembles those of frogs and salamanders.

be pedicellate. It is often dubbed the 'frogamander' and for good reason: a first blush, this animal looks like something close to the common ancestor of both frogs and salamanders! If correct, the initial split between frogs and salamanders goes back millions of years before the Triassic (Anderson et al., 2008). And this is where the controversy comes in. First, we have nothing definitive, frog or salamander, until the Triassic and Jurassic, so we must use caution in how we interpret and extrapolate from the *Gerobatrachus* chassis. Second, there is disagreement about whether or not the teeth of *Gerobatrachus* are actually pedicellate (Sigurdsen and Green, 2011). Finally, many amphibamids have relatively broad, flattened heads and elongate bodies. Therefore, it is difficult to know whether what we see in *Gerobatrachus* truly reflects close common ancestry with frogs and salamanders, or whether we're just seeing some convergence. As with everything in paleontology, time will tell.

It is also during the Early Jurassic that we find *Eocaecilia*, an early caecilian with legs (Fig. 11.14) (Jenkins, Walsh, and Carroll, 2007). The body of *Eocaecilia* is elongate like that of a salamander, and its head bears large interpterygoid vacuities and well-developed pedicellate teeth. Most recent trait state analyses place *Eocaecilia*, and thus all caecilians, among the temnospondyls (Sigurdsen and Green, 2011). However, as explained in chapter 12, some paleontologists would consider *Eocaecilia* and the caecilians to belong, instead, to another group called the lepospondyls. Finally, an odd group of salamander-like lissamphibians arose during the Jurassic and persisted into the Age of Mammals (~23 Ma; Card 100). These were the albanerpetontids, and they paralleled the body forms and habits of salamanders (Gardner, 2001).

Youngest
Exposed
Rocks

Younger
Rocks

Oldest
Exposed Rocks

Clothed,
Bipedal Ape

1. Stratigraphy illustrated. The author is standing among the brightly colored sedimentary rocks of the Morrison Formation in eastern Utah. As the early geologists recognized, in a typical sequence of rock layers (strata), the oldest rocks will be on the bottom whereas the youngest rocks will be at the top of sequence.

Lancelet Chordate

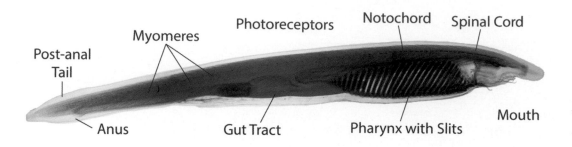

Post-anal Tail

Myomeres

Photoreceptors

Notochord

Spinal Cord

Anus

Gut Tract

Pharynx with Slits

Mouth

Juvenile Lamprey (Jawless Vertebrate)

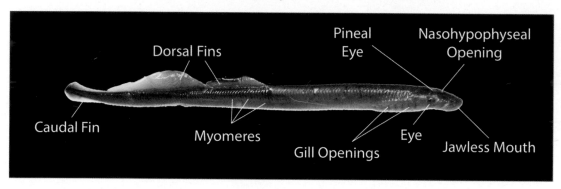

Dorsal Fins

Pineal Eye

Nasohypophyseal Opening

Caudal Fin

Myomeres

Gill Openings

Eye

Jawless Mouth

2. Chordate anatomy. The lancelet chordate shows to good effect the major anatomical trait states of all chordates, such as the notochord, pharynx with slits, and a post-anal tail. The juvenile lamprey, a primitive jawless vertebrate, demonstrates how similar the overall body plan of vertebrates is to that of other chordate animals.

3. Bone. This is a close-up of a cat femur with its proximal end sectioned longitudinally to expose the patterns of bone within. Note that dense, compact bone makes up the shaft where compression, tension, and torsion must be resisted, whereas spongy bone comprises the region associated with the joint surface (head of the femur) that must have some 'give' when stressed.

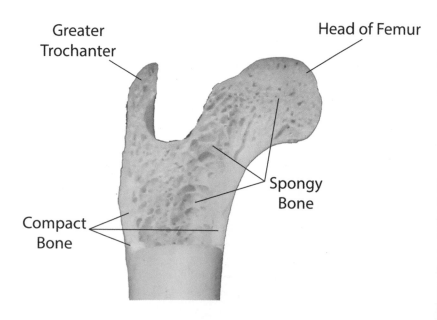

Greater Trochanter

Head of Femur

Spongy Bone

Compact Bone

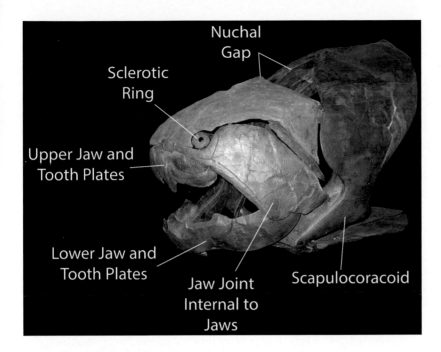

4. Placoderm skull. The skull of immense placoderm *Dunkelosteus,* which stretched nearly 9 meters in length from head to tail. Note the nuchal gap between the skull and scapulocoracoid that allowed the head to rock dorsally, greatly increasing the gape of this terrifying fish. Photo by the author, with permission of the Field Museum, Chicago.

5. The skull and jaws of the spiny dogfish shark, *Squalus acanthias.* The skull and jaws of a preserved specimen in lateral (A), dorsal (B), and ventral (C) views. See text for details and compare to Figure 6.6.

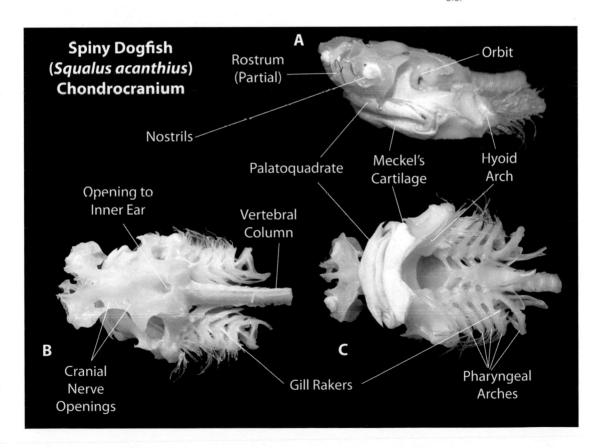

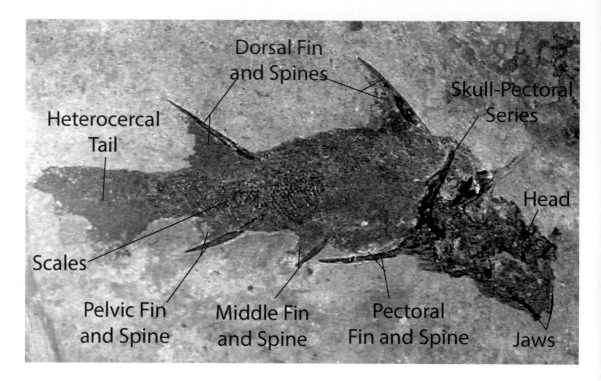

Dorsal Fin and Spines

Skull-Pectoral Series

Heterocercal Tail

Head

Scales

Pelvic Fin and Spine

Middle Fin and Spine

Pectoral Fin and Spine

Jaws

6. The body fossil of an acanthodian fish. Note the spines that form the base and leading edge of the dorsal, pectoral, pelvic, and middle fins. In this particular fossil, you can also make out the bony scales, the heterocercal tail, and skull-pectoral series of bones. Photo by the author, with permission of the Field Museum, Chicago.

7. The primitive actinopterygian fish called the bichir (*Polypterus*) and the African lungfish (*Protopterus*). Note the tough, scaly 'jacket' of the bichir that plays a crucial role in its respiration. In the African lungfish, notice that although fleshy pectoral and pelvic fins are present, they are relatively spindly. Despite the gracile nature of the fins, the African lungfish is capable of moving across land when necessary. The bichir is a preserved specimen, whereas the lungfish is a former denizen of the Stockton University animal labs.

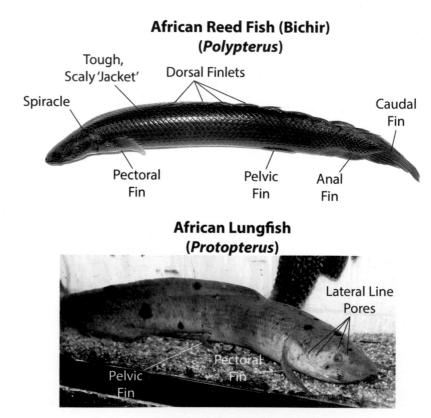

**African Reed Fish (Bichir)
(*Polypterus*)**

Tough, Scaly 'Jacket'

Dorsal Finlets

Spiracle

Caudal Fin

Pectoral Fin

Pelvic Fin

Anal Fin

**African Lungfish
(*Protopterus*)**

Lateral Line Pores

Pelvic Fin

Pectoral Fin

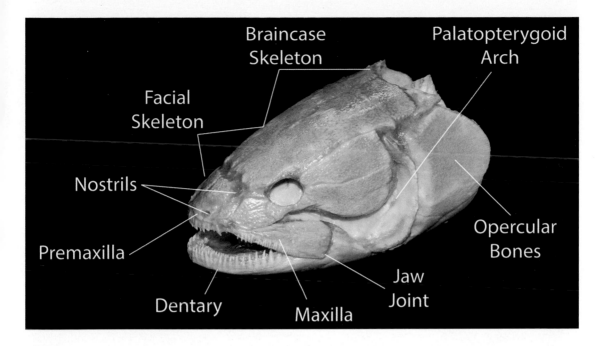

Braincase Skeleton

Palatopterygoid Arch

Facial Skeleton

Nostrils

Premaxilla

Dentary

Maxilla

Jaw Joint

Opercular Bones

8. The skull and jaws of the actinopterygian fish, *Amia calva* (the bowfin). See text for details and Figure 7.7.

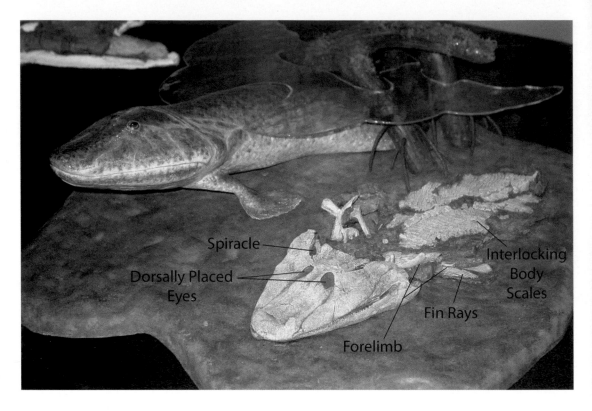

Spiracle

Dorsally Placed
Eyes

Interlocking
Body
Scales

Fin Rays

Forelimb

9. A display cast and model of the early 'fishapod' (transitional tetrapodomorph) *Tiktaalik* on display at the Field Museum in Chicago. See text and Figure 10.1 for more details. Photo by the author, with permission from the Field Museum, Chicago.

10. Early and modern amphibians compared. At left, a cast of the skull of the eryopid amphibian *Eryops*. At right, three blue poison dart frogs (*Dendrobates tinctorius*) in the animal labs at Stockton University.

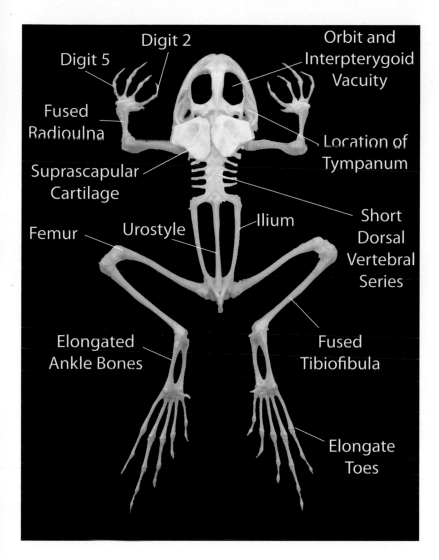

11. The skeleton of a modern bullfrog, *Lithobates catesbeianus*. See text and Figure 11.4 for more details.

12. The 'conger eel' salamander (*Amphiuma*). Note the tiny forelimbs and snakelike body of these lissamphibians. These animals are superficially similar to the completely limbless caecilians discussed in the text. Photo by the author, with permission of the Cape May Zoo, New Jersey.

13. Comparison of lissamphibian and reptile skin. Note the smooth, glandular texture of the skin in *Bufo guttatus* (smooth-sided toad) compared with the tough, water-resistant skin of a Savannah monitor lizard (*Varanus exanthematicus*). Photo of *B. guttatus* by the author, with permission of the Cape May Zoo, New Jersey.

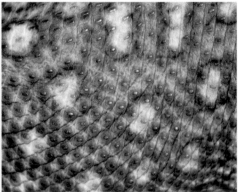

14. The parietal eye of a bearded dragon (*Pogona vitticeps*). The parietal eye directly senses light (but does not form images) and aids many lizard species in regulating their basking time. Photo of 'Blackbeard,' one of the author's research animals for locomotor studies at Stockton University. See text and figures in chapter 13 for more information.

15. Savannah monitor lizard (*Varanus exanthematicus*). Pictured is 'Hall,' the monitor lizard, one of the author's research animals for locomotor studies at Stockton University. See text and figures in chapter 13 for more information.

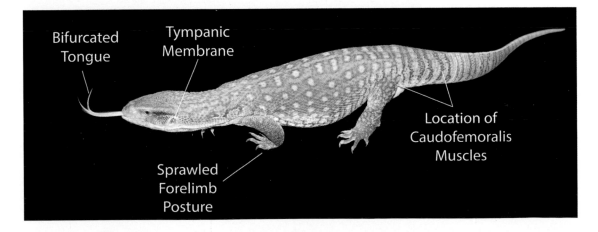

16. The tuatara reptile (*Sphenodon punctatus*). These reptiles are the last survivors of their lineage of lepidosaurs, and share a close common ancestor with lizards and snakes. Photo courtesy of Heinrich Mallison.

17. The common snapping turtle (*Chelydra serpentina serpentina*). Note how the shell influences every aspect of a turtle's posture. This female individual is a former member of the Stockton University animal labs. The author thanks Robert McKeage for help wrangling the turtle for this photo.

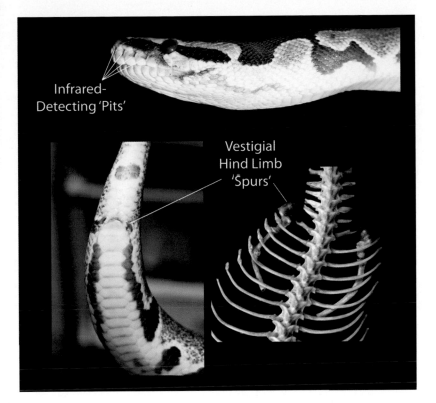

18. Python specializations. A male Burmese python (*Python bivittatus*) named 'Monty' in the Stockton University animal labs, showing the infrared sensory pits that adorn its face and its vestigial hind limbs used in copulation. The skeletal inset shows the reduced but still present pelvic girdle and limb bones that form the vestigial hind limbs in pythons. The author thanks Tom Gleason for help with photographing this snake.

Infrared-Detecting 'Pits'

Vestigial Hind Limb 'Spurs'

19. Skeleton and body outline of an ichthyosaur. Note the preserved body outline that shows the presence of a dorsal fin and a tuna-like caudal fin. See text in chapter 15 for more details. Ichthyosaur specimen at the Museum für Naturkunde, Berlin, Germany. Photo courtesy of Heinrich Mallison.

20. External crocodylian anatomy. An American alligator (*Alligator mississippiensis*) showing to good effect the 'waterproof' ear flaps, muscle 'jowls,' and bony scutes of these animals. See text in chapter 16 for more details. Photo by the author, with permission from the Cape May Zoo, New Jersey.

Ear Covered by 'Waterproof' Flaps

Scutes

Jaw Muscle 'Jowls'

Huge Caudofemoralis Muscle in Tail

Ankle

Ear

Metatarsus

Ankle

Digitigrade Foot

21. Some modern birds. At left, the wild turkey (*Meleagris gallopavo*) on the hunt for insects, and at right two ostriches (*Struthio camelus*) patrolling the Cape May Zoo grounds. Note the high ankle of birds and the digitigrade foot posture. See chapter 16 for more details about bird anatomy. Photo of wild turkeys courtesy of Heinrich Mallison. Photo of ostriches by author, with permission of the Cape May Zoo, New Jersey.

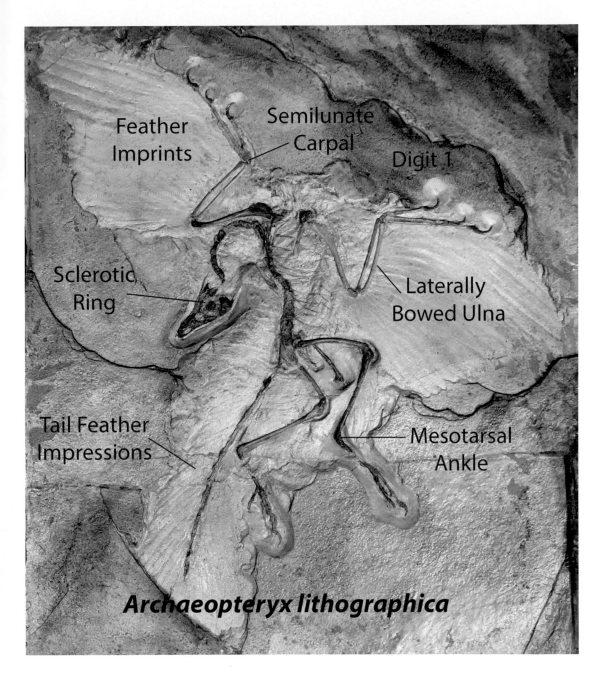

Feather Imprints

Semilunate Carpal

Digit 1

Sclerotic Ring

Laterally Bowed Ulna

Tail Feather Impressions

Mesotarsal Ankle

Archaeopteryx lithographica

22. *Archaeopteryx lithographica,* the 'first' bird. A cast of the famous Berlin specimen showing to good effect the feather impressions, the semilunate carpal of the wrist, and the mesotarsal ankle joint. The dinosaurian affinities of this animal can scarcely be overlooked.

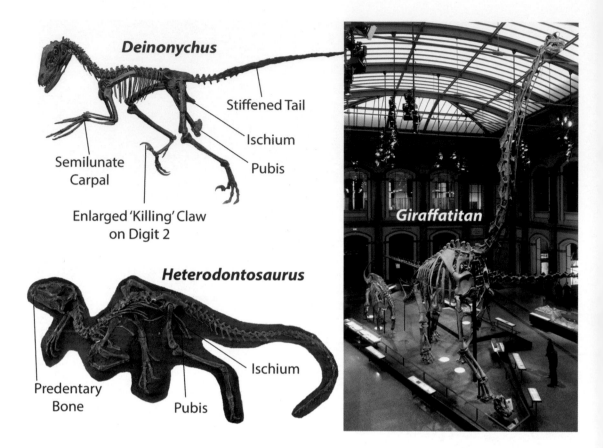

Deinonychus

Stiffened Tail

Ischium

Pubis

Semilunate Carpal

Enlarged 'Killing' Claw on Digit 2

Heterodontosaurus

Ischium

Predentary Bone

Pubis

Giraffatitan

23. Dinosaur diversity. *Deinonychus* is the predatory dinosaur that sparked the dinosaur 'renaissance' given its birdlike characteristics. A cast of *Heterodontosaurus,* a dog-sized herbivore from the Early Jurassic that shows the backswept pubic bone and predentary bone characteristic of all ornithischian dinosaurs. The skeleton of the giant sauropod *Giraffatitan* ('*Brachiosaurus*') at the Museum für Naturkunde in Berlin, Germany. Photo of *Deinonychus* by the author, with permission of the Field Museum, Chicago. Photo of *Giraffatitan* courtesy of Heinrich Mallison.

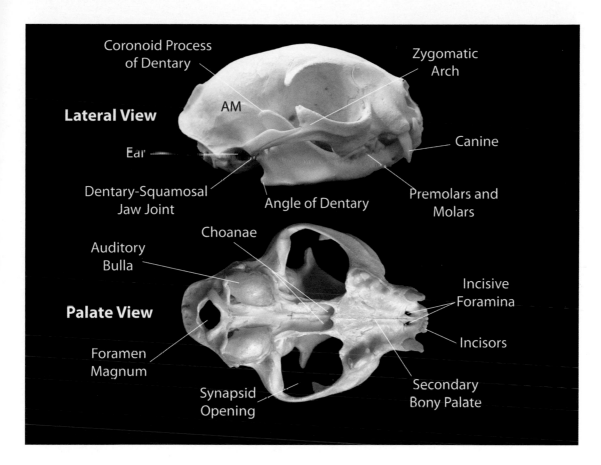

24. The skull of a domestic cat (*Felis catus*). Lateral and palate views of the domestic cat skull showing several mammalian trait states such as a single lower jaw bone (the dentary), a synapsid opening, and a secondary bony palate. See text for more details.

25. Synapsid diversity. *Dimetrodon* is more closely related to us than to reptiles. Note the prominent 'sail' composed of the elongated neural spines of the dorsal vertebrae. Cast of the skull of an herbivorous relative of *Dimetrodon*, *Edaphosaurus*, showing an array of crushing palatal teeth. Cast of the cynodont synapsid *Probelesodon* showing a skull in transition from the primitive amniote condition to that present in modern mammals. In *Probelesodon*, two jaw joints are present: (A) the original quadrate-articular jaw joint and (B) the new dentary-squamosal jaw joint. The quadrate and articular bones will eventually become the incus and malleus bones, respectively, in the mammalian middle ear. Photo of *Dimetrodon* by the author, with permission of the Field Museum, Chicago.

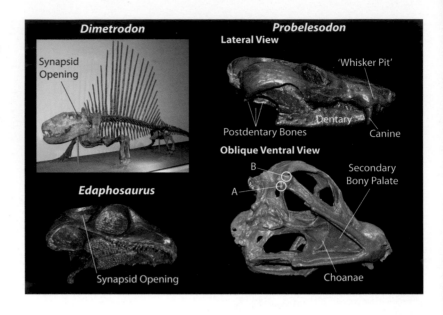

Dimetrodon

Synapsid Opening

Edaphosaurus

Synapsid Opening

Probelesodon

Lateral View

'Whisker Pit'

Postdentary Bones

Dentary

Canine

Oblique Ventral View

B

A

Secondary Bony Palate

Choanae

26. Modern mammal diversity. The most primitive living members of Mammalia are the egg-laying monotremes, as exemplified by the short-beaked echidna (*Tachyglossus*). Modern therian mammals consist of the Metatheria or marsupial mammals, as exemplified by a Bennett's wallaby (*Macropus rufogriseus rufogriseus*), and the Eutheria, as exemplified by the ring-tailed lemur (*Lemur catta*), a primate. Photo of short-beaked echidna courtesy of Heinrich Mallison. Other photos by the author, with permission of the Cape May Zoo, New Jersey.

Short-Beaked Echidna (*Tachyglossus*)

Bennett's Wallaby (*Macropus rufogriseus rufogriseus*)

Ring-Tailed Lemur (*Lemur catta*)

Skeletal Diagram of *Eryops*

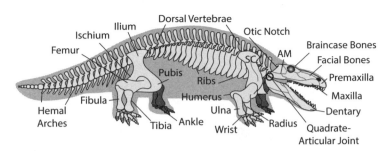

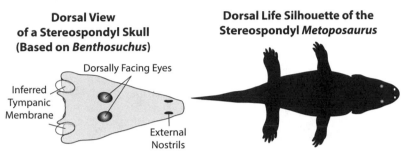

Dorsal View of a Stereospondyl Skull (Based on *Benthosuchus*)

Dorsal Life Silhouette of the Stereospondyl *Metoposaurus*

11.15. Eryopids and stereospondyls. *Eryops* was a large (nearly 2 meters long) amphibian that had many terrestrial adaptations (after Clack [2002] and Benton [2005]). In contrast, many stereospondyls were large, semiaquatic or aquatic predators with flattened, crocodile-like heads and broad bodies (after Benton [2005] and Sulej [2007]). AM, adductor mandibulae muscle group; SC, scapulocoracoid, cleithrum, and clavicle.

We end this chapter by briefly considering the large-bodied amphibians that branched off from the groups leading to lissamphibians some 310 Ma (Card 94). Technically called the eryopids and stereospondyls, some of these animals were large. For example, *Eryops*, known from numerous skeletons from the United States, grew nearly 2 meters long (Fig. 11.15) (Pawley and Warren, 2006) (Plate 10). The skull of *Eryops* was broad, tall, and long, measuring almost 30 cm from snout to braincase. The skull was robust and the bones were well sutured together, a feature shared by modern predators such as crocodiles and alligators that can clamp down on prey with tremendous force. The vertebral column, girdles, and limbs of *Eryops* leave no doubt that this large animal was capable of supporting itself and moving against gravity (Pawley and Warren, 2006). The vertebrae are large and robust with tall neural spines. Given that the neural spines are associated with the long back muscles and ligaments that hold up the vertebral column, the lengthening of these elements would provide excellent mechanical advantage in *Eryops*. The girdles and limbs of *Eryops* were massive. For example, the deltopectoral crest and epicondyles of the humerus were robust landmarks that must have served as strong attachment points for the forelimb protractors and wrist flexors and extensors, respectively (Carroll, 1988, 2009; Benton, 2005; Pawley and Warren, 2006). Moreover, some rare skin impressions of *Eryops* are fossilized, and these show that lozenge-shaped scales were embedded within its skin (Romer and Witter, 1941). The scales are similar in overall appearance to the ones that occur in the folds of skin on the limbless caecilians, but they were more broadly distributed across the body of *Eryops*. The scales appear to be related to the dermal scutes discussed earlier, because they contain some bone mineral content (Romer

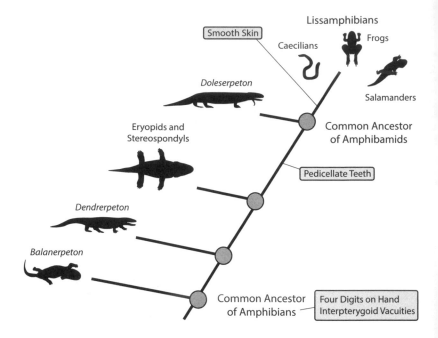

11.16. Pedigree of relationships among the fossil amphibians and lissamphibians discussed in this chapter. See text for details.

Smooth Skin

Lissamphibians

Frogs

Caecilians

Salamanders

Doleserpeton

Eryopids and Stereospondyls

Common Ancestor of Amphibamids

Pedicellate Teeth

Dendrerpeton

Balanerpeton

Common Ancestor of Amphibians

Four Digits on Hand
Interpterygoid Vacuities

and Witter, 1941). Overall, *Eryops* was a large, almost reptilian animal that could support its bulk against gravity. Given that *Eryops* lived contemporaneously with many early synapsids (mammal ancestors), it is not outside the realm of possibility that this large amphibian occasionally made lunch out of one of our long-lost relatives.

A group of fossil amphibians closely related to animals like *Eryops* were the stereospondyls, a diverse collection of tiny to large flat-headed, and in many cases crocodile-like, forms (Fig. 11.15) that persisted from the Permian into the Triassic (295–204 Ma; Cards 94–96) with a few stragglers making it through most of the Age of Dinosaurs (203–100 Ma; Cards 96–98) (Warren and Yates, 2000; Benton, 2005). Many of these amphibians had flexible growth strategies in which they were able to control the timing of when and how fast portions of their skeleton ossified (Schoch, 2009). This allowed some species to remain aquatic for longer periods of time when water was available, but also to accelerate the growth of their skulls and limbs if the environment became drier. Especially during the Triassic (250–204 Ma; Cards 95–96), the diversity of habitats occupied by stereospondyls was incredible, but alas none of these animals left living descendants.

Figure 11.16 presents a graphical summary of the major amphibian groups we have discussed and their relationships to one another. As the amphibians gained ground during the Carboniferous, another group of tetrapods took a different approach to leaving the water. These animals evolved the ability to lay eggs in terrestrial environments with tough shells to prevent embryo damage and desiccation. However, most importantly, the embryos of these animals produced four life-support membranes that allowed them to develop into a land-worthy chassis before hatching out onto dry land. These vertebrates were the common ancestors of the Amniota.

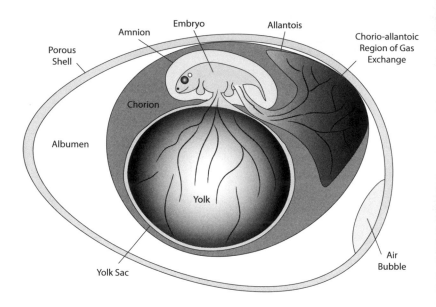

12.1. Diagram of amniotic egg based on the condition in birds and reptiles. Notice that the four life-support membranes (amnion, allantois, chorion, and yolk sac) are generated directly by the embryo. Note also the size of the yolk mass and region of highest gas exchange, where the chorion and allantois merge.

The Amniote Chassis: A Primer and the Lead-Up to True Amniotes

12

THE VACUUM OF SPACE IS A VERY HOSTILE ENVIRONMENT. HUMANS exposed to the vacuum of space suffer many life-threatening effects due to the lack of air pressure and oxygen (Thomas and McMann, 2011). Unlike the cartoonish Hollywood depictions of people rapidly inflating and exploding when exposed to space, the actual results are much less startling but equally unpleasant and lethal (Thomas and McMann, 2011). To ensure survival the space suit was developed to be a multipurpose life-support device (de Monchaux, 2011). First, it supports an astronaut's body with air pressure. Second, it supplies oxygen for inhalation and stores exhaled carbon dioxide in specialized tanks. Third, it cycles fluid around the wearer to keep the body cool within the confines of the suit. Fourth, it will absorb, store, and sequester liquid and solid waste in a so-called MAG (maximum absorbency garment), the outer space version of a diaper (Thomas and McMann, 2011).

One group of vertebrates, the amniotes, adapted to a purely terrestrial existence, and this entailed severing their development from watery environments (Shedlock and Edwards, 2009). Part of their success in becoming purely terrestrial was the development of a specialized, amniotic egg. Like a space suit, the egg containing the amniote embryos had to protect them from the harsh realities of this new environment until they were developed enough to survive. Just as the early tetrapods and amphibians had to adapt to gravity and desiccation, so too did the embryos of amniotes have to resist these environmental hazards and more.

But who are the amniotes? Today, the living amniotes are composed of two large, diverse groups. One group, which we will call the Reptilia, includes lizards (and the odd tuatara), snakes, turtles, crocodylians, and birds. The fossil record has revealed an even greater diversity among the Reptilia, and we will encounter fantastic examples of these animals such as the dinosaurs. The other group, the Mammalia, contains all the fur-bearing vertebrates that produce milk, including the egg-laying spiny echidna and duck-billed platypus, the pouched mammals called marsupials, and the so-called placental mammals (Shedlock and Edwards, 2009). As we will see, mammals themselves belong in a larger group that includes their nonfurry ancestors called the Synapsida. Some of the early synapsids were so reptile-like in general appearance that traditionally they were called mammal-like reptiles! These two lines, the Reptilia and the Synapsida, are both long-lived, having split shortly after the appearance of the first amniotes in the Carboniferous (~310 Ma; Card 94) (Benton, 2005). Despite the great differences between them, they are linked together by skeletal trait states and by the possession of some form of the amniotic egg.

In this chapter, we will first focus on the components of the amniotic egg, the skeletal adaptations of modern amniotes, and the soft tissues that are associated with the amniote chassis. These characteristics will again provide us with a search image for finding the ancestral amniotes in the fossil record. We then finish the chapter by discussing some fossil vertebrates that form a series of common ancestors leading up to the amniotes, as well as the earliest known amniotes themselves.

The Amniotic Egg

You may not appreciate it when you are cooking or eating chicken eggs, but you are observing and ingesting a life-support device that makes terrestrial existence possible for us and our amniote relatives. The evolution of an amniotic egg with its life-support capabilities is one of the key traits that allowed the amniotes to radiate into all the terrestrial environments they now dominate. All amniote embryos generate four life-support membranes during their development. The living members of Reptilia typically develop within a hard- or soft-shelled egg containing a large reservoir of yolk. However, although most mammals complete at least some of their development within the uterus of their mother and not in a-shelled egg, they still generate the four life-support membranes we see in the Reptilia (Gilbert and Raunio, 1997; Gilbert, 2010). Based on the distribution of trait states in living amniotes, we know that development within a shelled egg was the primitive condition in amniotes (Benton, 2005). Therefore, we will use the chicken egg and embryo as a template for describing the basics of amniote development, and detail the derivatives of this development in different amniote groups later.

Unlike sperm, which are tiny, are produced almost continuously, and require very little of a male's energy, eggs always require a greater investment of resources. The egg and its contents (the yolk, minerals such as calcium, and the proteins and molecular machinery to get the embryo up and running) all come from the mother (Wilt and Hake, 2003; Gilbert, 2010). In the fishlike vertebrates, early tetrapods, and amphibians (the so-called anamniotes), most embryos are deposited in large clutches of tiny eggs with small amounts of yolk. This means that the mother's energy investment per egg is rather small. Anamniote eggs are typically deposited within water or in damp conditions, and the eggs are permeable, allowing the embryo to stay moisturized and to exchange gases with the environment. Most anamniote eggs are surrounded by a thick jelly layer that insulates the embryo from bacterial and fungal invaders. In fishlike vertebrates, a small volume of yolk is present in a yolk sac connected to the embryo's gut, whereas in many lissamphibians the yolk sac is nearly vestigial and yolk proteins are incorporated into the cells of the gut directly. Development is rapid, and the embryo develops into a free-swimming larva, which then undergoes various transformations that yield the adult chassis (Gilbert and Raunio, 1997; Wilt and Hake, 2003; Gilbert, 2010).

The small, jelly-coated eggs of anamniotes would not last long on land. First, there is nothing to stop them from drying out. Second, the thick jelly coat would inhibit gas exchange in air. Third, there is nothing to support the embryo against gravity's relentless pull. Fourth, even if the embryo could make it to the larval stage, its fishlike body and gills would not allow it to live long once it left the egg. What we observe in amniotic eggs like those of a chicken are a number of 'solutions' to these problems.

First is the problem of fertilization, which cannot happen externally. Amniotes like chickens have internal fertilization, and male amniotes have a penis or an analogous organ that ejaculates sperm into the female's reproductive tract through her cloaca (Liem et al., 2001; Pough, Janis, and Heiser, 2002; Kardong, 2012). I did not grow up on a farm, and as a kid I could never figure out how the rooster's sperm got through the hard-shelled egg of the mother hen. The answer is it doesn't, because the sperm travel up the female's oviduct and fertilize the egg before any other layers are added onto it (Wilt and Hake, 2003; Gilbert, 2010). As the embryo develops as a small disc on top of a large ball of yolk, the hen's oviduct has specialized regions that add various outer components. Much like a car moving down an assembly line where different parts are added in different specialized areas, so the oviduct of a hen adds specialized layers as the embryo travels toward the cloaca. The white of the egg, the albumen, is added around the embryo and yolk, followed later by the addition of shell membranes and the hard, calcium-rich shell (Wilt and Hake, 2003; Gilbert, 2010). If as a child you were told not to lick the raw batter of sponge cake or a lemon meringue pie, there was a good reason. The batter of these treats is made by beating a large amount of the albumen from chicken eggs. The albumen contains antibacterial and antifungal properties that protect the developing chicken embryo from infection (Wilt and Hake, 2003; Gilbert, 2010). If you ingest a lot of raw egg whites, the antibiotic properties of this substance can kill off a lot of the good bacterial flora in your gut, causing you to become very ill. The calcium of the eggshell comes from the mother herself. In birds like chickens, specialized deposits of calcium within the marrow cavities of the mother's long bones are stripped away to shell the egg (Schweitzer et al., 2007). In other members of the Reptilia, such as lizards, turtles, and alligators, the calcium is stripped directly from the mother's bones: a large investment indeed (Wink and Elsey, 1986; Schweitzer et al., 2007)! Besides its providing protection, the calcium in the shell is absorbed by the developing embryo to help build its skeleton (Gilbert and Raunio, 1997).

While the mother's investment in each egg is substantial, amniote embryos earn their keep by producing the four life-support membranes we have alluded to. Much like the air supporting an astronaut in space, a membrane called the amnion (after which the egg and group are named) develops and surrounds the embryo with a fluid-filled chamber that supports it against gravity (Fig. 12.1) (Gilbert and Raunio, 1997; Wilt and Hake, 2003; Gilbert, 2010). A portion of the gut tract of the developing embryo expands outward and surrounds the large mass of yolk, becoming the yolk sac (Fig. 12.1) (Gilbert and Raunio, 1997; Wilt and Hake, 2003; Gilbert, 2010). This in and of itself is nothing too extraordinary because many fishlike vertebrate embryos do a similar thing. What is extraordinary is that the yolk sac must surround and engulf such a large yolk deposit. But why do amniotes require so much yolk? The answer is that unlike a fish or frog, the embryo has to develop for much longer, bypassing a swimming larval stage and generating a skeletal chassis capable of supporting itself against gravity. That requires more time and energy, which necessitates the large yolk mass (Gilbert and Raunio, 1997; Wilt and Hake, 2003; Gilbert, 2010).

The eggshell has pores and allows air into and out of the egg, but that alone would not be enough to keep the developing embryo alive. Moreover, the lungs of the embryo are not yet sufficiently developed to breathe air directly. Two other life-support membranes resolve this problem. One membrane, called the chorion, develops from the embryo and surrounds all the other membranes, eventually pressing up against the eggshell (Fig. 12.1). The chorion contains a mesh of blood vessels that pick up oxygen and release carbon dioxide through the pores in the eggshell (Gilbert and Raunio, 1997; Wilt and Hake, 2003; Gilbert, 2010). But another multipurpose membrane develops that not only improves gas exchange but also deals with the unpleasant task of sequestering waste products generated as the embryo develops. This membrane is called the allantois, and it develops from the location where the urinary bladder (if present) will eventually form (Fig. 12.1) (Gilbert and Raunio, 1997; Wilt and Hake, 2003; Gilbert, 2010). The allantois is extremely vascular, and most of the gas exchange in an amniote embryo occurs where this membrane expands and touches the chorion. It is also in this region that calcium salts from the eggshell are transported to the developing embryo to build its skeleton (Gilbert and Raunio, 1997; Wilt and Hake, 2003; Gilbert, 2010). Because the allantois is a large sac connected to the region that will develop into the urinary bladder and cloaca, it can act like the astronaut's MAG, absorbing, storing, and sequestering metabolic wastes generated by the embryo. After hatching, part of the allantois is retained as the urinary bladder (although not all amniotes have a urinary bladder) (Gilbert and Raunio, 1997; Wilt and Hake, 2003; Gilbert, 2010).

How the amniotic egg evolved is a more difficult matter to resolve. Unfortunately, we don't have a great fossil record of amniotic eggs: the earliest known amniotic egg fossils are from the Triassic (~220 Ma; Card 96) (Benton, 2005), but eggs must have been present prior to then. Moreover, the membranes and soft tissues of such eggs are not typically preserved even when the shell is. In essence, all that we can study directly from fossils are eggshells and occasionally embryonic bones. But all hope is not lost, because certain living frogs may provide us with clues about how the transition from a jelly-coated egg to a shelled egg occurred. A small (2–5 cm in length) terrestrial frog from Puerto Rico known as the coqui (*Eleutherodactylus coqui*) has a type of direct development that may approximate what was occurring in the common ancestor just prior to the emergence of true amniotes. Coqui frogs have larger eggs with more yolk than that typical of other frog species, and their fertilization is internal. As the embryo develops, distinct forelimb and hind limb buds appear after the establishment of the main body axis, in contrast to most frogs, where such structures develop during the free-swimming larval stage (Pough et al., 1998; Gilbert, 2010). The large yolk sustains embryonic growth and development to a point where their skeleton is robust enough to support and move them in air. Although the eggs do not have a hard, calcium-rich shell, the male coqui frog protects them by sitting on them and keeping them moisturized (Gilbert, 2010).

What is perhaps most interesting about coqui frog development is the size of the hatchlings, or 'froglets' as they are often called. Upon hatching, the tiny frogs are so minuscule that several of them could sit comfortably together on an American dime (Gilbert, 2010). What is fascinating about all of this is that the earliest fossils that are recognized as amniotes, as well as fossil vertebrates thought to be related to the common ancestor of amniotes, are also relatively small. This has suggested to some paleontologists that the initial development of the amniotic egg was tied in with small size (Laurin, 2004). Perhaps one of the first crucial steps in developing a fully functional hatchling is to rapidly bypass the ancient larval stage. It has been hypothesized that this rapid development restricted the size of the earliest amniotes and their ancestors until the full suite of amniotic features was firmly in place (Laurin, 2004). Although coqui frog development can't be a direct reflection of how amniote embryology evolved, these frogs and other lissamphibians with similar embryonic development at least show how such a transition from fishlike development might have occurred in broad strokes. Ultimately, the evolution of the amniotic egg remains a matter of continued study among paleontologists and developmental biologists.

Common Features of the Amniote Skeletal Chassis

Luckily for the paleontologist, we don't identify amniotes simply by their eggs. In fact, hard-shelled eggs are not typical for all amniotes, and many snakes, lizards, and turtles lay soft-shelled eggs, a feature that appears to be primitive (Pough et al., 1998; Liem et al., 2001). Thus, even though we are not finding fossil eggs with the first amniotes, it could be due to their possession of a soft-shelled egg that is less likely to be preserved. Moreover, you don't often find fossils with eggs tucked neatly inside of them (though there are some spectacular exceptions). However, amniote skeletons always and without exception have certain specialized traits that no other vertebrates possess. These calling cards of the amniote skeletal chassis are also functionally significant, and so they help illuminate how these animals moved about and interacted with their environments. Thus, we are identifying amniotes by their skeletons and not by their eggs (Benton, 2005).

At first glance, the skulls of amniotes are not all that much different from what we have encountered for tetrapods generally (Fig. 12.2). Many

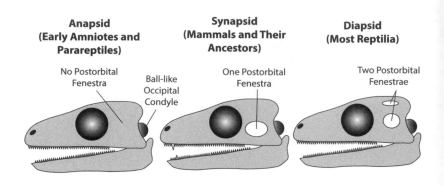

Anapsid (Early Amniotes and Parareptiles)

Synapsid (Mammals and Their Ancestors)

Diapsid (Most Reptilia)

No Postorbital Fenestra

Ball-like Occipital Condyle

One Postorbital Fenestra

Two Postorbital Fenestrae

amniotes have skulls that are tall and narrow; the teeth-bearing bones of the head are almost exclusively the premaxilla, maxilla, and dentary; and the jaw joint is still composed primitively of the quadrate and articular bones. However, the sides of the skull in the braincase region are commonly characterized by one or more openings called fenestrae (Fig. 12.2) (Liem et al., 2001; Pough, Janis, and Heiser, 2002; Benton, 2005). These openings are important both for interpreting the pedigree of an amniote and for functional reasons. The temporal fenestrae, as they are technically called, appear in the region of the skull (around bones such as the squamosal) to which the large adductor mandibulae muscles anchor. Openings like the temporal fenestrae appear to provide expanded ledges and lips onto which the adductor mandibulae muscles can expand, and the openings themselves allow room for larger muscles that can expand outward from the braincase (Liem et al., 2001). Certainly, loss of bone and the creation of openings in the skull may also be related to a combined weight-saving and force-directing strategy. As we discussed in chapter 4, because bone is typically placed where forces are greatest and stripped away from where they are small, the fenestrae may result from this bone cell tug-of-war, resulting in a lighter but strong skull (Liem et al., 2001; Pough, Janis, and Heiser, 2002). Intriguingly, there appear to have been two major ways of developing fenestrae because in the earliest amniotes, there were no temporal fenestrae. In the Synapsida line leading to mammals, only one temporal fenestra appears, whereas in most members of the Reptilia, two temporal fenestrae are the norm (Fig. 12.2). Thus, mammal skulls are referred to as synapsid, whereas those of most Reptilia are called diapsid (Fig. 12.2) (Benton, 2005). As we will see, one branch of the Reptilia, the Parareptilia, usually lacked temporal fenestrae (Fig. 12.2). The braincase itself tends to be more expanded compared with lissamphibians.

All amniotes have two specialized cervical vertebrae (Fig. 12.3). The first cervical vertebra is referred to as the atlas. Just as Atlas in Greek mythology held up the world, so the atlas of amniotes holds up the head (Liem et al., 2001; Benton, 2005). The atlas can be either a compound structure made up of many pieces or a single, fused bone, depending on the amniote. The atlas connects to the ball-like occipital bones of the skull, but primitively only slight dorsoventral nodding motions of the

Atlas-Axis Complex in Amniotes

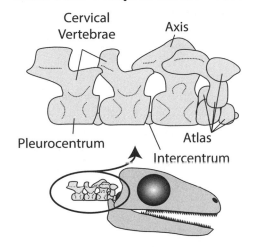

Cervical Vertebrae

Axis

Pleurocentrum

Atlas

Intercentrum

12.3. Postcranial traits of amniotes. All amniotes possess distinct atlas (which may be comprised of several parts or fused into one) and axis vertebrae; illustration based on the condition in *Paleothyris* (Carroll, 1969; Benton, 2005). Note the distinct astragalus and calcaneum that typify all amniotes. Also note that primitively the major axis of ankle flexion and extension passes between the astragalus and calcaneum proximally and the other ankle (tarsal) bones distally.

Dorsal View of Ankle   Lateral View of Ankle

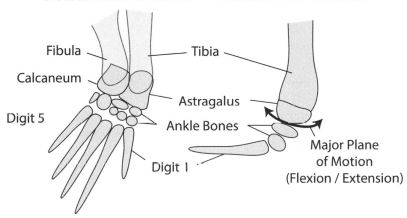

Fibula

Calcaneum

Digit 5

Tibia

Astragalus

Ankle Bones

Digit 1

Major Plane of Motion (Flexion / Extension)

head were possible at this joint, partly due to the presence of smaller more cranial bones called the proatlas, which helped prevent overstretching of the spinal cord (Carroll, 1988; Liem et al., 2001). The second cervical vertebra, the axis, articulates with the atlas, lends resistance that prevents overflexing of the spinal cord at the base of the skull, and allows some more dorsoventral movements of the head (Fig. 12.3). As we will see, the shapes and actions of these first two cervical vertebrae become more elaborated in mammals. Depending on the amniote, anywhere from 2 to more than 20 cervical vertebrae follow the atlas and axis.

A series of dorsal and caudal vertebrae are present in all amniotes, and the sacrum typically is composed of two or more vertebrae. The ribs are more prominent in amniotes than they were in most amphibians, and form a distinct ribcage. For most amniotes, lung ventilation is intimately linked to muscles that span and move the ribs, as described in chapter 9.

Primitively, the sternum, limb girdles, limbs, hands, and feet of amniotes are not all that much different than what we have already encountered in our basic description of the tetrapod chassis and for many amphibians. One difference occurs in the pectoral girdle, where there are primitively two coracoid bones, a cranial or procoracoid and a caudal or metacoracoid (Vickaryous and Hall, 2006). After years of confusion and debate, it seems likely that the majority of the coracoid in both the reptile and synapsid lines is the metacoracoid, whereas the procoracoid forms small remnants in both groups of interest to specialists but beyond the scope of this book (Vickaryous and Hall, 2006). Thus, for our purposes, we will still refer to a scapulocoraoid or a coracoid, and in both reptiles and synapsids, this will be equivalent to the metacoracoid. However, there are fewer bones in the ankle, and three proximal ankle bones have fused into a new ankle element called the astragalus (Fig. 12.3) (Rieppel, 1993). In amniotes, the astragalus is intimately associated with the distal end of the tibia, and usually bears complex articulating surfaces against which some of the distal ankle bones can rotate and flex. Another ankle bone, called the calcaneum, becomes enlarged and articulates with the distal end of the fibula. Together, the astragalus and calcaneum reduce the amount of sliding and rotation at the ankle, restricting the major movements of the foot to flexion and extension and improving the forward propulsion of the hind limbs (Benton, 2005).

Again, the skeletal chassis tells only part of the story, for a number of changes have occurred in the soft tissues and organs of amniotes compared with early tetrapods and amphibians. Although these tissues and organs are seldom preserved in amniote fossils, they are present in all their living descendants, and so we infer that these features were likely present in the common ancestor.

The Softer Side of Amniotes: Skin, Kidneys, and How Not to Dry Out

Moving to a terrestrial existence meant that amniotes had to develop ways to keep from drying out. To understand what amniotes have done to prevent excessive water loss through their skin, we need to briefly discuss the basic components of skin itself (Plate 13). In all vertebrates, the skin is composed of two major layers (Fig. 12.4). There is an outer layer called the epidermis, whose thickness varies considerably among different vertebrates (Fig. 12.4). The cells in this outer layer develop and multiply from a stem cell layer in the deepest part of the epidermis (Liem et al., 2001). As they multiply and pile on top of one another, the older cells begin to produce a waterproof protein called keratin, the same protein that gives fingernails their stiffness and makes up most of your hair. In fact, the older cells produce so much keratin that they eventually die as this protein crowds out their organelles. So, the top layer of the epidermis, the part you call your skin, is made up of layers of flat, dead epidermal cells (Liem et al., 2001; Wilt and Hake, 2003). As these dead cells wear and flake away, they are being replaced continuously by newer cells generated in the stem cell layer below. The deeper layer of skin is called the dermis, and it is much thicker than the overlying epidermis (Fig. 12.4). The dermis overall is enriched with blood vessels, contains the pigment

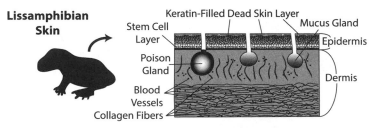

Lissamphibian Skin

Stem Cell Layer
Keratin-Filled Dead Skin Layer
Mucus Gland
Epidermis
Poison Gland
Dermis
Blood Vessels
Collagen Fibers

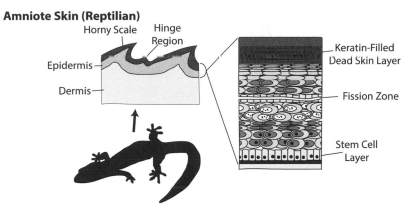

Amniote Skin (Reptilian)

Horny Scale
Hinge Region
Epidermis
Dermis
Keratin-Filled Dead Skin Layer
Fission Zone
Stem Cell Layer

12.4. Diagrammatic comparison of the skin layers and their organization in lissamphibians and an amniote such as a lizard. For all vertebrates, the lowest layer of the epidermis is a stem cell layer that generates new epidermal cells as the old ones are sloughed off or shed. For lissamphibians, the epidermis is very thin, allowing for easy exchange of gases and water, and is punctuated by various poison and mucus glands. In contrast, amniotes like lizards have a thick epidermis, in many covered by horny scales, containing a fission zone that allows old sheets of epidermis to be shed at once. Hinge regions between the horny scales allow the skin to flex and extend. The thicker epidermis of amniotes prevents excessive water loss. Diagrams based on illustrations and data in Liem et al. (2001), Kardong (2012), and Pough, Janis, and Heiser (2013).

cells that give vertebrates their color, and is flexible and stretchable due to a large number of collagen fibers (Liem et al., 2001; Wilt and Hake, 2003). Various glands associated with the skin make their home in the dermis, a network of nerves and nerve endings permeate this layer of skin, and muscles are commonly present in this region. Hair, scales, and feathers also take their root in the dermis (Liem et al., 2001; Wilt and Hake, 2003).

In fishlike vertebrates, amphibians, and presumably in the basal and early tetrapods, the epidermis is very thin, being only a few cell layers thick. The thin epidermis of these vertebrates allows moisture to easily move back and forth across the skin, which means it dries out quickly in air (Liem et al., 2001). In amniotes, the epidermis has become much thicker to prevent water loss (Fig. 12.4). The epidermis of animals like lizards and snakes is covered by horny scales separated by hinge regions (so the scales can move when the skin stretches and bends) that overlie a thick layer of dead, keratin-filled cells (Fig. 12.4) (Liem et al., 2001). This type of epidermis is nearly waterproof and dramatically reduces the amount of evaporative water loss typical of amphibians. As reptilians like lizards grow, the old layer of scales and epidermis must be shed. This occurs by the development of a so-called fission zone, a weakened layer between the upper, dead layers of epidermal cells and the lower, living and multiplying layers (Fig. 12.4) (Liem et al., 2001). We will return to the skin of mammals later, which is thicker than that of amphibians but more glandular than that of reptiles. It is unclear if the earliest amniotes had scaly skin, but it must have been thicker than that of amphibians.

In all vertebrates, kidneys process nitrogenous wastes (preparing them for excretion) and regulate the water content of the body (Liem et al., 2001; Pough, Janis, and Heiser, 2002; Kardong, 2012). Nitrogenous

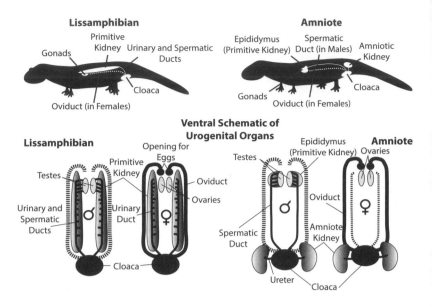

12.5. Differences in the urogenital systems of lissamphibians and amniotes. In most vertebrates, both sexes begin development with the precursors of the oviduct and a urinary duct. The oviduct degenerates during development of male vertebrates (broken lines in the figure). In lissamphibians, a primitive fishlike kidney that excretes copious amounts of water and/or urea is retained. In male lissamphibians, which lack an oviduct, the developing testes commandeer some of the tubules within the primitive kidneys and use the urinary duct to transport sperm (males usually develop an accessory urinary duct as a consequence), whereas female lissamphibians use the urinary duct only for excretion. In amniotes, new kidneys that are more water efficient evolved, and a new urinary duct called the ureter drains these organs. Part of the primitive kidney remains a sperm transport organ in male amniotes (the epididymus), and the old urinary duct is now solely a spermatic duct. Female amniotes lose nearly all of the primitive kidney and old urinary duct (broken lines in the figure) because they now transport urine through the new ureter. Diagrams based on illustrations and data in Liem et al. (2001), Wilt and Hake (2003), Gilbert (2010), Kardong (2012), and Pough, Janis, and Heiser (2013).

wastes are generated by the breakdown and digestion of food. One common nitrogenous by-product is ammonia, which is a great chemical for cleaning floors and disinfecting surfaces, but a deadly compound to vertebrates. The options available to a vertebrate are either to flush the ammonia out quickly with lots of water or to convert the ammonia to a less-toxic substance. The kidneys of freshwater fishlike vertebrates usually take the first option, excreting copious amounts of ammonia and water frequently (Fig. 12.5) (Liem et al., 2001; Pough, Janis, and Heiser, 2002; Kardong, 2012). In amphibians, the kidneys work in similar ways, but these animals commonly convert ammonia into water-soluble urea. The urea is less toxic than ammonia and can be sequestered in higher concentrations in tissues or in a urinary bladder (Liem et al., 2001; Pough, Janis, and Heiser, 2002; Kardong, 2012). When urea is excreted, less water is lost, and such a system is beneficial to amphibians spending their adult lives in terrestrial habitats. However, amphibians are still working with the same basic kidneys as fish, and as such they still lose a great deal of water during excretion.

Amniotes, on the other hand, need to retain all the water they can. Even though the skin prevents some evaporation of precious moisture, evaporation through the nose, mouth, and tears wicks valuable water away. Plus, amniotes still must rid the body of nitrogenous wastes, but they must do it in a more water-conserving way. The primitive vertebrate kidney possessed by fishlike vertebrates and amphibians wouldn't suffice anymore because ridding the body of ammonia and urea would waste a lot of water. Instead, amniotes developed a new kidney that was much more water efficient, and many amniotes developed the ability to convert ammonia into uric acid, a paste-like substance that is water insoluble (Fig. 12.5) (Liem et al., 2001; Pough, Janis, and Heiser, 2002; Kardong, 2012). As a consequence, the primitive kidney is either lost during development in amniotes or contributes to sperm transport (see Fig. 12.5). Although a detailed assessment of the amniote kidney filtration system is best left for other texts, we can say a few things here. First, the Reptilia appear to have retained the primitive amniote kidney, which mostly produces uric acid and a small amount of urea (Liem et al., 2001). By the way, the white pasty splotches called bird 'crap' on parked cars are actually bird 'pee,' the pasty substance being uric acid and not feces. Because such excretion does not require much water, the kidneys of these amniotes are rather small. Second, because so little water is used in their excretion, many reptilians either lack a urinary bladder entirely or hold on to only the vestiges of one (Liem et al., 2001; Kardong, 2012). Third, mammals have a derived excretory system and only convert ammonia into urea, so their kidneys are generally larger and crammed with millions of looping tubules that concentrate the urea and reabsorb as much water as possible (Liem et al., 2001; Kardong, 2012).

These physiological and skeletal features clearly set amniotes apart from their anamniote brethren. However, the transition from what we see in living members of the Reptilia and Synapsida (the living mammals) happened piecemeal over millions of years. Whereas we can't trace the soft tissue evolution of amniotes, we can trace their skeletal evolution from the fossil record. As with amphibians, much debate continues, with several conflicting hypotheses about which fossil vertebrates gave rise to the amniotes. Again, because our focus is on the major changes in the skeletal chassis of various vertebrate groups, I knowingly simplify what is a much more complex evolutionary history. For our purposes, we will assume that a group of amphibian-like animals called the lepospondyls were distant relatives of the amniotes, and that another group, the reptiliomorphs, was their closest common ancestor. These assumptions stem from the general consensus arising from trait state analyses (Ruta, Coates, and Quick, 2003; Ruta, Jeffery, and Coates, 2003; Shedlock and Edwards, 2009). We will again select a few specimens that appear to point the way toward changes leading up to the amniote chassis, and then briefly consider the skeletons of the first recognized amniotes.

Lepospondyls and Reptiliomorphs: On the Road to Amniotes

Lepospondyls are a diverse group of relatively small vertebrates that first appeared in the Carboniferous (~340 Ma; Card 93) and went extinct by the end of the Permian period some 250 Ma (Card 95) (Benton, 2005). Some of these animals led semiaquatic, amphibian-like lives, others were limbless and snakelike, and still others resembled small lizards (Benton, 2005). Not only is the relationship of these vertebrates to amniotes disputed, but the relationships among the lepospondyls themselves are not well resolved. Here, we will not concern ourselves with the interrelationships of these animals, and we simply focus on features that are derived in relation to what we saw previously for amphibians.

Lepospondyls appear to share with amniotes a relatively more elongate and taller skull, and there are reductions and fusions of the ankle

Pedigree Tetrapods Closer to Amniotes than to Amphibians

Date of First Appearance 340 Ma (Card 93)

Specialties of the Skeletal Chassis Some but Not All Amniote Traits

Eco-niche Aquatic to Terrestrial Predators, Omnivores, and the First True Herbivores

bones, including the development of an astragalus-like element (Benton, 2005). The originally two-piece vertebral body has become dominated by the section called the pleurocentrum, a pattern also observed in amniotes (Benton, 2005). However, in amniotes the pleurocentrum and neural arch remain separate for much of their growth, and when they fuse a scar or ridge develops that demarcates one part from the other. In contrast, the pleurocentrum and neural arch fuse early during development in lepospondyls into a single, solid piece with no trace of separation (Benton, 2005).

One of the most diverse lepospondyl groups was the Microsauria, so named because many of these animals superficially resemble tiny lizards (Benton, 2005). As with other lepospondyl groups, these animals appeared in the Carboniferous but went extinct in the middle of the Permian (340–270 Ma; Cards 93–95). Some microsaurs were aquatic animals, but others seem to have been well adapted to a terrestrial existence. One such animal, *Tuditanus*, has a skeletal chassis that closely resembles those of the common ancestors of the reptiliomorphs and amniotes that we discuss next (Fig. 12.6) (Carroll and Baird, 1968). The skull is tall and strong, with no temporal fenestrae. The stapes is stout, flat, and plate-like, suggesting that *Tuditanus* did not detect high-frequency, airborne sounds. The occipital region beneath the foramen magnum is unlike that of amniotes in that it is concave, and receives the ball-like centrum of the atlas. Although an atlas is present, there is no axis and the neck is relatively short, with the remaining cervical vertebrae being nearly identical to those of the dorsal series. The sacrum appears to be composed of two vertebrae, but it is unclear if this is the case based on the available specimens. The length of the tail is unknown, but the preserved caudal series is relatively thin and whiplike (Carroll and Baird, 1968).

Surprisingly, all that is preserved of the pectoral girdle in *Tuditanus* are the cleithra, clavicles, and elements called the interclavicles (Carroll and Baird, 1968). These bones are thin and presumably formed the cranial border of the pectoral girdle, but the scapulocoracoid is not preserved in any *Tuditanus* specimen. However, based on the trends we have already encountered, the scapulocoracoid was most likely the largest element in the pectoral girdle. The pelvic girdle consists of a vertical, sticklike ilium and a broad pubis and ischium surrounding a large acetabulum. Such features suggest that hind limb adduction was crucial to this animal in holding its body off the ground as it walked. Both the forelimb and hind limb are well developed but somewhat delicate in their construction (Fig. 12.6) (Carroll and Baird, 1968). The ulna has a discernible olecranon process. As you will recall, this region is the insertion point for the triceps muscles that resist collapse of the elbow (De Iuliis and Pulerà, 2011), and its elaboration suggests this function was well-developed in *Tuditanus*. Numerous wrist bones and a long hand with small, recurved claws complete the forelimb. The femur is the largest long bone, the tibia and fibula are approximately equal in size and length, a distinct astragalus-like tarsal and calcaneum are present, and the foot is elongate and tipped with

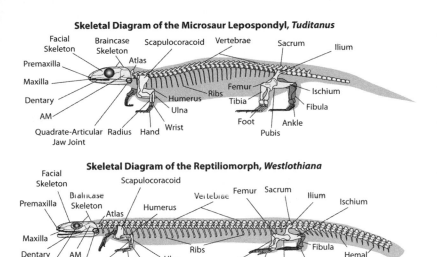

Skeletal Diagram of the Microsaur Lepospondyl, *Tuditanus*

Facial Skeleton · Premaxilla · Maxilla · Dentary · AM · Quadrate-Articular Jaw Joint · Radius · Hand · Wrist · Braincase Skeleton · Atlas · Scapulocoracoid · Vertebrae · Humerus · Ribs · Ulna · Femur · Tibia · Foot · Sacrum · Ilium · Ischium · Fibula · Ankle · Pubis

Skeletal Diagram of the Reptiliomorph, *Westlothiana*

Facial Skeleton · Premaxilla · Maxilla · Dentary · Sclerotic Ring · Quadrate-Articular Jaw Joint · Braincase Skeleton · Atlas · AM · Radius · Hand · Wrist · Scapulocoracoid · Humerus · Vertebrae · Ulna · Ribs · Femur · Sacrum · Tibia · Foot · Pubis · Ilium · Ischium · Fibula · Ankle · Hemal Arches

12.6. Skeletal diagrams of the microsaur lepospondyl, *Tuditanus* (based on Carroll and Baird [1968]), and the reptiliomorph, *Westlothiana* (based on Smithson et al. [1994]). Note that for both *Tuditanus* and *Westlothiana*, the body is elongate, the limbs are short, and an otic notch is absent, suggesting that both animals were deaf to high-frequency airborne sounds. Note that for *Westlothiana* the scapulocoracoid has become the dominant element of the pectoral girdle, a sign that terrestrial locomotion was the predominant form of movement. AM indicates the location of the adductor mandibulae muscles.

recurved claws. There is every indication that the forelimb and hind limb worked in crank-like fashion as they do in salamanders. Finally, some *Tuditanus* specimens show evidence of small, delicate, oval scales that covered the body and perhaps aided in preserving moisture (Carroll and Baird, 1968). Most likely, if we were to see a *Tuditanus* quickly scamper by, we would swear we had just seen a small lizard.

The reptiliomorphs are fossil vertebrates just this side of the common ancestor of the amniotes, sharing many of the same skeletal features, but lacking certain traits such as a ball-like occipital condyle and the presence of a distinct atlas and axis. For simplicity, we are including among the reptiliomorphs members of a group sometimes called anthracosaurs (literally, reptiles of the coal). One of the earliest known reptiliomorphs is *Westlothiana*, a small but elongate animal from the Carboniferous of Scotland (340 Ma; Card 93). This animal was discovered at the same site that has also yielded some early amphibians such as *Balanerpeton* (see chapter 11) (Smithson et al., 1994). Stretching almost 30 cm in length from head to tail, *Westlothiana* resembles a reptile-like dachshund (Fig. 12.6) (Smithson et al., 1994). The head is approximately 2 cm long, the skull is tall, and large eye sockets are present, containing a sclerotic ring. As in some modern amniotes, there are small, numerous denticles adorning the roof of the mouth. The temporal region of the skull lacks any fenestrae or any trace of an otic notch and related tympanic membrane. Like many salamanders, *Westlothiana* was probably deaf to high-frequency sounds.

The cervical vertebrae of *Westlothiana* are short and blend into the dorsal series. The dorsal vertebrae themselves bear well-developed short ribs that suggest this animal utilized rib movements to inhale and exhale (Fig. 12.6). The limb girdles of *Westlothiana* are similar to what we will encounter in true amniotes. The major pectoral girdle element

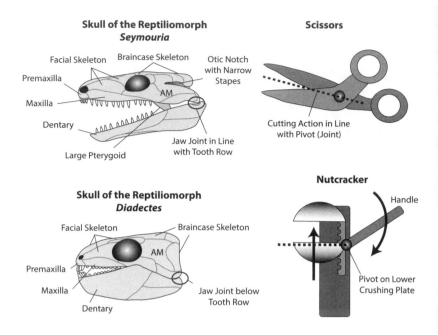

12.7. Reptiliomorph skulls compared. The skull of the predatory reptiliomorph, *Seymouria,* has expanded pterygoid bones that project below the palate. A similar situation occurs in alligators and crocodiles, and is associated with greatly expanded pterygoid muscles, which improve biting power. Note that *Seymouria* has an otic notch containing a delicate stapes, and this strongly indicates that this animal could hear and respond to high-frequency airborne sounds. The robust skull, incisor-like front teeth, rounded, molar-like back teeth, and expanded squamosal and cheek region of *Diadectes* strongly suggest an exclusively herbivorous diet. Moreover, the jaw joint of *Diadectes* is located below the tooth-line, allowing the teeth to come together simultaneously, a situation analogous to the crushing plates in a nutcracker, and unlike the scissors-like jaw alignment in *Seymouria.* Skull diagrams based on Laurin (1996), Berman et al. (2000), and Benton (2005).

is the scapulocoracoid, and the clavicles and cleithra are either poorly preserved or not present at all. This shift to a pectoral girdle dominated by the scapulocoracoid leaves no doubt that the forelimbs of *Westlothiana* were adapted to terrestrial support and movement. *Westlothiana* was definitely in rear-wheel drive: the hind legs are much longer than the forelegs (Fig. 12.6) (Smithson et al., 1994). As we will see, although differences between forelimb and hind limb length vary tremendously among amniotes, with rare exceptions, the forelimbs are always a bit shorter than the hind limbs. The humerus and femur are once again adorned in muscular scars associated with crank-like limb movements, such as broadly expanded epicondyles. The radius and ulna are relatively short but thin, and each articulates with the humerus at distinct condyles. The ulna bears an olecranon process, again suggesting that well-developed triceps muscles held the forelimb up against gravity. Several small wrist bones were present, and although only portions of the hand are preserved in the known specimens of *Westlothiana*, the shapes and lengths of the known parts strongly suggest that this animal had five digits (Smithson et al., 1994). The hind limb shows similar adaptations. The tibia and fibula are robust bones, and the former bears a crest, called the cnemial crest, for the insertion of the knee extensors. We will see a lot of variation in the cnemial crest among various amniotes related to hind limb function throughout the remainder of the book. The bones of the ankle are well preserved and closely resemble those of amphibians. The foot, like the hand, is elongate and lizard-like. All of these skeletal features suggest that *Westlothiana* moved in ways similar to many living species of long-bodied lizards with short limbs, such as skinks (Smithson et al., 1994).

Other reptiliomorphs are animals knocking at the door of true amniotes. One example is *Seymouria* (280–270 Ma; Cards 94–95), a predatory

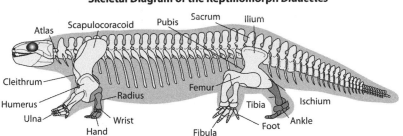

Skeletal Diagram of the Reptiliomorph *Seymouria*

Sacrum
Ilium
Atlas
Scapulocoracoid
Pubis
Femur
Humerus
Ischium
Wrist
Ankle
Radius
Tibia
Hand
Ulna
Fibula
Foot
Cleithrum

Skeletal Diagram of the Reptiliomorph *Diadectes*

Pubis
Sacrum
Ilium
Atlas
Scapulocoracoid
Cleithrum
Radius
Femur
Humerus
Tibia
Ischium
Ulna
Wrist
Ankle
Hand
Fibula
Foot

12.8. Reptiliomorph skeletons compared. The predatory *Seymouria* had a relatively stout but flexible skeleton with well-developed crank-like limbs. The presumably herbivorous *Diadectes* had a very heavy, robust skeleton and measured nearly 1.5 meters in total length. The large size of this reptiliomorph, combined with its skull and jaw construction, suggest a large digestive tract capable of breaking down vegetable matter. Skeletal diagrams based on Carroll (1988) and Benton (2005).

tetrapod just over 30 cm in length hailing from North America and Europe (Fig. 12.7) (Laurin, 1996; Berman et al., 2000). The skull of *Seymouria* shows that this animal had a serious bite (Fig. 12.7). The snout and facial skeleton are much longer than the braincase skeleton, the teeth resemble long, spiked stalactites, and the jugal and pterygoid bones are expanded (Laurin, 1996), presumably creating room and purchase for large jaw-closing muscles. In side profile, one can see the bottom portions of the pterygoids projecting below the skull itself (Fig. 12.7) (Laurin, 1996; Berman et al., 2000). This characteristic of the pterygoid is observed in modern crocodiles and alligators, where the pterygoid muscles are huge. A deep but narrow otic notch occupies the rear of the skull, and a slender stapes occupies this space, suggesting that *Seymouria* had a tympanic membrane and ear similar to that of amphibians (Laurin, 1996). The lower jaw is deep and full of pointed teeth. As with *Westlothiana*, the neck is short and the cervical vertebrae blend into the dorsal series (Fig. 12.8). Short ribs attach to the cervical and dorsal vertebrae, the ribs become shorter toward the sacrum, and *Seymouria* has a heavy tail almost a quarter of its total body length (Fig. 12.8). The open and flat joints (zygapophyses) between the vertebrae suggest lateral movements of the body axis were well developed (Berman et al., 2000).

The pectoral girdle is dominated by a large scapulocoracoid, although the cleithrum remains a rather large element clinging to the front of the shoulder (Fig. 12.8). In the pelvis, the ilium is tall and expanded, suggesting that the gluteal and knee-extending muscles were well developed in *Seymouria*. However, the pubis and ischium are broad and plate-like, and we infer that the hind limb adductors were still very prominent in this reptiliomorph. The humerus is thick, expanded proximally and distally, and has a well-developed deltopectoral crest and large epicondyles (Berman et al., 2000). Again, the proximal and distal ends are twisted in relation to one another, and the distal end of the humerus bears two distinct condyles for articulation with the radius and ulna. The forearm bones are slender and taper at their midpoints, but an olecranon process on the ulna again indicates the significance of the triceps muscles in stabilizing and extending the elbow against gravity. Multiple small wrist bones suggest there was an ability to twist and rotate the hand, and the hand itself is composed of five stout digits capped by blunted claws (Berman et al., 2000). The femur is large with prominent muscle attachment points and distinct condyles for the tibia and fibula. As with the forearm, the tibia and fibula are slender with narrow midpoints (Berman et al., 2000) (Fig. 12.8). The ankle does not bear a distinct astragalus and, as in *Westlothiana*, it is comprised of many small, rounded bones (Berman et al., 2000). The digits of the foot are both longer and stouter than the hand, and their asymmetrical arrangement is again suggestive of lizards. Overall, *Seymouria* was apparently an active, terrestrial predator that used its well-muscled, crank-like limbs and flexible vertebral column to pursue its meals (Fig. 12.8).

Finally, we turn to *Diadectes*, an animal that may deserve two titles: the closest fossil relative to the common ancestor of amniotes and the first terrestrial vertebrate herbivore. *Diadectes* was a large, bulky reptiliomorph approximately 1.2–1.5 meters long that populated regions of North America and Europe during the transition from the Carboniferous into the Permian (310–271 Ma; Card 94) (Figs. 12.7, 12.8) (Case, 1910; Sues and Reisz, 1998). The skull of *Diadectes* is broad and stout, and is overall much more boxy and robust than in reptiliomorphs like *Seymouria* (Fig. 12.7). The teeth and jaw joint of *Diadectes* are very unlike those of all other vertebrates we have considered so far. On the premaxilla and front portion of the dentary are pencil-like teeth that project outward, giving *Diadectes* a buck-toothed appearance (Fig. 12.7) (Sues and Reisz, 1998). These are certainly not teeth made for piercing flesh, but instead resemble the incisors of many herbivorous mammals that are developed for nipping and stripping plant material. The teeth that occupy the maxilla and the remainder of the dentary are large and blunt, and resemble in a general way the cheek teeth of many vertebrate herbivores. These features alone would be enough to convince some paleontologists that, at the very least, *Diadectes* had an omnivorous diet. But it is the construction of the jaw joint that strongly points toward plant-based feeding.

The quadrate-articular joint of *Diadectes* is not in line with the teeth but is instead situated far below them (Fig. 12.7) (Case, 1910; Sues and Reisz, 1998). This is significant for mechanical reasons. As you may recall from chapter 2, in herbivorous mammals we find that the jaw joint is located far above the tooth row. You will also recall from chapter 6 our discussion of lever mechanics in relation to jaws. Chewing plants is different from slicing meat. Meat is easier to slice, whereas plant material, being made of tough cellulose, requires more mashing and pulping. Just as you wouldn't cut walnuts with scissors, so as an herbivore you would not have scissors-like jaws with a jaw joint in line with the tooth row. Instead, like a nutcracker, you would place your jaw joint above or below the tooth row so that when you closed your jaws, all of the teeth would come together simultaneously with tremendous force (Fig. 12.7) (Liem et al., 2001). In herbivorous mammals, evolution has placed the jaw joint above the tooth row to accomplish this. However, in members of the Reptilia, this same feat is accomplished by placing the jaw joint below the tooth line. Either way you get the same outcome: lots of blunted teeth coming together at once to smash and crack open vegetable matter. The in-force for crushing plant matter comes from the adductor mandibulae and pterygoid muscles, and the expanded squamosal and jugal regions of *Diadectes* strongly indicate that these muscles were well developed.

Breaking down plant material takes more than a massive set of jaws: it takes powerful enzymes to speed up the breakdown of cellulose (Liem et al., 2001; Kardong, 2012). Surprising as it may seem, no animal on earth has ever developed the enzymes to break down cellulose, which is probably why it took so long for true vertebrate herbivores to make their appearance. Humans can't break down cellulose, so we tend to eat plant stems and leaves for the roughage, which helps regulate our digestive systems. If you can't break down cellulose, you can't get at the proteins, sugars, and nutrients you need inside the plant. Yet, we know from experience that there are many animals that make their living entirely off plants. How is this possible?

Often different organisms work together for mutual benefit, a process called symbiosis. It turns out that vertebrate herbivores have made a pact with bacteria to accomplish cellulose digestion. These bacteria thrive in the guts of horses, cows, iguanas, and all sorts of plant-eaters, and they have enzymes that will break down plant cellulose (Liem et al., 2001; Kardong, 2012). The bacteria get a home (the guts) and free food out of the deal, while their digestion of the cellulose frees proteins, sugars, and nutrients the herbivores need to survive. Because these bacteria are anaerobic, like the fermentation tanks of a brewery, they need space and time to accomplish their cellulose digestion. Therefore, one of many reasons driving size in herbivorous vertebrates may be to provide room for guts large enough to extract energy from the plants they ingest.

It is no surprise, therefore, that *Diadectes* would be one of the largest animals of its time. The vertebral column bears large and swollen neural spines, which, as we have discussed for *Eryops* in the previous chapter, increases the mechanical advantage of back muscles to hold up a heavy animal (Fig. 12.8). The ribs are also large and would have provided ample room for voluminous guts. The pectoral girdle is comprised of a massive scapulocoracoid, with a much smaller cleithrum and clavicle. The size of the scapulocoracoid in *Diadectes* is not surprising given that it is the anchoring point for the deltoid muscles that swing the arm cranially and laterally, as well as the trapezius muscles that support the shoulder. The pelvic girdle is equally robust, with a wide ilium, pubis, and ischium, suggesting that large hind limb rotators, adductors, and abductors moved and supported the heavy chassis of *Diadectes*. The humerus, radius, and ulna are thick bones with prominent muscle scars, including a large olecranon process on the ulna (Fig. 12.8). The wrist bones are short and rounded, and five stout fingers radiate from the hand. The femur is the longest and most robust limb bone, and the tibia and fibula are stout elements, with the tibia bearing a prominent cnemial crest. An astragalus-like bone and calcaneum occupy the proximal portion of the ankle, providing a hinge-like movement to the five-toed foot. Overall, *Diadectes* has a robust skeletal chassis that would have supported and moved a heavy gut tract from one clump of vegetation to the next (Fig. 12.8). All of these features of the skeletal chassis of *Diadectes* suggest it was one of the first true herbivores (Sues and Reisz, 1998).

The Earliest True Amniotes

Pedigree The First Tetrapods Possessing All Amniote Traits

Date of First Appearance 310 Ma (Card 94)

Specialties of the Skeletal Chassis Definitive Atlas, Axis, and Astragalus

Eco-niche Small Terrestrial Omnivores

The evolution of the first true amniotes occurred in a relative geological blink of an eye, occurring side by side with the evolution of later reptiliomorphs. By approximately 310 Ma (Card 94), around the same time animals such as *Diadectes* and *Seymouria* were coming onto the scene, we get fossils of animals that have the full complement of amniote traits. Two animals in particular, *Hylonomus* and *Paleothyris*, have shed much light on the anatomy and ecology of the earliest amniotes (Benton, 2005). Given their similarities to one another in anatomy, geography (both are from Nova Scotia), and time (they appear within 10 million years of one another) we simplify things here by focusing on the better known skull and skeleton of *Paleothyris*.

Paleothyris was a small animal, less than 30 cm long (Fig. 12.9) (Carroll, 1969). Its delicate skull was long, narrow, and tall, and the premaxilla, maxilla, and dentary bones bore small, sharp, pointed teeth. The squamosal, jugal, and pterygoid bones are expanded in relation to the facial skeleton when compared to most reptiliomorphs, and this would coincide with large jaw-closing muscles. There is no otic notch, and the stapes of *Paleothyris* is thick and probably could not have conducted high-frequency sounds to the inner ear (Carroll, 1969, 1988; Benton, 2005). Like many lepospondyls and reptiliomorphs, *Paleothyris* was likely deaf to airborne sounds. The palate was speckled with tiny denticles that prey could have been pressed against for additional crushing. Given its small size, the most likely prey for an animal like *Paleothyris* would have been insects and other small invertebrates. As a general rule, small vertebrates get the most bang for their buck, calorie-wise, by eating insects rather than eating plants or fruits (Modesto, Scott, and Reisz, 2009). For example, the smallest living placental mammal, the shrew, lives on a diet of insects to fuel its hot-blooded little body. Although it is doubtful that *Paleothyris* had a high metabolic rate, the small body size of this animal did not permit a gut capacity necessary to ferment and digest plant material. Moreover, the powerful jaws of this little amniote and its pointed teeth are precisely what we see in modern insectivores. This jaw and tooth setup allows

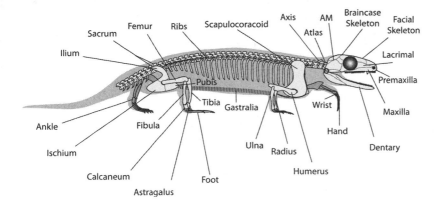

12.9. Skeletal diagram of one of the earliest true amniotes, *Paleothyris*. This little animal (less than 30 cm in length) possesses all of the major traits that denote amniotes, including a ball-like occipital condyle, a distinct atlas and axis, and a distinct astragalus and calcaneum. The small size and the combination of these traits are very close to predictions of what the common ancestor of all amniotes was like, and collectively suggest that *Paleothyris* was laying amniotic eggs. *Paleothyris* skeletal diagram based on Carroll (1969, 1988) and Benton (2005).

insectivores to puncture the hard outer chitinous shell of their prey to get at the tasty nutrients inside (Carroll, 1969, 1988; Benton, 2005).

But further direct evidence of insectivory comes from the fossil record. First, the earliest amniotes like *Hylonomus* and *Paleothyris* are well known and well preserved because they are found within fossil tree stumps. Large lycopod (club moss) trees of the genus *Sigillaria* were common constituents of Carboniferous forests, and were plentiful 310–300 Ma (Card 94) in Nova Scotia. When these large trees died and fell over, their stumps rotted out much more quickly than their bark, creating a hollowed-out home and shelter for insects, and a tasty attraction for insectivores. It is perhaps no coincidence that we find the fossils of *Hylonomus* and *Paleothyris* so well preserved within these *Sigillaria* stumps (Carroll, 1988; Benton, 2005). During heavy rains, flood waters would rise above the hollow stumps and inundate them with muddy sediments. Animals that had died or were trapped within the stumps would have been quickly buried and preserved (Carroll, 1988; Benton, 2005). More recent evidence from some early fossil members of the Reptilia shows that the link between the sharp teeth, jaw structure, and insectivory inferred for animals like *Paleothyris* is well supported. These Early Permian (280 Ma; Card 94) fossil reptilians from Oklahoma have well-preserved delicate skulls with pointed teeth and expanded cheeks for jaw-closing muscles (Modesto, Scott, and Reisz, 2009). Preserved inside the jaws of these animals are definitive insect parts (Modesto, Scott, and Reisz, 2009), from which we infer that the earliest amniotes like *Paleothyris* were almost certainly insectivores.

The atlas and the axis are well developed in *Paleothyris* (Fig. 12.9). The socket-like facet of the atlas articulates with the ball-like condyle on the skull, which would have allowed head nodding and some rotary movements. The axis is a large vertebra, and it locks onto the atlas from behind, stabilizing the neck at the skull-neck juncture; this probably allowed further nodding movements of the head. Four other smaller cervical vertebrae follow the atlas and axis, which are in turn followed by about 27 dorsal vertebrae. Long ribs that could be rocked craniocaudally for respiration are present, and these become smaller toward the sacrum.

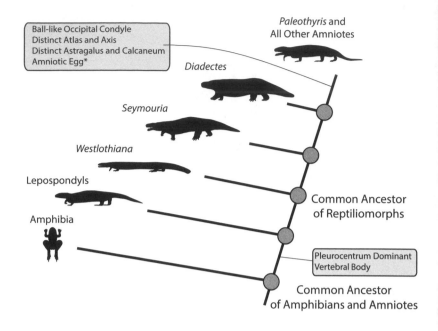

12.10. Pedigree of relationships among the Lepospondyls, Reptiliomorphs, and early true Amniotes discussed in this chapter. The asterisk (*) in the diagram indicates that we are inferring the presence of an amniotic egg in *Paleothyris* based on its skeletal trait states. See text for details.

Ball-like Occipital Condyle
Distinct Atlas and Axis
Distinct Astragalus and Calcaneum
Amniotic Egg*

Paleothyris and
All Other Amniotes

Diadectes

Seymouria

Westlothiana

Lepospondyls

Amphibia

Common Ancestor
of Reptiliomorphs

Pleurocentrum Dominant
Vertebral Body

Common Ancestor
of Amphibians and Amniotes

The sacrum itself contains two vertebrae that firmly unite with the ilium through their ribs, and some caudal vertebrae with delicate hemal arches are preserved in the known specimens. Belly ribs, known technically as gastralia, were embedded within the belly skin of *Paleothyris* (Fig. 12.9) (Carroll, 1969, 1988; Benton, 2005). In modern vertebrates with gastralia, such as alligators and crocodiles, these bony elements stiffen the abdomen and provide purchase for abdominal muscles that compress the belly during breathing (Carrier and Farmer, 2000). Perhaps the gastralia of *Paleothyris* accomplished a similar task.

The pectoral and pelvic girdles of *Paleothyris* take on shapes and arrangements that would now appear in various modified forms in all the amniote descendants. The pectoral girdle is composed of a substantial scapulocoracoid and tiny, sticklike cleithra and clavicles. An interclavicle is present as a bony, T-shaped rod that was probably embedded within a cartilaginous sternum, as in many modern lizards (Carroll, 1969, 1988; Benton, 2005). In the pelvic girdle, the ilium is long and the ischium and pubis are broad, suggesting that hind limb rotation, abduction, and adduction were well developed. As with other tetrapods we have encountered, *Paleothyris* was a rear-wheel-drive animal with hind limbs longer than its forelimbs. Both the forelimbs and hind limbs are slender, but still take on the form and muscle scars of the now familiar crank-like elements we have encountered in a variety of amphibians, lepospondyls, and reptiliomorphs. Distinct but rounded wrist bones probably contributed to rotational movements of the hands, whereas a large and distinct astragalus and calcaneum likely enhanced the flexion and extension of the foot in the direction of travel. Finally, the hands and feet are long, asymmetrical, and comprised of five digits each (Carroll, 1969, 1988; Benton, 2005).

Overall, *Paleothyris* is almost precisely what our anatomical hypothesis would predict for the form of the earliest amniotes (Fig. 12.9). We predicted that the first amniotes would be relatively small animals, in part because of trade-offs with direct development and the initial evolution of the amniotic egg. This small size would in turn require an insectivorous diet for the amniote to be a viable competitor in the new terrestrial world. *Paleothyris* has the hallmark characters we find in all modern amniotes: a tall and narrow skull, a specialized atlas-axis complex, and a definitive astragalus and calcaneum that improve flexion and extension of the foot against the ground in the direction of travel. The ribs are also well developed and could have been rocked back and forth to allow respiration. The slender limbs and robust limb girdles suggest *Paleothyris* and the common ancestor of the amniotes was an active, predatory animal. It is from this kind of a skeletal chassis that both the Reptilia and Synapsida (including Mammalia) have descended. To graphically summarize our discussion in this chapter, Figure 12.10 diagrams the pedigree of relationships among the lepospondyls, reptiliomorphs, and *Paleothyris*.

Not long after little animals like *Paleothyris* appeared, the amniotes split into their two great family branches: the Synapsida and Reptilia. Often, for chronological reasons, the Synapsida and Reptilia are introduced at the same time, and both lineages are followed simultaneously. Although it is historically accurate and instructive to understand that both of these groups have existed and coevolved since 310 Ma (Card 94), here we will take a different approach, and first follow out the line that led to the Reptilia. After exploring the diversity and evolution of the Reptilia chassis, we will return to the Synapsida branch later in the book so that the evolution of their skeleton can be traced in an unbroken pattern.

Deep Scaly I: Reptilian Chassis from Early Reptiles to Sea Monsters

5

And he who made kittens put snakes in the grass.

JETHRO TULL, *Bungle in the Jungle*

Modern Lizards and the Tuatara as an Introduction

13

AMONG MY PALEONTOLOGY FRIENDS, THERE IS A COMMON JOKE about the hot, dry environments we tend to work in. When your brain is cooking out of your head under the desert sun, someone is bound to say, "Well, at least it's a dry heat!" To which the reply is often, "Yeah, like your oven." I am the first to admit that I would rather put up with dry heat than the hot, humid air I have experienced living in the Midwest and now the mid-Atlantic. However, hot is hot, and just because the heat is 'dry' does not mean it will be comfortable. You also lose a lot of body water to the environment. In desert environments, you almost never feel yourself sweat because it evaporates off your skin so fast, often leaving a thin crust of body salts on your arms at the end of the day. It is also not uncommon to drink more than a gallon of water in a day and yet not pee at all! Drink early and drink frequently is the rule to live by (literally) when you are out in a desert.

As the Carboniferous closed and the Permian dawned (~300 Ma; Card 94), the continental landmasses were coalescing into a supercontinent called Pangea. What follows is a greatly simplified picture of what changed in the environment, and I encourage interested readers to immerse themselves in the popular and scientific literature on this pivotal time in earth history. Here, we will focus simply on overall climate changes. In some ways, the formation of Pangea was advantageous, because, as with the Internet, the world became more interconnected. We would be envious of land travel during this time: imagine being able to drive from North America into Africa, or take a train from South America into Antarctica and on to Australia! Not surprisingly we find similar groups of fossil vertebrates, other animals, and plants on continents across the world in Permian rocks (Redfern, 2003; Benton, 2005).

But what at first seems ideal became deadly. Not only did many of the continents come to lie over the equator, but as they converged water evaporated or drained away from the center of Pangea. Gone were the tropical coal forests. In their place, without intervening oceans between the continents to cool and moisturize the air, conditions became hotter and drier over the huge supercontinent in an escalating feedback cycle (Benton, 2005; Blakey and Ranney, 2008). Essentially, the center of Pangea became a large frying pan, and by the end of the Permian, most of the supercontinent was a gigantic, inhospitable desert with only damp fringes of coastline left. As a result of this and other complex factors (including rampant volcanism), one of the largest extinctions of all time, larger than the one that killed off the nonavian dinosaurs (66 Ma; Card 99), occurred at the end of the Permian Period (250 Ma; Card 95), and huge numbers of animals and plants died out (Benton, 2005).

During the Permian, the synapsids and reptilians vied for land supremacy, and initially synapsids were more successful. As we will see later, synapsids formed the first modern terrestrial vertebrate ecosystems during the Permian. However, although both groups were decimated by the end-Permian extinction, a quirk of physiological luck seems to have favored the reptilians, for they bounced back and dominated the following Mesozoic Era (250–66 Ma; Cards 95–99). As you will recall, reptilians solved the problem of excessive water loss associated with excretion of nitrogenous wastes by expelling uric acid, a pasty substance that requires only a tiny amount of water. In contrast, mammals and their synapsid ancestors excreted urea, which is more water-costly. As the Mesozoic Era dawned, the ability to produce concentrated uric acid in reptilians may have allowed them to gain the upper hand in exploiting newly opened environments that were still dry, whereas the ancestors of mammals could cope only by staying small and either remaining closer to water sources or becoming nocturnal. This again is an oversimplification of the very complex series of interactions that led up to rise of the reptilians during the Mesozoic. However, it is noteworthy that many reptilians continue to be very successful at exploiting desert habitats to this very day (Pough et al., 1998).

The earliest members of the Reptilia and Synapsida are often compared to lizards and called 'lizard-like' in many descriptions. However, the reptilians and synapsids of the Permian are an odd and unfamiliar bunch that is difficult to appreciate without some background knowledge of one of the great living reptilian branches that exists today, the Lepidosauria. This lineage includes lizards, snakes, and an odd lizard-like animal called the tuatara that exists only in some islands off New Zealand. We will spend this chapter focusing on modern lizards and the tuatara, and come back to discuss snakes in another chapter. By exploring the skeletal anatomy, biology, and diversity of lizards and the tuatara, we are then better prepared to understand the early evolution of the reptilian chassis during the Permian and beyond, as well as that of the early synapsids, which we will explore in later chapters.

Lepidosauria, Part 1: Lizards and the Tuatara

Today, lepidosaurs are a large and diverse amniote group comprising over 9,000 species. For the paleontologist, lepidosaurs provide both a window on the workings of the early amniote chassis and a natural experiment in amniote diversity. On the one hand, some lizards and the tuatara move about in ways not much different from salamanders and from the predicted locomotor movements of the common tetrapod and amniote ancestors (Reilly et al., 2006). On the other hand, some lizards can reproduce without sperm, others have developed placenta-like structures numerous times, still others have lost their limbs, and a plethora of species have developed venom glands (Sites, Reeder, and Wiens, 2011). In most

Pedigree Amniotes with Overlapping, Keratin-based Epidermal Scales

Date of First Appearance ~245 Ma (Card 95)

Specialties of the Skeletal Chassis Specialized Ankle Joints and Tails with Breakage Zones

Eco-niche Terrestrial to Aquatic Predators and Herbivores

discussions of lepidosaurs, the tuatara is examined first because it is the most primitive living member of the group, and relatives of these animals appeared earliest in the fossil record. However, because the tuatara is an unfamiliar animal, I have found it more instructive to discuss lizards first.

One of the standout features of lizard skeletons is that they have kinetic skulls: the bones of the skull move during feeding (Pough et al., 1998; Iordansky, 2011). As mammals, we are so used to the idea of chewing our food that the idea of a flexible skull would seem like a disadvantage: if your teeth and jaw bones slid around as you chewed, you wouldn't make much progress with anything harder than oatmeal. But lizards don't chew their food, at least not in a way recognizable to us. For example, anyone who has watched a carnivorous lizard feed knows that they grab prey with their toothy maws and through a series of head thrusts push the hapless victim down their throats with assistance from gravity (Pough et al., 1998). As we have discussed previously for many fishlike vertebrates, prey does not go down easily and tends to struggle. Wouldn't a flexible skull be a disadvantage in subduing struggling prey?

To get at what is advantageous about a flexible skull, I have to first recount a rather embarrassing chapter from my young adult life. When I moved into my first apartment, I bought a large bookshelf. After assembling the frame and shelves, I noticed this cardboard thing painted to look like a wood backing. Because I already was putting this against a wall, I figured that cardboard thing was an unnecessary decoration and pitched it into a closet. That was a major mistake. The book shelf stood okay for a few days, but then began to lean to one side under the weight of the various objects I had placed on it (a TV among them!). My solution was to redistribute the weight of things on the shelves, but this did little to stop the leaning. One day, my roommate took a look at the bookshelf and asked me if it had come with a backing. Pulling it out of the dust bunnies in the closet, I tacked it to the back of the bookshelf frame. Funny, that 'unnecessary decorative' backing helped the bookshelf stand tall and straight. This was my first lesson in the importance of crossbeams.

When you build a boxlike structure that will be loaded statically, it will support weight well. However, its weakness will come when you push on it from the side: the box will tend to distort easily into a parallelogram, and the once-vertical support beams will now be angled and experience potentially damaging bending stresses (Fig. 13.1). Many bridges and buildings are essentially statically loaded boxes, and engineers and architects have known for a long time that you must place crossbeams within the 'box' to resist sideways stresses. This is why many metal bridges have X-shaped steel girders in their sides (Fig. 13.1) (Bloomfield, 2006). That bookshelf I tried to build similarly needed crossbeams, this time in the form of a cardboard sheet tacked to the frame.

By comparison, the skulls of many amniotes are a boxlike arrangement of bony struts, and some of these elements act as crossbeams to resist distortion of the head during feeding. For example, the bones that lie between the temporal fenestrae we discussed in chapter 12 are essentially

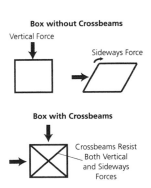

13.1. The importance of crossbeams. Without crossbeams, the box can be easily distorted into a parallelogram by sideways forces. With crossbeams, the box is able to resist sideways forces and maintain its shape.

crossbeams that provide lightweight strength to the back of the skull to resist the stresses of chewing or clamping down on prey. As we will later see, this kind of box-girder construction has been advantageous for crocodiles as well as mammals, but it is not always advantageous to have immobile skull architecture.

A lizard skull has evolved along somewhat the same principle as a set of adjustable pliers (Fig. 13.2). Typical pliers work great when you match the size of the tool to the bolt you are trying to loosen or tighten. However, if you have the wrong size of pliers, either the tool will not fit around the bolt or it will slip off. Adjustable pliers give you more versatility because you can adjust the size and grip of the tool for the problem at hand, applying the appropriate mechanical advantage to turn or tighten a bolt. Likewise, the kinetic skulls of many lizards can actively adjust to the shape of their struggling prey by sliding their bones in relation to one another (Herrel et al., 2007). Lizards have accomplished this, in part, by losing the lower temporal bar around their diapsid fenestrae. Without this brace, just like the box without the diagonal crossbeam, the square-like back end of the skull can now slide and distort parallelogram-like in relation to its normal shape (Fig. 13.2) (Herrel et al., 2007). More significantly, many of the bones in lizard skulls have joints that are similar to that of the quadrate-articular jaw joint. Although we haven't mentioned it previously, the quadrate-articular and limb joints are technically called synovial joints because they are surrounded by a capsule of soft tissues that holds the lubricating (synovial) fluid, which allows smooth movements and prevents direct rubbing of the bone ends. Many of the other joints (but not all of them) between the bony struts in lizard skulls are also synovial joints, permitting smooth, frictionless movements that allow the skull to distort and then return to its original shape (Holliday and Witmer, 2008; Payne, Holliday, and Vickaryous, 2011).

In many lizards, with the loss of the lower temporal bar, the quadrate bone is no longer connected in line with the upper tooth row and is free to rotate. In fact, the quadrate can rotate in relation to the squamosal bone above it, allowing it to act as a mobile, lever-like extension of the lower jaw (Fig. 13.2) (Herrel et al., 2007). In most non-mammalian tetrapods, the deeply situated pterygoid muscles pull on the lower jaw, causing it to pivot about the quadrate-articular joint to clamp the jaw closed with assistance from the outer adductor mandibulae muscles. Recall that the mechanical advantage for jaw clamping is increased if the functional jaw joint lies above or below the tooth row, as we saw for the herbivorous reptiliomorph, *Diadectes*. Similarly, the now mobile joint between the quadrate and squamosal bones in most lizards has effectively become a pivot point above the tooth row that the deeply situated pterygoids can act along. This is how something as seemingly counterintuitive as a kinetic skull can act to increase jaw clamping power in lizards. I should know— one of my lab monitor lizards has bitten my hand on several occasions. Although no harm was done (I do wear protective gloves), the squeezing of this lizard's jaws on my hand is quite sensational. In addition to the

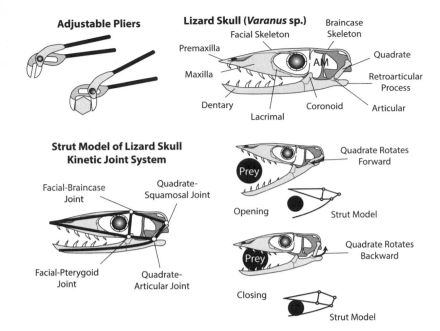

Adjustable Pliers

Lizard Skull (*Varanus* sp.)
- Braincase Skeleton
- Facial Skeleton
- Premaxilla
- Quadrate
- Maxilla
- Retroarticular Process
- AM
- Dentary
- Coronoid
- Articular
- Lacrimal

Strut Model of Lizard Skull Kinetic Joint System
- Facial-Braincase Joint
- Quadrate-Squamosal Joint
- Facial-Pterygoid Joint
- Quadrate-Articular Joint

Prey — Opening — Quadrate Rotates Forward — Strut Model

Prey — Closing — Quadrate Rotates Backward — Strut Model

13.2. Lizard skull mechanics. Like a set of adjustable pliers, the skull of many lizards can adjust to the size of its prey and maintain prey in the mouth once captured. Four joints within a typical lizard skull (exemplified here by *Varanus* sp.) allow the skull and mouth to change shape to suit the prey being eaten. In addition, the quadrate bone can actively rotate in relation to the skull and jaw joint, improving the mechanical clamping advantage of the pterygoid muscles. AM indicates the location of the adductor mandibulae and other jaw-closing muscles. Based on Smith and Hylander (1985), Pough et al. (1998), Metzger (2009), Montuelle et al. (2009), and Kardong (2012).

mobile quadrate, the gape of the jaws is increased substantially by two other joints. A hinge-like joint between the facial and braincase bones above, and another joint between the pterygoid and facial bones below, allows the snout to tip up when the mouth is opened (Fig. 13.2) (Pough et al., 1998; Herrel et al., 2007; Moreno et al., 2008). It should be noted that there is quite a diversity of skull kinesis across lizards, and this mobility is reduced in some species (Iordansky, 2011).

Jaw closing and opening are also enhanced by modifications in the lower jaw. First, the lower jaw of most lizards has a conspicuous pointed bone called the coronoid located just caudal and dorsal to the dentary bone (Fig. 13.2). The coronoid serves as the region of insertion for the adductor mandibulae muscles, and its size and shape provide a lever-like projection upon which these muscles can pull the jaw shut with improved mechanical advantage (Herrel et al., 2007). Caudal to the articular bone, a prominent retroarticular process has developed that acts as a lever upon which the depressor mandibulae muscle (discussed in chapter 9) can pull to open the jaws (Fig. 13.2).

Lizard teeth deserve brief mention here. All lizards can shed and replace their teeth on a regular basis, and in predatory lizards the upper jaw is wider than the more narrow lower jaw, allowing the teeth to slide past each other like scissors blades. In herbivorous lizards, the teeth are commonly blunted and contact one another to help pulp vegetation before it is passed on to the digestive tract. In many lizard species, the teeth are loosely cemented to the side of the premaxilla, maxilla, and dentary bones, and as they fall out or are broken they are replaced by new tooth buds growing in beneath them (Pough et al., 1998; Liem et al., 2001). In other lizard species (and in snakes) the teeth are instead perched atop the outer edge of the tooth-bearing bones and are replaced less often by a new tooth that rotates into the space left by the shed tooth (Pough et al., 1998; Liem et al., 2001).

Many lizards see and hear very well. In many lizards, a recess located behind the quadrate bone houses a taut tympanic membrane that transmits high-frequency airborne vibrations into the inner ear through a slender stapes (Pough et al., 1998; Liem et al., 2001). However, not all lizards have an external ear opening. For example, many burrowing lizards have secondarily lost this feature to prevent debris from lodging in their ears when they push through sediment and sand (Pough et al., 1998; Hildebrand and Goslow, 2001; Liem et al., 2001). Many lizards also have excellent color vision, and many of them see into the ultraviolet parts of the light spectrum that are invisible to us (Yokoyama, 1997). Various nerve cells within the light-detecting retina pick up different frequencies of light. One group of these cells, called rods because of their rodlike shape, detects differences between light and dark in the environment. In vertebrates that lead a nocturnal existence, rods are plentiful in the retina (Liem et al., 2001; Kardong, 2012). Rods are great at picking up images and movement in low light, but the images they form are slightly blurred and not sharply focused. In contrast, the cells called cones (again, due to their shape) perceive different frequencies of light that make up the color spectrum. Cones provide excellent, focused, fine-tuned color vision, but only work well in bright light. Several cones that detect different color frequencies collectively allow the brain to colorize the images transmitted from the retina (Liem et al., 2001; Kardong, 2012).

If you're like me and you geek out over large TVs, then you may have noticed that the individual pixels that make up the picture are one of three primary colors: red, blue, or green. In elementary school, many of us learned from our art teachers that the primary colors were red, blue, and yellow. We learned that combining these colors in different ratios gave us oranges, greens, and so on. If you wanted to make a dark brown or black blob, you combined all the primary colors together. However, in nature the primary colors of light are red, blue, and green. Unlike paint, when all three primary light colors are combined they give you white light, whereas the absence of light creates darkness. Other combinations of these three colors create the extensive Technicolor world we know so well. Combinations of red, blue, and green pixels enable televisions, computer screens, and cell phone displays to form bright and vivid pictures. Similarly, the retina of many vertebrates contains two or more types of cones for detecting the primary light colors. In humans, the retina contains three cone types that detect red, blue, and green, and combinations of signals from the cone cells allow the brain to discriminate the colors in your environment (Liem et al., 2001). Color blindness occurs when one or more of the cone types do not develop or cannot detect a particular part of the primary color spectrum.

Many lizards not only have the three cone types we possess, but also have additional cone types that can detect spectra of light in the ultraviolet. In many lizards, specialized colored oil droplets within certain cones act to more finely tune in various color tints, whereas others allow ultraviolet light to be detected and visualized (Loew et al., 2002; Fleishman, Lowe, and Whiting, 2011). Intriguingly, some lizard species have been shown to have scales and body coloration that reflect ultraviolet light, especially on parts of the body (such as the erectable skin flap under the chin called a dewlap) used for visual communication (Loew et al., 2002). Other lizards that lead a more nocturnal existence have done something bizarre: they have no rods in their retinas and instead rely on derived cones sensitive to light and dark (Pough et al., 1998; Fleishman, Lowe, and Whiting, 2011). Ultimately, most lizards have excellent color vision, and when they view a rainbow in the sky, they see colors beyond the violet ends of the spectrum. The remarkable color variation seen in lizards is intimately linked to their color vision: just as you don't spend energy making vocal calls if other members of your species can't hear you, you also don't bother with colorful body patterns unless others can see and interpret them.

Lizards also have a well-developed sense of smell and taste, as well as an enhanced ability to detect pheromones and other chemical signals in the environment via a prominent vomeronasal organ. Typical odor detection occurs as described generally for tetrapods in chapter 9. Tastes are picked up on both the tongue and roof of the mouth, and primitively lepidosaurs are hypothesized to have acquired prey using the tongue (Elias, McBrayer, and Reilly, 2000; Iwasaki, 2002). Although some lizard species have fleshy tongues, many have narrow tongues that are bifurcated or, in some groups, deeply forked. A bifurcated or forked tongue allows a lizard to taste chemosensory information in the air, on another lizard or animal, or on the ground. Because the tongue is bifurcated or deeply forked, it can discern which direction chemical signals are coming from in 'stereo' (Pough et al., 1998; Pough, Janis, and Heiser, 2002; Liem et al., 2001). The tongue draws in odor particles and places these into dual slots in the roof of the mouth where the odorants are passed on to the vomeronasal organ (Fig. 13.3) (Liem et al., 2001). This dependence on the tongue primarily to facilitate olfactory and pheromone detection would seem to render it relatively insignificant in processing food. However, surprisingly, the bifurcated tongue of lizards has actually been shown to assist significantly with food processing (Elias, McBrayer, and Reilly, 2000).

Based on extensive neurobiology research over the past 20 years, we know that the brains of lizards are a reasonable proxy for the primitive brain structure characteristic of the earliest amniotes. If we peer into the skull of a typical lizard, we find a brain that is somewhat larger and more complex than what we have encountered in actinopterygian fishes and lissamphibians (Fig. 13.4). One of the striking differences between the brain of a lizard and that of a frog is that the cerebrum is much more expanded (Northcutt, 2002; Butler and Hodos, 2005; Butler, Reiner, and Karten, 2011). It is so expanded, in fact, that it partially covers much of the caudal forebrain region. There is an enlarged part of the brain in reptilians (including birds) called the dorsal ventricular ridge (DVR)

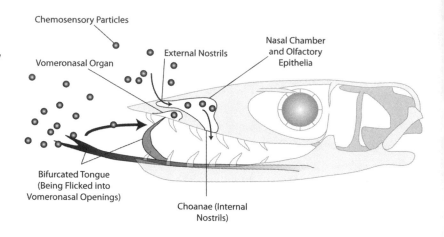

13.3. The bifurcated tongue and vomeronasal organ in lizards. Many lizards pick up odors and other chemosensory particles on a bifurcated or forked tongue. This allows lizards to smell in 'stereo.' The tongue is also placed into openings within the roof of the mouth where certain chemosensory particles can activate the vomeronasal organ, which is used to finely discriminate between other species members and prey. Based on Pough et al. (1998) and Kardong (2012).

Chemosensory Particles

External Nostrils

Nasal Chamber and Olfactory Epithelia

Vomeronasal Organ

Bifurcated Tongue (Being Flicked into Vomeronasal Openings)

Choanae (Internal Nostrils)

that develops beneath the dorsal cerebrum (Butler and Hodos, 2005). That is correct–reptilians invented the DVR. Previously, scientists had thought that the reptilian DVR was an expanded part of the deep cerebrum, which is an area that typically enhances and suppresses complex behavioral reflexes (Butler and Hodos, 2005; Reiner, 2009). In older texts and discussions of brain evolution, it was often suggested that most of the complex behaviors observed in reptilians (including birds) were a result of preprogrammed reflexes in the DVR. However, new neurological data show that this earlier hypothesis was incorrect and that the DVR is in fact related to the dorsal cerebrum (technically known as the dorsal pallium) (Reiner, 2009; Butler, Reiner, and Karten, 2011; Walsh and Milner, 2011). This means that the DVR is an expanded area for somatosensory integration, particularly for vision and hearing. These new data strongly suggest that lizards and other reptilians are probably smarter and much more aware of their world than previously suggested.

One of the lizard species I work with in my lab is *Pogona vitticeps*, known colloquially as the bearded dragon. I have found them to be personable, friendly, and clever, as do many people who care for them as pets. Fascinatingly, a recent behavioral study on these lizards shows just how wrong previous assessments of reptilian intelligence have been (Kis, Huber, and Wilkinson, 2015). In this study, some of the lizards were trained to open a trapdoor to get at tasty insects, while a control group was given access to the trapdoor but not shown how to open it. Not surprisingly, the bearded dragons trained to open the trapdoor learned how to use it to get treats, whereas the untrained beardies (as they're affectionately called) never did figure it out. Here's where it gets interesting: when beardies that had not been trained to open the trapdoor were placed in the same enclosure with those who were, they were able to observe and imitate the behavior of the trained lizards to get the insect treats. In other words, untrained beardies learned by watching and imitating trained individuals, an ability usually thought to be reserved only for mammals and birds (Kis, Huber, and Wilkinson, 2015)! So, those of you who learned in various biology or psychology courses that our thinking mammal

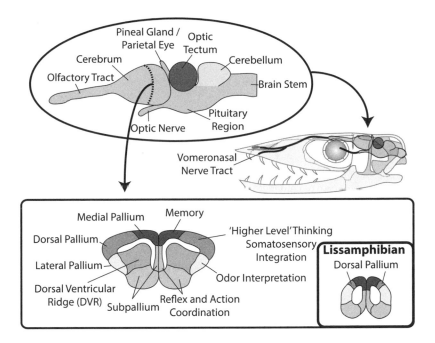

13.4. Schematic of the lizard brain. The brain of many lizards is a good proxy for the relative size and functional abilities of the ancestral amniote brain. Unlike in amphibians, the cerebrum of lizards is enlarged and covers part of the thalamus and caudal forebrain regions. A cross-sectional slice through the cerebrum of a typical lizard reveals the major functional regions of the pallium and subpallium. In amniotes, the dorsal pallium is greatly enlarged compared to this same region in lissamphibians (see inset box). In reptilians like lizards, there is also the development of a dorsal ventricular ridge (DVR) that is related to the dorsal pallium and boosts somatosensory integration, especially vision and hearing. Based on Liem et al. (2001) and Butler, Reiner, and Karten (2011).

brain overlies a primitive and brutish reptile brain may now deposit that hypothesis along with lizard droppings in the trash.

Most lizards retain the primitive pineal opening through which the median (parietal) eye detects light (Fig. 13.4 and Plate 14). Because lizards are ectothermic, they must be sensitive to daylight changes to efficiently shuttle among basking sites, their territories, and shaded regions for cooling. Therefore, it is not surprising that the light-detection system afforded by the parietal eye would remain prominent in these amniotes (Pough et al., 1998; Liem et al., 2001). The optic lobes are well developed, as we would expect given the keen eyes of most lizards. The cerebellum is also much more expanded than in lissamphibians, given the larger role of terrestrial locomotion and the need for greater coordination and refinement of motor impulses (Fig. 13.4) (Liem et al., 2001; Butler and Hodos, 2005).

The vertebral column of lizards resembles in many ways those of the reptiliomorphs we have previously discussed. A distinct atlas and axis is present, as well as a cervical series of varying lengths (Fig. 13.5). The dorsal vertebrae bear prominent ribs for protection and respiration, the cranial-most ribs connect to the sternum through cartilaginous extensions, and the sacrum is intimately connected with the pelvic girdle. The zygapophyses of the dorsal vertebral series allow the vertebral column to flex laterally, and these movements increase the step length of the forelimbs and hind limbs. Moreover, the centrum of most lizard vertebrae has a hollow socket in front and a ball-like process behind (Liem et al., 2001; Kardong, 2012). This allows the ball on the end of one centrum to articulate with the hollow socket on the following vertebral centrum. This ball-and-socket configuration further enhances the lateral movements of the vertebral column.

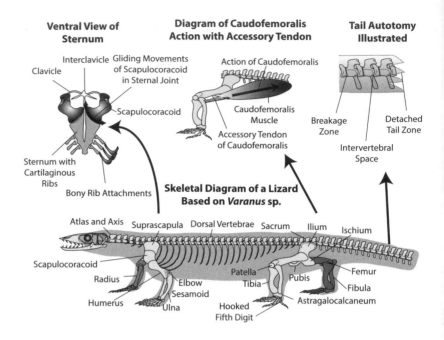

13.5. Skeletal anatomy of lizards as exemplified by *Varanus* sp. See text for details.

Ventral View of Sternum

Clavicle
Interclavicle
Gliding Movements of Scapulocoracoid in Sternal Joint
Scapulocoracoid
Sternum with Cartilaginous Ribs
Bony Rib Attachments

Diagram of Caudofemoralis Action with Accessory Tendon

Action of Caudofemoralis
Caudofemoralis Muscle
Accessory Tendon of Caudofemoralis

Tail Autotomy Illustrated

Breakage Zone
Detached Tail Zone
Intervertebral Space

Skeletal Diagram of a Lizard Based on *Varanus* sp.

Atlas and Axis
Suprascapula
Dorsal Vertebrae
Sacrum
Ilium
Ischium
Scapulocoracoid
Radius
Patella
Femur
Elbow
Tibia
Pubis
Fibula
Sesamoid
Humerus
Ulna
Hooked Fifth Digit
Astragalocalcaneum

One of the key features of lepidosaurs is that the caudal vertebrae have breakaway zones: the tail can be actively broken and detached when grabbed by a predator, a process called caudal autotomy (Zani, 1996; Bateman and Fleming, 2009). When many lizard species are grabbed by the tail, a series of rapid physiological responses cleaves the vertebrae apart at a breakage zone just in front of the point where the tail has been caught, and at the same time the severed blood vessels in the tail squeeze shut to prevent excessive blood loss (Fig. 13.5) (Bateman and Fleming, 2009). The tail muscles are segmented in such a way that they easily detach from one another once the tail is broken off. The detached tail end will jump and wriggle, perhaps distracting a predator while the rest of the lizard scampers off to live another day. For most lizards, the tail will regenerate after such an event, but the new tail replacement will lack vertebrae. Instead, the newly developed piece will be a rod of cartilaginous material sheathed in skin (Pough et al., 1998; Bateman and Fleming, 2009). Although many lizard species possess caudal autotomy, some have secondarily lost or reduced this ability (Zani, 1996; Pough et al., 1998; Bateman and Fleming, 2009).

Lizards have perfected the art of the sprawling, crank-like limb gait (Plate 15). In the pectoral girdle, the coracoid bones fit into slippery, cartilaginous grooves on the sternum that allow the scapulocoracoids to rotate and shift as the lizard walks (Fig. 13.5) (Jenkins and Goslow, 1983). These movements add significantly to the length of the forelimb's step. A T-shaped interclavicle is present in most lizards, and its ends typically articulate with the clavicles, which in turn link to the scapulocoracoids (Fig. 13.5). This interconnected system of chest bones is hypothesized to brace and enhance scapulocoracoid rotation during locomotion (Gray, 1968). This is not an unusual function for elements like the clavicle, and as we will see later, the clavicle plays a critical functional role in bracing the forelimb in some mammals. The pelvic girdle is similar to what we have discussed previously for reptiliomorphs and microsaurs, and the ilium, ischium, and pubis are all broad bones with ample regions for hind limb protractors, retractors, abductors, adductors, and rotators (Fig. 13.5). A space between the pubis and ischium known as the thyroid fenestra is present in all lepidosaurs, but this space is occupied by cartilage and membranous tissues in the living animal (Pough et al., 1998; Liem et al., 2001; Benton, 2005).

The humerus rotates, abducts, and adducts, the radius and ulna commonly rotate and slide independently of each other, and the multiple wrist bones provide a flexibility that allows the hand to remain pronated as the forelimb is retracted (Landsmeer, 1981, 1983). In the hind limb, the femur typically bears a prominent landmark called the fourth trochanter. This landmark is associated with the insertion point of a large muscle called the caudofemoralis, which arises from the tail. The action of the caudofemoralis is to forcefully retract the femur, and thus the hind limb. In fact, retraction of the hind limb in lizards accounts for a majority of hind limb propulsion (Gatesy, 1990). The caudofemoralis muscle in lizards not only attaches to the fourth trochanter on the femur, but a taut, tendon-like projection of the muscle inserts into the caudal side of the tibia just under the knee (Fig. 13.5). Experimental studies have shown that the caudofemoralis in lizards both retracts the femur and simultaneously flexes the knee (Reilly, 1998). However, this is not a trait unique to lizards, as it has also developed in the archosaurs we will discuss in later chapters (Gatesy, 1990). Despite how awkward such a posture and movement may seem to us, I have witnessed bearded dragons and monitor lizards alike bolt down my lab's treadmill with an unexpected speed and grace. A so-called sprawled posture need not be a disadvantage as it lowers the center of mass and can act to make lizards stable across a large variety of surfaces (Russell and Bels, 2001).

The ankle of lepidosaurs like lizards is unique in that the astragalus and calcaneum have fused together into a single unit called the astragalocalcaneum (Fig. 13.5) (Rewcastle, 1983; Russell and Bels, 2001). This bone articulates with the ends of the tibia and fibula and has a distal surface reminiscent of the threads on a screw that interconnect with the other ankle bones. This screwlike articulation allows mostly flexion and extension at the ankle, but also some controlled sideways rotation of the foot (Russell and Bels, 2001). The foot itself is asymmetrical, and lepidosaurs enhance this asymmetry by having a hooked outer fifth digit that is distinct and separate from the inner four. This foot shape appears to increase the proportion of the sole in contact with the ground as the lizard propels itself forward, especially because the joint shapes and mechanics of the ankle cause the foot to rotate outward as the hind limb is retracted (Russell and Bels, 2001).

Lizards are also one of the first groups of amniotes we will encounter to possess additional bones between their joints. Where tendons wrap

13.6. The mechanical advantage of ramps. Mechanical advantage comes down to force versus distance. If you (1) throw a refrigerator up to the second floor of a house, you cover less distance but (2) you require an inhuman amount of force. The more practical way to accomplish the same thing is to push the refrigerator up a gently sloping ramp, covering more distance but using far less force.

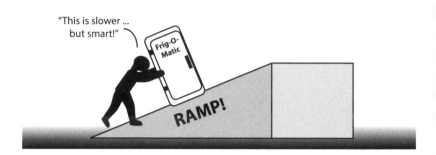

around joint surfaces, the constant tension in these regions induces the formation of cartilaginous and bony centers (Haines, 1969; Carter and Beaupré, 2001). However, recent studies on mice suggest some sesamoids may develop as portions of limb bones that 'defect' to a developing tendon (Eyal et al., 2015). These 'floating' bones within tendons reminded early anatomists of seeds, and they are now called sesamoid bones (Hildebrand and Goslow, 2001). Sesamoid bones improve the mechanical advantage of muscles pulling across joints in much the same way a rope and pulley system improves your ability to lift heavy objects off the ground (Hildebrand and Goslow, 2001; Liem et al., 2001).

To understand pulleys and sesamoid bone mechanics, it is practical to start with something simple. Let's say you want to move our heavy refrigerator introduced in chapter 6 from the ground floor to the second floor of a house (Fig. 13.6). There are two major things to consider here: distance and force. Certainly, throwing the refrigerator straight up to the second floor would cover less distance, but the force it would take to do this (let alone the destruction and mess it would create on landing) is practically inhuman. But if we increase the distance over which we move the refrigerator, most people of average build would have enough force to get the refrigerator up to the second floor. We increase the distance in this case either by moving in short increments up a series of steps or, better yet, pushing the refrigerator up one or a series of gently sloping ramps (Bloomfield, 2006). Although we will have to cover more distance to get the refrigerator up to the second floor than if we simply threw it, this is the practical and safe way to accomplish our goal (Fig. 13.6).

In a pulley system, we have a similar trade-off between distance and force. Let's say in this case we wanted to lift our refrigerator with a rope

The Mechanical Advantage of a Pulley System

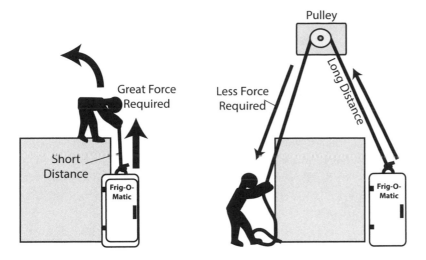

Great Force Required

Short Distance

Less Force Required

Long Distance

Pulley

Frig-O-Matic

Frig-O-Matic

13.7. Pulley systems and sesamoids. As with ramps, a typical pulley system augments the relationship between force and distance. If you lift the refrigerator with a short rope straight up to the second floor, you cover a short distance but require a huge amount of force. Using a pulley system, you trade force for distance: attaching a long length of rope to the refrigerator and pulling it over a wheel requires far less force to accomplish the same task. Sesamoid bones like the patella (kneecap) work with long muscle tendons to achieve pulley-like effects that require less muscular force for a given movement.

The Patella and Quadriceps as a Pulley

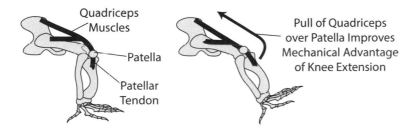

Quadriceps Muscles

Patella

Patellar Tendon

Pull of Quadriceps over Patella Improves Mechanical Advantage of Knee Extension

up onto a raised section of the floor (Fig. 13.7). Certainly, one way to do it would be to get above the refrigerator and pull vertically on a rope to raise it. Although this would require a shorter piece of rope, this exercise becomes difficult and impractical for numerous reasons. If instead we stand on the ground and pull a longer length of rope over a wheel, we will be much more successful. Thus, a pulley system is another trade-off between force and distance: improving mechanical advantage to lift an object attached to a rope requires a long rope angled over a pulley (Fig. 13.7) (Bloomfield, 2006).

Lizards develop sesamoids in between the elbow, knee, finger, and toe joints (Haines, 1969). The sesamoid that forms in the tendon that wraps over the knee is known as the patella, the bone we call our kneecap. Here, the knee extensor muscles are wrapping over the end of the femur and inserting into the tibia. Improving the mechanical advantage of this muscle group to enhance leg extension (kicking) is accomplished by the patella, which acts as a pulley, changing the length and angle over which these muscles act (Fig. 13.7) (Liem et al., 2001). Unlike a mechanical pulley system, the patella is not a wheel and cannot spin to decrease the friction of the tendon sliding over the knee. Instead, the patella has a slippery, cartilaginous side that slides in a groove on the end of the femur, reducing the friction of the tendon at the knee joint. Moreover, because the patella is restricted to sliding in a groove on the end of the femur, this prevents lateral slippage of the quadriceps tendon that could decrease the mechanical advantage of the knee extensors (Fig. 13.7) (Liem et al., 2001; Kardong, 2012).

Understanding the relationships among the living lizards has been complicated and controversial, although a general consensus has emerged in recent years based on both genes and anatomical traits (Fry et al., 2009; Hedges and Vidal, 2009; Vidal and Hedges, 2009; Sites, Reeder, and Wiens, 2011). It would be impractical to review the lizard family tree here, but a few groups can at least be very briefly highlighted to give an impression of the diversity in body shape and habits among the thousands of known lizard species. The most primitive living lizards, called dibamids, are nearly blind, insectivorous burrowers with small jaws. The geckos and their kin are nocturnal climbing specialists that form the outgroup to all more derived lizards. Specialized scales (called foot-hairs or setae) on the ventral surfaces of the fingers and toes in most geckos allow them to cling to and climb vertical surfaces. The tiny setae are so small they interact with attractive forces on a molecular level, something called Van der Waals forces (Autumn et al., 2000), allowing the geckos to stick and propel themselves at high speeds up vertical surfaces (Autumn et al., 2006).

The most derived lizards fall under two large groups within the so-called Bifurcata, and are characterized in part by their bifurcated tongues and expanded vomeronasal organs (Vidal and Hedges, 2009). Moreover, each of these groups includes numerous species that have independently evolved limb reduction or limblessness. One group, the skinks (Scincoformata), consists of numerous stout-bodied lizards with varying degrees of limb loss. The other group, Episquamata, characterized by their expansively developed vomeronasal organs and deeply forked tongues, includes two diverse branches. One branch, the Laterata, includes stocky lizards with short limbs called teiids and the weird amphisbaenians, a group specialized for burrowing with solid, shovel-shaped heads for compacting soil or pushing it out of the way. Given their burrowing habits, amphisbaenians lack external ears and have deeply recessed eyes covered by scales. In some ways, amphisbaenians resemble the caecilians we discussed in chapter 11: they, too, have a loose skin with which they can move by internal concertina motion (Pough et al., 1998).

The other branch of Episquamata is called Toxicofera, a diverse group united by their possession of toxin glands (Fry et al., 2005, 2009). Among these lizards are the infamous Gila monster of the southwestern United States, the iguanians, which include iguanas as well as chameleons, and the anguimorph lizards, which include the large, predatory varanid lizards (Plate 15) such as the giant Komodo dragon. Many of these lizards can run quickly, and several species assume a bipedal posture when doing so (Irschick and Jayne, 1999). Some even run on water:

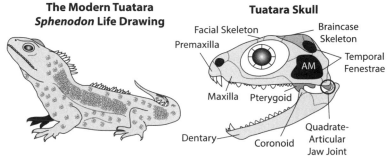

The Modern Tuatara
Sphenodon **Life Drawing**

Tuatara Skull

Facial Skeleton
Premaxilla
Braincase Skeleton
Temporal Fenestrae
AM
Maxilla   Pterygoid
Dentary
Coronoid
Quadrate-Articular Jaw Joint

Tuatara Skeletal Diagram in Lateral View

Dorsal Vertebrae
Scapulocoracoid
Humerus
Femur   Sacrum
Pelvis
Atlas-Axis Complex
Clavicle
Fibula   Tibia
Astragalocalcaneum
Radius
Ulna   Uncinate
Sternum and Rib Processes   Gastralia
Interclavicle

13.8. Skeletal anatomy and life portrait of the tuatara. Note that the skull is solid, retains the lower temporal bar, and is akinetic, unlike the more kinetic and lightly strutted skulls of lizards. AM indicates the location of the adductor mandibulae and other jaw-closing muscles. See text for more details. Skull and skeleton based on Romer (1956) and Liem et al. (2001).

among the iguanians is the famous 'Jesus Christ lizard' that can sprint across the surfaces of ponds with their large feet and rapid back-and-forth shifting of their body mass to maintain stability (Hsieh, 2003; Hsieh and Lauder, 2004). We will return to the dynamics of bipedal lizards in future chapters when considering the evolution of bipedalism in dinosaurs. Although not all toxicoferan lizards have toxic venom (your pet iguana is not a health hazard), they all possess salivary glands that produce complex enzymes that form the basis for toxins in many groups (Fry et al., 2005, 2009). One other toxicoferan group has become so successful that it is often given its own place among the Reptilia: the snakes. Essentially, snakes are limbless, eyelidless, earless toxicoferan lizards with megakinetic skulls and well-developed salivary glands that may produce venom (Fry et al., 2005, 2009). We will discuss these diverse, limbless vertebrates in chapter 15.

The diversity of lizards is quite impressive, and as you by now appreciate, calling a fossil amniote 'lizard-like' is a huge oversimplification. You may well wonder if there are any living amniotes that better resemble the early reptilians than lizards. Remarkably, there is a 'living fossil,' the tuatara, which, while specialized, may give us glimpses into what some of the early reptilians, or at least the earliest lepidosaurs, were like (Plate 16). The tuatara is the only remaining member of a once more diverse family of lepidosaurs called sphenodontians (Benton, 2005). Only two (or perhaps three) species of tuatara are left in the world, found on small islands off the coast of New Zealand today; they stretch about 60 cm long. At first glance, you would suspect the tuatara was a lizard. In fact, it looks kind of like an iguana, having a boxy head and bearing a spiny frill-like covering down its back (Fig. 13.8). But closer inspection of its skeleton and study of its habits reveal that this animal has some significant differences from its lizard relatives.

The skull resembles those of lizards in some respects, but the bones are immobile and do not form kinetic joints (Fig. 13.8) (Evans, 2003; Benton, 2005). Moreover, both the upper and lower temporal bars are present around the temporal openings, and so the back of the skull cannot distort parallelogram-like as in lizards. Canine-like teeth jut ventrally from the premaxilla and dorsally from the dentary, forming a beak-like chisel at the front of the mouth (Fig. 13.8). Robust, triangular teeth adorn the maxilla and the rest of the dentary, and a second row of pointed teeth projects from the palate. A prominent parietal opening houses a parietal eye for detecting light and modulating thermoregulation. Surprisingly, there is no external ear opening in tuatara, although they possess a delicate stapes and the ability to hear sounds (Pough et al., 1998; Benton, 2005). Because the tuatara is a lepidosaur, it shares with lizards breakaway zones in its tail, a fused astragalocalcaneum in its ankle, and a hooked fifth digit in the foot. However, its ribs are a bit more robust and develop overlapping processes reminiscent of birds and the transitional tetrapodomorph, *Tiktaalik*, we discussed in chapter 10 (Fig. 13.8). As you may recall, these rib processes provide leverage for intercostal muscles that rock the ribs craniocaudally during respiration (Tickle et al., 2007).

The habits of the tuatara are odd and unlike those of most lizards. These lepidosaurs are crepuscular, being most active during dusk and dawn. As ectotherms, they bask in the sun during the warmer parts of the day and store some of this thermal energy for their evening and early morning foraging. However, they are active at relatively low body temperatures compared to lizards, and seem to bask mostly to speed the digestion of their previous night's meal. Tuatara break open food using a specialized shearing action between their dentary and upper jaw bones. When the jaw closes, the dentary bone first clicks vertically into place between the tooth rows of the maxilla and palatal bones. Following this, the dentary bone slides forward, shearing and tearing the prey caught in its jaws (Jones et al., 2011). This shearing action is made possible by a flexible joint between the two dentary bones called mandibular symphysis (Jones et al., 2012). Tuatara are mostly insectivorous and appear to use the chisel-like teeth in their premaxilla and dentary bones to crack open invertebrate exoskeletons. However, they are also known to feed on seabirds, especially the chicks, and assorted small vertebrates. They tend to populate burrows they excavate near seabird colonies. This environment provides them with plenty of food options: the seabirds themselves are vulnerable at night, and the guano and dead chicks in the colonies attract hordes of insects and other tasty invertebrates (Pough et al., 1998; Benton, 2005). It would be tempting to correlate modern tuatara behavior with that of the common ancestor of lepidosaurs, but given what we know from the fossil record, it seems instead that these modern sphenodontians are highly derived in their habits.

The chassis of modern lizards and the tuatara have now provided us with a series of search images for the fossil record and some insight into the biology of both the earliest reptilians and the major radiations of the lepidosaurs and their close relatives. In the next chapter, we will first explore the chassis of the earliest branch of reptilians, the parareptiles, to see what is similar and different about these animals from our lizard friends. After an examination of turtles, we will follow this in chapter 15 by examining the great diversity of chassis in the lepidosaur pedigree, which range from the remarkable adaptations of snakes to the great sea dragons of the Age of Dinosaurs.

14.1. Skull and skeletal diagram of the procolophonid parareptile, *Procolophon*. Note the overbite at the front of the jaws, the well-developed pterygoid, and the prominent coronoid bone. The robust and squat skeleton and features such as the well-developed olecranon process suggest that animals like *Procolophon* were capable of burrowing. Note the large parietal opening shown in the dorsal silhouette of the skull. AM indicates the location of the adductor mandibulae and other jaw-closing muscles. Diagrams after DeBraga (2003) and Benton (2005).

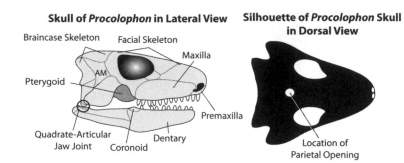

Skull of *Procolophon* in Lateral View

Braincase Skeleton
Facial Skeleton
Maxilla
Pterygoid
AM
Premaxilla
Quadrate-Articular Jaw Joint
Dentary
Coronoid

Silhouette of *Procolophon* Skull in Dorsal View

Location of Parietal Opening

Skeleton of *Procolophon* in Lateral View

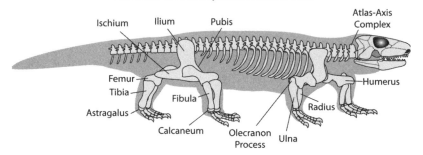

Ischium
Ilium
Pubis
Atlas-Axis Complex
Femur
Tibia
Fibula
Humerus
Astragalus
Radius
Calcaneum
Olecranon Process
Ulna

Early Reptiles and Turtles

14

GASOLINE-FUELED ENGINES REQUIRE THE RIGHT MIXTURE OF oxygen and fuel in their valves so that when a spark is applied, a small explosion can occur (Macaulay, 1988; Newton, 1999; Walker et al., 1999; Langone, 2004). This combustion moves valves in the engine that ultimately turn the wheels. Achieving the appropriate oxygen-fuel mixture was often controlled in the past by a carburetor: a winglike valve that adjusted the amount of air mixed with fuel as a car sped up or slowed down (Walker et al., 1999; Langone, 2004). As it turns out, better fuel economy could be achieved if fuel was injected directly into the engine valves in precisely controlled amounts, a process called fuel injection. Initially, there was little difference in fuel economy between carburetor and fuel-injected cars, and carburetors were often preferred as they were easier vehicles to maintain. However, changes in fuel efficiency standards in the 1970s and 1980s changed the economic landscape for car manufacturers. Now, it was more economical and desirable to build fuel-efficient cars, and this, combined with technological advances, eventually made the fuel-injected car the standard, with carburetor models almost completely phased out of existence by the early twenty-first century.

Similarly, two reptilian chassis types appeared approximately 310 Ma (Card 94), and neither one was initially more successful than the other. One group, the Parareptilia, consisted of a number of diverse chassis types, from small, aquatic predators to large herbivores over 2 meters in length (Benton, 2005; Ruta et al., 2011). The other group, the Diapsida, characterized by their primitively diapsid skulls, would eventually give rise to everything alive today that we call a 'reptile' or 'bird,' as well as a huge variety of forms now extinct. A huge extinction event at the end of the Permian shook the status quo, and a large number of vertebrate lines and skeletal chassis did not make it through to the beginning of the Mesozoic (Age of Dinosaurs). This change in the environmental and competitive landscape seems to have ultimately selected for the success of the Diapsida within the Reptilia, as the parareptiles did not make it past the Triassic period early in the Age of Dinosaurs. As discussed previously, the other major amniote group, the Synapsida, diversified initially during the Permian but suffered huge losses as well at the extinction event.

As we have already seen, it continues to be difficult to resolve and tease out the relationships of certain fossil vertebrates from one another, and the same can be said for the parareptiles. As of this writing, there is a controversy over whether the parareptiles are a completely extinct lineage or whether one part of their family tree lives on in the form of turtles. A turtle skull has the anapsid condition in which no temporal fenestrae are present. The only other amniotes that have an anapsid skull are the fossil parareptiles, and so it has often made sense based on this and several other features to place turtles with the parareptiles (deBraga and Rieppel, 1997). However, as we will discuss later, a new consensus is emerging from embryology, paleontology, and molecular biology that points very strongly toward turtles being members of the Diapsida (deBraga and Rieppel, 1997). Moreover, new research and new specimens of parareptiles show that some of these animals had temporal fenestrae (Sues et al., 2009). There is also the distinct possibility that parareptiles are not even a real group but an amalgamation of early diapsid reptiles. For our purposes, we will consider turtles to be a type of derived diapsid, and treat the parareptiles as a diverse but extinct branch of the early reptiles.

In this chapter we will briefly explore the parareptile and early diapsid chassis, followed by that of turtles. Having the lizard and tuatara chassis from the previous chapter as both a search image and a reference for soft tissues, over the next several chapters we can better appreciate the amazing diversity and adaptation of the reptilian skeleton.

The Parareptilia and Early Diapsida

Pedigree Earliest True Reptiles

Date of First Appearance ~302 Ma (Card 94)

Specialties of the Skeletal Chassis Solid Skulls, Larger Body Size

Eco-niche Terrestrial to Aquatic Predators and Herbivores

It used to be enough to say that parareptiles were anapsids, an odd group of early reptilians with no temporal fenestrae. Now, however, it is recognized that this group of early reptiles was a prominent part of the Permian landscape, and one diverse group, the procolophonids, survived for a time into the Age of Dinosaurs (Cisneros and Ruta, 2010; Ruta et al., 2011). Aside from primitively not having temporal fenestrae, their jaw joint is odd in that it is shifted to the level of, or just behind, the occipital region of the skull. This is somewhat like having your jaw joint just caudal to your ear. Moreover, there are some parareptile fossils that hint that some of these animals may have had some form of temporal fenestrae (Sues et al., 2009). As with so many groups, we cannot explore the entire diversity of the parareptile chassis. Here, we will focus on the most successful branch, the procolophonids.

Procolophonids were a successful parareptile group that diversified in the Late Permian (~260 Ma; Card 95) and lasted well into the Triassic period (nearly 200 Ma; Card 96) (Cisneros and Ruta, 2010). When in South Africa, I worked side by side in the field with paleontologist Juan Cisneros, who has traveled the world to study and unravel the evolutionary history of these peculiar parareptiles. We were digging in the Elliot Formation, which is right on the cusp of the Triassic-Jurassic boundary, and Juan was hopeful he would find a very early Jurassic or very late Triassic procolophonid among the dinosaur bones. Alas, the only thing we found out about procolophonids on our dig was my inability to pronounce 'procolophonid' effectively.

Most procolophonids were approximately 30–60 cm in length, possessed short, robust heads that were distinctly triangular in dorsal view, and had robust and squat bodies (Cisneros and Ruta, 2010). The best known procolophonid is *Procolophon* from the Early Triassic (~250 Ma; Card 95) of South Africa, Brazil, and Antarctica. In this animal, the snout is short, the orbits are large, and the premaxilla, maxilla, and dentary bones all bear robust, blunted peg-like teeth (Fig. 14.1) (Benton, 2005; Cisneros and Ruta, 2010). *Procolophon*, like many procolophonids, has a distinct overbite where the premaxilla juts out over the dentary. The quadrate-articular jaw joint lies slightly below the tooth row, and the coronoid bone in the lower jaw and the pterygoid bones in the skull are somewhat enlarged. All of these features strongly suggest *Procolophon* had a mostly herbivorous diet consisting of tough, fibrous plants that could be smashed between its peg-like teeth using powerful adductor mandibulae and pterygoid muscles (Benton, 2005; Cisneros and Ruta, 2010), possibly supplemented with bugs. Perhaps the overbite in these animals assisted them in nipping off plant parts. Again as in lizards, the skull of *Procolophon* bears a distinctive parietal opening through which the parietal eye could peek to detect light levels necessary to modulate its presumably ectothermic metabolism (Fig. 14.1). The braincase region of the skull is relatively solid, although some specimens of *Procolophon* and other procolophonids have what appears to be a small temporal fenestra. A small embayment behind the quadrate may have housed a tympanic membrane, and what is known about the stapes in *Procolophon* suggests this animal could hear some airborne sounds. Perhaps the most distinguishing features of the skull in *Procolophon* and other procolophonids were the expanded quadratojugal 'horns' that jutted out sideways (Fig. 14.1). Among the procolophonids these 'horns' vary greatly in size, shape, and orientation, and may have been used for everything from display to head butting (Cisneros and Ruta, 2010).

The vertebral column of procolophonids resembles in many respects that of the reptiliomorph herbivore *Diadectes* in having relatively large, expanded vertebrae (Fig. 14.1). Many of the dorsal vertebrae possess tall neural spines that we infer provided appropriate mechanical advantage for back muscles and ligaments to suspend the animal's squat body. Unlike lizards, intercentra are still present as distinct, stout wedges between the pleurocentra. An atlas-axis complex is present, the sacrum is composed of three stout vertebrae with short ribs, and the tail is short, tapering to a nub over a distance less than half the length of the ribcage. Well-developed ribs were present in *Procolophon* and its relatives, and respiration is inferred to have relied on rib movements (Fig. 14.1). The pectoral girdle is again similar to what is seen in lizards, except that the scapulocoracoid and interclavicle are more robust. The pelvic girdle is also lizard-like but more robust and lacking a thyroid fenestra (Fig. 14.1).

The forelimbs and hind limbs show most of the adaptations we have seen in lizards and other tetrapods for crank-like locomotor movements. However, especially in the forelimb, there are some aspects that may indicate burrowing habits in *Procolophon* and its kin (deBraga, 2003). The ulna is greatly expanded proximally and bears a large olecranon region around which the triceps muscles would have inserted (Fig. 14.1). Beyond enhancing the antigravity function of the triceps muscles, the large size of the olecranon region in *Procolophon* is intriguing because it would provide a large in-lever that could translate a great deal of force through the forearm and into the hand (deBraga, 2003). This would be useful for scratching away dirt and sediment. The radius is more slender than the ulna, and its shaft bows away from the midline of the arm so that the two forearm bones are not in contact distally (Fig. 14.1). Instead, both the radius and ulna articulate independently with the wrist. The wrist bones of *Procolophon* resemble a small cluster of rounded pebbles, whereas the hand consists of a stout array of five digits capped by large claws (Fig. 14.1).

All of these forearm and hand features are found in burrowing or digging reptiles (deBraga, 2003), such as the turtles we will discuss later in this chapter. In turtles that have these features, the radius and ulna can move and rotate about one another independently, the pebble-like wrist bones allow additional flexion, extension, and rotation, and the large claws are ideal for digging and clearing sediment (Hildebrand and Goslow, 2001). I have been able to demonstrate some of these movements to myself in fresh dissections of various turtle species. For example, when the muscles are removed from the forelimb of a turtle but its ligaments and joints are left intact to hold the bones together, the radius and ulna can easily be rotated back and forth and around one another, and this helps twist the wrist and hand. Similar features can be noted in the hind limbs, although the ankle bones are more robust and less pebble-like. Intriguingly, a number of fossilized Triassic burrows have been found in North America, South Africa, and Antarctica that are of the right proportions to have housed one or more procolophonids (Abdala, Jasinoski, and Fernandez, 2006). In one spectacular case, the skeletons of a procolophonid and a synapsid as well as the body of a fossil millipede are all preserved in one fossil burrow (Abdala, Jasinoski, and Fernandez, 2006).

Our necessarily short coverage of parareptiles has left out many other varieties of these animals. One other group deserves brief mention: the bolosaurids. These were small, lizard-like animals that had elongate hind limbs and a long tail. As we will see later for dinosaurs, these features usually signal that a reptilian is a biped, an animal capable of walking or running solely on its hind limbs. Studies of one bolosaurid in particular, named *Eudibamus*, strongly suggest that this parareptile was one of the very first bipedal vertebrates (Berman et al., 2000)! Unfortunately for the parareptiles, their line seems to have gone completely extinct by the beginning of the Jurassic period some 200 Ma (Card 96).

The diapsid line appears to have started with just a handful of species that eventually blossomed into the amazing reptile diversity we know today. Among the several different early diapsids, we will briefly review the earliest and best known here. Back at the end of the Carboniferous (~302 Ma; Card 94), the first known diapsid with a characteristically diapsid skull appeared. Called *Petrolacosaurus*, it was a 30 cm long, slender animal with a relatively small, delicate skull and thin, elongate limbs (Fig. 14.2) (Reisz, 1977). Viewed from above, the skull has a triangular

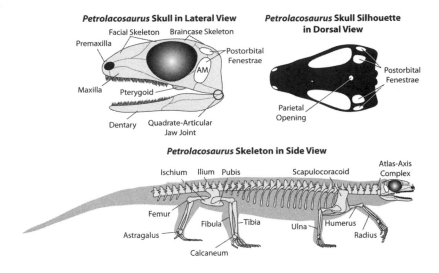

Petrolacosaurus Skull in Lateral View

Facial Skeleton — Braincase Skeleton

Premaxilla

Postorbital Fenestrae

AM

Maxilla — Pterygoid

Dentary — Quadrate-Articular Jaw Joint

Petrolacosaurus Skull Silhouette in Dorsal View

Postorbital Fenestrae

Parietal Opening

Petrolacosaurus Skeleton in Side View

Atlas-Axis Complex

Ischium — Ilium — Pubis — Scapulocoracoid

Femur

Fibula — Tibia — Ulna — Humerus

Astragalus — Radius

Calcaneum

14.2. Skull and skeletal diagrams of the earliest diapsid with a diapsid skull, *Petrolacosaurus*. Note the relatively small skull, a trend that continues in many diapsid reptiles The limbs of this early diapsid reptilian are very elongate and delicate, suggesting this vertebrate was an agile little insectivore. AM indicates the location of the adductor mandibulae and other jaw-closing muscles. Diagrams based on Reisz (1977) and Benton (2005).

shape with an expanded braincase region and a narrow snout (Fig. 14.2). The parietal opening is once again prominent, and must have housed a parietal eye to help regulate the animal's metabolism. Small, pointed teeth arise from the premaxilla, maxilla, and dentary bones, and smaller clusters of teeth adorn the palate. The pterygoid bones are well developed and project below the skull, a feature we have discussed previously as being associated with powerful pterygoid muscles (Fig. 14.2). The two postorbital fenestrae provide ample space and room for the adductor mandibulae muscles, and the quadrate-articular jaw joint is in line with the tooth row (Fig. 14.2). All of these skull features and the small size of *Petrolacosaurus* suggest this animal was an insectivore (Reisz, 1977).

The vertebral column and limb skeleton are very similar to those of other reptilians such as lizards. The vertebrae are somewhat small and delicately constructed, the ribs are short but well developed, the limb girdles are robust, and the elongate limbs bear all the hallmarks of a crank-like locomotor gait. Particularly interesting for our purposes is the ankle joint of *Petrolacosaurus*. When you buy chicken drumsticks at the grocer, or when you eat chicken legs at a restaurant, you have probably noticed that one end is thick with meat and the other end tapers into a thin bone with two knobs on the end separated by a groove (Fig. 14.3). The thick, meaty part of the drumstick is the knee end of the tibia, and the thin, tapered part with the two knobs is actually the end of the tibia plus the astragalus and calcaneum. In other words, the proximal bones of the ankle fuse with the end of the tibia in birds, and we technically call their fused tibia and upper ankle the tibiotarsus (Fig. 14.3). This is why the end of a chicken drumstick looks the way it does: the two knobs are the fused remnants of the astragalus and calcaneum, and the groove is a space in between that accepts a ridge from the distal ankle bones (Fig. 14.3). The ankle joint of a chicken is a hinge between the tibiotarsus above and the distal ankle bones below. The ridge-and-groove articulation at the ankle joint in chickens directs the foot caudally, and while

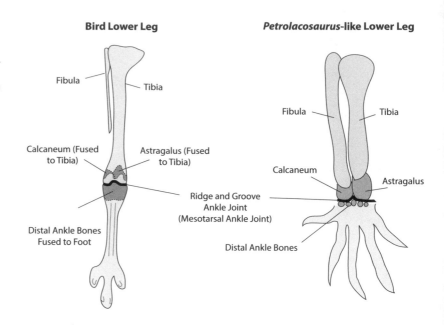

14.3. Schematic comparison of the ridge-and-groove ankle articulation in a bird and an early diapsid like *Petrolacosaurus*. In all cases, the functional ankle joint is situated between the proximal (astragalus and calcaneum) and distal ankle bones. Technically, such an ankle joint is described as mesotarsal (literally, between the ankle bones). In birds, the calcaneum and astragalus have fused with the tibia, and the distal ankle bones have fused with the foot.

Bird Lower Leg

Fibula

Tibia

Calcaneum (Fused to Tibia)

Astragalus (Fused to Tibia)

Ridge and Groove Ankle Joint (Mesotarsal Ankle Joint)

Distal Ankle Bones Fused to Foot

***Petrolacosaurus*-like Lower Leg**

Fibula

Tibia

Calcaneum

Astragalus

Distal Ankle Bones

allowing a wide range of flexion and extension, it prevents the ankle from slipping laterally or overly rotating. What does this all have to do with *Petrolacosaurus*? This early diapsid is one of the first known to have a ridge-and-groove ankle joint (Fig. 14.3) (Reisz, 1977; Benton, 2005). If you recall, the ankles of lizards have a ridge-and-groove screwlike articulation between the astragalocalcaneum and the distal ankle bones. By reference to all the living diapsids that possess this anatomical trait, we can infer that the foot of *Petrolacosaurus* could provide forward thrust with less outward turning. Such a joint is technically known as a mesotarsal joint: literally, an ankle joint between the ankle bones.

The hands and feet are also long and lizard-like in *Petrolacosaurus*. Intriguingly, there are trackways from the Carboniferous of Canada that resemble the hands and feet of animals like *Petrolacosaurus* (Falcon-Lang, Benton, and Stimson, 2007). These trackways show five long and distinct digits on the hand and foot, with the outer digit commonly set apart from the rest of the foot. The narrow spacing between the right and left hands and feet is also interesting in that it suggests that animals like *Petrolacosaurus* had a more upright and less sprawled gait than previous tetrapods and amniotes we have encountered (Falcon-Lang, Benton, and Stimson, 2007). Other details are preserved in the trackways, including the presence of small, elongate scales on the pads of the feet and a discernible tail drag line that is relatively straight. The last feature is intriguing because it suggests that the trackmaker was using far less lateral body movement (which would have resulted in a more sinuous tail drag impression) than many of the tetrapods we have so far encountered (Falcon-Lang et al., 2007). What is most remarkable about these footprints is that they are preserved on sedimentary rocks that were laid down in a dry riverbed, not in a swampy coal forest. Thus, the trackways provide additional evidence that animals such as *Petrolacosaurus* and other early

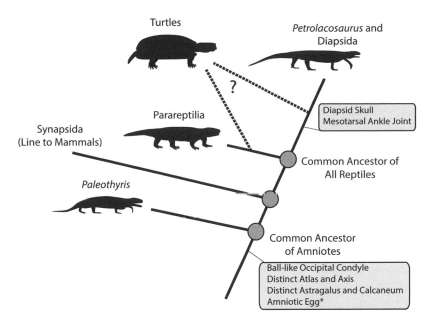

Turtles

Petrolacosaurus and Diapsida

Pararreptilia

Synapsida
(Line to Mammals)

Paleothyris

?

Diapsid Skull
Mesotarsal Ankle Joint

Common Ancestor of
All Reptiles

Common Ancestor
of Amniotes

Ball-like Occipital Condyle
Distinct Atlas and Axis
Distinct Astragalus and Calcaneum
Amniotic Egg*

14.4. Pedigree of relationships among the early amniotes, parareptiles, and diapsids discussed in this chapter. Note that *Paleothyris* is standing in as an example of an animal close to the common ancestor of all amniotes. Some new data suggest *Paleothyris* may lie at the base of the diapsid line. Note also how early the Synapsida (which eventually leads to mammals) branches from the reptiles. Parareptilia is shown as a separate and extinct branch of the Reptilia, but see the text for dissenting views. Turtles are placed at the end with a question mark to denote the uncertain relationship of these reptiles among the Reptilia. The asterisk (*) in the diagram indicates that we are inferring the presence of an amniotic egg in *Paleothyris* and all basal amniotes based on their skeletal trait states. See the text for more details.

diapsids had by the end of the Carboniferous broken their ties with water (Falcon-Lang, Benton, and Stimson, 2007).

To graphically summarize our discussion in this chapter, Fig. 14.4 diagrams the pedigree of relationships among the early amniotes, parareptiles, diapsids, and synapsids. Certainly, some diapsids were present in the Permian, but not in the diversity seen for parareptiles or (as we will later see) for the synapsids. Among the other diapsids that did populate the Permian landscape, some of the most exotic were small, arboreal gliders with ornamented skulls and thin, horizontally projecting ribs that presumably acted to support a thin skin membrane (Benton, 2005). These animals may have led lives similar to those of modern gliding lizards. By the end of the Permian (~254 Ma; Card 95), the diapsids had split into two major lines. One line, the one that we continue to trace in the next chapter, is the Lepidosauromorpha, the group that includes the Lepidosauria (the lizards, snakes, and tuatara), and the Sauropterygia, the marine reptiles. Very close relatives of the lepidosauromorphs were the dolphin-like marine reptiles called ichthyosaurs. The other major diapsid branch, the Archosauria, was perhaps the most successful reptilian group, giving rise to crocodylians and birds, as well as the dinosaurs and the flying reptiles known as pterosaurs. The archosaur chassis will be detailed further in subsequent chapters. Now we turn our attention to that of the enigmatic, shelled diapsids (?) known as turtles.

Turtles

Among the reptiles (probably the diapsids), turtles took an evolutionary gamble and decked out their chassis with a heavy armor option. Their iconic shell has made them well protected from predators and from

Pedigree Shelled Reptiles
(Probably Diapsids)

Date of First Appearance
~220 Ma (Card 96) and
Possibly Much Earlier

**Specialties of the Skeletal
Chassis** Shell, Pectoral Girdle
Placed inside Ribs

Eco-niche Terrestrial to
Aquatic Predators, Omnivores,
and Herbivores

paleontologists trying to understand their pedigree. In short, turtles are the most recognizable reptiles with the most exasperating family tree. All modern turtles have a shell, but that shell has been a true nuisance for biologists and paleontologists because it has literally covered over the evolutionary history of this group. Until recently, the earliest known turtles in the fossil record were . . . well . . . turtles, complete with shell. Even the assignment of turtles to diapsids, as well as their place among diapsids, continues to be fraught with debate and uncertainty (Lee, Gow, and Kitching, 1997; Joyce and Gauthier, 2004; Joyce et al., 2009; Lyson et al., 2010, 2011, 2013; Field et al., 2014). However, a combination of recent embryonic data and new fossil insights may finally be illuminating the origin of the turtle chassis (e.g., Sánchez-Villagra et al., 2009; Kuratani, Kuraku, and Nagashima, 2011; Lyson et al., 2013, 2014).

For about a month when I was five years old, I had a pet red-eared slider turtle I named Gamera after a Japanese monster movie starring a giant turtle. I was a bit disappointed to find out that my Gamera could not fly by spinning sideways while shooting fiery jets from its shell's leg cavities. Instead, my Gamera spent his days swimming in a terrarium, eating crickets and the occasional bologna I would throw in, and watching tv with me. After my parents grew weary of siphoning Gamera's 'leavings' from the tank, they convinced me that he was lonely and needed to be out among the turtles in the wild. I still remember going with my father to a local creek and setting him down on the muddy bank. Gamera looked around briefly in what I can only imagine was stunned surprise, and then burst forth faster than I ever thought something like a turtle could move, sliding quickly down the muddy bank and into the water. A few moments later, and Gamera had disappeared down the creek channel. I've always imagined he went on to sire generations of turtles and passed down his legendary story of captivity and escape.

The short time I spent with Gamera gave me an appreciation for turtle anatomy I hold to this day. First, that shell is hard and very difficult to damage. I ought to know: I accidentally dropped my poor turtle on our concrete patio, but he thankfully remained unharmed. It is harder to crack a convex shape than a flat one. All things being equal, if you try to smash a cube and a sphere made of the same material, the flat surfaces of the cube will dent and crush far more easily than the sphere. This is because the sphere's rounded shape doesn't provide a large enough surface at any one point to concentrate the blow of the hammer. The convex upper portion of many turtle shells acts in much the same way against a predator's tooth or kick: the energy expended is dispersed over the shell where it can't easily build up to form a devastating crack.

We don't often appreciate how weird turtle shells are until we stop to think about a more familiar pet like a cat. Those of you with cats have probably noticed that when they walk or run, you can clearly see their shoulder blades moving up and down along their sides. In fact, almost all tetrapods have the pectoral girdle bones lying above and outside the ribs. But you'll never a see a turtle's scapulocoracoid or other pectoral girdle elements unless you are a veterinarian with a fluoroscope (X-ray machine), because these bones are tucked inside their shell. When you realize that a turtle shell is made up of ribs and dermal armor, this is truly mind-blowing. It means turtles have an entire pectoral girdle inside their ribs! It gets even weirder when you realize that the pelvic girdle is similarly encased.

The shell quite simply affects every aspect of a turtle's anatomy. Imagine if you placed a heavy box around your body with holes for your head and limbs (Fig. 14.5 and Plate 17). More specifically, imagine if your shoulder and hips were inside this box, and unconstrained movement could occur only from your elbows and knees on down. On top of this, not only does this heavy box make your body incapable of flexing or bending in any way to help with movement, but imagine if the box were intimately connected to your ribs, preventing you from moving them as well. Such a situation would make breathing extremely difficult, especially if, like turtles, you do not have our internal diaphragm muscle! Because most of a turtle's anatomy and lifestyle is strictly dictated by its shell, there are a lot of skeletal and soft tissue 'workarounds' that are necessary to keep turtles alive and functional. The turtle chassis has a remarkable diversity despite the shell's constraint, but this feature has also prevented turtles from evolving galloping or flying forms. Then again, can you imagine being struck by a flying turtle?

Given its profound influence on turtle anatomy, it is probably best to start with the development of the shell itself and then work from there. Recent breakthroughs in studies of turtle embryology have demonstrated the truly remarkable way that the turtle shell develops and how the pectoral girdle ends up inside the ribs. To start with, the turtle's shell is made up of a combination of ribs and dermal bone. As with all tetrapods, the ribs develop from segmented blocks of tissue called somites, the same tissues that give rise to the segmented body muscles called myomeres common to all vertebrates (Wilt and Hake, 2003; Gilbert, 2010). Typically, the ribs of tetrapods grow out and down from the somites, eventually curving toward belly or sternum (Fig. 14.6). The neural crest cells we spoke of back in chapter 3, which generate much that is unique to vertebrate anatomy, are responsible for developing any dermal bone, such as the armor scutes of alligators. If you imagine the ribs of a turtle to be the rafters of a house, the shell develops from neural crest cells laid over the top of the ribs like a roof over a loft (Gilbert, 2010; Kuratani, Kuraku, and Nagashima, 2011). So, in turtles the shell is made from a combination of materials from the somites (the ribs) and the neural crest cells (the dermal armor or carapace) (Fig. 14.10). The ventral piece of the shell, the plastron, has a more mysterious origin, but it seems to be a combination of dermal bone, bones like the clavicles and the gastralia (belly ribs) (Fig. 14.6) (Gilbert, 2010; Kuratani, Kuraku, and Nagashima, 2011).

The first weird thing that happens during turtle development is the development of a ridge between the dorsal and ventral halves. This ridge, technically called the carapacial ridge (CR), will eventually form the outer

14.5. How being a turtle is akin to being trapped in a solid box. The shell of turtles is such a significant feature in their chassis that there have been numerous evolutionary 'workarounds' in turtle anatomy. Turtles don't seem to mind.

14.6. Rib and shell development in turtles. Unlike in all other tetrapods, the pectoral and pelvic girdles of turtles lie inside their ribs. The ribs, in turn, have become fused into the dermal armor that makes up the shell. During embryonic development, a region called the carapacial ridge forms and attracts the ribs and neural crest cells laying down the dermal armor of the carapace or upper shell. This causes the developing carapace and ribs to extend outward over the pectoral and pelvic girdles, which remain in their typical positions. Muscles that run from the ribs to the girdles or limbs are forced to fold in unique ways into the shell to accommodate this odd developmental pattern. However, this infolding of the skeletal muscles contributes to the unique respiratory patterns of turtles. Development diagrams based on information and illustrations in Kuratani, Kuraku, and Nagashima (2011).

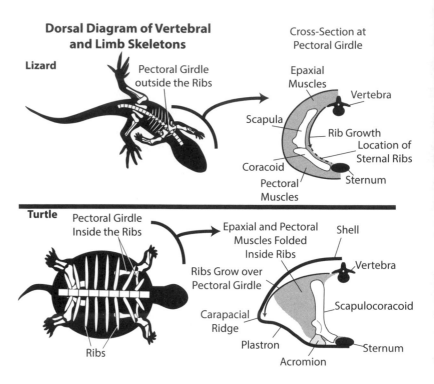

Dorsal Diagram of Vertebral and Limb Skeletons

Lizard
Pectoral Girdle outside the Ribs

Cross-Section at Pectoral Girdle
Epaxial Muscles
Vertebra
Scapula
Rib Growth
Location of Sternal Ribs
Coracoid
Pectoral Muscles
Sternum

Turtle
Pectoral Girdle Inside the Ribs
Epaxial and Pectoral Muscles Folded Inside Ribs
Shell
Ribs Grow over Pectoral Girdle
Vertebra
Carapacial Ridge
Scapulocoracoid
Ribs
Plastron
Sternum
Acromion

rind of the turtle shell (Fig. 14.6) (Sánchez-Villagra et al., 2009; Kuratani, Kuraku, and Nagashima, 2011). The CR arises from neural crest cells and possibly from other tissues as well, but it is unique to turtles. The second weird thing to happen is that the ribs don't grow out and down – instead, they grow out and sideways toward the CR! A cocktail of signals appears to prevent the ribs from arcing downward and instead sends them outward to the CR (Kuratani, Kuraku, and Nagashima, 2011). When the ribs reach the CR, they interact with signals in this region that cause their ends to broaden out, helping to form the lip of the upper shell (carapace). This leads us to the third weird thing to happen: because the ribs have grown out and sideways but the scapulocoracoid has remained on the side of the body as in all tetrapods, the pectoral girdle ends up being covered by the developing dorsal shell (Fig. 14.6) (Kuratani, Kuraku, and Nagashima, 2011). In other words, the outwardly growing ribs act somewhat like a patio awning that is pulled over the scapulocoracoid instead of curving down and behind this element. This is apparently how the pectoral girdle ends up inside the ribcage. Finally, a fourth weird thing happens: muscles associated with the shoulder girdle become folded inside the ribcage because of the difference in developing skeletal anatomy, and this has a profound effect on respiration (Fig. 14.6) (Kuratani, Kuraku, and Nagashima, 2011).

At this point, it is necessary to introduce or further detail four muscles, two associated with the forelimb and two with the abdomen, so that we can discuss respiration in turtles. In the forelimb, one such muscle is the serratus ventralis. This muscle is large and fan-shaped, and in a typical tetrapod it anchors to the ribs and inserts into the scapulocoracoid caudally (Fig. 14.7). Among other actions, when the forelimb is not

Generalized Amniote

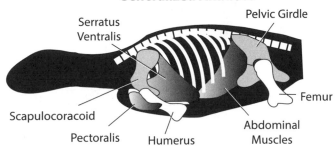

Serratus Ventralis

Pelvic Girdle

Femur

Scapulocoracoid

Pectoralis   Humerus

Abdominal Muscles

14.7. The odd placement of turtle girdle muscles compared with other amniotes. The serratus ventralis, pectoralis, and abdominal muscles (transversus abdominus and obliques), being folded inside the shell, take on new roles in respiration. Diagram based on Kardong (2012) and Lyson et al. (2014).

Turtle

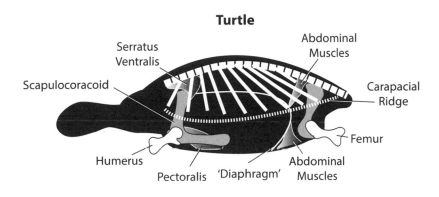

Serratus Ventralis

Abdominal Muscles

Scapulocoracoid

Carapacial Ridge

Femur

Humerus

Pectoralis   'Diaphragm'   Abdominal Muscles

supporting the body, contraction of this muscle helps to protract the forelimb by rotating the scapulocoracoid so that the humerus is swung craniodorsally (Liem et al., 2001; De Iuliis and Pulerà, 2011; Kardong, 2012). The humerus and forelimb are adducted to the body's midline by the pectoralis muscles, which we have previously discussed. Ventrally, as discussed in chapter 9, large sheetlike muscles encircle the body like layers of cummerbunds. One of these is the transversus abdominus (Fig. 14.7), and its constriction in a typical tetrapod squeezes the body core, forcing air out of the lungs (Liem et al., 2001). Another of these ventral muscles is actually a complementary pair that run at oblique angles to the belly from the ribs: hence their name, the abdominal obliques (Fig. 14.7). Acting together, these muscles also squeeze air out of the body in collaboration with the transversus abdominus (Liem et al., 2001).

To really get at how turtles breathe, we also need to get into the guts. Tissues called the mesenteries surround the internal organs of all vertebrates, and turtles are no exception. In a tetrapod like you, the mesenteries do a number of things, and one of these is acting as separators between various regions of the body. A sheetlike wall of mesentery divides the lung chamber (the pleural cavity) from the gut chamber (the peritoneum). There is also a tough, bowl-like mesenteric sheet that stretches across the inside of the pelvis, which supports the guts and urogenital organs and provides attachment points for various body core muscles. The mesenteric wall that separates the lungs from the guts we will casually refer to as the 'diaphragm,' and in most tetrapods the lungs rest against this wall. In mammals, the diaphragm has been invaded by muscles that actively pull down or passively bounce back to increase and decrease chest pressure, inflating or deflating the lungs (Perry et al., 2010).

In turtles, the lungs are large and pushed up against the ribs and carapace (Fig. 14.8). The 'diaphragm' does not have muscle tissue in it, but it is very tough and connected intimately with the guts, which hang between the lungs and the plastron. The weight of the guts pulling on the lungs naturally helps them to expand (Fig. 14.8) (Gans and Hughes, 1967). One can imagine the guts like a spring: when they are allowed to expand and 'stretch,' they pull the lungs open; when they are forced back, they recoil and squeeze air out of the lungs. And this is where the weird inward folding of the pectoral muscles has had an important effect on turtle respiration. The pectoralis, having been folded into the body cavity, now attaches to the inside of the plastron. When it contracts, it pulls the forelimbs into the shell, which push against the guts, which in turn compress the lungs, expelling air (Fig. 14.8) (Gans and Hughes, 1967; Lyson et al., 2014). In contrast, the serratus ventralis muscle, having been folded into the shell, now pulls the scapulocoracoid upward and outward toward the head, moving the forelimb cranially and increasing the gut space (Fig. 14.8) (Lyson et al., 2014). As the guts fill the newly opened space, their weight tugs on the diaphragm sheet, which in turn expands the lungs, allowing inhalation (Fig. 14.8) (Gans and Hughes, 1967). The aforementioned abdominal muscles complement the actions of the pectoralis and serratus ventralis in exhalation and inhalation, respectively (Lyson et al., 2014). These abdominal muscles pull on the bowl-like mesenteric sheet that covers the turtle's pelvic girdle (Fig. 14.8). When the transversus abdominus contracts, it pulls this mesenteric sheet inward, forcing the guts up into the lungs, assisting the pectoralis in causing exhalation. The abdominal obliques pull the pelvic mesenteric sheet caudally, allowing the guts to fall into the abdominal cavity, assisting the serratus ventralis with effecting inhalation (Gans and Hughes, 1967; Liem et al., 2001; Lyson et al., 2014). Imagine if every breath you took involved pulling your arms in, pushing them out, and doing stomach crunches. Turtles do this all the time, but it would make driving a car or playing a trumpet a bit awkward. How this came to be is discussed later in the chapter.

With an immobilized core and with the bases of the limbs tucked up inside the body cavity, turtles have a very odd way of getting about. The forelimbs cannot be splayed out sideways like those of a lizard because the shell is literally in the way. Instead, turtle forelimbs emerge through forward-pointing openings near the front of the shell (Fig. 14.9). This is because if turtle forelimbs simply projected out sideways like a lizard's, they would not be able to swing through a large enough arc to pronate the hand. Instead, turtles would have laterally facing hands but their shell would limit their ability to push against the ground, so they would sit there struggling to push their heavy bodies forward with the equivalent of ineffective paddles. The 'workaround' here is that the humerus of turtles is rotated so that its distal end points cranially. This puts the elbow out

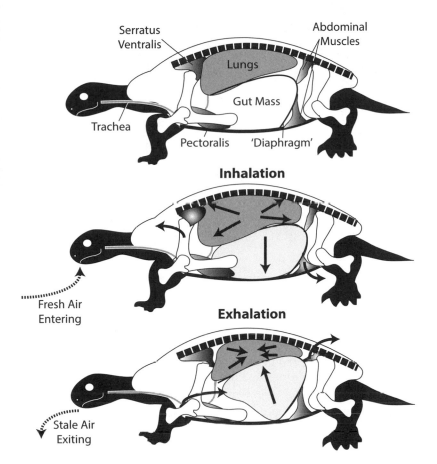

Serratus Ventralis

Abdominal Muscles

Lungs

Gut Mass

Trachea

Pectoralis   'Diaphragm'

Inhalation

Fresh Air Entering

Exhalation

Stale Air Exiting

14.8. How a turtle breathes with a rigid shell. The lungs are located dorsally against the carapace and are connected to the gut mass via diaphragm-like tissues. When a turtle inhales, the serratus ventralis muscle pulls the scapulocoracoid (and thus the forelimb) forward and outward, creating space inside the shell, while one set of the abdominal muscles pulls the 'diaphragm' caudally, also increasing space. These movements cause the gut mass to sink and expand, which in turn pulls on and increases the size of the lungs. During exhalation, the pectoralis muscles pull the forelimbs into the shell, and another set of abdominal muscles pulls the 'diaphragm' upward, pushing the guts into the lungs and pushing out stale air. Diagrams based on Kardong (2012) and Lyson et al. (2014).

in front, rather than pointing caudally or laterally, and this has the additional effect of rotating the radius medial to the ulna, causing pronation of the hand (Fig. 14.9). You can demonstrate this yourself by bending your elbows, placing them in front of you, and turning your hands to face forward with the palms down. The hind limb is somewhat less modified in turtles, and although it is also limited by the shell remains oriented more like that of lizards. Both the humerus and femur of turtles have a spherical proximal end that creates a ball-and-socket joint with the glenoid and acetabulum, respectively, and this allows the forelimbs and hind limbs to create rotational movements at their girdles (Fig. 14.9). Bones in the forearm (radius, ulna) and leg (tibia, fibula) are relatively free, and they can move independently of one another. These independent movements are translated to the pebble-like wrist and distal ankle bones, further imparting turning motions to the hands and feet.

When my pet turtle Gamera walked, he was surprisingly stable if not terribly graceful. As with all turtles, Gamera got around by moving limbs on opposite sides of the body and shifting his weight to one side then another as one combination of forelimbs and hind limbs switched between supporting weight and rotating into a supporting position. Because turtles cannot flex their shells, there comes a point where the leading limb has reached as far forward as it can go before being planted firmly on the

14.9. How cervical vertebra shape affects neck retraction in modern turtles. The centra (vertebral bodies) of turtle cervical vertebrae are saddle-shaped, which allows for a large range of rotational and swiveling movements that give the neck an extraordinary range of flexibility. In pleurodire or 'side-necked' turtles, the head is pulled back toward the shell because two key areas of flexion allow the neck to fold up in a horizontal S-shape. In cryptodire turtles, similar regions of flexion allow the neck to fold up into a vertical S-shape. In several modern cryptodire turtle species, the head can be tucked completely under the carapace, and commonly the limbs can be withdrawn into the shell as well. Neck folding diagrams based on Benton (2005). Cryptodire turtle skeleton is based on that of a bog turtle, *Glyptemys muhlenbergii.*

Turtle Cervical Vertebra Shape

3 4 5

Body (Centrum)

Posterior View of #3

Anterior View of #4

Schematic of Vertebrae #3 and 4 Articulated

Complex Movements

3

4

Turtle Neck Retraction

'Side-Necked' Turtle (Pleurodire)

'Typical' Turtle (Cryptodire)

Cryptodire Turtle Retracting Head and Limbs into Shell

Carapace

Plastron

ground. Turtles then simply fall forward toward the outstretched limb to make up for the difference that a flexible core would otherwise provide (Liem et al., 2001). So when you watch a turtle walk, you see the shell rise and fall as the animal pitches forward onto the leading hand or foot. Clumsy as it may seem, turtles have been doing this for at least 220 million years, and it obviously hasn't hurt them enough to cause their extinction.

With the dorsal vertebrae and ribs firmly united into the shell, the only mobile vertebrae are those of the neck and tail. The tails of turtles are relatively short and play only a small role in their locomotion. The cervical vertebrae of turtles, however, are a different story (Fig. 14.9). First, the centra of the vertebrae have saddle-shaped ends that fit together in such a way that a wide range of movements is available at each joint in the neck (Liem et al., 2001). To understand this, make an 'L' with your thumb and index finger on each hand, and then orient them perpendicular to one another. Now, put these two 'Ls' together so that the space between the thumb and index finger on each hand 'articulates.' You should be able to demonstrate to yourself that this configuration affords a lot of rotational and up-and-down movements. This wide range of movement is necessary in most turtles to allow them both to extend the head rapidly when snapping at prey and to retract the head into or under the shell when hiding from enemies (Liem et al., 2001).

There are two living groups of turtles, the cryptodires and pleurodires (Carroll, 1988; Benton, 2005). Both groups appeared at roughly the same time in the fossil record (~200 Ma; Card 96), but cryptodires have been more successful at populating the globe. Cryptodires include most of the familiar turtles, from red-eared sliders to tortoises to sea turtles. In contrast, pleurodires are restricted to the Southern Hemisphere, and are

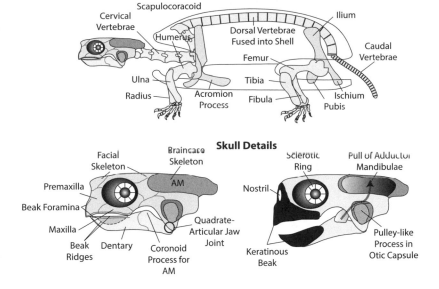

See-through View of a Cryptodire Turtle

Cervical Vertebrae · Scapulocoracoid · Humerus · Ilium · Dorsal Vertebrae Fused into Shell · Caudal Vertebrae · Femur · Ulna · Radius · Tibia · Acromion Process · Fibula · Ischium · Pubis

Skull Details

Facial Skeleton · Braincase Skeleton · Sclerotic Ring · Pull of Adductor Mandibulae · Premaxilla · AM · Nostril · Beak Foramina · Maxilla · Quadrate-Articular Jaw Joint · Beak Ridges · Dentary · Coronoid Process for AM · Keratinous Beak · Pulley-like Process in Otic Capsule

14.10. Aspects of the turtle skeleton and skull based on the bog turtle, *Glyptemys muhlenbergii*. Ribs are not shown in the see-through side view to aid in showing other aspects of the skeleton. Notice that the scapulocoracoid is tall and anchored into the plastron through its coracoid portion. Most turtles have a long acromion process that juts caudally. Given the constraints of the shell, the forelimbs project out with the elbows facing forward and the fore-arm flexed caudally. This odd orientation allows the hand to be pronated as the turtle walks. In the skull, a large embayment acts in an analo-gous fashion to the fenestra of other diapsids, giving purchase and space for the large ad-ductor mandibulae muscles (AM). The jaws of turtles are toothless and dotted with nutrient foramina that provide a blood supply and innervation to a keratinous beak. A large, pulley-like process develops within or near the otic region in turtles that changes the direction of pull from the ad-ductor mandibulae muscles, increasing their mechanical advantage when snapping the jaws shut on prey. Adductor mandibulae diagram based on illustrations in Benton (2005). AM indicates the location of the adductor mandibulae and other jaw-closing muscles.

represented by such forms as the Australian snake-necked turtle. Neck re-traction is present in most of these turtles, but the two groups accomplish this in different ways. In cryptodires, the cervical vertebrae are wide, with zygapophyses that are open vertically but narrow horizontally (Fig. 14.9). This zygapophysis shape allows a lot of movement in the vertical plane, but restricts horizontal movements (Pough et al., 1998; Pough, Janis, and Heiser, 2002; Benton, 2005). Overall, this configuration causes their neck to bend vertically into an S-shape when they tuck their heads into their shells. In pleurodires, the cervical vertebrae are narrow with zygapophyses that are flat, a configuration that allows the neck to bend sideways when it is tucked back against the body (Fig. 14.9) (Pough et al., 1998; Pough, Janis, and Heiser, 2002; Benton, 2005). There are typically eight cervical vertebrae in turtles, and the last three usually possess very well developed interlocking joints that cause most of the vertical or horizontal bending of the neck.

The skulls of turtles are also an interesting affair (Fig. 14.10). The premaxilla, maxilla, and dentary lack teeth altogether, and are instead covered by a sharp, keratinous beak (Pough et al., 1998). Like your finger-nails, the keratinous beak of turtles requires a blood supply, and a nerve network invades the beak to give the animal feedback on mouth pressure and biting forces. These blood vessels and nerves emerge from the edges of the premaxilla, maxilla, and dentary bones through tiny pores called foramina. Having prepared a number of turtle skulls, I can tell you that when you remove the keratinous beak you can see these foramina along the jaws. This correlation between the beak and the jaw foramina allows us to infer the presence of a beak in fossil turtles when we see these same features. The skulls of turtles are rather stoutly constructed, and they bear no temporal fenestrae. However, many turtle species develop a deep notch, sometimes called an emargination, on the dorsal part of the skull that may or may not be related to a previous diapsid condition (Fig. 14.10) (Carroll, 1988; Benton, 2005). A conspicuous notch behind the quadrate bone in all turtles houses the tympanic membrane for the ear, and a slender stapes is present that allows turtles to hear well.

In both lines of modern turtles, changes in the otic region of the skull have improved the mechanical advantage of the adductor mandibulae muscles in clamping the jaws shut (Fig. 14.10) (Sterli, Marcelo, and La Fuente, 2010). To put this in perspective, you can easily find your otic capsule because that's where your ears are. If you bite down while placing your hand over your temple, you can feel a portion of your adductor mandibulae contracting. Typically, the adductor mandibulae muscles run at a steep angle from the squamosal region of the head to the lower jaws cranial to the otic capsule. What turtles have done is expand their otic capsule so that it sticks cranially into the squamosal region of the skull. This forces the tendons of the adductor mandibulae muscles to bend around the otic capsule before inserting into the lower jaw. If you imagine the adductor mandibulae tendons as a rubber band that normally runs vertically, the otic capsule of turtles is acting like a finger that is pushing and bending that rubber band out of its vertical path (Fig. 14.10). Essentially, the expansion of the otic capsule in turtles has turned the tendons of the adductor mandibulae into a pulley that increases the mechanical advantage of these muscles on the lower jaw. The evolution of this pulley system seems to have arisen in conjunction with skull immobility (Sterli, Marcelo, and La Fuente, 2010). The earliest fossil turtles have somewhat loose and kinetic skulls (Sterli, Marcelo, and La Fuente, 2010), but the appearance of the otic capsule modification associated with the pulley-like adductor mandibulae tendons appears once the skull becomes akinetic. The two living turtle groups have expanded their otic capsules in slightly different ways, resulting in somewhat different but analogous ways of achieving the adductor mandibulae pulley effect.

Not surprisingly, given that the shell is such a constraining part of turtle anatomy, much of the variation that occurs in these reptiles happens in the head and limbs. For example, given their slow walking speed, the dietary adaptations of particular turtles tend to vary based on how they get around. In fully terrestrial turtles like tortoises, their slow pace makes them experts at hunting and catching plants. Turtles that are semiaquatic, like my old pet Gamera, tend to be omnivorous because they can move faster when in water to catch animal prey. Fully aquatic turtles run the gamut in terms of their diet, with some specializing in mollusks, others in jellyfish, and still others in sea plants (Pough et al., 1998). The limbs of turtles like tortoises tend to be heavy, club-like appendages with stubby feet and hoof-like claws. In contrast, the forelimbs of sea turtles tend to be elongate with greatly expanded hands, whereas their hind limbs are typically much smaller, possibly to improve streamlining when swimming (Pough et al., 1998; Pough, Janis, and Heiser, 2002). Many turtles excavate nests for their eggs and some burrow, and the hands and feet can take on a spade-like shape. Moreover, the claws of burrowing species tend to be enlarged and are commonly longer than most of the hand or foot so that they can act as effective levers for digging (Hildebrand and Goslow, 2001).

Regarding the shell, there are several variations on this portion of the chassis as well. Most terrestrial and semiaquatic turtles have rigid domed shells and tough plastrons, with flexible hinges that accommodate the retraction of the neck and limbs in many species (Pough et al., 1998). In some soft-shelled turtles, the shell is flattened dorsoventrally, its dermal bones are reduced, and the shell itself is covered in skin with no scutes. Some marine turtles retain a hard but more flattened and streamlined shell and plastron, whereas others, like the huge leatherback sea turtle (*Dermochelys*), have reduced much of their shelly armor to small polygonal dermal bones embedded within a thick skin over the shell (Pough et al., 1998). What is significant to note about all of these differences in turtle shells is that the plastron remains relatively unchanged. Intriguingly, during turtle development, the plastron develops prior to the development of the dorsal shell or carapace, which suggests that this piece of turtle anatomy appeared first during their evolution (Sánchez-Villagra et al., 2009; Kuratani, Kuraku, and Nagashima, 2011). Moreover, because the unique respiratory system of turtles depends on muscles anchored to the plastron, it has been argued that the presence of a plastron is an indication of marine origins for the group. In other words, because the muscles involved in turtle respiration are derived from body-supporting muscles, the buoyancy of water would provide an ideal environment in which this odd system could best develop without the crushing force of gravity (Rieppel and Reisz, 1999).

With a working knowledge of turtle shell development and a skeletal search image, we now turn briefly to the turtle fossil record. Until recently, the earliest known turtle was the nearly 1 meter long animal *Proganochelys* from the Triassic of Germany (210 Ma; Card 96) (Fig. 14.11) (Gaffney and Meeker, 1983; Gaffney, 1990). This animal is already very much a turtle, having a complete carapace and plastron that encapsulate the limb girdles. The head is solid and robust, and there is a notch behind the quadrate reminiscent of the tympanic notch in other turtles, suggesting *Proganochelys* could hear well, although the stapes is more robust than in modern turtles (Gaffney and Meeker, 1983; Gaffney, 1990). Unlike in most other more derived turtles, the otic capsule of *Proganochelys* is not expanded, and as a consequence the adductor mandibulae probably did not function as a modified pulley (Gaffney and Meeker, 1983; Gaffney, 1990; Sterli, Marcelo, and La Fuente, 2010). The premaxilla, maxilla, and dentary bones lack teeth and foramina adorn the edges of these bones, strongly suggesting a tough, keratinous beak was present, but unlike modern turtles some small denticles speckled the roof of the mouth. The cervical vertebrae have relatively flat articulations with one another, and as a consequence the neck could not be retracted into the shell (Gaffney, 1990). The forelimbs already assume the modified condition for turtles we have discussed with the elbow rotated cranially. Large, tough

14.11. Diagram of *Progano-chelys,* one of the earliest known turtles. Note the beginnings of beak ridges on the lower jaw in *Progano-chelys,* features that become more developed in modern turtles (see cryptodire skull for comparison). The skull of *Pro-ganochelys* is relatively robust and contains no fenestrae or embayments. The skeleton of this turtle resembles those of modern species in many ways, but it is armored and the cervical vertebrae lack saddle joints with which to retract the neck. The spade-like claws on the hands and feet suggest *Proganochelys* spent at least some of its time excavating mud and dirt for nesting or for rooting out small prey. AM indicates the location of the adductor mandibulae and other jaw-closing muscles. Based on Gaffney and Meeker (1983), Gaffney (1990), and Benton (2005).

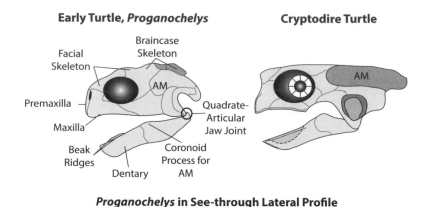

claws are present on the hands and feet, and these may have been used, as in modern turtles, to excavate nests (Gaffney, 1990). Knobby armor-like scutes adorn the neck, shell, proximal limb segments, and tail. In fact, the tail of *Proganochelys* bears numerous scutes that fuse into a sort of club (Gaffney, 1990). Fossils of this early turtle are associated with freshwater sediments, and this has suggested to some researchers that *Proganochelys* was at least semiaquatic in its habits (Rieppel and Reisz, 1999).

Prior to more recent discoveries, *Proganochelys* was the earliest known fossil turtle, and because it already possesses so many turtle features, it shed very little light on turtle origins. Some researchers pointed toward procolophonids and other parareptiles as close relatives of turtles, and not without good reason. Many of these parareptiles became increasingly encased in dermal armor throughout their evolution and had solid skulls like turtles that lacked any temporal fenestrae. Overall, the parareptile hypothesis of turtle origins favors a terrestrial origin for these vertebrates, in which the shelled chassis developed over time from a combination of expanded ribs and increasingly interlocking, scute-like armor (Lee, Gow, and Kitching, 1997; Joyce and Gauthier, 2004; Joyce et al., 2009; Lyson et al., 2010). However, an especially good candidate for a diapsid progenitor of turtles is an animal called *Eunotosaurus*, which has expanded ribs and a morphology eerily similar to that of turtles (Lyson et al., 2013) (Fig. 14.12). New research even shows that a juvenile specimen has diapsid openings in the skull, and the upper openings become filled in with bone in the adult (Bever et al., 2015)! Moreover, if *Eunotosaurus* is a turtle ancestor, then its T-shaped and expanded ribs may show how turtle respiration originated. *Eunotosaurus* does not have a shell, and

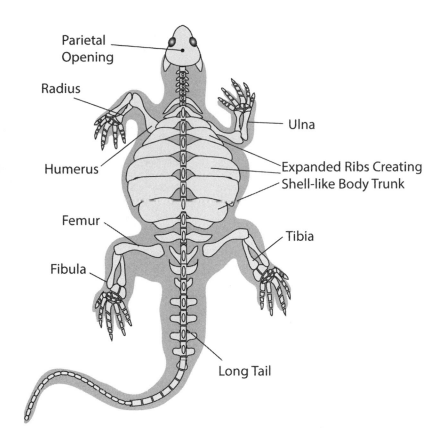

Parietal Opening

Radius

Ulna

Humerus

Expanded Ribs Creating Shell-like Body Trunk

Femur

Tibia

Fibula

Long Tail

14.12. The earliest (?) turtle *Eunotosaurus* in dorsal view. Note that the ribs of *Eunotosaurus* are broad and expanded, creating a shell-like body profile that is eerily reminiscent of turtles. Diagram based on drawings and information in Lyson et al. (2014).

presumably breathed using both its ribs and abdominal muscles (Lyson et al., 2014). However, the expanded ribs of this fossil reptile indicate more limited movements, suggesting that respiration was more reliant on abdominal muscles (Lyson et al., 2014). Moreover, muscles attach to the outer, living rind of bones (the periosteum) through connective tissues called Sharpey's fibers (Liem et al., 2001; Kardong, 2012). These connection points can be detected in extant and fossil bones microscopically. Curiously, Sharpey's fibers are diminished or absent in the ribs of *Eunotosaurus*, suggesting that its rib-spanning intercostal muscles were already reduced (Lyson et al., 2014).

Research into turtle development, likewise, has shown that many features that appear in turtles during their early embryonic stages are similar to those of lizards, including specialized traits like the fusion of the astragalus and calcaneum and the development of a hooked fifth digit, features absent in parareptiles (deBraga and Rieppel, 1997; Rieppel and Reisz, 1999). Shell development also shows that the shell armor itself, while dermal in origin, is not related to bones like scutes but is instead linked to the CR we discussed previously (Kuratani, Kuraku, and Nagashima, 2011). As we have discussed, the plastron develops first, and its placement and robust nature are similar to the expanded gastralia observed in the sauropterygian marine reptiles we will discuss in chapter 15. In fact, one group of sauropterygians, the placodonts, even have a plastron

14.13. *Odontochelys,* a Triassic turtle lacking a carapace. *Odontochelys* possesses an odd mixture of turtle and non-turtle features. Its mouth and palate are covered in small, pointed teeth, and its pectoral and pelvic girdles are free and not covered by a carapace. However, the ribs of *Odontochelys* still project outward as they do in turtles, and a well-developed plastron is present. Diagrams based on work by Li et al. (2008).

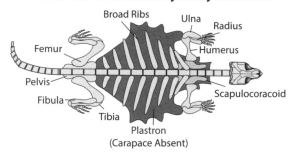

The Triassic Toothed Turtle, *Odontochelys*

Skull in Dorsal View

Braincase Skeleton

Facial Skeleton

Beak-like Premaxilla

Skull in Side View

AM

Quadrate-Articular Jaw Joint

Maxilla

Dentary

Cryptodire Turtle Skull for Comparison

AM

Dorsal View of *Odontochelys* Body Schematic

Broad Ribs

Ulna

Radius

Humerus

Femur

Pelvis

Fibula

Tibia

Scapulocoracoid

Plastron (Carapace Absent)

(Rieppel and Reisz, 1999). This has led to a marine origin hypothesis that places turtles near the sauropterygians on the lepidosauromorph family tree (deBraga and Rieppel, 1997; Rieppel and Reisz, 1999; Hill, 2005). Moreover, genetic traits have consistently placed turtles squarely among the diapsids, with some studies placing turtles close to archosaurs (crocodylians and birds) (Iwabe et al., 2005) or lizards (Lyson et al., 2011). The most robust genetic studies to date strongly suggest turtles share closer relationships with archosaurs than they do with lizards and snakes (Field et al., 2014). The signal coming from all of these studies suggests that turtles are diapsids, not parareptiles, and that features of the skull, such as lack of temporal fenestrae, are malleable and not always reliable indicators of pedigree.

Another fossil discovery provides further support for the diapsid affinities of turtles. Called *Odontochelys,* this animal from the Late Triassic of China (220 Ma; Card 96) predates *Proganochelys* by a good 10 million years (Fig. 14.13) (Li et al., 2008). It stretched just over 30 cm in length, and possessed a combination of traits that are quite unusual but telling. First, this animal has a narrow, pointed skull with teeth in its premaxilla, maxilla, and dentary bones, as well as across its palate. The skull also lacks temporal fenestrae. Second, *Odontochelys* has a plastron but lacks a shell, and yet the ribs are expanded sideways as they are in embryonic turtles before the shell begins to form! Even more interesting is that the ribs in *Odontochelys* have formed in such a way that the scapulocoracoid is still on the outside of the ribcage, as embryonic evidence would predict for turtle ancestors that predated the full development of the shell (Li et al., 2008). Overall, *Odontochelys* resembles much of what would be predicted for a turtle ancestor given recent genetic and developmental studies (Li et al., 2008), and it seems to share more derived traits with diapsids than with parareptiles. Just before this book went to press, another older turtle ancestor was discovered. Called *Pappochelys*, it existed 240 Ma (Card 95) in what is now Germany (Schoch and Sues, 2015). *Pappochelys* was a small animal with a skull approximately 20 cm in length. The skull of *Pappochelys* has temporal fenestrae like those of other diapsid reptiles (Schoch and Sues, 2015). Moreover, it has robust gastralia that resemble the overall shape of a turtle plastron (Schoch and Sues, 2015). This new fossil and new data on old fossils strongly suggest that turtles may fit comfortably among the diapsids (Schoch and Sues, 2015).

New information from the fossil record, genes, anatomy, and development continues to pour in about turtles, and it is certain that the debate over their origins and pedigree is far from over. A new consensus does seem to be emerging that turtles are diapsids, but their relationships with other reptiles, as well as their ecological origins (aquatic or terrestrial) remain challenging to unravel. The chassis of turtles, as odd as it is, has apparently served them well and has made them a very successful branch of the reptile family tree, even though a general headache for those attempting to unravel their relationships. Turtles are weird and have been nothing if not resilient. It is difficult not to respect these reptiles: some species can breathe underwater when times get tough . . . through their vascular cloaca (Flanagan, 2015)!

15.1. Lepidosaur skull evolution. Starting from a common ancestor close to animals like *Petrolacosaurus,* the diapsid condition in lizards and the tuatara became modified in different ways. Apparently, early during lepidosaur evolution the lower temporal bar (made up of the quadratojugal bone) was reduced and then lost. With the loss of the lower temporal bar, many lizards and snakes gained varying degrees of skull kinesis. In the lineage leading to the tuatara, early sphenodontids, such as *Gephyrosaurus,* had no lower temporal bar, but this gap was filled in during the evolution of the modern *Sphenodon* species. Skull illustrations after Reisz (1977), Carroll (1988), Evans (1994, 2003), Pough et al. (1998), Pough, Janis, and Heiser (2013), and Liem et al. (2001).

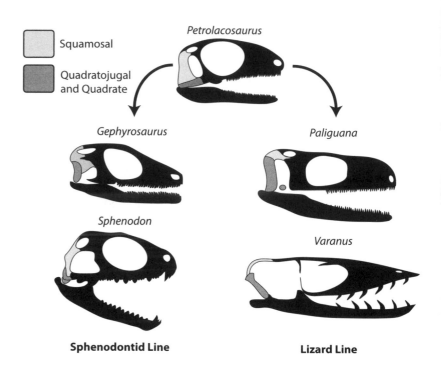

Squamosal

Quadratojugal and Quadrate

Petrolacosaurus

Gephyrosaurus

Paliguana

Sphenodon

Varanus

Sphenodontid Line

Lizard Line

Snakes and Sea Dragons

15

VERSATILITY OF USE IS THE KEY TO A SUCCESSFUL PRODUCT OR tool. In electronics, there is probably no better example of versatility than the transistor. Transistors are devices that can both amplify the power of electronic signals and act as a switch, turning power on or off to other components in a device or circuit (Langone, 2004; Bloomfield, 2006). Previous electronic amplifiers and switches, such as vacuum tubes, were large and ran hot. The invention of transistors in the mid-twentieth century was an important milestone for electronic devices because they were small, cheap to make, and generated less heat (Walker et al., 1999; Langone, 2004). Transistors opened the door for all the small, portable electronic devices we take for granted now.

Similarly, the versatility of the lepidosauromorph chassis has been remarkable. These reptilians have spawned an amazing variety of chassis types, and have adapted to environments ranging from arboreal gliders to bipedal runners to seagoing predators. We now return briefly to lizards and the tuatara to discuss fossil evidence for the evolution of skull kinesis, highlight the amazing skeletal chassis and evolution of snakes, and finish with the fossil diapsids often called sea dragons.

Far and away, the most noticeable trend among the menagerie of vertebrates that make up the Lepidosauromorpha is the gradual loss of bones in the skull, especially those around the braincase and in the cheeks (Fig. 15.1) (Evans, 2003; Benton, 2005). At first, one of the cheek bones, the quadratojugal, became a small, rudimentary structure that linked the jugal to the back of the skull via the quadrate. But very early during the evolution of lepidosauromorphs, the connection between the quadratojugal and the jugal was lost, leaving a small space that continued to widen over time, eventually culminating in the complete loss of the lower temporal bar (Fig. 15.1) (Evans, 2003; Benton, 2005). This loss, as you recall, is what allows lizards and (as we will soon see) snakes to have such kinetic skulls. In fact, some of the very earliest lepidosauromorphs had reduced or completely lost the quadratojugal bone itself. This loss, in turn, freed the quadrate to move as an independent lever as we have covered in some detail already for lizards.

However, the most primitive living lepidosaurs, the tuatara, retain a complete lower temporal bar and lack skull kinesis. Intriguingly, the lower temporal bar is open in the tuatara embryo, and it closes only when the jugal bone grows into the quadratojugal bone later in development (Fig. 15.1) (Evans, 2003). This embryology is supported by the first known

A Few Lepidosauromorph and Lizard Trends

fossil relatives of the tuatara, the sphenodontids. The best known, most primitive sphenodont from the fossil record is called *Gephyrosaurus* (it appears roughly 200 Ma [Card 96], although the sphenodont fossil record goes back at least 210 Ma) (Evans, 2003). This animal shows that its skull, like that of the developing tuatara, lacks a connection between the jugal and the quadratojugal (Fig. 15.1). In fact, many of the fossil sphenodontids lack a connection between the jugal and quadratojugal bones in the cheek (Evans, 2003). Therefore, the closed lower temporal bar observed in the modern tuatara is a secondary adaptation of the skull to be less kinetic. The sphenodontids comprised a fairly diverse group during the Late Triassic to Middle Jurassic (210–165 Ma; Cards 96–97) that evolved both terrestrial and aquatic forms. However, by the end of the Cretaceous (66 Ma; Card 99) most of these lepidosauromorphs had gone extinct, with stragglers isolated on the southern continents. It is from populations of these sphenodontid 'leftovers' that today's modern tuatara descended (Benton, 2005). The other lepidosauromorphs kept the lower temporal bar open and, as we have seen for lizards, this change in skull interconnectedness allows for a great deal of kinesis. We can infer that many of the extinct marine reptiles (Sauropterygia and Ichthyosauria – close lepidosauromorph cousins) probably had some skull kinesis as well.

The early fossil record of lizards has been patchy, although the earliest known true lizards appear in the fossil record during the middle of the Jurassic period (176–161 Ma; Card 97) (Evans, 2003). By the end of the Cretaceous period (66 Ma; Card 99), most of the major lizard groups are known from fossils. One group of fossil toxicoferan lizards is best discussed briefly at this juncture. Three families of varanid lizards became major aquatic predators some 100 Ma (Card 98), going extinct at the end of the Mesozoic (66 Ma; Card 99). The largest and best known of these aquatic varanids are the mosasaurs, with some members reaching over 9 meters in length (Fig. 15.2) (Lindgren, Polcyn, and Young, 2011). The skulls of mosasaurs were more robust than typical of many varanid lizards, and they lost the kinesis between the facial and braincase skeleton (Lingham-Soliar, 1995). On the other hand, this loss seems to have been offset by an extra joint in the lower jaw that presumably increased the gape of these animals (Fig. 15.2) (Lingham-Soliar, 1995). The teeth of many mosasaurs were pointed and cone-shaped, very similar to those of modern dolphins and whales that capture fish. Some mosasaurs had larger, blunted teeth that suggest they fed on shelled invertebrates. This inference is supported both by the fossil shells of ammonoids (squid relatives) with puncture marks that correspond to the size and placement of mosasaur teeth and by mosasaur fossils that contain invertebrate shells in their preserved stomach contents. As with many of the reptiles we have previously discussed, mosasaurs had palatal teeth that may have been useful at holding slippery prey in their mouths (Mahler and Kearney, 2006). The vertebral column was long and sinuous, and in some mosasaurs the tail was tall, deep, and almost fluke-like (Lindgren, Jagt, and Caldwell, 2007; Lindgren, Polcyn, and Young, 2011). The pelvic girdle of many

Mosasaur Skull Based on *Mosasaurus* and *Plotosaurus*

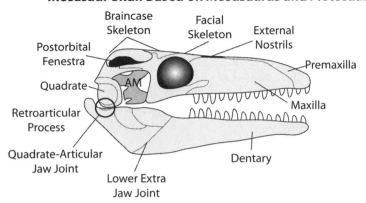

Braincase Skeleton
Facial Skeleton
External Nostrils
Postorbital Fenestra
Premaxilla
Quadrate
AM
Retroarticular Process
Maxilla
Quadrate-Articular Jaw Joint
Dentary
Lower Extra Jaw Joint

Mosasaur Skeleton with Side Body Profile Based on *Mosasaurus* and *Plotosaurus*

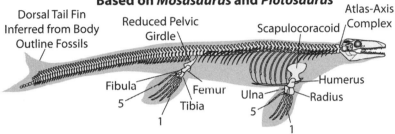

Dorsal Tail Fin Inferred from Body Outline Fossils
Reduced Pelvic Girdle
Scapulocoracoid
Atlas-Axis Complex
Fibula
Femur
Ulna
Humerus
5
Tibia
Radius
1
5
1

15.2. Skull and skeletal diagrams of the marine varanid lizards known as mosasaurs. Notice that, as with many marine tetrapods, the premaxilla in mosasaurs has expanded back and dorsally, displacing the nostril. Although the skulls of mosasaurs have some reduced kinesis, this may have been compensated for by an extra joint in the lower jaw. AM indicates the location of the adductor mandibulae and other jaw-closing muscles. Numbers indicate digits of the same number. Skull and skeletal restorations after Carroll (1988), Lingham-Soliar (1995), Lindgren, Jagt, and Caldwell (2007), and Lindgren, Polcyn, and Young (2011).

mosasaurs was not attached to the vertebral column, a situation reminiscent of early, mostly aquatic, tetrapods like *Acanthostega* (Lindgren, Jagt, and Caldwell, 2007; Lindgren, Polcyn, and Young, 2011). Moreover, the forelimbs and hind limbs were modified into paddle-like organs, with the humerus, femur, forearm, and shin all foreshortened, whereas the fingers and toes were elongated and splayed (Fig. 15.2). Finally, mosasaur skin impressions have revealed that at least some of these lizards had small, overlapping scales that formed channels along the back that would have reduced drag when the animal was swimming (Lindgren et al., 2009). Other features of this group appear in the sauropterygian aquatic reptiles and in dolphins and whales, and we will save our discussion of these aquatic adaptations for later. I encourage interested readers to consult Everhart (2005) for more about these amazing lizards.

Lepidosauria, Part 2: Snakes

Snakes, extremely specialized toxicoferan lizards, represent the end of a long trend in skull kinesis among the lepidosauromorphs. Snake skulls take kinesis to the extreme, losing both the upper and lower temporal bars and freeing not only the quadrate but the squamosal bone as well to rotate and move about (Fig. 15.3) (Pough et al., 1998; Pough, Janis, and Heiser, 2002; Benton, 2005). Sharp, needle-thin and hook-like teeth adorn the maxilla and dentary as well as the palatal and pterygoid bones, but the premaxilla lacks teeth completely (Fig. 15.3). Many of the bones

Pedigree Limbless Lizards to the Extreme

Date of First Appearance ~130 Ma (Card 98)

Specialties of the Skeletal Chassis Extremely Kinetic Skulls

Eco-niche Terrestrial to Aquatic Predators

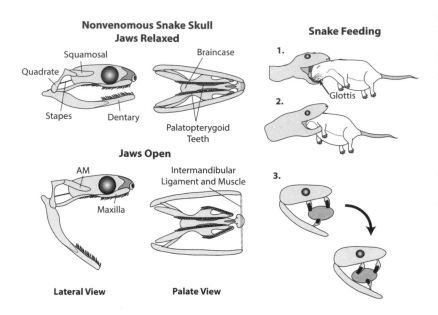

15.3. Nonvenomous snake skull mechanics. Snake skulls are incredibly kinetic because they possess multiple skull and jaw joints coupled with the loss of both the upper and lower temporal bars. Note that both the quadrate and squamosal have become mobile levers that help to increase the gape of the mouth, and that the stapes is an immobile strut braced against the braincase. In the lower jaws, the mandibular symphysis is absent and occupied by a flexible ligament and muscle that can stretch, allowing the lower jaws to bow apart to consume prey. Extra sets of palatopterygoid teeth help snakes pull prey into their mouths. The inset shows a snake walking its mouth (1, 2) along a recently dead mouse, and a diagram (3) shows how the snake mouth operates like a set of independently movable forks that draw its head over the prey. In part 1 of the inset, note the forward position of the glottis, which enables snakes to breathe snorkel-like while they are consuming their prey whole. AM indicates the location of the adductor mandibulae and other jaw-closing muscles. Diagrams after Pough et al. (1998) and Pough, Janis, and Heiser (2013).

in the upper and lower jaws are loosely interconnected to one another by ligaments and muscles, allowing independent movements (Liem et al., 2001). About the only bones in a snake skull that do not move are the small cluster that make up the braincase. Again, prey does not wish to be eaten and struggles vigorously to escape. If the snake braincase were mobile or had large openings, one good kick from its prey would result in traumatic brain injuries. If you're going to be eating your food whole and have to distort your skull to do it, you also would not want your brains to fall out of your head!

I have had the privilege of keeping several corn snakes (*Elaphe guttata guttata*). These animals have beautiful red and orange patterns on a silvery background, with belly scales that form a black-and-white checkered mosaic, and are nonvenomous (I'm crazy but not that crazy). Keeping such snakes is not for everyone because they require a diet of mice. I fed dead (rather than live) mice to my snakes for reasons I indicated in chapter 5, but it gave me a weekly opportunity to watch how these vertebrates eat their food. Before striking the dead mouse, there was a lot of tongue flicking that allowed the snake to hone in on its prey's scent. Next, a quick strike, and the needle-thin, hook-like teeth firmly anchored its head on to the mouse while the snake quickly wrapped its coils around the mouse's body. After several minutes (the snake did not know the mouse was already dead, of course), the snake would release the mouse and search for the head, its scaly face poking around the body and its tongue flicking incessantly.

What followed next was always quite fascinating to watch (Fig. 15.3). First, the snake would open its jaws and grip the nose of the mouse. Next, the head of the snake would literally begin to 'walk' slowly down the head of the mouse toward the body. One side of the skull and jaws would grip and pull the mouse while the other side would release its previous hold and shift forward. When I could see the lower jaw, it was apparent that the dentary bones were beginning to spread apart, widening the mouth several times larger than the snake's head. You could also see that the region of the skull containing the quadrate and squamosal bones was expanding as well. As more of the mouse passed into the esophagus of the snake, it began to use its neck and body muscles to forcibly push its meal further into its gut. Eventually, only the skinny tail of the mouse would be poking out of the snake's mouth, and with a few more turns and twists of its neck, this too would disappear. Finally, the snake would make a few movements of its jaws and skull, and 'reset' its head back to its original shape.

Forget all the myths perpetuated out there about snakes 'dislocating' or 'unhinging' their jaws to swallow prey. Instead, the series of head and jaw movements I have just described in my former pets are possible because, on each side of a typical snake's head, there are eight independently movable skull joints (Pough et al., 1998; Liem et al., 2001). The maxilla and palatopterygoid bones (palate and pterygoids) can be advanced and retracted independently of one another, and in this way the snake can literally walk its head over its prey. Imagine if your upper jaw bore four independently movable forks that could take turns pulling steak into your mouth, and you get some idea of how a snake is feeding (Fig. 15.3). In most jawed vertebrates, the dentary bones are tightly connected or fused at the front of the mouth in a region dubbed the mandibular symphysis. In snakes, the dentary bones can be moved independently of one another and splayed because in place of the mandibular symphysis are muscle, ligament, and skin that can be stretched apart (Fig. 15.3) (Pough et al., 1998; Liem et al., 2001).

Snakes don't have arms or anything to hold down or grab onto struggling prey. Not surprisingly, there has been selective pressure among different snake groups to develop ways to dispatch prey before feeding. Venom in various formulations has evolved as one means of subduing and even predigesting prey before ingestion. The vipers, such as rattlesnakes, have modified their maxilla so that it bears a single, hollow fang that can be folded against the palate when the mouth is closed (Fig. 15.4) (Pough et al., 1998; Liem et al., 2001). As the snake opens its mouth to strike, the maxilla with its fang is rotated downward, and like a hypodermic needle, is driven deep into prey. The specialized venom glands situated behind the eyes are connected to the fang by a modified salivary duct. Toxins are squeezed out of the glands by modified branches of the adductor mandibulae muscles as the snake bites down (Fig. 15.4) (Pough et al., 1998; Liem et al., 2001). Not only are toxins introduced into the prey's body but, just as you might deeply inject meat with tenderizers, so enzymes that break down proteins are introduced deep into the victim (Warrell, 2010). The prey may be released after being bitten, and the toxins and enzymes do their work killing the animal, with the snake following the scent trail to its now safely immobile and tenderized lunch. Other venomous snakes such as cobras have small, fixed fangs that deliver fast-acting neurotoxins that

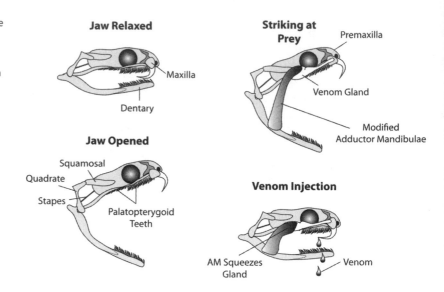

15.4. Fanged venomous snake skull mechanics, based on pit vipers. Note that the maxilla is a mobile lever bearing one large, hypodermic fang. When a snake like a viper opens its mouth, the maxilla swings forward and readies the fang for venom injection. As the mouth closes down on prey, a modified portion of the adductor mandibulae squeezes a venom gland, forcing the potent cocktail of enzymes and toxins through the fang. AM indicates the location of the adductor mandibulae and other jaw-closing muscles. Diagrams after Pough et al. (1998) and Liem et al. (2001).

quickly paralyze prey and stop the heart (Warrell, 2010). Nonvenomous snakes, such as boas and pythons, neutralize their prey through constriction. Constrictors commonly have many short vertebrae and powerful, overlapping body muscles that can provide tremendous squeezing power. As the prey is squeezed, respiration becomes more and more difficult, and the heart is compressed (Pough et al., 1998). Through a combination of suffocation and cardiac arrest, the prey succumbs to the snake's coils.

Snakes no longer have eyelids; instead the eye is covered with a clear scale (called a spectacle) that is shed with the skin on a regular basis. The takeaway lesson here: never enter a blinking contest with a snake! Snake eyes are also odd in how they focus on objects in their field of vision, and operate, strangely, like those of fish in some respects. For example, the lens of the eye is spherical, not disc-like as it is in most tetrapods (Caprette et al., 2004; Davies et al., 2009). Stranger still, they squeeze expanded iris muscles (the muscles that control how much light enters the eye), which increases the internal pressure of the eyeball, forcing the spherical, rigid lens forward to allow the snake to focus on near objects (Caprette et al., 2004; Davies et al., 2009). When these iris muscles relax, the spherical lens glides back to its initial position. The cornea itself is also very spherical, again like what is observed in many aquatic vertebrates.

The eyes of snakes seem to have evolved from a more degenerated version of the typical reptilian eye in their common ancestor (Caprette et al., 2004; Foureaux et al., 2010). This would explain why the lens, the cornea, iris, and focusing system are all so different from those of lizards. Essentially, snakes reinvented the wheel, taking the leftover parts of a degenerate eye and adapting them to again create focused images. However, this odd anatomy has led to ongoing debates about what these eyes are telling us about the history of snakes (Caprette et al., 2004). On one hand, burrowing vertebrates and those that live underground tend to show evolutionary trends toward eye degeneration and simplification

(Foureaux et al., 2010). Thus, some scientists have taken this to mean that the common ancestor of snakes was a burrowing animal. On the other hand, the aquatic-like features of the spherical lens and its method of focusing suggest to others that snakes have an aquatic or semiaquatic common ancestor (Caprette et al., 2004). Currently, there is no strong consensus as to which scenario is more likely, and the pedigree and fossil record of snakes has not been the most helpful in this regard, either.

Adding to the oddity of snake vision, certain groups of snakes have developed ways to detect the heat generated by their prey in infrared. Any object that is warmer than absolute zero emits some form of infrared light (Bloomfield, 2006). We humans can only feel infrared as heat, but many vertebrates like snakes not only can finely detect its strength and direction, but can see it as well. Some snakes, like boas and pythons, have specialized infrared detecting cells beneath their lip scales (Plate 18), whereas numerous viper species have finely tuned systems where infrared detecting membranes are suspended in specialized pits on the face (Moon, 2011). Back in chapter 4, we learned that the fifth cranial nerve from the brain (called the trigeminal) is complex and both controls the jaw-closing muscles and provides sensation to the face and teeth. In snakes, the trigeminal nerve innervates the infrared detectors in the skin or pits (Moon, 2011). When scientists have followed these branches of the trigeminal nerve back to the brain, they have found that it not only sends signals to the cerebrum for processing but also sends information to the optic lobes. This means that the infrared signals detected in the face or pits are overlaid as visual cues onto what the snake sees with its eyes (Moon, 2011). Therefore, several snakes can see both their prey and its heat trail.

Snakes would never appreciate good music: they lack external ear openings, and the stapes is firmly braced between the quadrate and the braincase (Figs. 15.3, 15.4) (Pough et al., 1998; Liem et al., 2001). Various experiments have shown that snakes can hear some low-frequency sounds, and other evidence indicates that some snakes can pick up vibrations through the ground using their jaws, which transmit these sounds into the inner ear. The burrowing hypothesis of snake origins suggests that the loss of the external ear and most of the hearing in snakes was related to keeping debris out of the ears of their subterranean ancestors. However, another explanation for the loss of the external ear and the brace-like stapes has to do with how snakes feed. The very mobile quadrate bone of snakes allows the skull to deform and expand in the region where an external ear would be (Berman and Regal, 1967). Thus, snakes could have also lost their ability to hear as a compromise related to their highly kinetic skulls (Berman and Regal, 1967).

As anyone who has had a bad cold knows, it is hard to eat when your nose is stuffed with mucus. We forget that we breathe through our nose while we are chewing, and when this air passage is obstructed with snot we have to stop and breathe around our food. If you think that's a problem, now imagine slowly swallowing something like an entire loaf of

15.5. Diagram of a snake vertebra and skeleton. Snake vertebrae possess joints in addition to the typical zygapophyses, which prevent the vertebral column from twisting on itself as the snake moves about. The amazing flexibility of the snake skeleton arises from a combination of hundreds of vertebrae spanned by complex body wall muscles. In all snakes, the pectoral girdle does not form, but in some species, a small, vestigial pelvic girdle and hind limb form. Snake vertebra diagram based on Liem et al. (2001).

Snake Cervical Vertebra (Cranial View)

- Neural Spine
- Zygapophyses
- Additonal Joint
- Centrum
- Additional Joint
- Hypapophysis

Flexible Snake Skeleton

- 120+ Dorsal Vertebrae and Ribs
- Lack of Pectoral Girdle
- Pelvic Remnant

bread without being able to breathe through your nose. This is precisely the problem snakes have when engulfing large prey. Although they have internal nostrils (choanae) like we do, theirs open from the roof of the mouth, and air entering the external nostrils would be blocked by the presence of large prey backed up against the palate. The remarkable solution to this problem is that snakes have shifted their glottis cranially so that it opens just caudal to the front of the dentary bones (Fig. 15.3) (Pough et al., 1998; Liem et al., 2001). If you recall from chapter 8, the glottis is the valve that allows air into the trachea and lungs. By shifting the glottis forward, the snake trachea now has a sort of air snorkel with which to take in oxygen while the rest of its mouth is stuffed with prey. This shift in the glottis has also resulted in the loss of anything like vocal cords, but, as most people know, snakes hiss. Much like the hissing sound created by partially squeezing open the valve on a car or bike tire, so too a snake hisses by forcing air out of the glottis when it is partially closed (Hildebrand and Goslow, 2001; Liem et al., 2001).

Snakes have a well-developed sense of smell. As with many lizards, snakes have a deeply forked tongue that is flicked about to 'taste' the air and capture minute chemosensory molecules. Similarly, the tongue can be inserted into the roof of the mouth, where two pores lead to the vomeronasal organ, which, as we have previously discussed, is extremely sensitive to various odors and pheromones (Pough et al., 1998; Hildebrand and Goslow, 2001; Liem et al., 2001). A snake's tongue does not have many taste buds. Instead, a variety of taste buds are spread across the palate and insides of the mouth.

The vertebrae of snakes are numerous and almost all are dorsal vertebrae. A typical snake chassis consists of an atlas and axis followed by at least 120 dorsal vertebrae. The sacrum and caudal vertebrae are typically small and short, and most snakes do not have long tails or breakaway zones. Anyone who has ever had to use a long, metallic measuring tape to get the dimensions of a room for some home project knows the all too common frustration of having it twist and bend on itself, usually just before you write down the information you need. Whereas this is an inconvenience for us, twisting of the vertebral column on itself in an elongate vertebrate like a snake would be catastrophic. Therefore, in addition to the zygapophyses that interlink the vertebrae of most tetrapods, snakes have additional intervertebral joints that prevent such body axis twisting (Fig. 15.5).

As is well known, snakes slither to get around. Snake locomotion is biomechanically complex, and what follows here is a very simplified overview of how these limbless lepidosaurs get around (based on Gray, 1968; Jayne, 1988; and Hu et al., 2009). Snakes bend their bodies into a series of curves, and push themselves forward against the ground at points where the body is least curved (Fig. 15.6). For example, if the snake is curved into a simple S-shape, then the points of greatest force and forward propulsion will occur at the straightest parts of the 'S' (i.e., just behind the head and just in front of the tail) (Fig. 15.6). The rough, overlapping belly scales of most snakes provide ample friction for gripping the terrain and moving forward, even over relatively smooth surfaces. As with caecilians, if snakes are confined to small spaces, such as a burrow, they employ concertina movements where one part of the body is pushed against the ground while another part is stretched out to move forward. Once the stretched-out piece is placed on the ground, it becomes still and the rest of the body is drawn forward (Fig. 15.6). Other snakes that live in areas such as deserts where the ground over which they travel is loose (e.g., sand grains) have evolved alternative forms of serpentine movements such as sidewinding. In sidewinding, at any given point in time only a small portion of the snake's belly is briefly in contact with the sand or loose gravel, and just like a smart swimmer knows to swim perpendicular to a rip tide, so the snake's body moves perpendicular to the slope of the sand dune (Fig. 15.6).

Because the vertebral column of snakes is so elongate and complex, the dorsal and ventral muscles associated with the vertebrae play a major role in locomotion (Gray, 1968; Jayne, 1988; Hu et al., 2009). Some dorsal and ventral muscles simply attach to and move adjacent vertebrae, but others have long tendon-like extensions that span across multiple vertebrae. When you grab a string at two ends far apart from one another and pull those points together, the portion of string between the two converging points bends. Similarly, when long tendons pull one vertebra several segments away toward another, it causes the intervening portion of the vertebral column to bend, throwing the snake into a series of curves. In snakes that constrict their prey, there tend to be shorter vertebrae and more short tendon-like connections between the vertebral segments so that the snake can throw its coils around prey and apply sufficient pressure to kill it (Pough et al., 1998). However, these snakes tend to be bulkier and less able to move quickly. Conversely, in snakes that chase their prey and move rapidly, the vertebrae tend to be longer and the longer tendon-like connections are emphasized (Pough et al., 1998). This enables these snakes to move quickly, but hampers their ability to make tight coils.

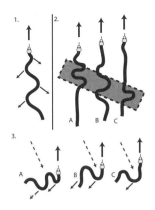

15.6. Generalized movement patterns in snakes. The common form of locomotion for most snakes consists of slithering (1) such that where the body curves it pushes down and back to propel the snake forward. In concertina locomotion (2), one portion of the body is anchored against the ground or the inside of a tunnel while part of the snake's body stretches ahead. In the diagram, the dashed box shows the portion of the snake being used as the anchor, whereas the rest of the snake is stretching forward or catching up to the anchor point. During sidewinding, the snake slithers perpendicular to the slope and movement of a sand dune, which allows forward progress on an angle. Large arrows indicate the direction of travel, whereas small arrows show regions of the snake pushing on the ground. Dashed arrows in (3) indicate the direction of sand grain movement. Diagrams patterned after Gray (1968) and Pough et al. (1998).

But how do you get animals that have 120 or more dorsal vertebrae and no limbs? It turns out that a simple but profound difference in the timing of the expression of the *HOX* genes we discussed in chapter 10 renders snakes limbless, whereas an increase in the frequency of another set of segmenting genes generates their amazing number of vertebrae (Cohn and Tickle, 1999; Caldwell, 2003; Vonk and Richardson, 2008; Woltering et al., 2009). In most tetrapods, there is a restricted region where three *HOX* genes that produce the rib-bearing dorsal vertebrae are active simultaneously. These three genes are expressed together only after the pectoral limb forms, and end their simultaneous expression just before the pelvis. In snakes, these three *HOX* genes are instead switched on just after the axis vertebra forms in the neck and continue their simultaneous expression right up to the pelvis (Cohn and Tickle, 1999). Therefore, there is never a chance to produce forelimbs, and the pelvis and hind limb may be rudimentary or nearly gone (Plate 18). In addition, another set of genes that regulates vertebral segmentation is sped up in snakes, so that multiple vertebrae will form rapidly along the body's axis (Vonk and Richardson, 2008). As with fins and limbs, although the same *HOX* genes are at work, differences in the timing and duration of their expression have profound effects on anatomy.

The habits of the common ancestor of snakes continue to be disputed. One of the largest problems with reconstructing the early history of snake evolution is that we have so few snake fossils. The first known snakes appear in the middle of the Cretaceous period some 130 Ma (Card 98) and already they have most of the features of snakes like pythons and boas (Rage and Escuillié, 2003; Rieppel et al., 2003; Zaher, Apesteguía, and Scanferia, 2009). In fact, modern pythons and boas retain a small pelvic girdle and limb-like spur that is used during copulation. The earliest known snake fossils bear a striking resemblance to boas and pythons, but tiny hind limbs complete with toes are present in these specimens (Rage and Escuillié, 2003; Zaher, Apesteguía, and Scanferia, 2009). The odd and reworked degenerate eyes, lack of external ears, and limb loss have suggested to some scientists that the common ancestor of snakes was a burrowing animal (Foureaux et al., 2010). Several lizard groups burrow, and many of these species develop smaller, scale-covered eyes, closure of the outer ear openings, and loss or dramatic reduction of the limbs in some combination (Foureaux et al., 2010). Moreover, the most primitive living snakes are small, burrowing insectivores. If a burrowing ancestry is correct, then the skeletons of legged snakes from the middle of the Cretaceous must have already diverged greatly from their early ancestors.

Some tantalizingly new support for a burrowing origin for snakes comes from discoveries of skull pieces from a primitive Cretaceous snake named *Coniophis* (Longrich, Bhullar, and Gauthier, 2012). The dentary and maxilla of this snake are partially preserved, and these skull bones more closely resemble the less kinetic skulls of lizards, whereas the teeth are hooked like snakes'. Moreover, known vertebrae of *Coniophis* are very snakelike. This snake with a lizard-like head combined with snakelike teeth and vertebrae is found in terrestrial flood plain sediments, not the marine sediments predicted for an aquatic snake ancestry. It consistently comes out close to the common ancestor in snake pedigrees, further supporting a burrowing, terrestrial origin for snakes (Longrich, Bhullar, and Gauthier, 2012). Just prior to publication of this book, a four-legged snake from the Cretaceous of Brazil was discovered (Martill, Tischlinger, and Longrich, 2015). The snake, named *Tetrapodophis*, has diminutive limbs similar to burrowing lizards, and may bolster a burrowing origin for snakes. However, its collection and provenance from Brazil are controversial (Perkins, 2015), rendering its place in snake evolution difficult to resolve.

Other larger primitive snake fossils are known from the end of the Cretaceous to the early Paleogene (~70–50Ma; Card 99). One fossil snake from India, *Sanajeh indicus*, was very boa-like, measured up to nearly 3.5 meters in length, and was preserved in the act of plundering a nest of baby sauropod dinosaurs (Wilson et al., 2010). The largest known snake, a fossil boid aptly named *Titanoboa*, hails from South America. *Titanoboa* is estimated to have been over 11 meters in length, had a midsection nearly a meter wide, and may have weighed over a metric ton (Head et al., 2009). Clearly, large predatory terrestrial snakes were already terrorizing the countryside soon after their initial evolutionary origin.

An alternative hypothesis to snake origins has been that snakes have an aquatic ancestry and that they share a close, common ancestor with the mosasaurs. This hypothesis is supported in part by new molecular data that show that snakes and varanid lizards belong to the lizard group Toxicofera (see chapter 13) (Fry et al., 2005, 2009). Furthermore, the sinuous vertebral column and foreshortened limbs of mosasaurs have been marshaled as evidence of a close aquatic ancestor for snakes and these varanid lizards (e.g., Lee, 2005). Recall, too, that the eyes of snakes share some anatomical similarities with those of fish and other aquatic vertebrates (Caprette et al., 2004). Future fossil discoveries, molecular data, and anatomical studies will undoubtedly shed more light on this issue. For now, we can marvel at how snakes took the lizard chassis and modified it into an elegant and often deadly skeletal framework.

Sauropterygia and Ichthyosauria: The Great Sea Dragons and Fish-Lizards

The Sauropterygia and Ichthyosauria are diapsid groups that appeared before the first dinosaurs, and some of the sauropterygians survived right until the end of the Mesozoic (66 Ma; Card 99) (Benton, 2005). According to some sources, sauropterygians may have survived even longer in the depths of Scottish fjords. As with the other lepidosauromorphs we have discussed, there was loss of the lower temporal bar and the apparent development of cranial kinesis. Sauropterygians, despite the differences among their various groups, share a number of derived trait states. In particular, their premaxilla is long, their temporal fenestra is larger than their orbits, and their limb bones have reduced muscle attachment sites

Pedigree Marine-Adapted Lepidosauromorphs and Kin

Date of First Appearance ~250 Ma (Card 95)

Specialties of the Skeletal Chassis Specialized Diapsid Skulls

Eco-niche Aquatic Predators

distal to the elbow and knee (Carroll, 1988; Benton, 2005). These features go part and parcel with the aquatic habits of these lepidosauromorphs. A long premaxilla shifts the position of the nostril caudally and dorsally, allowing such animals to breathe through their nose when the rest of the head is submerged. The expanded temporal fenestra lightens the skull, provides room for kinesis, and expands areas of jaw muscle attachment. The reduced muscle attachment sites distal to the elbow and knee are characteristic of aquatic amniotes in that these animals are no longer using their lower limbs for pushing against the ground. Instead, the lower limb segments are being utilized as paddle-like structures to propel and steer the body through the water, and most of these movements are now accomplished almost solely at the shoulder and hip. If you think about it, this is similar to rowing a boat. The power for rowing the boat comes from the muscles in your body, which is sitting in the boat, and the paddles themselves just need to be stiff and rotatable. Likewise, the shoulder and hip muscles are well-developed (especially at the shoulder) to swing the limbs back and forth and rotate them like paddles. Having extra muscle mass on their ends, or making the digits or wrist mobile, would interfere with the ability to have efficient underwater propulsion. In many sauropterygians, the ends of the long bones are roughened and flat, and presumably were capped by thick cartilaginous epiphyses, a trait very common in marine-adapted amniotes (Benton, 2005).

The Sauropterygia consisted of two major chassis types. One short-lived group, known as the placodonts, was turtle-like in appearance with squat, sometimes armored bodies (Carroll, 1988; Benton, 2005; Motani, 2009). The other group of sauropterygians is known technically as Eosauropterygia. It includes three major branches of lizard-like animals with small heads, rather elongate necks, paddle-shaped limbs, and short tails. Among these groups, we will focus on the most successful and widespread chassis, that of the plesiosaurs. The first plesiosaurs appeared around 210 Ma (Card 96), and quickly diversified throughout the remainder of the Mesozoic (Benton, 2005). They were typically large animals that could measure from 1.5 to over 12 meters in length. Several times during their evolutionary history, the plesiosaurs repeatedly developed two chassis types (O'Keefe, 2002). One common variant were animals with extremely elongate necks with small heads (Fig. 15.7). In these forms, the skull was very small, perhaps less than one-sixth the total length of the body. As with all sauropterygians, the premaxilla was long and displaced the nostrils dorsally and close to the eyes. In some long-necked forms, sharp, pointed teeth projected from the premaxilla, maxilla, and dentary bones in such a way that when the plesiosaur closed its mouth, the teeth interlocked with one another to form a toothy trap. In others, the teeth were tiny and needlelike, perhaps an adaptation for catching small, fast moving fish or invertebrates (O'Keefe, 2002). Some long-necked plesiosaurs were apparently bottom-feeders: gut contents in some specimens show a preponderance of benthic mollusks (McHenry, Cook, and Wroe, 2005). A well-developed retroarticular process on the lower jaw would

Long-Necked Plesiosaur Skull Based on *Cryptoclidus*

Postorbital Fenestra

Premaxilla

AM

Quadrate-Articular Jaw Joint

Dentary

Maxilla

Short-Necked Plesiosaur Skull Based on *Liopleurodon**

Postorbital Fenestra

Premaxilla

AM

Quadrate-Articular Jaw Joint

Dentary

Maxilla

Long-Necked Plesiosaur Skeleton Based on *Cryptoclidus*

Femur

Humerus

Atlas-Axis Complex

Platelike Scapulocoracoid

Platelike Pelvis

Robust Gastralia

Short-Necked Plesiosaur Skeleton Based on *Liopleurodon**

Femur

Humerus

Atlas-Axis Complex

Platelike Scapulocoracoid

Platelike Pelvis

Gastralia

15.7. Skulls and skeletons of plesiosaurs. Skull details and skeletal reconstructions of two different plesiosaur chassis. *Cryptoclidus* represents the long-necked forms, whereas *Liopleurodon* exemplifies the short-necked chassis type. Notice in both plesiosaurs the large platelike pectoral and pelvic girdles, which probably anchored massive adductor muscles for the flippers. Well-developed gastralia were also present in these marine reptiles, adding to overall body stiffness. AM indicates the location of the adductor mandibulae and other jaw-closing muscles. *This particular *Liopleurodon* is not magical (cf. Charlie the Unicorn). Skull and skeletal diagrams patterned after Benton (2005) and Motani (2009).

have acted as a lever for the depressor mandibulae muscles to quickly and efficiently open the jaws. The necks of these plesiosaurs could be quite long, and one species, *Elasmosaurus*, had 72 cervical vertebrae (Benton, 2005)! The competing chassis type among plesiosaurs was characterized by possession of a relatively large, long skull and a short neck (Fig. 15.7) (O'Keefe, 2002). The skulls of the short-necked plesiosaurs were quite large, many measuring over a meter or more in length from nose to braincase, and thick, coarse teeth studded the premaxilla, maxilla, and dentary bones in these animals. As with the long-necked plesiosaurs, the short-necked varieties had a long premaxilla that displaced the external nostrils dorsally and close to the eyes. Again, a prominent retroarticular process was present. Incidentally, one species of short-necked plesiosaur, *Liopleurodon*, has become popular in an Internet cartoon!

All the plesiosaurs shared a number of chassis features (O'Keefe, 2002). The entire vertebral series had relatively flat vertebral bodies (centra) and tightly interlocking zygapophyses and additional intervertebral joints (Fig. 15.7). These features alone suggest that the vertebral column of plesiosaurs was not sinuous and snakelike but rather rigid. Before considering the functional implications for this type of vertebral architecture, it is important to examine the rest of a typical plesiosaur skeleton. Adding to the body's rigidity and stiffness were a robust ribcage and a long plate of gastralia that stretched from the pectoral girdle back to the pelvis (Fig. 15.7). The pectoral and pelvic girdles were greatly expanded ventrally into platelike bones (Fig. 15.7). The coracoids in the pectoral girdle and the pubes and ischia in the pelvic girdle were especially large. Given that these bones are associated with limb adductors and protractors, their large size suggests adduction and protraction of the limbs were crucial to locomotion in plesiosaurs (Benton, 2005). The limbs themselves were highly modified into paddle-like flippers. The humerus and femur were the largest and longest bones, and projected laterally from the body wall. The bones of the forearm, wrist, shin, and ankle were extremely reduced, squat, rectangular elements that all together comprise maybe one-third the length of the humerus or femur. However, the digits of the hand and feet are greatly elongated, and commonly the number of finger and toe segments is significantly increased. For example, whereas most tetrapods have two to five finger or toe segments on their digits, plesiosaurs may have digits with five or more such segments (Benton, 2005). The entire forelimb and hind limb gently arc backward, and the curvature and articulation of the fingers and toes is such that they form a paddle that tapers to a point (Fig. 15.7) (Benton, 2005). Essentially, the forelimbs and hind limbs are flippers that are shaped in some ways like aquatic wings.

In a strange way, the plesiosaur chassis gives us a sneak preview for what we will encounter in flying vertebrates like birds. The relatively stiffened chassis of plesiosaurs would have ensured that when the flippers propelled the body forward it wouldn't bend or twist, movements that would cause additional drag and make the animal unstable. As you may recall, some fast swimmers have additional joints between their vertebrae that make the vertebral column rigid and only allow the tail to beat against the water. The flippers of plesiosaurs show adaptations to swimming that are eerily similar to how airplanes, birds, and bats modify their wings for flight. There is a concept in aviation science called the aspect ratio, which is the number you get when you divide length of a wing by its breadth (Fig. 15.8) (Bloomfield, 2006). This means that airplane wings that are long and narrow have a high aspect ratio, and those with short and stubby wings have a low aspect ratio. You can make the same measurements of bird and bat wings, and retrieve similar aspect ratios. But what do the numbers mean? Airplanes, birds, and bats that have high aspect ratio wings are most efficient at cruising: that is, flying straight over long distances and turning slowly (Fig. 15.8). In contrast, aircraft and vertebrates with low aspect ratio wings tend to be more maneuverable (Fig. 15.8). In World War II, the combat planes that were involved in the 'dogfights' had low aspect ratio wings and were very maneuverable. The shapes of most plesiosaur flippers resemble airplane wings with different aspect ratios (O'Keefe, 2001; O'Keefe and Carrano, 2005). Intriguingly, the long-necked plesiosaurs tend to have high aspect ratio flippers, suggesting they were open ocean cruisers that were efficient at traveling long distances

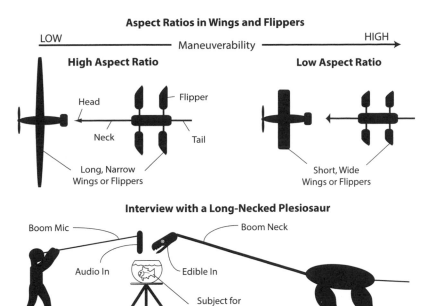

Aspect Ratios in Wings and Flippers

LOW ——————— Maneuverability ——————— HIGH

High Aspect Ratio

Head
Flipper
Neck
Tail
Long, Narrow
Wings or Flippers

Low Aspect Ratio

Short, Wide
Wings or Flippers

Interview with a Long-Necked Plesiosaur

Boom Mic
Boom Neck
Audio In
Edible In
Subject for
Discussion/Digestion

15.8. Flipper aspect ratios and plesiosaur necks. Airplane wings and flippers contribute to maneuverability in similar ways. Wings and flippers with a high aspect ratio are long and thin, making them less maneuverable but more efficient at soaring or cruising. Low aspect ratio wings and flippers show opposite trends: greater maneuverability but less efficient long-distance travel. At bottom, the stiffened necks of long-necked plesiosaurs may have functioned as a boom mic does, precisely placing and holding the head near prey. The generalized silhouettes of plesiosaurs are based on work by O'Keefe and Carrano (2005).

(Fig. 15.8) (O'Keefe, 2001; O'Keefe and Carrano, 2005). In contrast, most of the short-necked plesiosaurs have low aspect ratio flippers, indicating that these animals were very maneuverable in the water (O'Keefe, 2001; O'Keefe and Carrano, 2005). These results suggest that the short-necked plesiosaurs with their maneuverability likely gave chase to fish and other animals, twisting and turning to catch their quarry much as modern seals and sea lions pursue their prey.

But why would the long-necked plesiosaurs have relatively stiff necks? The answer may be demonstrated by the important but relatively unsung heroes of the movie industry, the boom mic operators. Anyone who has attempted to record video and audio has known the frustration of getting a great shot but having wind or other background noise swamp out the voice of the person you are filming. A boom mic solves this problem: it places a small microphone on a long, stable rod (the boom), which can be placed, often with great precision, near the person whose voice needs to be recorded for the video (Fig. 15.8). In a certain sense, the long, stiff necks of some plesiosaurs may have operated under similar principles: the neck acting as a boom, the small head acting as a microphone, and the fish or other animal being the thing that must be precisely 'recorded' (Fig. 15.8). Because the aspect ratios of long-necked plesiosaur flippers suggest that these animals were long-distance cruisers, one can imagine them swimming through schools of fish or clusters of invertebrates, precisely orienting their tiny heads on their long necks toward prey as their stiff bodies followed behind. In some senses, long-necked plesiosaurs may have been capable of some surprise depending on the clarity of the water: with the tiny head on a long neck boom ahead of the larger body, fish and other animals may have been caught up in their jaws before the larger body was detected (Massare, 1988).

Very early during the Triassic (~250 Ma; Card 95), another group of diapsids appeared that were either very close relatives of the lepidosauromorphs or were close to the common ancestor of both the lepidosauromorphs and the archosaurs (Liu et al., 2011). Called ichthyosaurs, these were the original 'fish-lizards,' and their chassis converged in many ways with those of fishes and marine mammals like dolphins and whales. Were it not for some amazing preservational circumstances, however, we might never have been aware of how fishlike the ichthyosaurs truly were. For example, many ichthyosaurs had a well-developed C-shaped tail, much like a tuna, and a large dorsal fin, but neither of these features was skeletal (Fig. 15.9) (Carroll, 1988; Benton, 2005). Instead, the dorsal fin is completely constructed from flesh, and the upper lobe of the tail fin was likewise derived from soft tissues. The body outline itself is teardrop-shaped, giving these animals a streamlined body form. As we have discussed in previous chapters, all of these traits are common in fast-swimming vertebrates that need to cut down on drag (streamlined body form), prevent roll (dorsal fin), and generate powerful tail thrust (C-shaped, homocercal tail).

But how do we know that these animals had these features at all? It turns out that bacteria acted as forensic documentarians, outlining the body form of ichthyosaurs much as you might make a chalk outline of a crime scene victim. As the bacteria digested the carcasses of deceased ichthyosaurs that lay on the ocean bottom, their waste products generated a type of biofilm or 'gunk' that stained the seabed all around the animal's body. This biofilm stain was preserved, along with the skeleton, and it traces the outline of the ichthyosaur's body (Benton, 2005) (Plate 19)! This phenomenon occurs even today: whales that die and fall to the ocean floor are feasted upon by similar bacteria, and months later a ghostly biofilm outline of the whale is generated around the bones (Mascarelli, 2009).

The skulls of many ichthyosaurs were porpoise-like, with a long, tooth-lined snout and a short, thick braincase (Fig. 15.9) (Benton, 2005). The premaxilla was very long in ichthyosaurs and comprised a large portion of the upper jaw. Its growth subsequently pushed the nostrils far back toward the eyes and into a more dorsal position, where they presumably aided the animals in taking a breath at the water's surface. As a consequence of the 'overgrowth' of the premaxilla, the maxilla in ichthyosaurs was a small bone sandwiched between the former bone and a preorbital bone (Fig. 15.9) (Benton, 2005; Motani, 2005). Overall, the combined premaxilla, maxilla, and dentary bones formed a long, scissors-like set of jaws that contained many pointed, narrow, conical teeth. The length and shape of the jaws strongly suggests that many ichthyosaurs were excellent fish and squid predators, quickly snapping up their fast-moving prey in their toothy mouths. Moreover, the gut contents of several ichthyosaurs show both fish and squid remains (Pollard, 1968; Massare et al., 2014). The orbits of ichthyosaurs are both relatively and absolutely huge, and ichthyosaurs hold the record for the largest eyes in a vertebrate animal

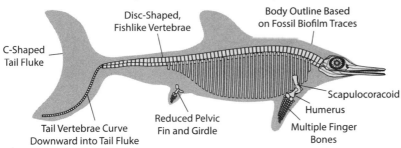

Ichthyosaur Skull

Postorbital
Fenestra
Sclerotic
Ring
AM
Premaxilla
Quadrate-Articular
Jaw Joint
Dentary
Maxilla

F-number Illustrated

High F-Number
Bright Light

`i am a camera`

Lens   Iris

Low F-Number
Dim Light

`i am a camera`

Lens   Iris

Ichthyosaur Skeleton with Body Outline

Disc-Shaped,
Fishlike Vertebrae

Body Outline Based
on Fossil Biofilm Traces

C-Shaped
Tail Fluke

Tail Vertebrae Curve
Downward into Tail Fluke

Reduced Pelvic
Fin and Girdle

Scapulocoracoid

Humerus

Multiple Finger
Bones

15.9. Ichthyosaur anatomy and the F-number concept. Note how long the premaxilla is and the effect it has on displacing the nostril backward. The eyes of most ichthyosaurs were absolutely huge, and well-preserved sclerotic ring bones enable relatively accurate information on the dimensions of these organs. The enormous eyes of ichthyosaurs suggest that, like a camera with a large lens set to a low F-number (see inset), these animals could see and detect movements in low-light conditions. The skeletal chassis is very fishlike, with a reduced pectoral girdle and tiny and detached pelvic girdle that floated in the body wall musculature. Fossil biofilms have preserved the outlines of several ichthyosaur species, and we know that some of them, such as *Ichthyosaurus*, had a well-developed dorsal fin and C-shaped tail fluke. AM indicates the location of the adductor mandibulae and other jaw-closing muscles. Ichthyosaur skull and skeleton based on *Ichthyosaurus* and close relatives, after work by Motani, Rothschild, and Wahl (1999) and Motani (2009) and images in Carroll (1988) and Benton (2005).

(Fig. 15.9) (Motani, Rothschild, and Wahl, 1999; Motani, 2005). In some ichthyosaurs the presence of a sclerotic ring of bones has allowed fairly accurate estimates of eye size. These data indicate that some ichthyosaurs sported eyes nearly 30 cm in diameter (Motani, Rothschild, and Wahl, 1999). But why did ichthyosaurs have such huge eyes? There appears to be a strong correlation between eye size and aquatic adaptations in these marine reptiles, a subject to which we will return shortly.

The earliest ichthyosaurs were not fish-shaped animals but had a more sinuous body form like those of aquatic lizards such as mosasaurs (Frobisch et al., 2013). The currently earliest known ichthyosaur, *Cartorhynchus*, was apparently a semiaquatic reptile no longer than 38 cm in length (Motani et al., 2014). The mouth lacks teeth and large hyoid elements are present, which may indicate this small ichthyosaur was a suction feeder (Motani et al., 2014). The forelimbs and hind limbs are already modified into flipper-like elements, and the wrist of *Cartorhynchus* was enlarged and cartilaginous. In fact, this type of flexible flipper is present in modern sea turtle juveniles that handily traverse the beach to the surf after hatching (Motani et al., 2014). The full tail is not preserved in *Cartorhynchus*, but in other basal ichthyosaurs the tails were long and bent in such a way that a small tail fin may have adorned the dorsal curve in the tail (Fig. 15.9) (Motani, 2005). Moreover, in *Cartorhynchus* and other early ichthyosaurs, the pelvis was still united to the vertebral column through the sacrum, and this suggests that these animals were capable of hoisting themselves onto land at least occasionally, perhaps to lay their eggs much as modern sea turtles do. However, ichthyosaurs were animals that from the earliest beginnings had an intimate relationship with the water. In fact, not long after the first ichthyosaurs appear in the fossil record (Early Triassic, ~250 Ma), we already have examples of live birth among their more aquatic descendants in the Middle Triassic

(245–235 Ma; Card 95) (Benton, 2005; Motani, 2009). Some spectacular fossils from the Jurassic of Germany show a small, articulated ichthyosaur skeleton emerging beneath the pelvis of an adult of the same species. Giving live birth allows aquatic amniotes to remain at sea their entire lives without having to return to land to lay eggs (Benton, 2005).

Live birth may seem an unusual mode of reproduction in diapsids, but it is actually relatively common in modern lepidosaurs (Sites, Reeder, and Wiens, 2011). In various lizard and snake species, the egg remains unshelled but the remaining membranes and yolk deposit still form. Commonly, the yolk sac, allantois, or both membranes will make some sort of connection with the expanded portion of the mother's oviduct, the uterus. Given that the yolk sac and allantois are both highly vascularized, they can easily transport gases from the capillaries lining the walls of the uterus into the developing embryo (Sites, Reeder, and Wiens, 2011). Even if, as in some lepidosaur species, there is not a firm connection between the mother and offspring, a close association of the embryo with the uterine walls is enough for the embryo to develop to term within the mother. So, ichthyosaurs giving live birth is actually not entirely unexpected. Even more intriguing is that fetal ichthyosaurs are preserved such that we can see they exited the birth canal tail first, just as modern dolphins and whales do (Benton, 2005), and there is a good reason for this. Because the birthing process can take some time, you want the head of the baby to come out last so that it does not suffocate underwater and can immediately surface for its first breath of air (though it should be noted that some viviparous sharks are born tail first). It should be mentioned at this point that there is also fossil evidence of certain plesiosaurs giving live birth as well, and several species have been preserved with small, fetal skeletons within the body cavity (Smith, 2008).

The forelimbs and hind limbs became highly modified into flippers early during ichthyosaur evolution, and trends we discussed in plesiosaurs are taken even further in these diapsids. First, only the humerus and femur are still recognizable as those bones because the radius, ulna, tibia, and fibula have been reduced to small, quadrangular bones that are almost indistinguishable from the bones of the wrist and ankle (Fig. 15.9) (Benton, 2005; Motani, 2005). Second, as in plesiosaurs, the only areas of muscular scar development are situated near the proximal ends of the humerus and femur, and their distal ends are relatively smooth and unremarkable. The joint surfaces are very pitted and rugose, again a condition related to the possession of thick cartilaginous epiphyses. Third, the bones of the hands and feet are greatly extended and expanded (Fig. 15.9), with not only multiple digit segments but in some cases more than 5 digits per hand or foot! In one ichthyosaur species, 10 fingers adorn each hand (Benton, 2005). All of these features enhanced the paddle-like shape of the fins and made the distal end of the fin a stiff but bendable element ideal for steering and maneuvering through the water.

Returning to the gigantic eyes of ichthyosaurs: there is a significant correlation between the evolution of aquatic adaptations and the size of their eyes. To understand the significance of this, it is first important to understand the concept of the F-number on a camera. When taking a picture you typically want the image to be focused and bright enough to see the details. The aperture of a camera is an adjustable opening, somewhat like an eye's pupil, that controls how much or how little light enters through the camera lens when a picture is snapped (Fig. 15.9). The size and shape of the glass lens in the camera in combination with how open or closed down the aperture is determines how well a distant object will be captured (Fig. 15.9). The F-number is the ratio of the focal length (determined by the size and shape of the lens) and the size of the aperture opening (Bloomfield, 2006). A low F-number value means the aperture is wide open so that most of the lens is used to take in all available light. A high F-number value tells you that the aperture is closed down a great degree, letting only some light in through the lens. In a standard photography camera with a 35mm lens, the F-number ranges from 2 to 22. So, if you want to take a picture of a landscape in the evening, you opt for low F-numbers, because this opens up the aperture a great deal, letting in enough light for a focused, discernible photograph. In contrast, if you want to take a picture on a very bright day, much higher F-numbers are recommended, because now the problem becomes 'white out' or overexposure.

Eyes work on principles similar to those of cameras, and so F-numbers can be calculated for the visual systems of many vertebrates (Motani, Rothschild, and Wahl, 1999). In humans, our F-number values fall just above 8 on a bright day, and down to nearly 2 when dark. Cats and other nocturnal vertebrates do far better at night, achieving F-numbers hovering around 1. We are fortunate that not only are many ichthyosaurs found reasonably intact, but their bony sclerotic rings are often well-preserved. Because the opening in the sclerotic ring represents the largest the pupil could possibly get, some paleontologists have been able to calculate approximate F-numbers for these aquatic reptiles. As it turns out, the more fishlike that ichthyosaurs become in their body form over time, the further their F-number values drop (Motani, Rothschild, and Wahl, 1999). For example, in some of the earliest ichthyosaurs with more lizard-like bodies, the lowest F-numbers approach 2. However, in the later, very fishlike ichthyosaurs, the F-numbers fall to almost 1 (Motani, Rothschild, and Wahl, 1999). Given that their eyes and lenses were huge, these low numbers suggest that ichthyosaurs could see distant objects very well in the dark.

Studies of ichthyosaur body form combined with F-number data over deep time suggest that larger eyes with lower F-numbers are correlated with the evolution of a more fishlike body form (Motani, Rothschild, and Wahl, 1999). But why should this be? One certainly could not argue that ichthyosaurs were primitively nocturnal animals that simply became better at seeing in the dark. Instead, the large eyes and low F-numbers in combination with fishlike bodies suggest that something about the water itself was driving this evolution. Available light decreases with depth, so

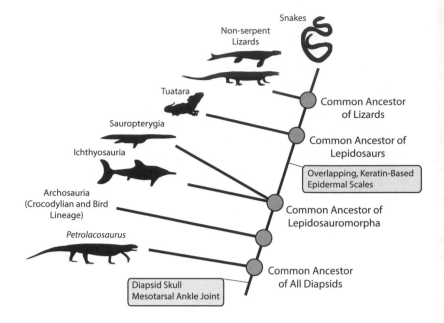

15.10. A pedigree of lepido-sauromorph relationships. See text for details.

Snakes

Non-serpent Lizards

Tuatara

Sauropterygia

Ichthyosauria

Archosauria (Crocodylian and Bird Lineage)

Petrolacosaurus

Common Ancestor of Lizards

Common Ancestor of Lepidosaurs

Overlapping, Keratin-Based Epidermal Scales

Common Ancestor of Lepidosauromorpha

Common Ancestor of All Diapsids

Diapsid Skull Mesotarsal Ankle Joint

if you dive deeply, your surroundings become very dark. Curiously, the earliest ichthyosaurs are more lizard-like and probably spent most of their lives swimming in shallow waters where there was ample light to see prey and navigate. Given that ichthyosaurs became more fishlike over time, they would not only get to a point where leaving the water was impossible, but they would also be more likely to dive to greater depths in search of prey. The deeper the dive, the less available light and the more advantageous it would be to have huge eyes with large lenses and wide pupils. In essence, the huge eyes of ichthyosaurs made it all the better to see their next meal (Motani, Rothschild, and Wahl, 1999).

For reasons still not clear, the ichthyosaurs went extinct some 25 million years before the end of the Mesozoic, or around 90 Ma (Card 99) (Benton, 2005; Motani, 2009). For a long portion of the Cretaceous period (145–90 Ma; Cards 97–99), many areas, including the middle of North America, were inundated by shallow seas. As the Cretaceous period began to wind down (90–66 Ma; Card 99), events such as the development of the Rocky Mountains in North America and other mountain-building events around the world raised the continental elevations, and the inland seas began to drain. It is possible that such environmental changes could have devastated the last populations of ichthyosaurs. However, it is also important to note that by the Late Jurassic (161–145 Ma; Card 97) ichthyosaur diversity shrank considerably, and perhaps competition from predatory fishes and later the aquatic mosasaur lizards and sauropterygians eventually did them in.

To review the remarkable pedigree of the lepidosauromorphs and their kin, see Figure 15.10. Against the backdrop of all of this lepidosauromorph and diapsid diversity, the archosaurs came to the fore. Perhaps the most successful terrestrial vertebrates to have ever lived, the archosaurs developed a body plan and a diverse range of habits that sent many of them to the top of the food chain. The archosaurs spawned the dinosaurs, birds, pterosaurs, crocodylians, and many other groups now long extinct. In the next chapters, we will set the stage and explore the chassis of these ruling reptiles.

Deep Scaly II: The Archosaur Chassis, Those Ruling Reptiles

Hey mighty *Brontosaurus,* don't you have a lesson for us?

THE POLICE, *Walking in Your Footsteps*

16.1. Archosaur skull traits. Unlike other diapsids, the skulls of archosaurs are characterized by a specialized opening, the antorbital fenestra, between the nose and orbit, and a lower jaw fenestra.

Non-Archosaur Diapsid Skull

Diapsid Fenestrae

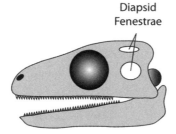

Archosaur Skull

Antorbital Fenestra

Diapsid Fenestrae

Lower Jaw Fenestra

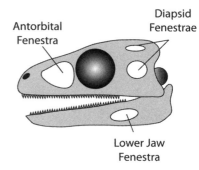

16

IF YOU HAD AN INTEREST IN RACE CARS AND WANTED TO LEARN HOW they operated, you could be bold and jump right into the technical details of these specialized machines and their chassis. However, such an approach would often leave you bewildered and confused, and you certainly would have difficulty determining what about the race car chassis and engine were specialized and which parts were common to cars generally. Such is the habit of many children and adults interested in those most recognizable of fossil archosaurs, the dinosaurs. To avoid a case of the pot calling the kettle black, I must admit that as a youngster on through high school I was often focused on dinosaurs and remained a bit disinterested in other vertebrates until I began to realize the benefit of seeing the bigger picture. Simply focusing on dinosaurs and other fossil archosaurs without the context of their living relatives is why I believe many people come away thinking dinosaurs were essentially 'monsters' with no real connection to the modern world.

Given that many fossil archosaurs had a mosaic of croc- and birdlike traits, I have found it more informative to the uninitiated to examine the chassis of modern crocodylians and birds first before exploring the large variety of skeletal forms found among fossil archosaurs. In this chapter, we explore the chassis of the two modern archosaur groups, the crocodylians and the birds. Despite their differences, understanding the skeletal anatomy and inner workings of these archosaurs provides us with search images and tools for understanding the variety of chassis forms we will encounter among their fossil relatives in chapters 17 and 18.

Archosauria: The Ruling Reptiles

The archosaurs were the most successful terrestrial vertebrates by almost any measure. Their ancestral chassis spawned a wide diversity of predators, herbivores, and omnivores; they gave rise to two of the three flying vertebrate groups; and their living members, the crocodylians and birds, have survived the onslaughts of both nature and humans . . . at least, so far. The archosaur chassis was rare when it first appeared early in the Triassic (~245 Ma; Card 95), but by the end of that same period (~200 Ma; Card 96) a plethora of archosaurs dominated both the land and the sky (Benton, 2005). As with so much in this book, because we are tracing the major patterns of the vertebrate skeletal chassis, I am simplifying things to show broader patterns. Technically, the term Archosauria applies only to the two major groups that have living descendants today: the crocodylian group and the bird group. However, we now recognize that there

Pedigree Crocodylians, Birds, and Their Relatives

Date of First Appearance 245 Ma (Card 95)

Specialties of the Skeletal Chassis Antorbital Fenestra

Eco-niche Terrestrial and Aquatic Predators, Omnivores, and Herbivores

are a number of diapsids that share a common ancestor with these two groups, and yet belong to neither. These animals are often called the Archosauromorpha, and they share most but not all of their trait states with the other archosaurs (Benton, 2005). To avoid becoming overly technical in our discussion, we will simply call these animals 'basal archosaurs' to distinguish them from the crocodylian and bird lines.

Basal archosaurs and the two major archosaur groups all share a number of trait states, but two are the easiest to spot. The most telling trait state is the possession of a fenestra situated between the nostrils and the orbits (Fig. 16.1). Called an antorbital fenestra, in modern archosaurs this opening is filled by what are called paranasal sinuses: air-filled sacs that line the snout (Witmer, 1997). It has now been shown that there probably are several competing factors that selected for the evolution of this unique archosaur trait, but the most likely explanations seem to be related to weight reduction, efficient channeling of biting forces through the head, and the tendency of air sacs to invade bones (Witmer, 1997; Rayfield and Milner, 2008). It seems that the bending stresses experienced by the bones of an archosaur skull would be concentrated along its margins rather than its middle. The antorbital fenestra, then, would be the equivalent of the marrow cavity in a long bone, arising because it is in the region where bending stresses are limited or negligible. Whatever the ultimate cause, or causes, of this trait, the antorbital fenestra is a sure sign that you have an archosaur on your hands.Another trait state of basal and derived archosaurs is the possession of a fenestra in the lower jaw, situated between the dentary and postdentary bones (Fig. 16.1). This opening is situated near the insertion point for portions of the adductor mandibulae muscles that close the jaws (Holliday et al., 2013).

Archosauria: The Crocodylia

Pedigree Crocodiles, Alligators, and Their Relatives

Date of First Appearance 90 Ma (Card 99)

Specialties of the Skeletal Chassis Robust, Dorsoventrally Flattened Skulls with a

Complete Secondary Bony Palate

Eco-niche Terrestrial Predators and Omnivores

The living crocodiles, alligators, and gavials, technically called the Crocodylia, give us a glimpse, if somewhat derived, of the basal archosaur chassis (Brochu, 2003; Brochu et al., 2009). The anatomy, biology, and behavior of these archosaurs can also inform us of the ancestral condition inherited in dinosaurs and their kin. One of the most common, well known crocodylians is the American alligator, *Alligator mississippiensis* (alligator hereafter). I have had the privilege of dissecting a great many of these archosaurs, and have even studied their embryonic development. It is fortuitous that alligators not only are common (thanks to concerted cooperative efforts by United States governmental conservation agencies and private farms) but happen to be generalists among the living crocodylians (Dodson, 2003; Meers, 2003). This means that alligators are in some senses the 'gold standard' against which other crocodylians can be compared and contrasted (Dodson, 2003). For our purposes, we will describe the alligator chassis and then briefly consider modern and fossil variants on the crocodylian body plan.

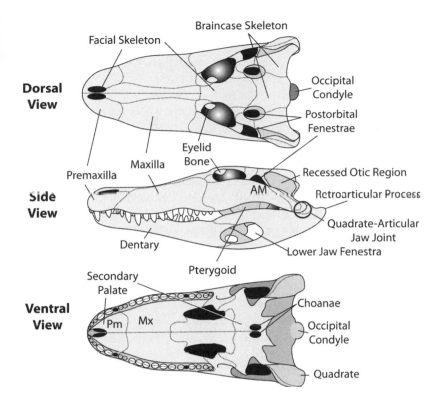

Facial Skeleton

Braincase Skeleton

Dorsal View

Occipital Condyle

Postorbital Fenestrae

Eyelid Bone

Maxilla

Premaxilla

Recessed Otic Region

Side View

AM

Retroarticular Process

Quadrate-Articular Jaw Joint

Dentary

Lower Jaw Fenestra

Pterygoid

Secondary Palate

Ventral View

Pm   Mx

Choanae

Occipital Condyle

Quadrate

16.2. American alligator skull. Note the robustness of the alligator skull, especially the secondary palate as seen in ventral view. Also note the retracted choanae that make it possible for alligators and other crocodylians to breathe through their external nostrils while their mouth is submerged. AM indicates the location of the adductor mandibulae and other jaw-closing muscles.

As we have encountered in several other vertebrates, the skulls of alligators are broad and flattened dorsoventrally (Fig. 16.2). In addition, the eyes, nostrils, and ears are placed dorsally on the skull. None of these trends should be surprising given that alligators spend a great deal of time in the water stalking their prey. The goal of an alligator is to float with just the eyes and nostrils poking above the surface of the water, its long, scaly body resembling a floating log to its potential prey. In fact, the caudal braincase bones overhang the otic region in alligators and provide a point of attachment for a fleshy lid that closes over the ear when these archosaurs are in the water (Benton, 2005) (Plate 20).

Unique among reptiles, the nasal passages in alligators are divided from the mouth by a hard, bony palate (Fig. 16.2). This region of bone is formed by outgrowths of the premaxilla, maxilla, and palatal bones, and is technically known as a secondary palate because it forms over the original fleshy palate (Benton, 2005). As we will see, a similar anatomical development has occurred in mammals. The secondary palate allows these archosaurs to remain almost completely submerged while inhaling fresh air through their dorsally placed nostrils. The inhaled air subsequently flows through the walled-off nasal chamber all the way to the glottis. The choanae of alligators are shifted far caudal to the end of the secondary palate, and the base of the tongue bears a special flap that blocks off the choanae from the mouth cavity when the animal is submerged (Fig. 16.2). This prevents water from going up into the nasal cavity and ensures that fresh air arriving through the choanae from the secondary palate is shunted into the glottis.

The secondary palate of alligators also serves an important mechanical function, but before we can discuss this aspect of the alligator chassis, it is important to survey the major aspects of skull function in this species. First, the alligator skull is essentially analogous to a flat, thick, heavy board of wood. Just as cracking someone over the head with a heavy board would shock, stun, and injure them (please avoid this behavior), so the massive skulls of alligators smash into their prey with quick and deadly force. In fact, much like the placoderms we encountered way back in chapter 6, when alligators open their mouths to slam them into their victims, it is the skull that rotates up and back with help from strong epaxial muscles (Fig. 16.3). Whereas the lower jaw is certainly depressed, the gape of the mouth is largely expanded through dorsal rotation of the skull. Second, adding to our board analogy, the upper and lower jaws are studded with conical teeth that bear small, pointed ends, much like large construction nails. When these teeth make contact with their prey, they not only hold it firmly but also puncture and crush soft tissues and bones as the skull slams back down onto the lower jaws (Pough et al., 1998; Benton, 2005).

As if all this weren't enough, both the pterygoid and adductor mandibulae muscles are massively enlarged in alligators (Holliday and Witmer, 2007; Holliday et al., 2013). So enlarged, in fact, are the pterygoids that they create the appearance of jowls on either side of the head against the neck (Fig. 16.3 and Plate 20). No, that large alligator does not have a weight problem: it has such huge jaw-closing muscles that they spill out of the skull and onto the neck! Moreover, a sesamoid called the cartilago transiliens further augments the mechanical advantage of deeper jaw muscles (Tsai and Holliday, 2011). The mechanical advantage of these jaw muscles in imposing an incredible amount of force onto their prey is augmented by the shape and orientation of the quadrate bone that forms the upper part of the jaw joint. Because the quadrate is long, it acts like a lever around which the deep pterygoid muscles pull, enhancing their mechanical advantage (Schwimmer, 2002). Once the upper and lower jaws start approaching each other, the huge adductor mandibulae muscles are in a mechanically effective position to pull the mouth shut with tremendous force (Fig. 16.3). Studies of alligator bite forces show that their jaws close with a force rivaled only by that of great white sharks (Erickson, Lappin, and Vliet, 2003). One 3.5 meter-long alligator's bite was recorded at nearly 10,000 N (Erickson, Lappin, and Vliet, 2003): this would be equivalent to having a heavy car or light pickup truck dropped on top of you! So an alligator's head is like a combined blunt-force weapon, a battering ram, and a studded vise that few animals can escape with their lives, let alone in one piece.

Returning to our consideration of the secondary palate, with all that force going through the skull, it is necessary to have structures that resist the powerful twisting and bending forces generated by biting. Therefore,

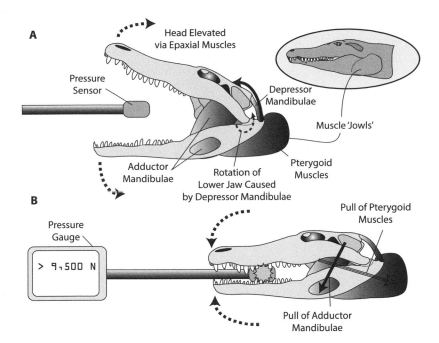

16.3. Alligator jaw mechanics. The large adductor mandibulae and pterygoid muscles of alligators and other crocodylians give them tremendous biting power. Pressure gauge data from Erickson, Lappin, and Vliet (2003).

the secondary palate not only separates the nasal cavities from the mouth, but is in an ideal position to act as a brace when the skull is slammed down onto prey. In addition to the resistance that the secondary palate provides when biting, there has been the loss of the ancestral antorbital fenestra (Rayfield and Milner, 2008). If you imagine the snout of an alligator as a beam, then any perforations along its middle might be a source of weakness. It seems that due to biomechanical selective pressures, the antorbital fenestra was 'phased out' in modern alligators and their kin (Rayfield and Milner, 2008).

Snapping the jaws open quickly involves epaxial muscles and the depressor mandibulae muscle we have come to know for tetrapods in previous chapters (Fig. 16.3). As with our early jawed friends the placoderms, epaxial muscles running from the neck pull on the skull to lift it up. Moreover, as with many lizards, there is a retroarticular process that juts off the back of the lower jaw that acts as a lever through which the depressor mandibulae can effectively depress the lower jaw (Fig. 16.2) (Pough, Janis, and Heiser, 2002). As with everything we have discussed, there is a tradeoff between force and speed. Because opening the jaws quickly is more important than the force associated with this motion, the depressor mandibulae in alligators is relatively small (Pough, Janis, and Heiser, 2002). Numerous nature television personalities have demonstrated the poor mechanical advantage of the depressor mandibulae in alligators by taping their mouths shut or holding them closed with their hands. For my own research involving alligator anatomy, I have helped to tape alligator mouths shut as well, but I can't say it's high on my list of things to do. Alligators have no lips, so even with their mouths taped shut, the teeth of these archosaurs still protrude outside their jaws. Alligators are

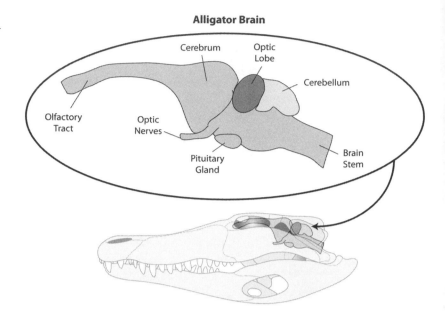

16.4. The brain of the American alligator. See text for details.

Alligator Brain

Cerebrum

Optic Lobe

Cerebellum

Olfactory Tract

Optic Nerves

Pituitary Gland

Brain Stem

powerful animals that can shake their closed mouths side to side in such a way as to injure and deeply cut you with those teeth, so I am happy to let others play this 'game' of alligator roulette.

It is now a good time to peer inside the skull of an alligator and consider its brain (Fig. 16.4). In many ways, the brain of an alligator is similar in overall structure to that of lizards, which we discussed in chapter 13. The cerebrum is much more expanded than in lissamphibians, and a dorsal ventricular ridge (DVR) associated with the dorsal cerebrum and its integration of somatosensory information is well developed in these archosaurs (Butler and Hodos, 2005). A prominent olfactory tract, somewhat reminiscent of what we encountered for sharks, provides a long and sensitive pathway for the detection of odors. A pineal gland is present, but unlike in many of the reptiles we've previously encountered, it does not poke dorsally through the skull as a parietal eye. Why alligators and other crocodylians should not possess a parietal eye is unclear. The optic lobes and cerebellum are also well developed for visual and motor coordination (Fig. 16.4). Delicate stapes bones rest against the otic capsule and convert a wide array of airborne sound frequencies into pulses interpreted by the inner ear. Despite claims made in some recent movies, the medulla oblongata of alligators is not especially large nor does it make them angry.

Alligator behavior is well studied and fascinating. As one would expect for a predator, their keen senses of odor detection and vision are crucial for them to capture prey. It is often remarked that alligators bite first and ask questions later, and their olfactory and visual centers are a testament to this predatory behavior. However, a series of pores along the jaw bones supply pressure-sensitive sensation to the mouth (Leitch and Catania, 2012). Somewhat as with the lateral line system in fishes, pressure waves stimulate the nerve endings in alligator jaws and seem to cue

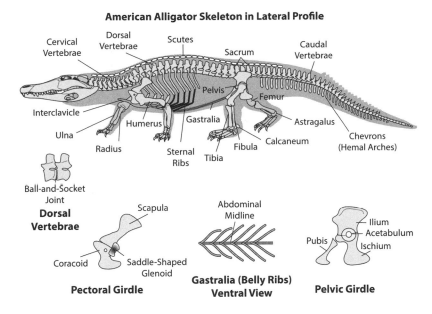

American Alligator Skeleton in Lateral Profile

Cervical Vertebrae

Dorsal Vertebrae

Scutes

Sacrum

Caudal Vertebrae

Interclavicle

Pelvis

Femur

Humerus

Gastralia

Astragalus

Ulna

Radius

Sternal Ribs

Tibia

Fibula

Calcaneum

Chevrons (Hemal Arches)

Ball-and-Socket Joint
Dorsal Vertebrae

Scapula

Coracoid

Saddle-Shaped Glenoid

Pectoral Girdle

Abdominal Midline

Gastralia (Belly Ribs) Ventral View

Pubis

Ilium

Acetabulum

Ischium

Pelvic Girdle

16.5. The American alligator skeleton. Note the scutes (dermal bones) dorsal to the vertebral column and embedded in the skin. Also note that the pubis is not directly attached to the pelvis–this bone plays a significant role in alligator respiration in conjunction with the gastralia. Based on skeletal pictures provided to the author courtesy of Heinrich Mallison.

them to when prey is close enough to bite (Leitch and Catania, 2012). Alligators are territorial, and males will guard areas of a swamp or waterway, mating with several females within that area (Lance et al., 2009). Alligators have rather elaborate mating rituals in which males make a variety of vocalizations and movements to attract females (Vergne, Pritz, and Mathevon, 2009). They are also very protective of their young, and mothers typically excavate a bowl-shaped pit into which they lay their eggs, which are incubated by heat generated from rotting vegetation piled on top. Mother alligators guard their nests fiercely and help their young make their way to the water when they begin to hatch. Hatchling alligators make chirp-like cries while emerging from their eggs or when in danger, and this vocal cue summons the mother quickly to their aid (Vergne, Pritz, and Mathevon, 2009). In fact, mother alligators will scoop up hatchlings gingerly in their mouths and carry them to safety (Pough et al., 1998; Pough, Janis, and Heiser, 2002).

The vertebral column of alligators is well developed and allows for a variety of movements on land and in water (Fig. 16.5). Nearly all alligator vertebral centra have a cranially facing socket and caudally facing ball (Fig. 16.5). These ball-and-socket articulations lend an extraordinary amount of flexibility to alligators. Alligators and nearly all crocodylians have nine cervical vertebrae in their relatively short necks that give purchase to various epaxial and neck muscles that power the movements of their skulls (Benton, 2005). The dorsal vertebrae allow a wide range of lateral movements as well as some dorsoventral flexing, and bear well-developed ribs. The sacrum is solidly anchored to the pelvic girdle, and the caudal vertebrae are numerous, with tall neural spines and deep chevrons (Fig. 16.5). The tail of alligators is tall, narrow, and thickly muscled, all attributes that make it an ideal sculling organ for pushing its heavy body through water (Fig. 16.5).

16.6. American alligator respiration. In alligators and other crocodylians, a 'diaphragm' muscle works in combination with the abdominal muscles and the liver to collapse and expand the lungs during respiration. Diagrams are based on Liem et al. (2001).

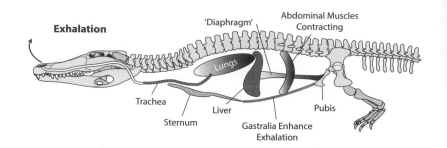

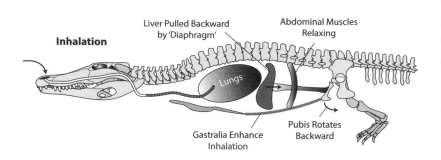

The pectoral girdle is comprised of a scapulocoracoid in which the coracoid end articulates within a slippery groove on a cartilaginous sternum, increasing, to a modest degree, the range of motion in alligator forelimbs (Baier and Gatesy, 2013). Clavicles are absent, but an interclavicle remains embedded within the sternum. In the pelvic girdle, the ilium and ischium are united and form the acetabulum, but the pubis is almost tacked on as an afterthought in front of the ilium, and articulates with the rest of the pelvis only through cartilaginous connections (Fig. 16.5). This loose connection of the pubis with the pelvic girdle not only is a unique trait state of crocodylians, but plays an important role in their respiration in combination with numerous V-shaped gastralia embedded in their ventral muscles. The gastralia overlap one another and can slide and slip past each other (Claessens, 2004). These movements allow the abdominal cavity to expand and contract as the alligator breathes. Several muscles from the pelvis attach and invest themselves among the gastralia, and it is the contraction of these abdominal muscles in concert with the moving gastralia that causes changes in the abdominal capacity of an alligator (Claessens, 2004). Second, the pubis bone, being united to the rest of the pelvis only through cartilage, is capable of swinging forward and backward, decreasing and increasing the size of the abdominal cavity, respectively (Carrier and Farmer, 2000; Claessens, 2004). Such gastralia and pubic movements help to draw air into the lungs and force it back out (Fig. 16.6). These movements work in conjunction with the liver and a diaphragm-like muscle to actively pump air into and out of the lungs (Liem et al., 2001; Pough, Janis, and Heiser, 2002) (Fig. 16.6). The skin along a good portion of an alligator's back is embedded with a series of dermal bones called scutes (Fig. 16.5 and Plate 20). The scutes appear to serve numerous functions in these animals ranging from stiffening of the

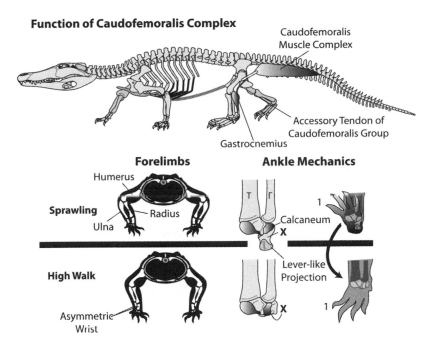

Function of Caudofemoralis Complex

Caudofemoralis
Muscle Complex

Accessory Tendon of
Caudofemoralis Group

Gastrocnemius

Forelimbs

Humerus

Sprawling

Radius

Ulna

High Walk

Asymmetric
Wrist

Ankle Mechanics

Calcaneum

X

Lever-like
Projection

X

1

1

16.7. American alligator loco-motion and posture. In alliga-tors, other crocodylians, and archosaurs broadly, the caudo-femoralis muscle complex pro-vides the power that retracts the hind limb. In particular, the caudofemoralis longus muscle has an accessory tendon that joins the gastrocnemius on the lower leg, increasing the mechanical advantage of this muscle across the entire hind limb. Alligators and other crocodylians can also assume two distinct postures—a sprawl-like posture where the forelimbs and hind limbs are splayed out to the side as in many reptiles, or a high-walk posture in which the forelimbs and especially the hind limbs are more adducted close to the midline of the body. In the forelimbs, the mechanics of the asymmetrical wrist appear to help pronate the hand during the high-walk, whereas the independent, rotational movements of the calcaneum in the hind limb maintain a forward-directed foot posture during this gait. The 'X' in the figure represents the dorsal surface of the calcaneum—you can see where the 'X' rotates to when the calcaneum pivots on the astragalus. The '1' indicates the location of the first digit on the foot. Ankle mechanics based on work by Parrish (1987, 2012).

body trunk, to protection from predators, to thermoregulation by shed-ding or absorbing additional heat from the blood supply that runs into and out of these elements (Farlow, Hayashi, and Tattersail, 2010).

Alligator limbs have all the hallmarks of sprawling reptiles like liz-ards and turtles. The general shapes of the humerus, radius, ulna, femur, tibia, and fibula are reminiscent of what we have encountered in lizards and even salamanders (Fig. 16.5). But first impressions can be deceiving. When I worked for the U.S. Geological Survey in Florida, I was able to see a lot alligators in action, and the truly stupid risks tourists would take to have their pictures with these 'slow' reptiles. I have literally witnessed a man stand within arm's length of a nearly 3-meter-long alligator lying on a swamp bank so that his family could photograph him! To demonstrate how foolhardy this is, keep in mind that alligators have two distinct gaits. Although alligators may adopt a slow, sprawling gait, they can also adopt what is called a high walk in which the forelimbs and hind limbs are ad-ducted and the body is held aloft, improving their prey-catching ability (Schwimmer, 2002; Benton, 2005) (Fig. 16.7). Last but not least, some cro-codylians (but not alligators so far as we know) can gallop (Zug, 1974; Webb and Gans, 1982; Allen et al., 2014)! This is no joke: some of these predators can adopt a bounding gait in which the forelimbs and hind limbs work in opposing pairs, bouncing the archosaur forward at a high rate of speed for a brief period of time (Zug, 1974; Webb and Gans, 1982; Allen et al., 2014). However brief, these more upright gaits last long enough for a high-walking alligator or galloping crocodylian to grab a tourist with its impressive jaws and take home its own souvenir! This more upright posture presages what we will encounter in dinosaurs (however, the 'sprawled' gait of crocodylians may be a reversion from an upright posture: see chapter 17).

Although the forelimb remains relatively sprawled even in the high-walk, in the hind limb, a wide acetabulum allows the femur to reorient itself into a more upright orientation. Moreover, the fourth trochanter of the femur is well developed, and, as in lizards, is associated with a caudofemoralis muscle complex that powerfully retracts the hind limb, propelling the alligator forward in this more upright posture (Fig. 16.7 and Plate 20) (Gatesy, 1990). Additionally, specialized joints in the ankle allow the foot to assume various postures. As in lizards, the major ankle joint is situated between the astragalus and calcaneum proximally and the remaining tarsals distally, forming a ridge-and-groove hinge joint. However, the astragalus and calcaneum have a peg and socket joint, respectively, that allows the calcaneum to rotate around the astragalus (Fig. 16.7) (Parrish, 1987, 2012). These movements collectively allow the ankle to change its orientation depending on the stance of the alligator (Fig. 16.7) (Parrish, 1987, 2012).

Before leaving alligators and exploring crocodylians more broadly, we should briefly discuss specialized aspects of their digestive, respiratory, and circulatory systems. The guts of alligators do much more of the breakdown of foodstuffs into digestible bits than their teeth and jaws. Alligators don't chew, of course, but rend from their prey large bolts of meat and bone that they swallow whole. Alligators may use their infamous 'death roll' to twist off stubborn pieces of their prey, but in the end nothing we would recognize as chewing occurs. Instead, the large packages of meat and bones slide down the esophagus into a specialized stomach chamber called the gizzard. Having dissected many an alligator digestive tract, I can tell you that the gizzard is rock hard with thick muscles. In essence, the gizzard of an alligator works like your teeth to pulp up large food bits before sending them on their way (Pough et al., 1998; Pough, Janis, and Heiser, 2002). But alligators do one better and swallow small stones and grit to help them process food in their gizzards. In fact, opening the gizzard of an alligator typically reveals a treasure trove of pebbles, rocks, and sand. These foreign objects are hypothesized to act like teeth, in combination with the powerful, muscular walls of the gizzard, to pulp food. However, it should be noted that how much the swallowed stones contribute to food breakdown is currently debated (Fritz et al., 2011).

Alligators are the first vertebrates we encounter in this book that have a completely separated, double-pump heart. A solid wall of tissue separates the right and left sides of the heart, just as occurs in ourselves and in birds. This means that low-pressure, oxygen-poor blood returning to the right side of the heart is completely separated from the high-pressure, oxygen-rich blood leaving the left side of the heart after returning from the lungs (Pough, Janis, and Heiser, 2002). As we have discussed in previous chapters, this separation of blood streams allows low-pressure blood to go to the lungs, which will not damage the delicate capillaries in the alveoli. At the same time, it allows the left side of the heart to be more muscular and pump oxygen-rich blood that has returned from the lungs at high pressure to the brain and body. Although alligators can no longer shunt blood from the right to left sides of their heart as many lissamphibians and diapsids can, they have retained the primitive circulatory plumbing exiting the heart. Thus, as in most diapsids, alligators can shunt blood around their lungs, but accomplish this just outside their heart using a pressure-sensitive valve between their bifurcated aortae (Pough, Janis, and Heiser, 2002). When an alligator is on land and breathing normally, the lungs are open and there is little resistance to the blood traveling to the lungs, so the valve remains closed. However, when an alligator dives, holds its breath, and closes down its lungs, blood traveling to the lungs backs up and forces open the valve in the bifurcated aorta, allowing blood to bypass the lungs and flow directly back into the body (Pough, Janis, and Heiser, 2002). The lungs of crocodylians have recently been demonstrated to function in a pattern eerily similar to those of birds: allowing air to flow through in a looping pattern that provides a continuous stream of oxygen across the lung tissues (Farmer and Sanders, 2010). We will discuss this in the next section of this chapter.

The alligator has served as a good introductory model for the primitive archosaur chassis, and with this knowledge in mind we will now briefly explore modern and fossil crocodylian diversity. Modern crocodylians are divided into three major groups, consisting of the crocodiles, the alligatoroids, and the gavials (Fig. 16.8) (Pough, Janis, and Heiser, 2002; Schwimmer, 2002; Brochu, 2003). Members of the crocodile group tend to have rather narrow, triangular snouts with sharp, pointed teeth. These crocodylians typically have glands in their mouths and around their cloaca that help rid their bodies of excess salts, allowing them to occupy both fresh- and saltwater habitats. Body size and habits vary greatly across crocodiles, and they occupy a wide variety of eco-niches. The largest living crocodylians are found among the crocodiles, with saltwater crocodiles from Australia and New Guinea and the African Nile crocodile measuring in as some of the largest on record (over 6 meters recorded for some individuals). Alligatoroids have broader heads than crocodiles, usually possess more blunted teeth, and lack salt-secreting glands. Modern alligatoroids are more restricted in their distribution than crocodiles, and they typically inhabit freshwater environments. Alligatoroids include the North American and Chinese alligators as well as the variety of South American forms known as caiman. The gavials are an odd group of crocodylians restricted to Asia that possess elongate and very narrow jaws. Given their tweezer-like jaw construction, gavials are excellent fish predators. Conflicting fossil, molecular, and anatomical data have made the placement of gavials within the crocodylian pedigree difficult, with some studies suggesting they are close relatives of crocodiles, and others placing them as an outgroup to other modern crocodylians (see Brochu [2003] for an overview).

The earliest true crocodylians first appeared in the Late Cretaceous some 90 million years ago (Card 99) and have had an unbroken record of success ever since (Schwimmer, 2002; Benton, 2005). Alligatoroids

16.8. A comparison of modern crocodylian skulls. Some early modern crocodylians, like *Deinosuchus,* were huge animals that approached 9 meters in length! *Deinosuchus* skull based on Schwimmer (2002).

Comparative Dorsal Skull Shape in Modern Crocodylians

Alligatorid Skull

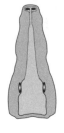
Crocodylid Skull

Gavial Skull

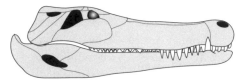

Deinosuchus Skull
Approximately 1.2 Meters in Length
Estimated Body Length = ~8.2 Meters

Alligator Skull
About 0.6 Meters in Length
Body Length of ~4–4.3 Meters

appear to have branched off from the rest of the Crocodylia early on, and some grew to fantastic sizes. One of the best known of these early alligatoroids is *Deinosuchus* from the Late Cretaceous rocks of the United States (~75 Ma; Card 99) (Fig. 16.8) (see Schwimmer [2002] for a detailed examination of this crocodylian). *Deinosuchus* may have reached sizes of nearly 10 meters in length, and was more than capable of securing full-grown dinosaurs for its meals. The middle and rear teeth of *Deinosuchus* were rather blunted and possessed deep roots. Fossil turtle shells associated with *Deinosuchus* localities show impressive blunted crushing and puncture marks that could only have been made by the teeth of this large alligatoroid. This has suggested to some paleontologists that *Deinosuchus* often fed on turtles. This is actually not all that unusual an eating habit among modern alligators, some of which catch smaller turtles, position them at the back of their jaws for maximum jaw closing advantage, and crack open their shells (Schwimmer, 2002). Within the Alligatoroidea, the fossil caimans are probably among the weirdest of all crocodylians. Some fossil caimans had extremely long, flat snouts studded with hundreds of tiny teeth, but it remains unclear what these animals were eating. Other fossil caimans had flat, broad, robust skulls with huge nostrils. Further study and newer specimens may reveal what these oddball caimans were up to (Brochu, 2003). Fossil crocodiles appear to have been more conservative in their overall diversity, with most possessing variations on the long, triangular snout theme.

Archosauria: The Modern Birds (Neornithines)

Flying is energy-intensive. The Boeing 747 passenger plane burns approximately 136,000 liters of fuel on a 10-hour international flight, which translates to 19 liters per mile! Of course, it is carrying over 500 people and their luggage, so on a per-person basis it is more efficient for each person to fly than to drive, but this doesn't change the fact that an airplane must burn a significant amount of fuel to stay aloft. No matter how you look at it, flying requires a lot of energy and fuel. The same goes for vertebrates that fly. They not only require enormous muscles to get them airborne, but they need to sustain rapid contractions of those muscles to keep them so. Birds are endothermic, internally regulating their body temperature within a narrow range, and maintain high body temperatures (often above 38° C) that sustain active flight (Attenborough, 1998). They also require a more complex command and control center, a larger brain, that itself requires a huge investment in calories.

In many ways, birds seem to contrast completely with crocodylians, and yet underneath the heat and feathers, both archosaurs have quite a bit in common. Traditionally birds were defined as those amniotes possessing feathers. However, spectacular dinosaur fossils that preserve unmistakable feathers and associated structures have shown that this body covering is more widespread among the Archosauria than previously recognized (Witmer, 2009; Holtz, 2012; Naish, 2012). Moreover, an overwhelming consensus of data supports birds as direct living, feathered descendants of dinosaurs (Fastovsky and Weishampel, 2012; Holtz, 2012; Naish, 2012). Saying birds are dinosaurs used to be very controversial, but for modern paleontologists this is the equivalent of placing humans among mammals. That birds are recognized as the surviving members of the Dinosauria highlights the exciting prospect of discovering how dinosaurs ticked by understanding the daily guests at your birdfeeder. But more importantly, birds are a living and breathing example of how a diapsid group evolved a high metabolic rate and the ability to fly. Here we will focus on Neornithines, the group of modern flightless and flying birds that fill our lands and skies. For simplicity, we will hereafter refer to members of the Neornithines as birds.

Bird skeletons are instructive for understanding features that appear earlier in other archosaur groups. Moreover, because we can dissect and examine bird soft tissues in relation to their bony structures, we can form hypotheses about what might have been present in a wide variety of fossil archosaurs (Witmer, 1995). The variety of birds is great, and it would be folly to suggest any one species represents all birds equally. However, we have to start somewhere, and turkeys are at once familiar and tasty. In particular, we will focus on the wild Eastern turkey of North America, known scientifically as *Meleagris gallopavo* (Plate 21). It comes as a surprise to many U.S. citizens that wild turkeys can both fly and be rather cunning animals (of course, this is not news to turkey hunters) (Dickson, 1992). We are used to thinking about turkeys solely in terms of

Pedigree The Common Ancestor of Modern Flightless and Flying Birds

Date of First Appearance ~65 Ma (Card 99)

Specialties of the Skeletal Chassis Loss of Teeth, Mobile Quadrate, Air Sac Spaces throughout Skeleton

Eco-niche Terrestrial Predators, Omnivores, and Herbivores

their consumable domestic cousins, which have been bred to grow very large, flightless, and meaty, but who have brains 30% smaller than their wild cousins (Ebinger and Röhrs, 1995).

Adaptations for flight have been the major selective pressure on birds, especially upon their skeletons. To this end, one of the major trends in bird skeletons has been weight reduction (Attenborough, 1998; Benton, 2005). The vertebral column and limb skeleton are remarkably light and hollow, strengthened by strategically placed cross-struts of bone (trabeculae) that resist the forces applied on these elements during flight and ground movements (Liem et al., 2001; Pough, Janis, and Heiser, 2002). You can feel the remarkable difference this hollowing-out has made in birds by holding the femur of a 4.5 kg turkey in one hand and that of a small dog of similar mass in the other. Although the dog femur has some heft to it and that of the turkey feels almost weightless, both support animals of equal overall mass! I have challenged students to break a fresh chicken limb bone by bending it, and to their astonishment they rarely can. Despite their hollow and light construction, bird bones are surprisingly tough.

Lightening of the skeletons starts at the head. Upon examining the skull of a wild turkey (turkey from here on), one can immediately see that the facial skeleton has become a series of jointed struts connected together through ligaments and muscles (Fig. 16.9). Of the facial bones, perhaps the largest are the premaxilla and maxilla, but even that is saying a lot. The premaxilla ends in a beak, and this bone is somewhat more developed than the maxilla. As we saw for turtles, the premaxilla of the turkey (like that of most birds) is pocked with tiny holes (foramina) that transmit the blood and nerve supply to a keratinous beak (Fig. 16.9). An elongate nostril stretches from the premaxilla nearly back into the orbit. Most other facial bones are thin and delicate, including the jugal and lacrimal bones. The braincase bones are more or less fused into a tight unit to protect the brain (Fig. 16.9). In turkeys, the antorbital fenestra characteristic of archosaurs is present but much smaller, and tucked somewhat beneath the caudal extent of the large nostril (Fig. 16.9). The orbits are large to accommodate their eyes, and the diapsid fenestrae behind the large orbits are tiny. Small, rounded openings just behind the quadrate bones provide openings for the ears, and delicate stapes bones are present to conduct airborne sounds. The lower jaw is likewise a delicate, narrow affair, and the dentary possesses a pointed tip with blood and nerve foramina that supply the lower beak (Fig. 16.9). A stick-like retroarticular process is present that accepts the jaw-opening depressor mandibulae muscles.

Weight reduction has become so critical in birds that the turkey skull, like that of all living birds, lacks any trace of teeth. Remarkable as it may seem, even the weight of teeth can be enough to make a difference in terms of mass savings (Attenborough, 1998). Instead, the keratinous beak of these birds allows them to grasp, manipulate, and sometimes smash up food before swallowing it. Many birds, including the turkey, have a degree

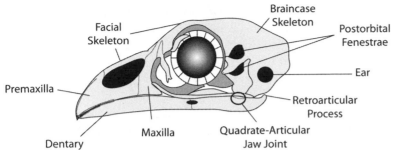

Wild Turkey Skull in Lateral View

Facial Skeleton

Braincase Skeleton

Postorbital Fenestrae

Ear

Premaxilla

Retroarticular Process

Dentary

Maxilla

Quadrate-Articular Jaw Joint

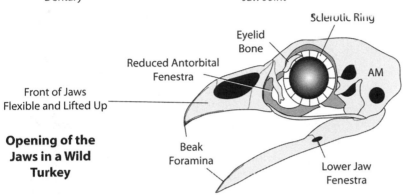

Sclerotic Ring

Eyelid Bone

Reduced Antorbital Fenestra

AM

Front of Jaws Flexible and Lifted Up

Opening of the Jaws in a Wild Turkey

Beak Foramina

Lower Jaw Fenestra

16.9. Skull of a wild turkey. Note that the skull, like that of many other birds, is kinetic, and that the front of the jaw is flexible and lifts dorsally when the bird feeds. Illustration of skull mechanics in the turkey based on work by Dawson et al. (2011) on ducks, and figures in Liem et al. (2001).

of skull kinesis somewhat similar to what we have previously described in lizards (see chapter 13) (Dawson et al., 2011). Synovial joints between numerous bones in the facial skeleton, and between the quadrate and the braincase, allow the beak of the turkey and many birds to flex dorsally as the mouth is opened, and to flex ventrally as the mouth closes (Fig. 16.9) (Dawson et al., 2011). In this way, the beak can accommodate objects of various sizes and actively adjust to struggling prey. As you are probably aware, differences in the size and shape of a bird's beak make it adept at grasping and processing certain foods, whereas other potential food items will be off-limits or very difficult to deal with. For example, many predatory birds have strongly curved beaks that make great meat hooks and slicers, but that would be disadvantageous if they tried to crack open seeds and nuts (Pough, Janis, and Heiser, 2002).

The large orbits of many birds house equally large eyes, which allow them to see very well. For comparison, human eyes account for approximately 5% of the total volume of the skull, whereas many birds have eyes that comprise nearly 50% of their skull's volume (Walsh and Milner, 2011)! Birds rely to a great extent on the sclerotic bones to squeeze and distort their lenses to adjust their focus (Walsh and Milner, 2011). As we saw for lizards, the retina of many birds possesses more than three cone types and commonly bears cones with oil droplets that magnify and enhance specific portions of the color spectrum. As with many diapsids, most bird species can detect colors across the spectrum and into the ultraviolet, and their often colorful body and feather patterns are a testament to their excellent vision (Walsh and Milner, 2011). In fact, many bird species rely on colorful displays to attract mates and show up rivals, sometimes to the detriment of the often flashy males that are noticed equally well by hungry predators (Attenborough, 1998). Many bird species also hear exceptionally well (Walsh et al., 2009). This is especially true for owls, which have increased the density of their sound-detecting cells to hear their small mammalian prey at quite a distance (Walsh et al., 2009; Walsh and Milner, 2011).

'Bird brain' is an insult without merit. A revised, modern understanding of their brains has only recently revealed in stark detail just how wrong earlier hypotheses about their supposed lack of intelligence were (Köppl, 2011; Northcutt, 2011; Walsh and Milner, 2011). Birds take the diapsid brain to the extreme (Fig. 16.10) (Charvet, Owerkowicz, and Striedter, 2009). The cerebrum of birds is greatly expanded, to a degree similar to what we will encounter in mammals, and the forebrain covers most of the midbrain and part of the brainstem. The optic lobes of most birds are huge and process copious amounts of visual data streaming from the eyes, and the DVR (the region for expanded somatosensory processing) is huge (Fig. 16.10) (Butler and Hodos, 2005; Charvet, Owerkowicz, and Striedter, 2009). If you are flying through the air at high speed you want eyes that respond to rapidly changing conditions lest you make an unexpected stop! The parietal eye no longer pokes out of the skull but instead is tucked comfortably within the cerebrum as a pineal gland (Fig. 16.10). It still serves its original function in regulating wake and sleep cycles, but given that birds are endothermic, a direct pipeline to sunlight falling on the head is no longer necessary. Instead, the pineal gland detects changes in daylight indirectly through signals relayed to the visual centers in the brain. The cerebellum is also greatly expanded in birds, which is not very surprising given that flight requires excellent motor coordination (Fig. 16.10) (Walsh and Milner, 2011).

Parental care is also diverse among birds. In turkeys, the females scratch out a shallow depression, lay their eggs, and cover them over with some vegetation and dirt. However, unlike alligators and other crocodylians, the females sit on and incubate the eggs, leaving them only for brief periods to drink, defecate, and gobble up some insects (Dickson, 1992). Females also turn the eggs, reaching underneath their feathery bellies to move and roll the eggs in the nest with their beaks. Egg turning is necessary for birds for the same reasons you rotate food in a microwave oven: even heating on all sides (Dickson, 1992). Turning the eggs can also prevent the unfortunate tendency of the developing chick to 'stick' to the side of the shell if it is left in one position for too long. This 'sticking' can have devastating consequences as the embryos grow, causing them to eventually rip away from the shell, hemorrhaging blood and organs. In many bird species, the hatchlings are born helpless and hungry and must be tended and fed almost constantly by one or both parents until they fledge and leave the nest (Attenborough, 1998). Other species, such as the Australian brush-turkey (*Alectura lathami*), have a parenting style that relies on communal egg laying and chick rearing. A typical group of

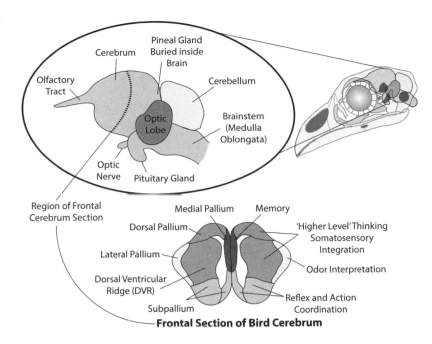

16.10. The brain of a bird. Note that in the cerebrum, the dorsal ventricular ridge (DVR) of birds is quite enlarged, and new studies on avian intelligence indicate that this region plays a significant role in somatosensory integration (i.e., smarts). Bird brain diagram and cerebral cross-section based on figures in Liem et al. (2001) and Kardong (2012).

brush-turkeys will excavate several pits into which eggs from a variety of females are laid, and these pits are subsequently covered with rotting vegetation for incubation. Here, the males guard and manage the vegetation at the nests (Birks, 1997).

Returning to the skeleton: the necks of birds are among the most flexible and mobile of any vertebrates (Fig. 16.11). There is no set number of cervical vertebrae for birds, which can range from as few as 13 to as many as 25 (Liem et al., 2001). In turkeys, the number of cervical vertebrae is on the low end, with a total of 13 from the head to the start of the dorsal series (Dickson, 1992). Most of the cervical vertebrae have saddle joints, and these allow a wide range of movements, as we saw for turtles (Kardong, 2012). In some birds like owls, a thick coat of feathers disguises complex movements of the neck so that their heads appear to turn completely around! The flexible neck also allows the head to remain stable while the rest of the body turns and moves during flight (Warrick, Bundle, and Dial, 2002). This is somewhat akin to the principles underlying the attitude indicator in an airplane, which shows the airplane's orientation relative to the horizon no matter how the plane is turning, rotating, or pitching in space. Similarly, the long, flexible neck of a bird allows the animal to keep its head oriented toward the ground or prey while the body adjusts itself however best for flying and landing (Attenborough, 1998; Warrick et al., 2002; McArthur and Dickman, 2011).

In contrast to the neck, the dorsal vertebrae and sacrum are largely immobilized by ligaments and fusions between the bones (Fig. 16.11). In fact, a good portion of the dorsal vertebrae and sacrum are fused together with the pelvis into a solid unit called the synsacrum (Fig. 16.11) (Liem et al., 2001; Kardong, 2012). In turkeys and many other bird species the ligaments that surround the vertebrae become ossified into bony rods

16.11. The skeleton of a bird, based on the wild turkey. Inset is the wild turkey skeleton with a body outline silhouette. The * indicates the uncinate processes on the ribs that provide additional mechanical leverage for the inspiratory muscles. See text for details.

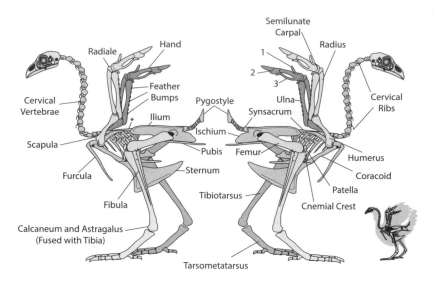

that further stiffen the vertebral column. This stiffening of the dorsal vertebrae, sacrum, and pelvis stabilizes the body core and prevents extraneous movements or wobble that would be disastrous for flying. As a consequence of the stiffening, the dorsal muscles are thin and reduced in birds, a benefit from an energy perspective because few calories are expended keeping the back rigid during flight (Kardong, 2012). From the dorsal vertebrae project a series of stout, well-developed ribs, many of which possess uncinate processes that act as levers for intercostal muscles to expand the ribcage (Fig. 16.11) (Tickle et al., 2007; Klein and Codd, 2010). These vertebral ribs connect with ossified sternal ribs, and a joint between these dorsal and ventral elements allows the ribcage to expand and contract in conjunction with movements of the sternum (Claessens, 2009). The caudal vertebrae are puny and few in number, and the last several are typically fused into a terminal vertebra known as the pygostyle (Fig. 16.11). In airplanes, the tail rudder is a stabilizing flap that can be moved in the horizontal plane to prevent the aircraft from drifting too far left or right (Fig. 16.11) (Langone, 2004; Bloomfield, 2006). Similarly, the pygostyle can be moved on the end of the tail to orient tail feathers like a rudder for similar aerodynamic reasons.

Flying would not occur in a turkey or other birds were it not for the possession of wings, and the pectoral skeleton has been highly modified for getting your future Thanksgiving meal airborne. Recall that the pectoralis muscles are major adductors and are ideal for forcing the wings downward. It is not surprising that pectoralis muscles in birds are huge, accounting for up to 35% of total body mass in some flying birds (Gill, 2006). Given that the sternum is the anchoring point for the pectoralis, the large size and tall, central ridge (the carina) of the sternum provide many points of purchase for this powerful forelimb adductor in birds. Another chest muscle, the supracoracoideus, helps raise the wings of birds (Fig. 16.12). The tendon of this muscle inserts into the deltopectoral crest with the pectoralis, but its tendon has become 'trapped' and inverted

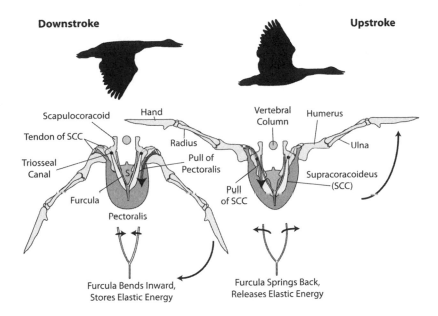

Downstroke

Scapulocoracoid
Hand
Tendon of SCC
Radius
Triosseal
Canal
Pull of
Pectoralis
Furcula
Pectoralis

Furcula Bends Inward,
Stores Elastic Energy

Upstroke

Vertebral
Column
Humerus
Ulna
Supracoracoideus
(SCC)
Pull
of SCC

Furcula Springs Back,
Releases Elastic Energy

16.12. Basic mechanics of a bird flight stroke. In birds, the old supracoracoideus muscle that originally complemented the actions of the pectoralis in forelimb adduction has been modified to serve as a wing abductor. An opening formed between the scapula and coracoid, called the triosseal canal, entraps the tendon of supracoracoideus and deflects it to the caudal side of the humerus, where its pull is now dorsal rather than ventral. The furcula, or fused clavicles, of birds act like a spring, repeatedly storing elastic energy and releasing it during flight. SCC stands for supracoracoideus in the diagram. Based on information and figures in Jenkins, Dial, and Goslow (1988) and Liem et al. (2001).

through a specialized opening in the scapulocoracoid called the triosseal canal. This inversion of the tendon causes the supracoracoideus to abduct the humerus instead of adducting it, making it a powerful, complementary muscle to the pectoralis. Cranial slots on the sternum accept boomerang-shaped scapulocoracoids that hug the body wall and allow passive movements of the shoulders when the wings are depressed (Fig. 16.12). In birds, the clavicles have become fused into a V- or U-shaped bone called the furcula. This bone connects the scapulocoracoids to the cranial end of the sternum, where it acts as a compressible spring that stores and releases energy as the bird flaps its wings (Jenkins, Dial, and Goslow, 1988) (Fig. 16.12).

The glenoid of the scapulocoracoid is saddle-shaped, whereas the proximal end of the humerus is hemispherical. The combination of these expansive surfaces allows a significant degree of powerful adduction, abduction, and mediolateral rotation of the humerus about the shoulder. A tough, strap-shaped ligament prevents the humerus from slipping out of the glenoid (Baier, Gatesy, and Jenkins, 2007). The radius and ulna of most birds (except for nonflying species) may be nearly as long as, or longer than, the humerus (Fig. 16.11). The ulna is always larger than the radius, and its caudal edge is studded with a series of small bumps that correspond to where the major wing feathers insert (Fig. 16.11). This occurs because the primary feathers of the wing have quills that insert so deeply into the arm flesh that they make ligament-like connections with the ulna (Turner, Makovicky, and Norell, 2007). The relatively open joints of the radius and ulna combined with the bulbous condyles on the distal end of the humerus allow the forearm bones to swivel into advantageous orientations during flight. Moreover, the radius and ulna can slide in parallel to each other, which in turn deflects or straightens out the hand in automatic ways that control flight and especially landing (Vazquez, 1994). In particular, a

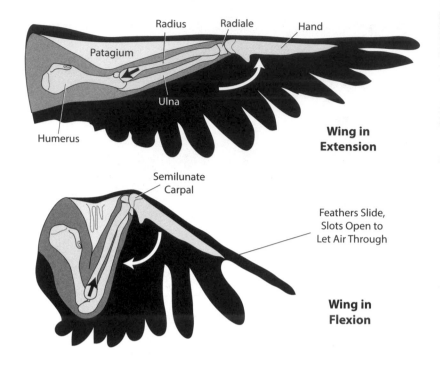

16.13. Folding actions of the bird wing. In bird wings, the radius can slide parallel to the ulna. When the elbow is straightened out, the radius slides proximally, drawing the radiale wrist bone back. This action, in turn, reorients the semilunate carpal bone such that the hand becomes extended. When the elbow is flexed, the radius slides distally, pressing on the radiale, which causes the semilunate carpal to pivot the hand, allowing the wing to fold up. Diagrams based on work by Vazquez (1994).

specialized wrist bone called a semilunate carpal (literally, a 'half-moon' wrist bone) prevents the hands from flexing or extending but allows them to pivot so they fold against the ulna (Naish, 2012). In fact, the entire bony apparatus of the bird forelimb allows it to fold up neatly against the body wall when not actively engaged (Naish, 2012) (Fig. 16.13).

Birds are different from other flying vertebrates we will discuss, such as the pterosaurs and bats, because their hind limbs are 'decoupled' from their forelimbs. Unlike a bat, in which both the forelimbs and hind limbs are united into the wing, avian hind limbs are in no way tied in with flight. The hind limbs of a bird are drawn toward the midline of the body into an upright, vertical orientation known as a parasagittal gait (Fig. 16.11). This gait was inherited from their dinosaurian ancestors as detailed in the following chapters. For now, suffice it to say that the upright limbs of birds allow them to cover more ground with each stride than a lizard of similar size, and allow many birds to move efficiently when on the ground (or when permanently grounded such as ostriches).

The ilium is a large, broad bone in most birds that anchors the femoral abductors and knee extensors, whereas the pubis and ischium are much more slender elements (Fig. 16.11). The pubis in birds has rotated caudally against the ischium, and the acetabulum is formed as a deep, open socket that points laterally (Fig. 16.11). The evolution of both a retroverted pelvis and an open acetabulum in birds are related to postural accommodations. In birds, most of the hind limb joints operate much like simple hinge joints without a great deal of rotation allowed. However, recent X-ray reconstructions of three-dimensional knee movements in guinea fowl show that a significant amount of rotation can occur between

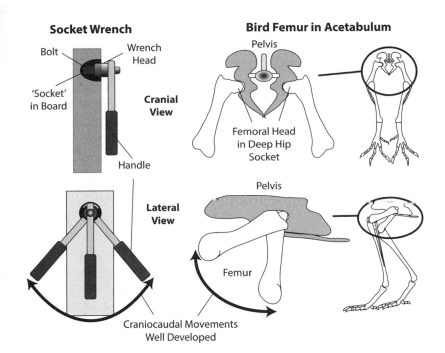

Socket Wrench

Bolt

Wrench Head

'Socket' in Board

Handle

Bird Femur in Acetabulum

Pelvis

Cranial View

Femoral Head in Deep Hip Socket

Pelvis

Lateral View

Femur

Craniocaudal Movements Well Developed

16.14. The mechanics of the bird hind limb. The cylindrical femoral head and deep, perforated acetabulum of the bird pelvis act in ways like a socket wrench turning a bolt in a board. See text for details.

the femur and the tibiotarsus (Gatesy, Kambic, and Roberts, 2010; Kambic et al., 2014). It is possible that long-axis rotation at the knee (turning of the tibiotarsus on the distal end of the femur) may compensate for the apparently more limited movements available at the hip and especially in the ankle (see below). The head of the femur is shaped like a cylinder, with the smoothest joint surfaces along its dorsal half, and it projects medially into the pelvis, fitting snugly into the deep, laterally facing acetabulum. This setup allows the femur to rotate craniocaudally quite well but restricts its ability to abduct away from the body (Fig. 16.14) (Hutchinson and Gatesy, 2000; Fastovsky and Weishampel, 2012). In fact, a simple mechanical analogy to this situation is how a socket wrench works. A typical socket wrench has a long handle that can be used to pivot an inwardly projecting metal cylinder with teeth that grip a bolt or nut (Fig. 16.14). The wrench handle held in a vertical position can rotate back and forth quite easily, rotating the metal cylinder on its long axis to apply torque (turning power). Now, imagine if you modified the socket wrench so that the metal cylinder was solid, and you placed this end into a deep socket in a board of wood. You could easily turn the wrench about its cylindrical end with the handle in a vertical position (Fig. 16.14). Replace the metal cylinder with a cylindrical femoral head, the handle of the wrench with the shaft of the femur, and the deep socket in the board of wood with the deep acetabulum, and you have in essence the basic mechanics of a bird hip joint (Fig. 16.14) (Fastovsky and Weishampel, 2012).

The femur of birds, although strongly adducted into a vertical plane, is naturally held nearly horizontally (Fig. 16.11). In fact, in most birds the femur never develops the columnar posture we are used to seeing in ourselves (Hutchinson and Gatesy, 2000; Abourachid et al., 2011). Instead, it is the tibia that is longer, larger, and held vertically (Fig. 16.10). In their typical walking, running, and standing postures, birds get about with more or less permanently flexed knees (Hutchinson and Gatesy, 2000; Biknevicius and Reilly, 2006; Abourachid et al., 2011). These changes in femur orientation are related to a bird's odd anatomy. First, birds are naturally front-heavy: a large sternum, huge pectoral muscles, and a majority of the body's mass are all in front of the acetabulum. Unlike those of their dinosaurian ancestors, the tail of modern birds is too small to counteract the mass of the chest and guts. This would be somewhat equivalent to having a wheelbarrow with the wheel near the handle rather than toward the front. In this situation, your legs are providing the forward thrust, and the pivot point (the wheel) lies behind most of the weight (Fig. 16.15). You would find such a wheelbarrow frustrating to use because it would tend to tip forward often, spilling its contents and you to the ground (Fig. 16.15).

The solution to this mechanical imbalance would be to move the wheel away from your legs and under the load you are carrying (Fig. 16.15). Similarly, if birds had a vertical femur, the feet would be placed underneath the hips, leaving most of the mass ahead of this pivot point. Just like the wheelbarrow, the bird would tend to tip forward, and walking would be clumsy and cumbersome. Instead, the horizontal femur acts to shift their lower legs cranially, which places their feet underneath the mass of the chest and guts (Fig. 16.15). In this way, birds can walk with a parasagittal gait without constantly having to prevent themselves from tipping forward. This is a nice story, but how can we test that this change in femur orientation is linked to the loss of a balancing tail in birds? In an ingeniously simple study, a group of researchers attached a prosthetic tail to chickens to see what would happen to their femur posture and center of mass (Grossi et al., 2014). The results of this study confirm what many paleontologists had suspected: the femur was held more vertically and the center of mass shifted caudally closer to the hips (Grossi et al., 2014).

Another advantage of having a more horizontally oriented femur is that it counteracts the tendency of a rigid body core to cause the body to jolt with each footfall. The movements of the femur act like 'shocks' so that a walking or running bird has a more sinuous, bouncing gait instead of one that is jerky, stilted, and more wasteful of mechanical and elastic energy.

Given the flexed posture of the knee, both the knee extensors and flexors have become enlarged in birds (Hutchinson and Gatesy, 2000). In fact, a lever-like projection called the cnemial crest points forward just below the articular surface of the tibiotarsus (Fig. 16.11). The combined tendon of the knee extensors (the quadriceps) inserts into the cnemial crest and, as with all lever systems we have discussed, gains mechanical advantage that allows efficient knee extension. Moreover, the tendon of the quadriceps is bent around the knee more or less permanently, which causes tensile stresses within it that commonly lead to the formation of one or more patella-like sesamoids (Chadwick et al., 2014; Regnault, Pitsillides, and Hutchinson, 2014; but see Eyal et al., 2015)!

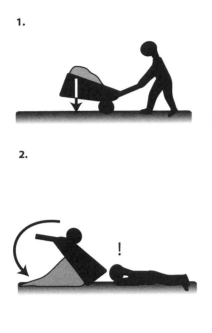

1.

2.

3.

Bird 'Solution'

16.15. Physical mechanics of walking with flexed knees. In birds, the knees are flexed during locomotion, and this places the feet beneath the center of the bird's mass. If birds did not walk with flexed knees, this would be equivalent to pushing a heavy wheelbarrow with the wheel near the handle and away from the center of mass (1). It would be difficult to control such a machine, and it would be in danger of tipping forward and spilling its contents (2). Instead, a typical wheelbarrow has the wheel placed forward under the center of mass (3). The flexed knees of birds accomplish the same thing: placing the feet beneath the forward-shifted center of mass created by loss of a bony tail and the expansion of the chest muscles.

In birds, the tibia has fused with the calcaneum and astragalus into a single bone, the tibiotarsus. The fibula itself becomes very thin and sticklike, and it has essentially become a part of the tibiotarsus. Because the calcaneum is now associated with the tibiotarsus, only the proximal portion of the fibula remains in birds where it completes the knee joint (Fig. 16.11). The two knobs at the distal end of a bird tibiotarsus (the fused astragalus and calcaneum) and the grooved space in between them accept a ridge from the distal ankle bones. This ankle shape ensures that the foot has a great range of flexion and extension that aids in terrestrial locomotion while nearly eliminating ankle rotation. The metatarsals and distal ankle bones have fused into a tarsometatarsus that further restricts movements beyond flexion and extension (Fig. 16.11). The foot of birds is technically tridactyl and digitigrade: birds stand on their central three toes (digits 2–4) and keep their tarsometatarsus and ankle up off the ground (Fig. 16.11 and Plate 21). This is equivalent to you standing on your tiptoes like a ballerina with your heels up in the air. By walking only on the toes, birds extend the length of their lower leg by effectively adding the length of the tarsometatarsus to the tibiotarsus, which increases their stride length and often their ground speed. We will return to this topic in the next chapter.

Flight would not be possible without feathers. The feathers of a bird are remarkable structures that help create the wing surface itself (technically, the airfoil) and interlock with one another over the body surface to create a smooth profile, and are capable of adjusting their orientations to modify the shape of the wings and body. The most familiar type of feather is the primary or flight feather (Fig. 16.16). It consists of a central, hollow tubelike quill from which extend filaments called barbs that themselves

16.16. The physics and anatomy of flight feathers. Flight feathers have an asymmetrical vane that, in cross-section, resembles an airplane wing with a flattened underside and a gently convex top side. This causes air to rush over the top of the feather faster than under the bottom, which produces lift. A typical flight feather is composed of a shaft or quill and a vane comprised of barbs and barbules held together by little hooklets. Based on diagrams and illustrations in Radinsky (1987), Liem et al. (2001), Kardong (2012), and Pough, Janis, and Heiser (2013).

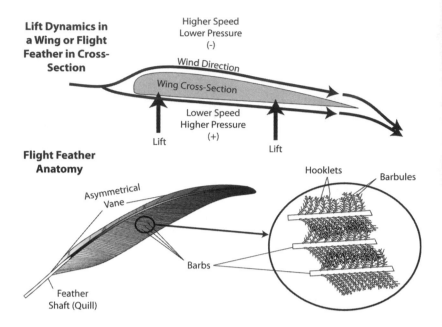

Lift Dynamics in a Wing or Flight Feather in Cross-Section

Higher Speed
Lower Pressure
(-)

Wind Direction

Wing Cross-Section

Lower Speed
Higher Pressure
(+)

Lift   Lift

Flight Feather Anatomy

Asymmetrical Vane

Hooklets   Barbules

Barbs

Feather Shaft (Quill)

possess multiple smaller projections called the barbules. The combined surface formed of the barbs and barbules is called the vane of the feather (Fig. 16.16). Moreover, the vane is adjustable and fixable because each barbule has numerous Velcro-like hooks (hooklets) (Fig. 16.16) that allow the feather vane to be unzipped, tended to by the bird using its beak, and then rezipped back into a competent flight structure (Attenborough, 1998; Gill, 2006). The vane of a flight feather is asymmetrically situated around the quill: the part of the feather that faces into the wind as the bird is flying (the leading edge) is short compared to the lengthier portion of the vane facing downwind (Fig. 16.16). This asymmetry in the flight feather's vane tapers it like the wing of an airplane, causing air to rush faster over its dorsal surface, eventually helping to generate lift (Attenborough, 1998; Bloomfield, 2006; Gill, 2006). As the quill bumps on the ulna testify, the flight feathers attach deeply into the skin around the caudal edge of a bird's forelimb. Special sensory nerves associated with the deep-rooted feather quills send signals to the bird's brain to provide feedback that informs the bird on how it should adjust its wings during flight (Gill, 2006).

Other feathers augment and improve flight in various ways. Across the cranial edge of a bird's forelimb stretches a fleshy strip between the shoulder and wrist, called the patagium, which 'fills in' the elbow. The patagium by itself creates a triangular airfoil that provides definition to the wing, whereas a smooth hump of shorter feathers (called coverts) runs along and across the patagium to contour the leading edge of the entire wing. This results in the coverts across the patagium being gently arced and tapered caudally toward the flight feathers. This feather-created profile again acts again like an airplane wing, directing faster air over the top of the bird's feathery forelimb to increase lift. Numerous soft contour

Wing with No Slat

Wind

Air Pressure above Wing Drops

Separated, Turbulent Air

Lift Weakens

Angle of Attack

Wing with Slat

Wind

Wing Slat Helps Counter Stall

Lift Maintained

Angle of Attack

Bird Alula = Wing Slat

Air Passing through 'Slat' Created by Lifting Alula

Alula

Wind

Perch

16.17. Preventing stall. In birds, as in airplanes, coming in for a landing is dangerous because the up-tilt of the wings can cause stalling, which weakens lift. As with modern aircraft, a modern bird wing has an adjustable slat called the alula. When birds approach a perch or come in for a landing, they lift their alula, creating a slat in the wing that maintains lift. Airplane wing diagrams based on Bloomfield (2006)

feathers run from the back of the skull down to the tail, and these help to smooth out the overall profile of the bird so as to cut down on drag. In flightless birds such as ostriches and emus, the flight feathers have lost their hooklets and tend to be bushier with less well defined vanes. The origin of these marvelous structures will be briefly considered in chapter 18.

When a typical bird begins takeoff, the wings are extended and adducted such that the feathers push air down, creating equal but opposite forces that launch the bird into the air. Once airborne, a bird's body tilts cranially and its wings flap up and down in a rowing sequence that generates air vortices that propel the animal forward and upward (McGowan, 1999; Liem et al., 2001; Vogel, 2003). This is somewhat reminiscent of helicopter flight: when a helicopter pilot wants to direct his flying machine forward, the nose of the aircraft tilts downward and the rotors now push the air down and back, forcing the helicopter forward and up (Vogel, 2003; Langone, 2004). Landing is a trickier affair, as shown by modern aircraft. First, when readying to land, an airplane's speed slows, decreasing lift and causing the aircraft to begin descending. Second, to stay aloft and not rapidly lose lift, the plane begins to tilt its nose up so that the angle at which the wings hit the oncoming air is increased. The increased wing angle (called the angle of attack) keeps enough air rushing over its top portion so that lift can still be generated (Fig. 16.17). However, there will come a point where the angle of attack becomes too steep and the low-pressure air over the top of the wing will separate and become turbulent (Fig. 16.17). This situation, called stalling, disrupts lift and has caused aircraft crashes (Bloomfield, 2006). Many modern airplanes get around the dangers of stalling through the use of slats and flaps, devices that can be extended and lifted to increase the wing's airfoil surface, and that generate gaps through which air traveling under the wing can escape to the topside (Fig. 16.17) (Langone, 2004; Bloomfield, 2006). This allows more air to rush over the tilted wing than would otherwise be possible

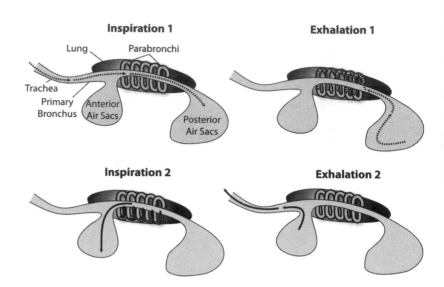

16.18. Airflow schematic of a bird lung. Each breath or 'packet' of air that a bird breathes in makes a circuitous trip through the respiratory system that takes two cycles of inspiration and exhalation to complete. In this diagram, we follow one 'packet' of air from when it is initially inhaled to when it is finally expelled. On the first inspiration, fresh, oxygen-rich air (indicated by the dotted line) bypasses the lungs and is instead drawn caudally into the posterior air sacs. On the first exhalation, the oxygen-rich air stored in the posterior air sacs is pumped into the parabronchi or air capillaries of the lungs, where oxygen is extracted and carbon dioxide is released. On the second inspiration, the now oxygen-depleted air 'packet' is pushed into the anterior air sacs. Finally, on the second exhalation, the stale air 'packet' is pushed up the trachea and out of the mouth. Diagrams based on Liem et al. (2001).

at slower speeds, increasing its lift, limiting turbulence, and preventing stall at high angles of attack. In birds, feathers associated with parts of the hand, called the alula, create similar flaps that allow most birds to land in the gentle, graceful, and demure way we are accustomed to seeing (Liem et al., 2001; Gill, 2006). Some of the caudal wing feathers also lift up and act to displace the turbulence formed during landing away from the front of the wing, which also counteracts stall (Liem et al., 2001; Gill, 2006).

Now we want to briefly consider some of the specializations of the bird digestive, circulatory, and respiratory systems. As with crocodylians, birds do little oral processing of their food. Nuts and grains may be crushed up to a point, but typically birds swallow a lot of their food whole or in large chunks. In fact, birds such as owls are famous for swallowing their small mammalian prey whole and later regurgitating the indigestible hair and bones of their prey as so-called pellets. In birds, the upper portion of the esophagus is commonly expanded into a holding chamber called the crop. Recent work on digestive efficiency shows that diapsids possessing a gizzard may have food-processing power equivalent to what mammals do with their specialized batteries of chewing teeth (Fritz et al., 2011).

As with alligators and other crocodylians, birds have four-chambered, double-pump hearts. In contrast to those of their crocodylian cousins, the hearts of birds are relatively large, and no mixing of oxygen-poor and oxygen-rich bloodstreams occurs within or outside the heart. With no valves or openings, this setup ensures that the brain and body of a bird are continuously supplied with high-pressure, oxygen-saturated blood (Liem et al., 2001; Gill, 2006; Kardong, 2012). For such endothermic, energetic, flying vertebrates, there simply is no other solution.

Just as in an internal combustion engine, oxygen is required to burn fuel in vertebrates, and the high-energy lifestyles of most birds demand as much oxygen as can be delivered quickly as possible. To do this, the lungs of birds have become evolutionarily modified into extremely efficient oxygen-extracting machines. Unlike those of mammals and other reptiles, the lungs of birds are actually quite small and fairly rigid. A series of enlarged air sacs is pumped by movements of the sternum and ribs so that fresh air flows in a circular pattern that pushes a continuous stream of air across the lungs (Attenborough, 1998; Liem et al., 2001; Gill, 2006; Claessens, 2009) (Fig. 16.18). In this way, the lungs are always ventilated, and thousands of tiny tubes called air capillaries can extract an extraordinary amount of oxygen from each breath. The air sacs of birds invade their vertebrae and long bones, adding to the hollowness of these elements and typically leaving tell-tale openings called pneumatic hiatuses (Wedel, 2009). We now know that a simpler, circuitous air flow pattern is also present in crocodylians and monitor lizards, suggesting that this unique way of extracting oxygen may have antecedents in the common ancestor of most diapsid reptiles (Schachner et al., 2011; Schachner, Hutchinson, and Farmer, 2013; Schachner et al., 2013).

The relationships of modern birds are currently difficult to understand and complex at best. We can say with some certainty that the large ground birds, such as the ostrich and emu, are more primitive than modern flying species but must have descended from flying ancestors. These modern, large ground birds also have a 'bulkier' palate that more closely resembles that of their dinosaurian ancestors (Benton, 2005). All of the flying lineages of modern birds have a much more reduced skull and palate structure. Turkeys, ducks, and many of the game fowl familiar to hunters appear to belong to one super group of more primitive flying birds, whereas the dozens of other more derived bird species familiar to most of us are harder to place with certainty. Perhaps the most terrifying birds among the modern lineages were the phorusrhacids. These flightless birds are related to falcons, pigeons, and parrots (Jarvis et al., 2014). Standing between 1 and 3 meters tall, they roamed South America 62–2 Ma (Cards 99–100) eating about anything they could kill, which often included mammals (Benton, 2005). For a time, some of them migrated as far north as the southern United States. The beaks of many phorusrhacid species were axe-like or wickedly hooked, and studies of neck anatomy reveal that many of these so-called terror birds would have wielded these weapons with astonishing speed and brutality (Degrange et al., 2010; Tambussi et al., 2012). We know from the preservation of the semicircular canals of one of these terror birds, *Llallawavis*, that some of these flightless predators were probably agile runners that could twist and turn in pursuit of their prey (Degrange et al., 2015). Sadly (or gladly, depending on your point of view), the last of the terror birds died out about 2 Ma (Card 100). Their closest living relatives appear to be South American flightless birds known as cariamids (Jarvis et al., 2014), which are nowhere near as tall or terrifying. The relationships and intricacies of modern bird relationships could fill a book, and as such go far beyond the scope of our coverage here (Benton, 2005; Naish, 2012).

At this juncture, you may well wonder how the starkly different chassis of crocodylians and birds evolved from a common ancestor. Why don't crocodylians have an erect, parasagittal posture? Where did feathers come into the picture, and why have the long tails present in the crocodylian relatives of birds not persisted? Now, armed with a working knowledge of the crocodylian and bird chassis and their associated soft tissues, we can review the amazing diversity and adaptations of several extinct archosaur groups during the Mesozoic Era. In turn, these fossil archosaurs inform us of the origins and evolution of the distinct crocodylian and avian chassis.

17.1. A demonstration of static stability with a coffee mug and a curious cat. As long as the center of mass is not pushed past the base of support, the coffee will not spill. However, a persistent force (in this case, a feline friend) that knocks the center of mass past the base of support will cause a spill.

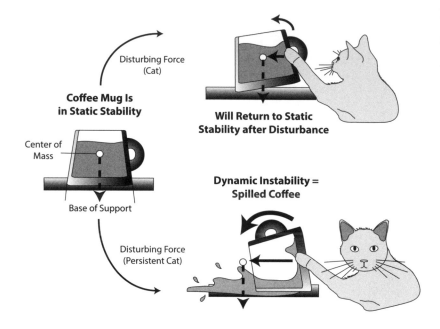

Disturbing Force (Cat)

Coffee Mug Is in Static Stability

Center of Mass

Base of Support

Will Return to Static Stability after Disturbance

Dynamic Instability = Spilled Coffee

Disturbing Force (Persistent Cat)

AS MY STUDENTS WILL TELL YOU, I THOROUGHLY ENJOY COFFEE. Most of the coffee mugs I drink from have a wide base, and there is a very good reason for this: stability. You don't want your hot beverage easily tipping over to spill on your lap or destroy a set of student papers you are grading (then again . . .). A wide-based mug prevents many spills because it obeys what physicists call static stability. Any object will remain stable as long as its center of mass remains over its base of support (Langone, 2004; Bloomfield, 2006). In objects with a wide base, small bumps or jostles will likely not push the center of mass (in this case the coffee and mug) far enough from the base of support to make it unstable (Fig. 17.1). Technically, pushing on the side of a coffee mug imparts a turning force or torque to its center of mass because as you push the coffee mug one way, the force of friction is pushing back in the opposite direction (Bloomfield, 2006). If this torque moves the mug's center of mass far enough away from its base of support, it's spill city.

Cars are built around the same concept of static stability: with all four wheels on the concrete, a car has a wide base of support and will remain stable when stationary or moving in a straight line (Newton, 1999; Langone, 2004; Bloomfield, 2006). Things begin to change, however, when a car starts to make sharp turns or to attempt tight cornering. As a car enters a sharp turn, the force of friction from the road keeps pushing it in its former straight-line direction. This force imparts a torque to the car that, if successful in displacing its center of mass far enough from its base of support, will cause the vehicle to flip (Fig. 17.2). This is why racecar chassis are built low to the ground: with a low center of mass positioned over a wide base of support, the car is resistant to flips during turns at speeds that would cause standard road vehicles to roll end over end (side over side?). This is also why many SUVs and vans are prone to flipping: their center of mass is located high above their base of support, and so it takes less of a disturbance to make the vehicle unstable.

Bicycles, on the other hand, do something different: they sacrifice static stability for something physicists call dynamic stability: a bicycle is more stable when it is on the move than when it is stationary (Bloomfield, 2006). A bicycle's base of support is very narrow, consisting of just two thin tires arranged in a straight line, above which lies the combined center of mass for the bike and the person riding it (Fig. 17.2). A nonmoving bicycle is unstable because the center of mass is perched precariously over a narrow base of support, and any small breeze or nudge will send it careening over. In contrast, a moving bicycle is dynamically stable: as long as it's moving forward, it is much less likely

17.2. Comparing static and dynamic stability. A car provides a good example of static stability – it has a wide base of support and as long as the center mass is not perturbed beyond this region (such as by oversteering into a turn), the car remains stable and unlikely to tip over. In contrast, a bicycle is dynamically stable – although it has a narrow base of support, it naturally leans and turns under its center of mass.

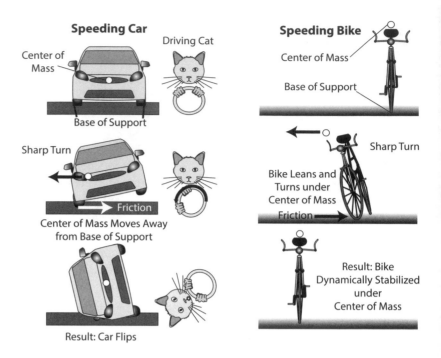

to pitch over. But why should this be? The answer is that as a bicycle is moving forward, any lean to the left or right causes the front wheel to turn in the same direction, which moves the tires (and hence the base of support) back under the center of mass, stabilizing the bike (Fig. 17.2) (Bloomfield, 2006). This is why, as a child, you could say, "Look, ma! No hands!" and not immediately crash: even without your input, the front tire was automatically rolling under the wobbling center of mass! Only if you slowed down or allowed the bike to wobble violently would this neat 'trick' backfire. Now, to be complete, we should note that the physical reason for the automatically correcting front wheel of a bicycle lies in its gyroscopic motion and something called precession (Bloomfield, 2006). In other words, a spinning wheel has a forward momentum that is difficult to perturb, and the leaning bicycle simply alters the direction of the forward-spinning tire (Fig. 17.2).

Certainly, wheels have never evolved in any archosaurs, nor other vertebrates! However, archosaurs were one of the first groups of vertebrates to do something new with their hind limbs: they stood upright, and assumed a parasagittal gait (Benton, 2005; Brusatte et al., 2010; Fastovsky and Weishampel, 2012; Parrish, 2012). As you will recall, birds have a parasagittal gait in which the hind limbs are positioned underneath the pelvis and swung like vertical pendulums. When archosaurs first pulled their hind limbs up underneath their pelvis, they effectively became reptilian bicycles by increasing both the height of their center of mass and their dynamic stability (Fig. 17.2). As with a gyroscopically moving bike tire, the long, alternately swinging hind limbs could constantly reestablish a narrow but temporarily stable base of support over which the body mass could move (Fastovsky and Weishampel, 2012). Moreover, like a speeding

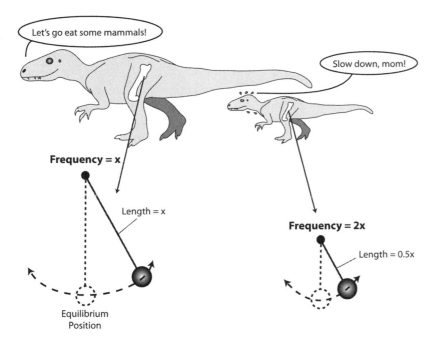

17.3. How an upright, parasagittal posture is like a pendulum. In dinosaurs and many archosaurs, the upright stance essentially transforms the hind limb into a pendulum that swings (oscillates) with a frequency correlated with femur length. A longer femur covers more ground per swing, but has a lower frequency, whereas a shorter femur covers less ground with each stride but has a naturally higher frequency.

bicyclist, they could navigate tighter turns by leaning while continuously repositioning the hind limbs beneath the body mass. Certainly, unlike a bicycle, an archosaur didn't need a kickstand because all tetrapods have postural muscles in the limbs and back to keep them upright when standing still (Dilkes et al., 2012).

Before getting too carried away with our bicycle analogy, there are some things more fundamentally significant about a parasagittal posture, and those are mobility and weight support. With vertically oriented long bones, most of the hind limb's length contributes to the stride, unlike the crank-like movements of many diapsids, where long axis rotational movements are more pronounced (Liem et al., 2001; Kardong, 2012). All of us can remember that as children, we found keeping pace with an adult difficult because mom and dad had longer legs and we were stuck with shorter ones. Thus, we had to move our legs much faster to keep up with an adult's leisurely pace. We also know that, if two people have approximately the same leg lengths and one of these people straps on a pair of stilts, the stilt-walker can easily cover more distance with the same pace simply because their stride length has increased (Leurs et al., 2011). In fact, the upright hind limbs of early archosaurs (and mammals as we will see) are reminiscent of a pendulum (Biewener, 1989, 2005; Liem et al., 2001).

In a classic grandfather clock, a large pendulum swings back and forth, turning gears that help mark the passage of time (Fig. 17.3). What many of us learned as children is that if you changed nothing else but the length of a clock pendulum, you could change the frequency (period of oscillation) with which it swung back and forth. A long pendulum has a long period of oscillation (low frequency), whereas a short pendulum has a short period of oscillation (high frequency) (Fig. 17.3) (Bloomfield, 2006). Simply put, long pendulums swing slowly and short pendulums swing rapidly. By the same token, long, upright hind limbs have a longer period of oscillation, whereas shorter limbs oscillate more frequently to cover the same distance (Fig. 17.3). Hence your childhood problem of keeping up with mom or dad: they could cover more ground more efficiently than you could because their longer hind limbs had lower oscillation periods coupled with a longer arc. Thus, standing upright and evolving a parasagittal stance would have been extremely advantageous to the common ancestor of most archosaurs, allowing these animals to cover more ground efficiently in search of prey, water, mates, and shelter.

This upright posture also has implications for speed. If you ever played an instrument in school, you are probably very familiar with the classic mechanical metronome (Fig. 17.4).This tool consists of an inverted pendulum that audibly clicks back and forth to help nascent musicians learn to keep time. A small weight on the metronome's pendulum can be slid to its tip, to its base, or anywhere in between. When you wanted the tempo to decrease, you slid the metronome's weight toward the end of the pendulum, which slowed its ticking. This happened because the pendulum had a displaced mass at its end that required more work to overcome, slowing it down. In contrast, if you wanted to speed up the metronome's pace, you slid that weight closer and closer to the base, where the gears that turned the pendulum were located. When the mass was placed near or at the base, it was mechanically more efficient for the metronome's pendulum to swing its distal, lighter end side to side, increasing its frequency. This is somewhat similar to wearing weights on your ankles versus on your belt. If you wear ankle weights, it is mechanically more challenging for your hip muscles to swing the distal ends of your legs, and you are slowed as a consequence. Instead, if you wear a weight belt, although you are heavier, the added weight is in line with your large hip muscles and therefore it remains mechanically 'easier' to swing the distal ends of your legs.

If we now think of the upright limbs of an archosaur as an inverted metronome, the hip joint represents the base, the femur represents the weight, and the tibia, fibula, and foot are the distal end of the pendulum (Fig. 17.4). Given that the largest and heaviest muscles that move the hind limb are located on the femur, the length of this bone indicates how this weight is distributed. Because the distal hind limb and foot have much lighter muscles, the combined length of these elements contributes to stride length. If the femur is shorter than the distal hind limb, the heavy hip muscles are bunched near the pelvis, making it mechanically more efficient to swing a longer, lighter distal hind limb. However, if the femur is longer than the distal hind limb, the mass of the hip and femur muscles is more displaced from the body, and this makes it more mechanically challenging to swing the shorter distal hind limb. Thus, in archosaurs and other vertebrates with upright, parasagittal postures, a short femur and long distal hind limb typically indicate a cursorial creature (capable

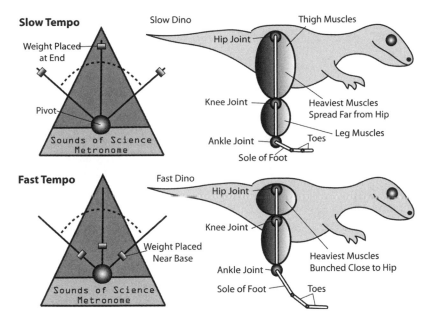

Slow Tempo

Weight Placed at End

Pivot

Sounds of Science Metronome

Slow Dino

Hip Joint

Thigh Muscles

Knee Joint

Ankle Joint

Sole of Foot

Heaviest Muscles Spread Far from Hip

Leg Muscles

Toes

Fast Tempo

Weight Placed Near Base

Sounds of Science Metronome

Fast Dino

Hip Joint

Knee Joint

Ankle Joint

Sole of Foot

Heaviest Muscles Bunched Close to Hip

Toes

17.4. The metronome of speed. In a musical metronome, the speed of the ticking pendulum is controlled by a weight on its end. In this case, a slow tempo results from placing the metronome's weight far away from the pivot, whereas placing the weight close to the pivot allows it to tick faster. Similarly, a hypothetical dinosaur with a long femur and short leg and foot segments would be relatively slow because the heavy muscles that move the thigh are spread far from the hip joint, much like a metronome weight displaced far from the pivot. In contrast, a hypothetical dinosaur with a short femur and long leg and foot segments would be relatively fast because now the heavy thigh muscles are bunched near the hip joint, much like a metronome weight placed close to the pivot.

of running or moving quickly), whereas a long femur and short distal hind limb are usually a sign that the animal in question should move to the slow lane.

Second, the mechanics of weight support change significantly with an upright hind limb posture. We should recall that long bones (humerus, femur, etc.) are strongest when loaded in compression: when a long bone is held vertically and supports the body's weight as a column, it is least likely to break, fracture, or buckle (McGowan, 1999; Carter and Beaupré, 2001; Currey, 2002). In nature, because of these properties very few long bone injuries occur due to crushing or buckling. However, long bones are less resistant to noncompressional forces such as bending or twisting. It is more likely that a long bone will break due to forces acting at angles to it. Long bones are even more susceptible to fracture due to shearing forces generated by twisting motions when both ends of the bone are forced to rotate in opposite directions. In sports injuries, broken leg bones usually occur when a planted limb is struck at an angle or twisted. Thus, because of these biomechanical properties, as vertebrates grow larger and increased stress is placed on their limbs, it is mechanically most efficient to orient the bones vertically to take advantage of their strength under compression (Schmidt-Nielsen, 1984; McGowan, 1999). These properties explain why there are no gigantic, sprawling vertebrates – the angular and twisting stresses placed on the long bones of such animals would make their limbs very susceptible to fracture.

Certainly, this scenario is an oversimplification of a complex series of evolutionary starts and stops along the way to a fully erect, parasagittal gait. For example, one of the earliest known archosaurs, *Proterosuchus* (~250 Ma; Card 95), was a Komodo dragon–sized sprawling tetrapod with crank-like limb mechanics (Parrish, 2012). Moreover, the evolution of an erect posture and parasagittal locomotor style may have occurred independently several times in the different major archosaur lines (Brusatte et al., 2010). However, despite these complexities, archosaurs were among the first amniotes to achieve such an erect stance, and this posture was present in early members of both the modern crocodylian and bird groups.

For reasons that remain ripe for scientific exploration, the forelimbs in archosaurs never seem to have 'caught up' with the hind limbs (Romer, 1956). Whereas the functional dynamics of the hind limbs changed, forelimb movements remained relatively unaltered and crank-like. Indeed, this can still be appreciated in modern crocodylians when they assume a more upright posture during their high-walk. Although the hind limbs are adducted significantly and operate in many ways like a pendulum, the forelimbs remain relatively less erect, with the elbows pointing out sideways (Fig. 17.5). If you think about it, the wings of a bird have changed little from the crank-like forelimb movements of other diapsids as well. As we discussed in the previous chapter, the wing can rotate and swivel in ways reminiscent of a sprawling tetrapod.

This decoupling of the hind limb and forelimb in terms of their overall function appears to have had significant evolutionary effects on the archosaur lineage. Let us consider a modern alligator sunning itself and waiting for prey. One of the things you might notice is that the length and mass of the head and body in front of the hind limbs is roughly equivalent to that of the tail trailing behind. You might imagine that the pelvis and hind limbs represented a pivot point on a seesaw and the head to tail axis represented a board (Fig. 17.5). You would also note that the forelimbs are shorter than the hind limbs. When an alligator or crocodile spots a meal and assumes a high-walk, its movements are reminiscent of the wheelbarrow race you may have played as a child where one standing person holds the legs of another person so they are walking on their hands. In this situation, the standing person is moving his/her legs like pendulums, but the person acting as the wheel has to move his/her shorter arms much more quickly (Fig. 17.5). Often, the standing person in this two-person race overtakes the struggling 'wheel' and both people fall over each other.

Now let's combine these analogies into a single chassis. If you are an alligator or crocodile in the high-walk going faster and faster, you will come to a point where the somewhat pendulum-like strides of your hind limbs will overtake your shorter, more crank-like forelimbs (Fig. 17.5). At this point, either you tumble end over end (which is not very becoming of a crocodylian) or you change your gait, such as the gallop in some crocodylians we discussed in chapter 16. But another option was common in the ancestral archosaurs of both the crocodylian and bird lines: removing the forelimbs entirely from the equation and progressing solely on the hind limbs (Fastovsky and Weishampel, 2012). Thus, there was strong selective pressure to become a biped (Fig. 17.5). Given that the relative length and mass of body or tail on either side of the pelvis were approximately equivalent, the center of mass would have been situated

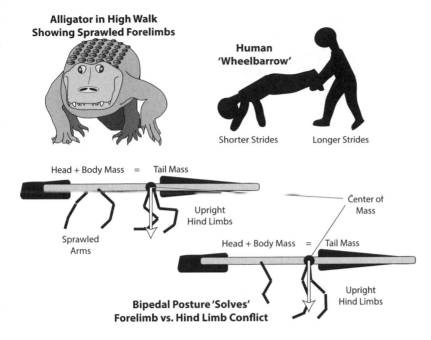

Alligator in High Walk Showing Sprawled Forelimbs

Human 'Wheelbarrow'

Shorter Strides   Longer Strides

Head + Body Mass = Tail Mass

Upright Hind Limbs

Center of Mass

Sprawled Arms

Head + Body Mass = Tail Mass

Bipedal Posture 'Solves' Forelimb vs. Hind Limb Conflict

Upright Hind Limbs

equidistant from either end of the body underneath the hind limbs. Like a seesaw, the body and tail would have acted as children of equal mass on either side of a pivot whose balancing forces would keep an archosaur upright and bipedal (Fig. 17.5) (Fastovsky and Weishampel, 2012). This solution to the decoupled mechanics of the forelimb and hind limb apparently occurred several times in various archosaur lineages and in gradual ways.

But is this all just guesswork? Apparently not: there are many species of modern lizards that become bipedal when they begin to run. Surprisingly, these lizards show that our wheelbarrow to seesaw concept actually occurs in reality. In fact, it has been documented that some lizards often trip over their own forelimbs when accelerating into a run (Irschick and Jayne, 1999)! This observation bolsters the suggestion that shorter forelimbs get in the way of longer hind limbs as running speed increases and that there are locomotor 'benefits' associated with eliminating those pesky forelimbs once a high-enough speed is reached. Moreover, the long tails of many lizard species act as a counterbalance to the body's mass, and the tail may act in part as a rudder to steer or balance the animal as it careens along on its hind limbs (Aerts et al., 2003; Olberding, McBrayer, and Higham, 2012). Of course, lizards are not archosaurs, and even when going full tilt, the hind limbs of bipedal lizards have a wide arc of rotation and retain many of the crank-like characteristics we are now familiar with (Irschick and Jayne, 1999).

In some of the earliest archosaurs, such as *Euparkeria*, locomotion was probably a bit like that of bipedal lizards. *Euparkeria* is a small archosaur from the Early Triassic of South Africa (251–245 Ma; Card 95) stretching at most 60 cm in length from head to tail (Fig. 17.5) (Ewer, 1965; Parrish, 2012) (Fig. 17.6). The skull of *Euparkeria* is rather sturdy

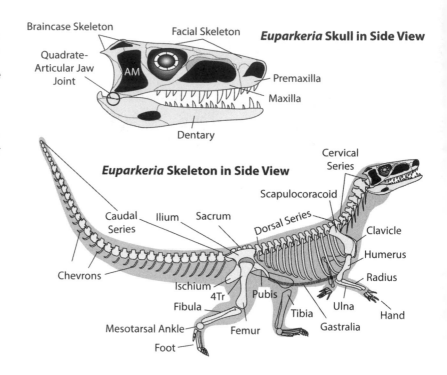

17.6. The early archosaur *Euparkeria*. This animal probably assumed a bipedal posture on occasion and bears a number of skeletal hallmarks that were inherited and modified in all more derived archosaurs. AM indicates the location of the adductor mandibulae and other jaw-closing muscles. 4Tr indicates the fourth trochanter on the femur where the large caudofemorales that pulled the hind limb caudally inserted. Skull and skeleton diagrams after Ewer (1965).

with a boxlike braincase and a triangular snout, and the premaxilla, maxilla, and dentary bones sport well-developed blade-like teeth. In particular, the teeth are serrated along their edges like steak knives, a trait that continued into many archosaur groups including the dinosaurs. The orbits are large, oval openings containing well-developed sclerotic rings, which may indicate keen eyesight and perhaps adaptations to nocturnal habits (Ewer, 1965; Benton, 2005; Parrish, 2012). The neck and trunk are rather short and stout, whereas the tail is long and heavy. A series of bony scutes runs the length of the body, indicating that *Euparkeria*, like modern crocodylians, utilized these dermal bones for trunk stabilization, calcium stores, and possibly thermoregulation. The pectoral girdle and limb are slender affairs with joint surfaces indicating crank-like forelimb mechanics. The pelvis and hind limb are also rather slender, and the hind limb measures almost one and half times the length of the forelimb. This forelimb to hind limb ratio is similar to those of bipedal lizards, and lends some support to the hypothesis that it was mechanically feasible for *Euparkeria* to assume a bipedal posture when running. However, the acetabulum was wide and outward-facing, and the head of the femur, like those of bipedal lizards, was located on the end of the femur where it could presumably enact various rotary movements related to a sprawling posture (Ewer, 1965). The long, heavy tail and well-developed fourth trochanter on the femur leave little doubt that *Euparkeria* had a large caudofemoralis muscle that could propel it forward rapidly, whether on all fours or as a biped. Such a long, stout tail would also conceivably act as a strong, lever-like counterbalance to the front of the body during bipedal locomotion.

Perhaps as early as 240 Ma (Card 95), the common ancestor of all modern archosaurs appeared, and very quickly after that point there was a major split in their family tree. One group was called the Pseudosuchia (also called Crurotarsi), and encompasses the radiations that lead to modern crocodylians (Brochu, 2003; Benton, 2005; Brusatte et al., 2010; Fastovsky and Weishampel, 2012; Parrish, 2012). The other group, called Ornithodira, contains the modern bird line as well as the flying pterosaurs and the dinosaurs (Brusatte et al., 2010; Benton, 2012; Fastovsky and Weishampel, 2012). In both groups, the pelvis became narrow, the acetabulum became deeper, and the head of the femur became offset medially from the shaft. This change in the shape of the femur is significant because it shows that when the head of the femur nestled into the deep acetabulum, the shaft of the femur could remain straight and more column-like. In other words, the cylinder-like femoral head reminiscent of a socket wrench was beginning to form (see chapter 16).

Moreover, the knee and ankle joints in pseudosuchians and ornithodirans became more hinge-like and were oriented such that these joints could extend and flex relative to the direction of travel (Brusatte et al., 2010; Fastovsky and Weishampel, 2012). As you will recall, the major ankle joint in archosaurs is a hinge situated between the astragalus and calcaneum proximally and other ankle bones distally. For pseudosuchians, the calcaneum and astragalus articulated with one another through a peg-and-socket joint just as in their modern descendants (Parrish, 1987, 2012). This suggests these ankle bones were capable of independent rotary movements in relation to one another, enhancing their ability to switch between a more sprawling and upright posture while keeping their foot pointing forward. Several Triassic archosaur footprints also show that many of the early pseudosuchians and ornithodirans were already using the high-walk or, in some cases, assuming a more permanent upright posture (Kubo and Benton, 2009; Klein et al., 2010). This can be inferred because of the narrow spacing and angles between the hand and foot placements in these trackways. Certainly, this is an oversimplification, and several pseudosuchians and ornithodirans had less erect hind limbs.

Pseudosuchians: Archosaurs with Crocodylian Skeletal Hallmarks

Oddly enough, among the most primitive members of the pseudosuchians were the phytosaurs, which bore an uncanny resemblance to the narrow-snouted fish-eating crocodylians called gavials, which we discussed in chapter 16. Currently, there is some uncertainty about the position of phytosaurs in the archosaur family tree (Stocker and Butler, 2013), but for our purposes we will regard them as basal members of the pseudosuchians. Phytosaurs had large heads with narrow, elongate jaws lined with many sharp and pointed teeth (Fig. 17.7) (Parrish, 2012). Like crocodylians, their heads were relatively flattened dorsoventrally and their orbits and external nostrils were positioned atop the head. As you might

Pedigree The Common Ancestor of Crocodylia and Their Distant Relatives

Date of First Appearance ~240 Ma (Card 95)

Specialties of the Skeletal Chassis Specialized Rotational Joint between Astragalus and Calcaneum

Eco-niche Predators, Omnivores, and Herbivores of the Land and Waterways

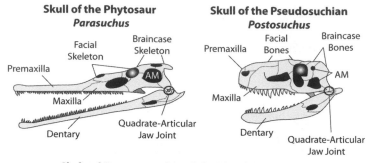

Skull of the Phytosaur
Parasuchus

Facial Skeleton
Braincase Skeleton
Premaxilla
AM
Maxilla
Dentary
Quadrate-Articular Jaw Joint

Skull of the Pseudosuchian
Postosuchus

Facial Bones
Braincase Bones
Premaxilla
AM
Maxilla
Dentary
Quadrate-Articular Jaw Joint

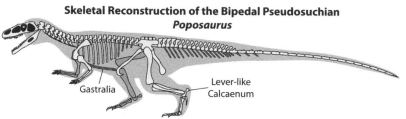

Skeletal Reconstruction of the Bipedal Pseudosuchian
Poposaurus

Gastralia
Lever-like Calcaenum

17.7. A sampling of Pseudosuchian archosaurs. Note the very crocodylian-like skull of *Parasuchus,* and the eerily dinosaur-like skeleton of *Poposaurus*–compare to *Marasuchus* (Fig. 17.9) and figures in chapter 18. AM indicates the location of the adductor mandibulae and other jaw-closing muscles. Diagram of *Parasuchus* and *Postosuchus* skulls based on Benton (2005); skeletal diagram of *Poposaurus* based on Gauthier et al. (2011).

anticipate, this head shape suggests phytosaurs were ambush predators that lay motionless in the water with only their eyes and nose poking above. In fact, two phytosaur specimens have been found with intact stomach contents consisting of the bones of small tetrapods that were living at the water's edge (Parrish, 2012). Unlike in crocodylians, the nostrils of phytosaurs were retracted far back toward the eyes and elevated on a 'hump' of nasal bone. The best known phytosaur, *Parasuchus* from India, measured over 2 meters in length and was covered in dermal scutes. As with modern crocodylians, its limbs are typically sprawled, but, based on their joint geometries, the hind limbs would have been capable of assuming something akin to the high-walk. *Parasuchus* also has a large, thick tail that must have housed a substantial caudofemoralis muscle capable of generating plenty of locomotor force. Phytosaurs were relatively short-lived and went extinct by the opening of the Jurassic period some 200 Ma (Card 96). It is often hypothesized that the eco-niche for waterside ambush predator left vacant by the phytosaurs was later filled in by their distant crocodylian relatives (Benton, 2005).

A variety of predatory members of Pseudosuchia flourished during the Triassic. These archosaurs have variously been placed near the common ancestors of what would eventually lead to the modern crocodylians, but their exact relationships are currently debated. These archosaurs had adaptations in their hind limbs that suggest many were capable of occasional and sometimes full-time bipedalism. They commonly had robust skulls and long jaws lined with dagger-like, serrated teeth (for example, *Postosuchus* [Fig. 17.7]). In some cases, the skulls of these pseudosuchians bear an eerie resemblance to those of predatory dinosaurs. Moreover, some of these predators, such as *Poposaurus* (235–204 Ma; Card 95–96), have bird- and dinosaur-like body forms (Fig. 17.7) (Gauthier et al., 2011; Parrish, 2012). For *Poposaurus* in particular, the forelimb is less than half the length of the hind limb, the femur has an in-turned head that articulates with a deep, open acetabulum, and the pelvis is narrow (Gauthier et al., 2011). In addition, the foot is symmetrical, with the central three toes forming its base. This is unlike the asymmetrical feet we have encountered in lizards, where the fourth and fifth toes are commonly elongate. Instead, the foot of *Poposaurus* is similar to those of birds, where the symmetrical central digits ensure that the foot faces forward when the hind limb is in an upright, parasagittal posture (Fig. 17.7). Like many of the mammals we will soon encounter, the calcaneum of *Poposaurus* and many other Triassic pseudosuchians forms a lever-like projection into which the foot-extending gastrocnemius muscle inserted via the Achilles tendon (Gauthier et al., 2011) (Fig. 17.7). The large size of the calcaneal process in *Poposaurus* strongly suggests a great deal of mechanical leverage was possible that drove the foot backward into the ground, pushing this pseudosuchian forward efficiently on its hind limbs. Again, the large tail would have housed a caudofemoralis muscle that connected to the prominent fourth trochanter on the posterior side of the femur. Such a large tail would again provide a counterbalance to the body and skull, further enhancing bipedal locomotion.

Herbivorous archosaurs also made an appearance during the Triassic. Aetosaurs, pseudosuchians closely related to the Crocodylia, were broad, stout-bodied, and covered head to tail in armor scutes (Benton, 2005; Parrish, 2012). Their heads were relatively small, and they possessed upturned snouts that may have been useful at digging in the soil for plant roots and tubers. Perhaps their armored skin was a useful defense against the wide variety of Triassic predatory archosaurs.

Among this plethora of pseudosuchians, the ancestors of what would eventually become the modern Crocodylia appeared. Called crocodylomorphs, many of these archosaurs look so little like anything related to crocodylians that they can be easily mistaken at first pass for dinosaurs or earlier predatory pseudosuchians. As an example, the crocodylomorph *Saltoposuchus* was a small (just about 50 cm head to tail) archosaur with a thin, delicate skull, a sleek vertebral column, and elongate limbs (Fig. 17.8) (Benton, 2005; Allen, 2010). The tiny, needlelike teeth that adorn its premaxilla, maxilla, and dentary bones suggest *Saltoposuchus* was an insectivore. Moreover, its hind limbs were far longer than its forelimbs, suggesting that, along with its lengthy tail, this little crocodylomorph could run bipedally after insect prey. Despite its very un-crocodylian appearance, we know that little *Saltoposuchus* was a distant relative of modern crocodylians partly from its wrist. Like modern crocodylians (see chapter 16), the proximal wrist bones have become elongate and asymmetrical. Unlike those in its living relatives, the wrist bones of *Saltoposuchus* appear to have helped to lengthen the forelimb, much as the tarsal bones of a frog elongate its hind leg (see chapter 11) (Fig. 17.8) (Benton, 2005).

By the Early Jurassic period some 195 Ma (Card 96), the first crocodylomorphs with crocodylian characteristics appear. One of these is *Protosuchus*, a meter-long crocodylomorph from North America and South Africa (Fig. 17.8) (Colbert and Mook, 1951). The skull of *Protosuchus*

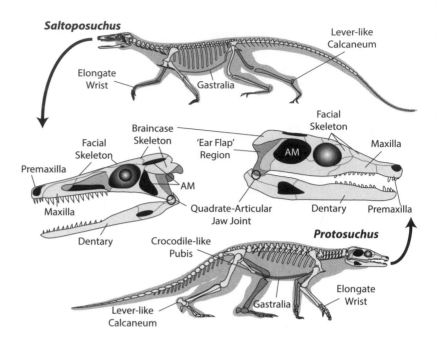

17.8. A comparison of crocodylomorphs. Note the very gracile and long-legged *Saltoposuchus* from early in the crocodylomorph radiation, and compare this to the skeleton of *Protosuchus,* which is a heavier animal and which shows the harbingers of the modern crocodylian skeleton. In particular, note the antorbital fenestra, so characteristic of the archosaurian ancestor, has disappeared in *Protosuchus,* presaging the evolution of the heavy, prey-crushing skull of modern crocodylians. AM indicates the location of the adductor mandibulae and other jaw-closing muscles. *Saltoposuchus* diagrams after Benton (2005) and Allen (2010); *Protosuchus* diagrams after Benton (2005) and Colbert and Mook (1951).

bears a number of features we have already encountered in modern crocodylians. For example, much of the skull is rugose and dimpled, a feature correlated with bumpy, closely adhering skin in crocodylians. In addition, the posterior braincase bones overhang the otic region in *Protosuchus,* suggesting that a fleshy ear flap was present in this animal as it is in modern crocodylians (see chapter 16) (Benton, 2005). Although an antorbital fenestra is still present in the snout, it is rather small (so small it is not pictured in Fig. 17.8), and this appears to go hand in hand with the evolution of a secondary palate in this archosaur (Colbert and Mook, 1951; Benton, 2005). These developments would suggest that the powerful jaw-closing muscles and crushing action typical of crocodylian jaws was already beginning to take shape in animals like *Protosuchus.* Recall that by approximately 100 Ma (Card 98), the antorbital fenestra had disappeared and a prominent secondary palate was already in place in the earliest true crocodylians. In other respects, the skull of *Protosuchus* differed from modern crocodylians in that the eyes were large and laterally facing, and the external nostrils were not dorsally placed (Fig. 17.8).

Protosuchus also shows the harbingers of other crocodylian features in its postcranial skeleton (Fig. 17.8). An elongate, somewhat asymmetrical wrist is present, the forelimbs are shorter than the hind limbs, and the body was covered in scutes. Crocodile-like belly ribs (gastralia) are present, and the pubis has already become isolated from the pelvis as in modern crocodylians (Colbert and Mook, 1951; Benton, 2005). As you will recall, these are skeletal hallmarks of the diaphragm-based respiratory apparatus of crocodylians, where the pubis and gastralia move to increase and decrease the size of the abdominal cavity (Carrier and Farmer, 2000; Claessens, 2004). The tail of *Protosuchus* is massive and likely indicates the presence of both a large caudofemoralis muscle and perhaps occasional bipedalism. The calcaneum bears a lever-like projection for the gastrocnemius and related foot-extensors, and the elongate foot is narrow, features that suggest *Protosuchus* was a rather fleet-footed archosaur. Whereas *Protosuchus* was much more crocodile-like than animals like *Saltoposuchus*, it was likely a mostly terrestrial predator that consumed small vertebrates.

Various lineages of terrestrial and even marine crocodyliforms appeared and diversified throughout much of the Mesozoic. Some small terrestrial crocodyliforms have even been implicated in raiding dinosaurs nests for eggs and babies (Kirkland, 1994). However, as the Cretaceous period waxed and then began to wane, the chassis of crocodyliforms became more and more like that of their long lost phytosaur relatives. Whereas their heads and bodies became large and muscular, their forelimbs and hind limbs became relatively shorter by comparison. This trend can still be seen today during the growth and development of alligators from embryo to adult: the overall proportions of the limbs change very little (Livingston et al., 2009) and their growth lags behind that of the body and skull (Dodson, 1975, 2003). Part of this 'shrinking' of the limbs in relation to body size must be tied to their semiaquatic habits. Larger alligators and crocodiles are known to spend more time in the water than juveniles, and tail-based propulsion becomes a more significant locomotor repertoire in adults (Willey et al., 2004). In adaptation to life as a semiaquatic ambush predator, selection for an upright hind limb posture apparently diminished as well. We may speculate that the ability to flatten the body and head as much as possible to hide from potential prey may have favored the selection of the currently sprawled resting posture in modern crocodylians. However, the femur, pelvis, and ankle still retain the articulations and mechanical relationships to allow the high-walk and occasional gallop, skeletal reminders of the more fleet-footed ancestors of these still impressive reptilian predators.

Ornithodirans: Archosaurs on the Line to Birds

Returning to the deep split in the archosaurs nearly 240 Ma (Card 95), the early members of the ornithodirans early on distinguished themselves from pseudosuchians in further modifications to their hind limb skeleton. There are a number of early ornithodirans known from the Triassic period, and the interrelationships of these animals to groups such as the dinosaurs or the pterosaurs are complex (Brusatte et al., 2010). Some of these animals led to the flying pterosaurs, whereas others, called dinosauromorphs, were, as you may guess, on the line to the dinosaurs. For simplicity, we will briefly focus on the early dinosauromorph *Marasuchus* from Argentina (230 Ma; Card 95) as an example of what was present in the common ancestor of the ornithodirans (Fig. 17.9) (Sereno and Arcucci, 1994; Benton, 2012).

Marasuchus was a small, predatory dinosauromorph, just a little over 30 cm in length. The skull of *Marasuchus* is a delicate collection of bony

Pedigree Pterosaurs, Dinosaurs, and Their Ancestors

Date of First Appearance ~240 Ma (Card 95)

Specialties of the Skeletal Chassis Specialized, Hinge-like Ankle Joint

Eco-niche Predators, Herbivores, and Omnivores of the Land and Sky

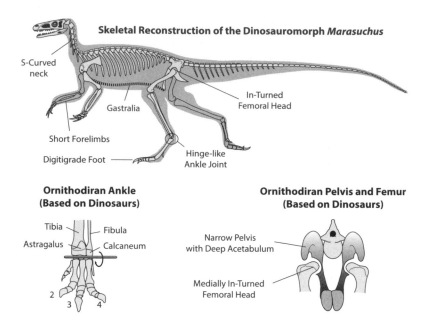

17.9. *Marasuchus* and the ornithodiran skeleton. *Marasuchus* shows a number of skeletal hallmarks of the ornithodiran chassis including an S-curved cervical series, forelimbs much shorter than the hind limbs, a deep acetabulum combined with a medially in-turned femoral head, and a hinge-like ankle. At the bottom of the figure, a dinosaur foot and pelvis (based on *Tyrannosaurus*) show the basic ornithodiran ankle shape, foot posture, and pelvic anatomy that collectively contribute to an erect stance in these archosaurs. The diagram of *Marasuchus* is based on Sereno and Arcucci (1994); the *Tyrannosaurus* foot and pelvis are adapted from Benton (2005) and Fastovsky and Weishampel (2012).

struts, and the premaxilla, maxilla, and dentary possess small, sharp, and recurved teeth. There are 9 cervical vertebrae that, when articulated, give *Marasuchus* a subtly S-shaped neck. A horizontal dorsal series bearing stout ribs is present, a short sacrum articulates with the pelvic girdle, and an elongate tail is over half again the length of the animal. Gastralia were also present, and probably aided *Marasuchus* in breathing by stiffening the body wall during inspiration, a suggestion proposed for pterosaurs as well (Claessens, 2004; Claessens, O'Connor, and Unwin, 2009). The pectoral girdle consists of a long but broad scapulocoracoid that presumably articulated into a small, cartilaginous sternum. The forelimb consists of a slender humerus, radius, and ulna, nubbin-like wrist bones, and a hand with five digits. The radius and ulna, as in crocodylians and birds, parallel each other and would have prevented *Marasuchus* from actively pronating its hand into a palm-down orientation. The hand itself is dominated by digits 1–3, with digits 4 and 5 being small and tucked into the palm. You can demonstrate this to yourself by flexing your ring and pinky fingers inward toward your palm—like *Marasuchus* and other early ornithodirans, you now have a functionally three-fingered hand.

The pelvic girdle has a modestly expanded ilium and long but narrow pubis and ischium bones. These features suggest that hip rotators, abductors, and knee extensors were well developed in *Marasuchus*, whereas the midline adductor muscles were relatively smaller, probably owing to the upright, parasagittal posture of the hind limb. The pelvis contains a deep, perforated acetabulum that accepted the crudely cylinder-like head of the femur. Overall, the hind limb is significantly longer than the forelimb, demonstrating without a doubt that *Marasuchus* was a biped. The femur sports a modestly sized fourth trochanter, which again suggests that a reasonably powerful caudofemoralis musculature was present to propel

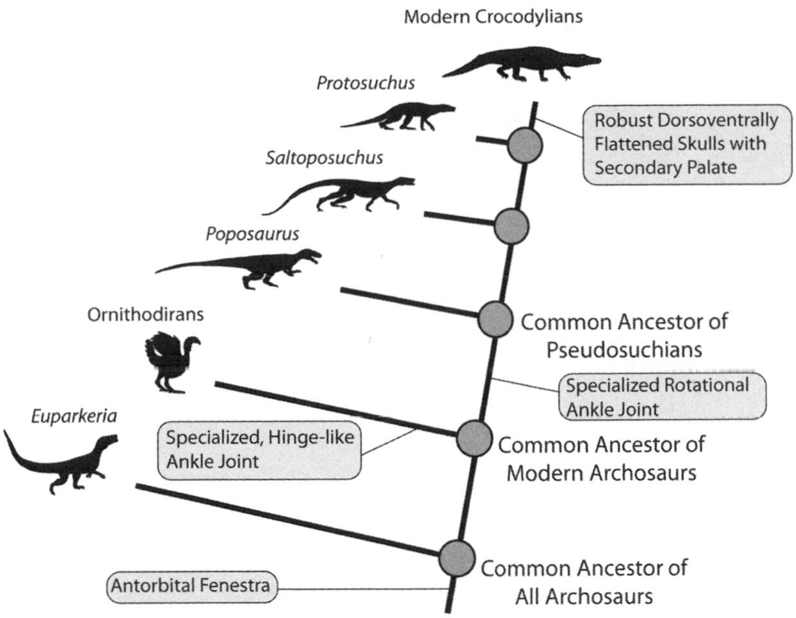

Modern Crocodylians
Protosuchus
Robust Dorsoventrally Flattened Skulls with Secondary Palate
Saltoposuchus
Poposaurus
Ornithodirans
Common Ancestor of Pseudosuchians
Specialized Rotational Ankle Joint
Euparkeria
Specialized, Hinge-like Ankle Joint
Common Ancestor of Modern Archosaurs
Common Ancestor of All Archosaurs
Antorbital Fenestra

17.10. A pedigree of archosaur relationships. See text for details.

this little dinosauromorph. The femur is shorter than the elongate tibia and fibula, and this suggests, based on our pendulum and metronome analogies, that *Marasuchus* was capable of being speedy. In *Marasuchus* and other early ornithodirans, the ankle became simplified into a strictly hinge-like joint between the astragalus and calcaneum proximally and the remaining tarsals distally (Fig. 17.9). In fact, the astragalus and calcaneum together formed a continuous roller-like surface against which the remainder of the ankle and foot could easily flex and extend (Sereno and Arcucci, 1994; Fastovsky and Weishampel, 2012). This roller-like ankle joint persists to this day in modern birds, as we have detailed in chapter 16. Unlike modern birds, the astragalus and calcaneum are not fused into the tibia, but articulate tightly with the tibia and fibula, respectively.

Furthermore, the foot of *Marasuchus*, as in all ornithodirans, was digitigrade, as seen in modern birds (Fig. 17.9). In particular, the bones of the metatarsus became narrow and more 'bunched' together, allowing them to act mechanically as a single unit (Benton, 2012; Fastovsky and Weishampel, 2012). Elevation of the metatarsus and walking on the toes also acted to further extend the length of the hind limb by adding an additional segment to its pendulum-like anatomy. In *Marasuchus* as in many ornithodirans, the toes are symmetrically arranged around the central or third digit, and digits 1 and 5 have become reduced and removed functionally from the foot proper. This pattern of digits and foot posture appears to have evolved early on, and may have been present at least 248 Ma (Card 95) (Brusatte, Niedzwiedzki, and Butler, 2010).

Figure 17.10 summarizes the pedigree of archosaurs we have discussed in the past two chapters. Putting this all together, little *Marasuchus* gives us insight into the basic chassis from which the pterosaur and dinosaur lineages arose. We should expect the earliest members of the pterosaurs and dinosaurs to have had relatively delicate skulls composed of birdy, strut-like elements and a carnivorous dentition. The cervical vertebrae should form a gentle S-curve, the dorsal vertebrae should be held horizontally and bear elongate ribs, and a rather long tail should initially be present. The forelimbs should be shorter than the hind limbs and have a hand with digits 1–3 being the most prominent. A pelvis with a deep acetabulum that accepts a medially in-turned femoral head should be present, which would allow for a parasagittal pendulum-like hind limb posture, and the femur will be somewhat shorter than the tibia and fibula. The ankle joint should be hinge-like and the foot itself will be digitigrade, with the early dinosaurs and pterosaurs supporting their weight on the central three toes. We now turn to the chassis of dinosaurs and pterosaurs to test this hypothesis.

18.1. The Jurassic rhamphorhynchoid pterosaur, *Rhamphorhynchus*. Note how the long fourth digit creates the leading edge of the wing. *Rhamphorhynchus,* like other primitive pterosaurs, had a rudder-like element on the distal end of its tail. That these archosaurs were powerful fliers is evidenced, in part, by the possession of a very large deltopectoral crest on the humerus for attachment of wing adducting muscles and a well-developed sternum. AM indicates the region of the adductor mandibulae and other jaw-closing muscles. Figure based on illustrations from Wellnhoffer (1991).

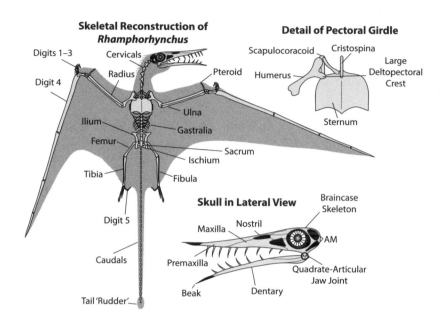

Skeletal Reconstruction of *Rhamphorhynchus*

Digits 1–3
Cervicals
Radius
Digit 4
Pteroid
Ulna
Ilium
Gastralia
Femur
Sacrum
Ischium
Tibia
Fibula
Digit 5
Caudals
Tail 'Rudder'

Detail of Pectoral Girdle

Scapulocoracoid
Cristospina
Large Deltopectoral Crest
Humerus
Sternum

Skull in Lateral View

Braincase Skeleton
Maxilla
Nostril
AM
Premaxilla
Quadrate-Articular Jaw Joint
Beak
Dentary

THE ORNITHODIRANS WERE THE MOST SUCCESSFUL GROUP OF archosaurs. Here, we will explore all too briefly the evolutionary diversity of their chassis. In addition, we will return to birds to consider the origin of their chassis and the evolution of feathers.

Pterosaurs: The Original Flying Reptiles

The song "Everything Old Is New Again" by Peter Allen could equally well apply to the convergences witnessed between pterosaur and bird skeletons. Indeed, one could make the argument that pterosaurs, being the pioneers of active vertebrate flight, were plagiarized evolutionarily by their later distant relatives the birds. Birds can be forgiven, of course, because there are only so many ways to adapt the vertebrate chassis to active flight through air. We shouldn't be surprised, therefore, to find many skeletal similarities among pterosaur and bird chassis.

The earliest pterosaurs appeared roughly 230 Ma (Card 95), nearly 80 million years before the first birds came onto the scene. Whereas many of the earliest pterosaurs were the size of sparrows, later forms achieved gigantic sizes: *Quetzalcoatlus* was the size of a small aircraft and had a wingspan approaching 9 meters across (Wellnhoffer, 1991; Witton and Naish, 2008)! As you might imagine, weight reduction was as much of a priority for pterosaurs as for birds, and the skull, vertebral column, and limb skeleton were commonly hollow and pneumatic. Like birds, the limb bones were hollow but supported internally by bony cross-struts (trabeculae) that resisted the powerful twisting and bending forces of flapping wings or ground landings (Habib, 2008; Witton and Habib, 2010). From the very beginning, pterosaurs had a specialized combination of features that defined their chassis. It would be impossible to adequately document the fantastic variety of these flying archosaurs in this book (I encourage interested readers to read Witton [2013]). Instead, we will focus on two well-known pterosaurs that show the two major versions of the chassis of these ancient fliers: *Rhamphorhynchus* and *Pterodactylus* (Figs. 18.1, 18.2). Once we have investigated the anatomy of these two pterosaurs, we will then briefly explore the evolution and diversity of these flying archosaurs.

As with all the vertebrate groups we have encountered, the relationships among the pterosaurs are somewhat complex. For simplicity, we will distinguish between the more primitive, long-tailed pterosaurs traditionally called 'rhamphorhynchoids' and the more derived, short-tailed pterosaurs called 'pterodactyloids.' Technically, it should be noted that

Pedigree The First Actively Flying Archosaurs

Date of First Appearance ~228 Ma (Card 95)

Specialties of the Skeletal Chassis Specialized Wing Consisting of an Elongated Digit 4 and Modified Wrist Bone

Eco-niche Predators of the Sky

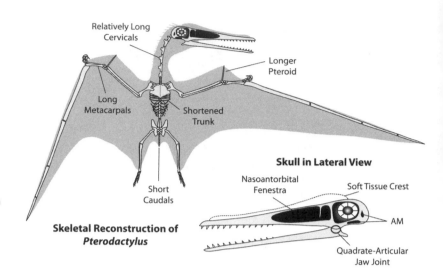

18.2. The Jurassic pterodactyloid pterosaur, *Pterodactylus*. Compared to *Rhamphorhynchus,* the skull and neck are longer, the body trunk is shortened, and the tail is tiny. In pterodactyloids, the nostrils and antorbital fenestra have merged into a nasoantorbital fenestra, but the functional significance of this feature is unclear. AM indicates the region of the adductor mandibulae muscles. Figure based on illustrations and data from Wellnhoffer (1991) and Bennett (2012).

the rhamphorhynchoids are not a natural grouping, and that some of these animals actually form transitional steps toward the pterodactyloids (Hone, 2012). However, this traditional division makes it easier for our purposes to discuss the major evolutionary changes from early pterosaurs to later species. It is paleontological serendipity that *Rhamphorhynchus* and *Pterodactylus* are both very well known pterosaurs and found together in the Solnhofen limestone, a famous rock unit from the Late Jurassic (~150 Ma; Card 97) (Wellnhoffer, 1991). This fine-grained limestone, laid down in a Jurassic lagoon in what is now Germany, preserves remarkable details of fossil anatomy. In fact, the Solnhofen limestone quarry is the source of *Archaeopteryx,* a small predatory dinosaur known as the 'first bird' the world over because this extraordinary fossil preserved in detail the outline and structures of wing and tail feathers. We will return to *Archaeopteryx* later in this chapter.

Given that *Rhamphorhynchus* and *Pterodactylus* are members of the long-tailed and short-tailed pterosaurs, respectively, and that they are both found in the Solnhofen limestone, we will compare the two animals side by side. Both *Rhamphorhynchus* and *Pterodactylus* were on average relatively small animals. *Rhamphorhynchus* typically measured 60–90 cm in length and had a wingspan of approximately 90 cm, whereas *Pterodactylus* ranged in size from small individuals with a wingspan of about 50 cm to larger forms known from fragmentary remains that may have had wingspans exceeding 2 meters (Wellnhoffer, 1991). Both pterosaurs have delicately constructed skulls with many small teeth set into their premaxilla, maxilla, and dentary bones (Figs. 18.1, 18.2). Both *Rhamphorhynchus* and *Pterodactylus* had a long face and short braincase, giving these pterosaurs heads with a scissors- or tweezers-like appearance, the sort of skull we have encountered in many insect and fish eaters (Figs. 18.1, 18.2). In *Rhamphorhynchus,* the ends of the premaxilla and dentary bones bend dorsally and have a roughened and pitted texture that strongly suggests a beak was present. Such a feature may have further enhanced fish-eating.

In fact, the skeletal remains of ingested fish were found within the fossil body cavity of a *Rhamphorhynchus* specimen (Wellnhoffer, 1991). In *Pterodactylus*, the roughened ends of its premaxilla and dentary bones suggest a small beak was present, but not as well developed as in *Rhamphorhynchus*. In both pterosaurs, the nostrils were somewhat retracted from the tip of the snout, but there is a difference in both the position and relationship of the nostrils to the skull. The primitive condition seen in *Rhamphorhynchus* is that the nostrils are retracted but separated from the antorbital fenestra by a bony bar. In contrast, *Pterodactylus* shows the derived state where the nostrils are retracted to a point where they are confluent with the antorbital opening, forming what some paleontologists call the nasoantorbital fenestra. The biological significance of this difference in nostril position and relationship to the antorbital fenestra is unclear. The orbits are large with well-developed sclerotic rings, an indication that they housed big and visually acute eyes (Figs. 18.1, 18.2). Interestingly, the quadrate-articular jaw joint in many pterosaurs was screwlike so that both sides of the lower jaw rotated outward as the mouth was opened, greatly increasing the gape (Wellnhoffer, 1991). This maw-expanding mechanism may have been correlated with storing food in the mouth while on the wing: a throat pouch structure, reminiscent of seabirds such as pelicans, is preserved in several pterosaur specimens including *Pterodactylus* (Bennett, 2012). Additionally, *Pterodactylus* also seems to have sported a low but distinctive crest on its head, a trait very common among pterodactyloids (Bennett, 2012) (Fig. 18.2).

The excellent preservation of many pterosaur specimens has meant that the braincase and its endocast have been preserved for several species. Dr. Tilly Edinger was a pioneering paleoneurologist who studied fossil vertebrate brains via their endocasts in the mid-twentieth century, and she is notable for her work on pterosaurs (Wellnhoffer, 1991). What Edinger found, and what has since been confirmed with the advent of CT-scanning, is that pterosaur brains were relatively large and similar in overall organization to those of birds (Witmer et al., 2003) (Fig. 18.3). In fact, the braincases of *Rhamphorhynchus* and another pterodactyloid were studied by Larry Witmer and colleagues (Witmer et al., 2003), and these show that the pterosaur cerebrum was enlarged, the optic lobes of the midbrain were big, and the cerebellum and inner ear regions were absolutely huge. As we saw for birds, these are all hallmarks of actively flying animals that require complex motor coordination (large cerebellum), keen eyesight (large optic lobes), and rapid higher-level somatosensory processing (expanded cerebrum). But the information gleaned from CT-scans of *Rhamphorhynchus* and a pterodactyloid shows further that the flocculus, the area of the cerebellum involved in keeping the head focused on prey while the body is moving, is larger than the optic lobes (Witmer et al., 2003) (Fig. 18.3). Moreover, the semicircular canals in the inner ear are likewise large and in fact relatively twice as big as in modern birds (Fig. 18.3). These data suggest that *Rhamphorhynchus* and other pterosaurs had a lot of space in their brains devoted to maintaining

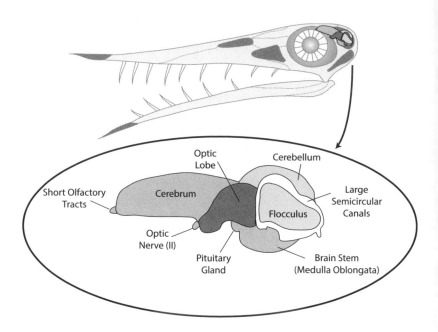

equilibrium, and this finding alone overturns long antiquated hypotheses that pterosaurs were poor or clumsy fliers.

The cervical series was mobile, as in birds, and consisted of nine ver-tebrae (Wellnhoffer, 1991). In primitive pterosaurs like *Rhamphorhynchus*, each vertebra was relatively short and compact with well-developed ribs. In contrast, derived pterodactyloids like *Pterodactylus* had longer necks by virtue of their elongated cervical vertebrae, and the vertebrae themselves had either very reduced ribs or none at all (Wellnhoffer, 1991) (Figs. 18.1, 18.2). These differences suggest that more primitive pterosaurs had stron-ger but less flexible necks, whereas derived pterodactyloids had gracile but mobile necks upon which to orient their heads (Lu et al., 2009).

As we have encountered previously for birds, the trunk skeleton of pterosaurs was short and relatively immobilized, likely an adaptation for stabilizing the body core during flight. The pelvis was commonly fused to the sacrum (creating a synsacrum) and, in some large pterosaurs, to the more caudal dorsal vertebrae as well. In some of the larger pterosaurs, the neural spines of the dorsal vertebrae have fused into a single unit called the supraneural plate, which effectively created a single trunk unit called the notarium (Wellnhoffer, 1991). Rhamphorhynchoid pterosaurs had long tails that could move freely at their bases but were stiffened through-out most of their length by bony rods (Wellnhoffer, 1991) (Fig. 18.1). It is hypothesized that some of these early pterosaurs were using their tails as rudders and balancers during flight, and in *Rhamphorhynchus* a diamond-shaped tail rudder was present (Fig. 18.1). The derived ptero-dactyloids such as *Pterodactylus* had relatively short tails but larger skulls that typically sported crests and other cranial adornments that may have assumed the rudder-like adjustments lost with the tail (Lu et al., 2009). A prepubic bone lay in front of the fused pelvis, and may have functioned to support the guts, aid in respiration, or both, in combination with a set of gastralia much like in crocodylians (Claessens, O'Connor, and Unwin, 2009). As with birds, the ribs of pterosaurs are completely ossified, and a joint between their dorsal and sternal portions would have allowed the ribcage to expand and contract with movements of the sternum (Claessens, O'Connor, and Unwin, 2009). Moreover, the pterosaur sternal ribs support numerous processes that probably enhanced the leverage of rib-expanding muscles in much the same way that uncinate processes on the vertebral ribs of birds do (Claessens, O'Connor, and Unwin, 2009). The sternum itself is a large and broad breast plate that connected to the ribs and sported a cranial cristospina or keel reminiscent of the bird sternal carina. The large and cranially keeled sternum leaves little doubt that pterosaurs had large pectoral muscles (Figs. 18.1, 18.2).

The scapula and coracoid are typically fused into a single, boomerang-shaped bone that articulated with a socket on the sternum ventrally and hugged the ribs dorsally (Wellnhoffer, 1991; Bennett, 2001, 2003) (Figs. 18.1, 18.2). No clavicles or anything resembling a furcula is present in any known pterosaurs, but the V-shaped arrangement of the coracoids leaves little doubt that these elements could resist stresses associated with active wing-beating. The glenoid fossa provides a large, shallow, oval depression into which the head of the humerus could articulate.

The forelimb of pterosaurs, as in birds, has become specialized into a wing. However, unlike in birds, the humerus tends to be much more robust and is one of the shortest elements in the forelimb (Figs. 18.1, 18.2). This short but robust humerus is observed even in the more primitive rhamphorhynchoids such as *Rhamphorhynchus*. The short pterosaur humerus is all the more curious compared with birds, especially in large, soaring species, which tend to have a long humerus (Witton and Habib, 2010). This enlarged humerus would have had the effect of placing the major muscles that lift and depress the wings near the body core (Witton and Habib, 2010), thereby keeping the remainder of the wing light. The humerus has a saddle-shaped joint proximally that would allow it rotate, abduct, and adduct in several planes (Wilkinson, 2008), a necessary adaptation in flying animals. The deltopectoral crest into which the chest and shoulder muscles that control flight would insert is huge, providing more than adequate leverage for flight movements (Bennett, 2003; Wilkinson, 2008) (Figs. 18.1, 18.2). Moreover, the distal end of the humerus bears two separate and well-developed condyles for articulation with the radius and ulna. The radius and ulna are both elongate sticklike bones that articulate with a blocky wrist (Figs. 18.1, 18.2). A specialized wrist bone called the pteroid projects above the wrist, resembling a spindly finger (Figs. 18.1, 18.2). The function of this bone has been debated (Wilkinson, 2008), but it likely served important functions in modifying the wing surface during flight (see below).

The hand is highly modified in more primitive pterosaurs like *Rhamphorhynchus* as well as in *Pterodactylus* (Figs. 18.1, 18.2). The metacarpals and fingers of the first three digits are relatively gracile, and the fingers of the first three digits are fairly short with grappling hook–like claws that may have aided pterosaurs in climbing, hanging, or moving about when on the ground (Wellnhoffer, 1991). However, the fourth digit is massive by comparison: its metacarpal is robust as well as long, and it terminates in a well-developed grooved condyle. Articulating with the distal condyle of the fourth metacarpal is the first in a series of long finger bones that form much of the wing's leading edge (Wellnhoffer, 1991; Bennett, 2003; Wilkinson, 2008). In essence, the majority of the pterosaur wing is formed by the fourth digit and is shown to good effect in *Rhamphorhynchus* and *Pterodactylus* (Figs. 18.1, 18.2). It would be as if your thumb, index finger, and middle finger all remained the same size, but your ring finger was massive and stretched out at least as far away from your hand as the length of your forelimb. Not only would this look amazingly weird, but it would be difficult to put a ring on that!

The hind limbs of pterosaurs are typically shorter than the forelimbs. The femoral head is medially offset from the shaft, and the tibia and fibula are usually longer and more gracile than the femur (Wellnhoffer, 1991) (Figs. 18.1, 18.2). There has been considerable debate over the hind limb posture of pterosaurs, but it is clear that they had a mostly upright, parasagittal stance (Hone, 2012). In most pterosaurs the acetabulum into which the head of the femur articulates faces laterally and a bit caudally. The head of the femur is also more vertically oriented than in birds, and the femur shaft is relatively straight (Witton, 2013). Traditionally, pterosaurs were assumed to have had a less-than-upright hindlimb (Wellnhoffer, 1991). Previous work had indicated that the femur could not completely adduct against the pelvis because the acetabulum was oriented a bit dorsally, which would give pterosaurs a wider stance than birds (Wellnhoffer, 1991). However, current research and reinvestigation of the acetabulum and pelvic architecture in pterosaurs shows that the acetabulum was probably oriented laterally but faced caudally. Thus, when the vertically oriented head of the femur is articulated with the outward- and downward-facing hip socket, a parasagittal, upright posture is effected (Witton, 2013). The ankle of pterosaurs was hinge-like, an ornithodiran trait, and this shows as well that these archosaurs were certainly not sprawling (Kellner, 2004). Debate has persisted over pterosaur posture and locomotion on the ground (Unwin, 1999; Mazin et al., 2003; Mazin, Billon-Bruyat, and Padian, 2009; Padian, 2003; Witton, 2013). Recent analyses of known pterosaur footprints suggest that the limbs were held in a relatively narrow, parasagittal orientation, and that these animals were 'comfortable' walking quadrupedally on the ground (Witton, 2013). Moreover, biomechanical simulations of pterosaur locomotion on the ground show convincingly that the center of mass in these animals was positioned far forward of the pelvis, so that standing upright would have been difficult if not impossible (Palmer and Dyke, 2011).

If we did not have the spectacular and well-preserved fossils of several pterosaur species, it would be very difficult to figure out just how these archosaurs flew. Fortunately, numerous pterosaurs, including

Propatagium

Brachiopatagium

Uropatagium

18.4. The major portions of the pterosaur wing. See text for details.

Rhamphorhynchus and *Pterodactylus*, were preserved in silty or muddy sediments that not only preserved their skeletons in detail, but preserved impressions and traces of their soft tissues as well. Such fossils have revealed that pterosaur wings were composed of a main triangular leathery membrane of skin that stretched from behind the shoulders, ran along the entire length of the forelimb to the tip of the fourth (wing) finger, and then angled caudally to insert near the knee (dashed outlines in Figs. 18.1 and 18.2; Fig. 18.4) (Unwin, 2003; Palmer and Dyke, 2011; Hone, 2012). This main wing is technically called the brachiopatagium. An accessory wing or propatagium also stretched from the humerus to the fingerlike pteroid bone, and a small tail membrane (uropatagium) was also present (Fig. 18.4). Although the wing membranes of pterosaurs often traditionally were portrayed as flimsy structures as delicate as tracing paper, we now know that they were anything but fragile. Detailed microscopic examination of the preserved wing membranes of numerous pterosaur species shows that they were embedded with thousands of collagen fibers and an extensive layer of well-developed wing muscles. Intricate rivers of capillaries fed and oxygenated the wing membrane, and numerous tendon-like fibers reinforced and held its shape, particularly toward the edges (Kellner et al., 2009; Hone, 2012). All of these tissue layers were stacked into a wing membrane only a few millimeters thick! In other words, pterosaur wings were dynamic, versatile structures that, in combination with a large and complex brain, would have given these animals exquisite control while in the air (Fig. 18.3) (Unwin, 2003).

Several species of pterosaurs have been preserved such that hairlike structures, called pycnofibers, have been detected radiating across their bodies (Wellnhoffer, 1991; Kellner et al., 2009; Hone, 2012). One of the most famous fossils displaying these pycnofibers is the Russian pterosaur *Sordes pilosus* (essentially, 'hairy devil') (Wellnhoffer, 1991; Hone, 2012). Some pycnofibers on this pterosaur have turned out under rigorous investigation to have been displaced wing fibers, but there are also many genuine pycnofibers that covered the body from the back of the head to the base of the tail. Other fossils, such as *Jeholopterus* from China, show similar patterns (Kellner et al., 2009). Curiously, these bristle-like

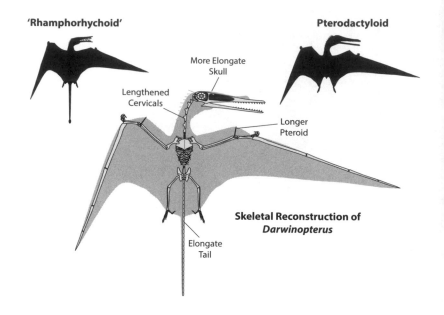

18.5. A transitional pterosaur, *Darwinopterus*. This pterosaur retains a mixture of rhamphorhynchoid and pterodactyloid features. Whereas *Darwinopterus* has a long neck and a more elongate skull, it still has primitive features such as an elongate tail. Based on figures in Lu et al. (2009).

'Rhamphorhychoid'

Pterodactyloid

More Elongate Skull

Lengthened Cervicals

Longer Pteroid

Skeletal Reconstruction of
Darwinopterus

Elongate Tail

pycnofibers may very well be a different version or anatomical structure homologous to feathers (Witmer, 2009), a topic we will address later in this chapter. What is remarkable is that most pterosaurs were probably covered in a bristly overcoat of pycnofibers (Kellner et al., 2009) that may have helped them regulate their body temperature and keep warm as they soared high in the air (Wellnhoffer, 1991; Witton, 2013).

Returning to the wing, just as we have discussed at length with birds, pterosaurs were probably able to actively adjust for a variety of typical flight situations. The propatagium that stretched between the pteroid wrist bone and the shoulder must have acted in similar ways to the fleshy patagium of birds (see chapter 16). In particular, it would have rounded out the cranial profile of the wing, and this would have enabled air to rush over the top of the wing faster than underneath, causing lift (Wellnhoffer, 1991; Palmer and Dyke, 2011). It also seems likely that the pteroid bone could be actively adjusted, and it may have allowed the propatagium to change its shape in various ways to prevent stalling and allowing for safe landings (Wilkinson, 2008).

A recently discovered pterosaur, *Darwinopterus*, sheds additional light on the evolutionary transition between the more primitive and long-tailed forms like *Rhamphorhynchus* and the derived and short-tailed forms such as *Pterodactylus*. *Darwinopterus* is another small pterosaur from the Middle Jurassic of China (~160 Ma; Card 97) with a wingspan of approximately 1 meter across and a skull measuring just barely 15 cm in length (Lu et al., 2009) (Fig. 18.5). The skull of *Darwinopterus* is remarkably like that of *Pterodactylus*: its nostril and antorbital fenestra have combined into a nasoantorbital opening, and a small crest graces its head from nose to braincase (Lu et al., 2009). The cervical vertebrae are also very pterodactyloid in being elongate and lacking ribs, suggesting

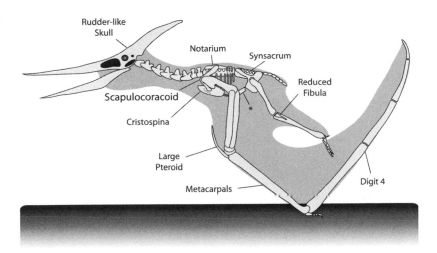

Rudder-like
Skull

Notarium

Synsacrum

Scapulocoracoid

Reduced
Fibula

Cristospina

Large
Pteroid

Metacarpals

Digit 4

18.6. The large pterodactyloid pterosaur, *Pteranodon*. In this figure, *Pteranodon* is shown launching itself into the air using its robust and powerful forelimbs, based on data from Habib (2008). This figure also shows to good effect the fused dorsal vertebrae (notarium) and fused pelvis and vertebrae (synsacrum) characteristic of many pterodactyloids. The figure itself is based on the artwork of Scott Hartman, copied by kind permission of the artist.

Darwinopterus had a much more flexible neck than typically encountered in more primitive pterosaurs like *Rhamphorhynchus* (Lu et al., 2009). Yet the body that follows the skull and neck of this pterosaur is very similar to *Rhamphorhynchus*, including a long, rudder-like tail stiffened by bony rods along most of its length (Lu et al., 2009). What the remarkable *Darwinopterus* fossils show is that the evolution of the pterodactyloid chassis took place in modular stages, and that changes in the skull and neck that increased head mobility came before trunk and tail shortening (Lu et al., 2009). The combination of a flexible neck and mobile head with a rudder-like tail suggests that *Darwinopterus* may have been an aerial predator, taking its meals on the wing and perhaps even preying on other pterosaurs (Lu et al., 2009).

Given that pterosaurs spent most of their lives airborne, their mode of reproduction has often been a source of speculation. Would they have laid eggs or could they have given live birth much like modern bats? Recent pterosaur fossils, including *Darwinopterus*, have resoundingly shown that pterosaurs were egg-layers (Hone, 2012). In fact, the eggs of several pterosaur species show that their shells were porous, in many cases leathery (although some were calcified like in birds, crocodylians, and lizards), and thin, all characteristics that suggest pterosaurs did not sit on their eggs to incubate them (Hone, 2012). Instead, pterosaurs probably covered their eggs in rotting vegetation, much as we have noted for crocodylians and some bird species (see chapter 16) (Hone, 2012). Embryos have been found associated with some of the eggs, and these pterosaurs-to-be had well-developed skeletons, suggesting that many of these flying archosaurs were precocial: ready to move about and fly shortly after hatching (Hone, 2012).

What is perhaps most remarkable about pterosaurs is their size range (Figs. 18.6, 18.7). Whereas it has been difficult enough to study and model the small and medium-sized pterosaur species (e.g., Palmer and Dyke, 2011), it has been incredibly challenging simply to understand how the largest pterosaurs became airborne (if they did at all) and made their

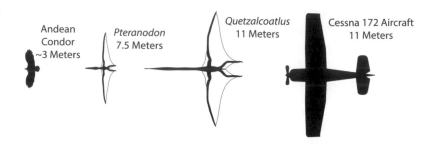

18.7. Size of the largest ptero-saurs. In this figure, the large pterodactyloids *Pteranodon* and *Quetzalcoatlus* are shown in scale in relation to one of the largest flying birds (the Andean condor) and a small aircraft, the Cessna 172. Bird and pterosaur silhouettes based on Witton (2010).

living. In *Pteranodon*, the most completely known large pterodactyloid (wing span: over 7 meters), the head is longer than the dorsal vertebral series, and a rudder-like crest extended almost as far caudally behind the skull as the jaws did in front (Fig. 18.6). Flight experiments with *Pteranodon* models suggest that the head may have played a significant role in steering and stabilizing the animal (Wellnhoffer, 1991; Roberts, Lind, and Chatterjee, 2011).

Although there seems to be little doubt that *Pteranodon* could fly, its much larger relative *Quetzalcoatlus* is another matter. How did such a huge animal get off the ground, stay aloft, and land without shattering its bones? As you might anticipate, debate has surrounded the mode of life and flight capabilities of this giant pterosaur. On the one hand, it has been hypothesized that *Quetzalcoatlus* was indeed capable of flight, and that it did so by vaulting itself into the air using its amazingly long but robust forearm and wrist (Habib, 2008; Witton and Habib, 2010) (see Fig. 18.6, showing *Pteranodon* launching itself). In this scenario, it is hypothesized that the chassis of *Quetzalcoatlus* indicates that this giant animal could still become airborne. In particular, it has been pointed out that the humerus of *Quetzalcoatlus* is incredibly robust and was supported internally by a lattice of light but tough bony struts (Witton and Habib, 2010). However, most recent biomechanical studies that take into account various mass estimates and the power required to launch this giant pterosaur into the air suggest that *Quetzalcoatlus* did not fly, at least not as an adult (Witton and Naish, 2008; Henderson, 2010). Imagine an animal with a 10 meter wingspan (the size of a Cessna airplane!) that would tower over a giraffe, and you begin to understand what flight would require. It is conceivable, perhaps, that as a juvenile the animal flew like other pterosaurs, but once it reached a critical adult size, virtual models indicate it was just too heavy to achieve flight (Henderson, 2010). Moreover, footprints from *Quetzalcoatlus*-sized pterosaurs show that these animals had padded feet, somewhat like a camel, which would have allowed them to move more effectively over the ground than other species (Witton and Naish, 2008). The acetabulum and femur shaft shape also suggest the hind limbs had a more upright posture than typical for other pterosaurs (Witton and Naish, 2008). Even the cervical vertebrae of pterosaurs like *Quetzalcoatlus* were relatively stiff, providing these animals with something more like a feeding boom and less like the flexible, gyroscope-like necks of birds and other, smaller pterosaurs. The large head and beak-like mouth of

Quetzalcoatlus and its kin also suggest ground-feeding habits similar to modern hornbill storks, birds that spend most of their time walking and snapping up small prey. Imagining an aircraft-sized *Quetzalcoatlus* strutting across the Cretaceous landscape and snatching 'small' prey off the ground (Witton and Naish, 2008) is at once arresting and chilling.

Pterosaurs flew their last at the close of the Cretaceous period (~66 Ma; Card 99). Yet, pterosaurs were the first of two archosaur groups that took to the air. Birds suffered extinctions but did not succumb to the end-Cretaceous die-off. Birds inherited their skeletal undergirding from a chassis of the most successful archosaurs and terrestrial vertebrates to have ever lived: the dinosaurs. It is to the dinosaurian chassis that we turn next.

Dinosaurs: The Original Ruling Reptiles

From chicken-sized predators and insectivores to herbivorous giants weighing over 20 metric tons to their modern feathery descendants, dinosaurs were and continue to be a vertebrate success story in every sense of the word. Entire entertainment and children's industries have been built on these amazing archosaurs, and nothing draws children to science faster than the mere utterance of their name. Dinosaurs are the very reason for my career and the writing of this book. It would simply be impossible to give a deserved nod to all the dinosaurs and the various branches of their family tree in this book. For our purposes, I will focus on a mere handful of examples among the major dinosaur lineages to give but the slightest taste of their skeletal diversity. Interested readers are strongly encouraged to immerse themselves in the nearly limitless volumes on the Dinosauria for more information and insight (e.g., Brett-Surman, Hotz, and Farlow, 2012).

As we have seen so often with other vertebrate groups, debate centers on the origins and branching patterns of the early dinosaur family tree. Most of our earliest dinosaur fossils can already be assigned to the two major lineages of this group (Benton, 2012; Fastovsky and Weishampel, 2012; Nesbitt et al., 2012), and this has made it extremely difficult to point to a particular fossil specimen and say, "Here is a basal dinosaur." Moreover, it has become more apparent that several radiations of 'almost-dinosaurs' called dinosauromorphs had a brief moment of diversification during the Triassic. For example, some recently discovered fossil dinosauromorphs, dubbed silesaurs, appear to have formed a brief but diverse and closely related group of animals that coexisted alongside the early dinosaurs (Nesbitt et al., 2010; Benton, 2012; Fastovsky and Weishampel, 2012). However, all of the earliest known dinosaur fossils share a number of features that shed light on what the overall body form and posture of these animals was. Without exception, all of the earliest dinosaurs were bipeds with short forelimbs, parasagittal hind limbs, a deep and fully perforated acetabulum, and stout tails. From such a bipedal and most likely predatory ancestor, all dinosaurs, both biped and quadruped, descended (Fastovsky and Weishampel, 2012).

Pedigree The Common Ancestors of Birds

Date of First Appearance ~235 Ma (Card 95)

Specialties of the Skeletal Chassis Fully Perforated Acetabulum

Eco-niche Predators, Omnivores, and Herbivores of the Land

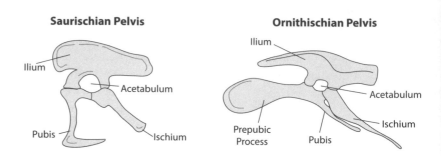

18.8. The pelves of saurischian and ornithischian dinosaurs. The saurischian pelvis is based on *Allosaurus* whereas the ornithischian pelvis is based on *Iguanodon*. See text for details. Based on illustrations in Fastovsky and Weishampel (2012).

Dinosaurs can be divided into two major lineages, called the Saurischia and the Ornithischia (Fig. 18.8). The 'ischia' portion of these names refers to the ischium bone in the pelvis, and it turns out that the shape of the pelvis, in combination with other trait states, helps paleontologists distinguish between these two groups (Fastovsky and Weishampel, 2012). Saurischian dinosaurs have a pelvis in which the pubis projects cranially and the ischium projects caudally, reminiscent of other reptilians and hence the 'sauria' portion of this group's name (Fig. 18.8). In contrast, ornithischian dinosaurs have a pelvis where both the pubis and ischium are rotated caudally as in birds, and thus the 'ornith' part of their name (Fig. 18.8). However, these names, coined long before birds were recognized as dinosaurs (Fastovsky and Weishampel, 2012), are not helpful in deciphering which branch of dinosaurs led to the birds. Confusing as it may seem, it is the Saurischian branch that led to the birds, whereas the Ornithischian branch ultimately went extinct. Thus, the caudally oriented pubis and ischium characteristic of birds evolved independently and in somewhat different ways in the saurischian ancestors of birds and the Ornithischia (Benton, 2012; Fastovsky and Weishampel, 2012). Yet, there are other traits that help us distinguish between the two dinosaur lineages. For Ornithischia, a unique lower jaw bone situated cranial to the dentary, called the predentary, formed a beak. In Saurischia, a grasping hand and long, flexible (and birdlike) cervical series distinguish them from ornithischians.

The only living descendants of the dinosaurs are the saurischians we call birds (see chapter 16), and an overwhelming amount of data points to these archosaurs having descended from the predatory or theropod dinosaurs (Holtz, 2012; Naish, 2012) (Plate 22). In fact, theropod dinosaurs in general are characterized by the possession of extremely hollow vertebrae and long bones (Fastovsky and Weishampel, 2012; Holtz, 2012). Mounting evidence from theropod and other saurischian skeletons very strongly suggests that their hollow vertebrae may have housed birdlike air sacs (Wedel, 2009; Yates, Wedel, and Bonnan, 2012). The presence of such air sacs would indicate a birdlike respiratory system, and it may indeed have been a primitive trait state of the saurischian lineage (Yates, Wedel, and Bonnan, 2012). Given this, we will first focus on the theropod dinosaurs because much of their skeletal anatomy is reminiscent of what we have seen in birds, but also in crocodylians. One can think of theropod

Early Theropod *Tawa* Skull in Lateral View

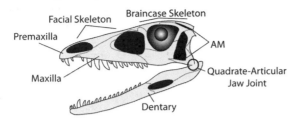

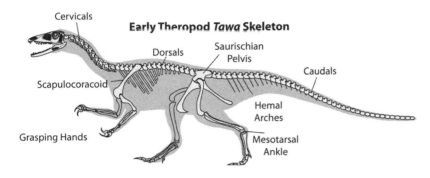

Early Theropod *Tawa* Skeleton

18.9. *Tawa,* an early theropod dinosaur. *Tawa* shows to good effect the major anatomical organization of both theropods in general and many early dinosaurs, including a bipedal posture, a digitigrade foot, and grasping hands. AM indicates the region and extent of the adductor mandibulae and related jaw-closing muscles. Figure based on illustrations in Nesbitt et al. (2009).

skeletons as a sort of hybrid chassis containing elements of crocodylians and birds.

The earliest theropods were medium-sized bipedal carnivores with relatively short forelimbs. No known theropods were quadrupeds, although new material and interpretations of the large Cretaceous predatory dinosaur *Spinosaurus* suggest it was an adept swimmer and may have been capable of quadrupedal locomotion (Ibrahim et al., 2014). For all other known theropods, the body mass was balanced over their hips by a long tail. Here, we start with the chassis of a primitive theropod whose characteristics carried over into all more derived predators and, in modified form, into the birds. The recent discovery of the primitive theropod *Tawa* shows the ancestral theropod chassis and has helped clarify the relationships of the predatory dinosaurs (Nesbitt et al., 2009). *Tawa* hails from the Late Triassic of New Mexico (215 Ma; Card 96) and was approximately 2 meters in length, standing just shy of 60 cm off the ground (Fig. 18.9). This small carnivore had a long snout filled with sharp, serrated and recurved teeth (premaxilla, maxilla, dentary), and overall its skull was shaped like an isosceles triangle with the nose as the apex and the braincase as the base. Although the teeth adorning the dentary are relatively small, many of the teeth protruding from the maxilla are quite pronounced, and the braincase region of the skull appears to have had space to house modest adductor mandibulae and pterygoid muscles. These variations in tooth size combined with its slender snout suggest *Tawa* was capable of feeding on small, fast-moving prey, whether insects or small vertebrates. The jaw joint was in line with the teeth, and the lower jaw is narrower than the upper jaw, allowing the teeth to slide past one another in scissors-like fashion.

A slender, birdlike neck comprised of 10 cervical vertebrae complements a relatively short but narrow trunk sporting delicate ribs. In *Tawa*, the sacrum is apparently composed of only 2 vertebrae, and an elongate caudal series of over 25 vertebrae stretches nearly a meter distal to the pelvis. A gracile scapulocoracoid with a birdlike, saddle-shaped glenoid would have received the hemispherical head of the humerus, and presumably would have allowed at least some rotation, adduction, and abduction of the forelimb. The humerus itself is very birdlike, having a pronounced deltopectoral crest that would have anchored shoulder and chest muscles for moving the forelimbs, and a sticklike radius and slightly bowed ulna lie parallel to one another beyond the elbow. The hand of *Tawa* is long, being about half the length of the forearm, and digits 1–3 bear stout, recurved claws for grasping. The pelvic girdle is characterized by a modestly expanded ilium, a slender pubis and ischium, and a large, open acetabulum. These features suggest that the knee extensors and hip abductors were quite well developed but that the femoral adductors were relatively small. This is not surprising given that *Tawa* and other dinosaurs already had a bone geometry that allowed them to stand with greatly adducted femora in a parasagittal stance. Thus, it was not mechanically necessary to utilize large femoral adductors to maintain this body posture. The femur is gently sinuous in lateral profile, and is slightly shorter than the tibia and fibula. The fourth trochanter is slender and located proximally, suggesting that the pull of the caudofemoralis muscles acted near the femoral head. If you recall from our discussion in chapter 6 on levers, forces applied close to the fulcrum of the lever increase the overall range of movement but require more force. Given the relatively small size and mass of *Tawa*, the pull (force) of the caudofemoralis muscles near the hip socket (fulcrum) would impart enough force to rapidly retract the femur. The proximal end of the tibia possesses a small cnemial crest for the insertion of the knee extensors, and the astragalus and calcaneum collectively form a distal roller-like surface that would have restricted ankle movements to hinge-like craniocaudal flexion and extension as in modern birds (chapter 16). The metatarsus was bunched and probably held vertically, increasing the overall length of the hind limb, and digits 1–4 were fairly large, with the central three (2–4) bearing the bulk of *Tawa*'s weight. Overall, *Tawa* shows a trend that continued through most of the evolutionary history of the theropod dinosaurs: the femur is typically shorter than the combined distal hind limb and foot. Given our pendulum and metronome analogies earlier in this book, these proportions indicate that *Tawa* and many theropod dinosaurs were cursorial or at least more than slow shufflers.

The basic chassis we have described for *Tawa* morphed in various lineages, yet still retained its core elements, including a bipedal stance (Farlow et al., 2000). Many theropods became larger animals (some over 12 meters in length), commonly in combination with reduced forelimbs, long hind limbs, and heavy tails (Paul, 1988; Holtz, 2012). Some became

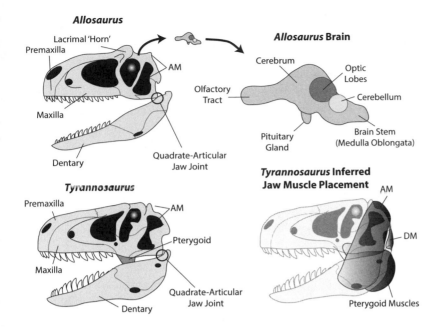

Allosaurus

Lacrimal 'Horn'
Premaxilla
AM
Olfactory Tract
Maxilla
Dentary
Quadrate-Articular Jaw Joint

Allosaurus Brain

Cerebrum
Optic Lobes
Cerebellum
Brain Stem (Medulla Oblongata)
Pituitary Gland

Tyrannosaurus

Premaxilla
AM
Pterygoid
Maxilla
Dentary
Quadrate-Articular Jaw Joint

Tyrannosaurus Inferred Jaw Muscle Placement

AM
DM
Pterygoid Muscles

18.10. Skull, brain, and jaw muscles of larger theropods. The skulls of two well-known theropods, *Allosaurus* and *Tyrannosaurus*. Note that these skulls are elaborations of the patterns seen in *Tawa* (Fig. 18.9), but that *Tyrannosaurus* has a very robust skull with much larger teeth than in *Allosaurus*. The brains of theropods, as exemplified here by *Allosaurus,* share many similarities with those of crocodylians. The jaw muscles of *Tyrannosaurus* were massive and probably delivered bone-crushing forces to its prey. AM indicates the location of the adductor mandibulae muscles whereas DM indicates the location of the jaw-opening depressor mandibulae. Skull reconstructions based on illustrations in Fastovsky and Weishampel (2012); brain diagram after Rogers (1998); and jaw muscles of *Tyrannosaurus* based on data and illustrations in Dilkes et al. (2012).

omnivorous, and a few even evolved into strictly herbivorous forms (Paul, 1988; Holtz, 2012; Novas et al., 2015).

One of the most common predators of the Late Jurassic (~150 Ma; Card 97), *Allosaurus*, shows to good effect many adaptations that occurred in larger theropods (Fig. 18.10). *Allosaurus* measured nearly 9 meters in length and was characterized by a hatchet-shaped, narrow skull, an S-curved neck, and shortened forelimbs boasting three large, grappling hook–like claws (Paul, 1988; Holtz, 2012). In the hind limbs, the femur remains shorter than the combined length of the tibia, fibula, and metatarsus, suggesting *Allosaurus* could at least on occasion adopt a more cursorial locomotor pattern. Like *Tawa's,* the skull of *Allosaurus* was birdlike and composed of multiple struts of bone; the premaxilla, maxilla, and dentary bones all bore recurved, serrated teeth; and the jaw joint was in line with the tooth row. In *Allosaurus*, as in many theropods, the lacrimal bones (recall that these are the bones associated with the tear ducts) have become enlarged and triangular, and may have served as display structures. Rayfield et al. (2001) demonstrated, using engineering software called finite element analysis (FEA), that the skull of *Allosaurus* was not robust but instead acted as a gracile but rapidly slashing knife. Calculations suggest that the bite force of *Allosaurus* was somewhere in the range of 2,000 N (compare this to great white sharks [>18,000 N] and American alligators [>9,000 N]; see chapters 6 and 16), which places its bite in the vicinity of wolves (Rayfield et al., 2001; Erickson, Lappin, and Vliet, 2003). However, it remains unclear whether the multijointed and strut-like skulls of *Allosaurus* and many other theropods were kinetic, and if so, how kinetic, given that we currently lack evidence for the presence of permissive linkages in their skull bones (Holliday and Witmer, 2008).

More recent mechanical analyses of neck posture and the attachment sites for head-flexing muscles in *Allosaurus* support a rapid downward biting and stripping motion, a feeding style similar to many raptorial birds (Snively et al., 2013).

With so many specimens of *Allosaurus* available, we can again peer inside the skull and get a look at the brain. What we find is a brain remarkably like that of an alligator (see chapter 16) (Rogers, 1998) (Fig. 18.10). The cerebrum is somewhat expanded and has a dorsal ventricular ridge (DVR) associated with the dorsal cerebrum and its integration of somatosensory information. Once again, there is a prominent olfactory tract for the detection of odors, and the optic lobes and cerebellum are modestly developed for visual and motor coordination. However, the overall impression of the inferred brain structure from *Allosaurus* is something much more like a crocodylian than a bird (Rogers, 1998). One gets the impression that most predatory dinosaurs were attuned predators but certainly not mathletes.

Perhaps the most famous theropod was *Tyrannosaurus*, an apex predator from the Late Cretaceous (~70–66 Ma; Card 99) of North America that was nearly 15 meters in length (Fastovsky and Weishampel, 2012; Holtz, 2012) (Fig. 18.9). The skull of *Tyrannosaurus rex* was one of the most robust of any theropod, and the braincase had large regions of attachment for adductor mandibulae and pterygoid muscles (Holliday and Witmer, 2007; Holtz, 2012). Such a skull must have generated an amazing amount of force. To be more exact, try a bite force in the range of 35,000 to 57,000 N (Bates and Falkingham, 2012)! Compare this to the calculated bite forces for a great white shark (see chapter 6, >18,000 N) and the American alligator (see chapter 16, >9,000 N), and the bite of *Tyrannosaurus* seems unworldly and terrifying. Even the teeth of *T. rex* are different from those of most other theropods, being thick rather than blade-like, and more reminiscent of crocodylian teeth or even those of hyenas, both vertebrates that specialize in crushing bone (Holtz, 2012). The snout of *T. rex* is narrow and the orbits were turned more cranially than in most theropods, suggesting that this predator had a good degree of binocular vision with which to judge the distance to its prey (Holtz, 2012). CT-scans of the well-preserved braincase of the *T. rex* known as 'Sue' show a brain similar to *Allosaurus* but with greatly expanded olfactory bulbs (Brochu, 2000; Witmer and Ridgely, 2009). Given the massive jaws and well-developed sense of smell, it would not be too far from the truth to say *Tyrannosaurus rex* was, to paraphrase Greg Paul, a land shark (Paul, 1988). In *Tyrannosaurus*, the cervical vertebrae are robust, the forelimbs have become small (about the size of your own arms) with but two digits, and the hind limbs are massive. What *T. rex* was doing with those small forelimbs is still unclear, but it seems clear that this theropod and its relatives dispatched prey with their massive jaws (Holtz, 2012). As with many predatory dinosaurs, the femur remains somewhat shorter than the combined tibia, fibula, and metatarsus, but given the size and estimated mass for *T. rex* (4–8 metric tons) (Erickson et al., 2004), this

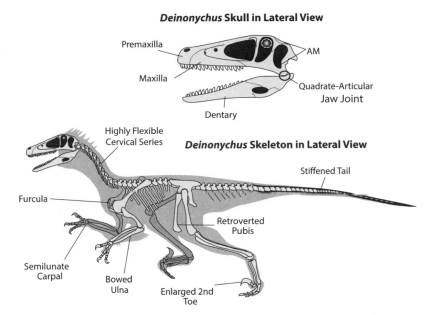

Deinonychus Skull in Lateral View

Premaxilla

AM

Maxilla

Quadrate-Articular Jaw Joint

Dentary

Highly Flexible Cervical Series

Deinonychus Skeleton in Lateral View

Stiffened Tail

Furcula

Retroverted Pubis

Semilunate Carpal

Bowed Ulna

Enlarged 2nd Toe

18.11. The birdlike theropod *Deinonychus.* Discovered in the 1960s by paleontologist John Ostrom, this theropod helped revitalize dinosaur science. Among the birdlike features of this predator are the presence of a furcula, a semilunate carpal, a bowed ulna, and a retroverted pubis. Note the enlarged second toe with a sickle-like claw that was probably used for dismembering and stabilizing prey, and the stiffened tail that may have been used for balance. *Deinonychus* reconstruction and skull based on illustrations in Fastovsky and Weishampel (2012) and Holtz (2012).

has led to controversy over its ability to pursue prey at high speed. In fact, Hutchinson and Garcia (2002) have shown with sophisticated computer models that the hind limbs of *T. rex* would have required abnormally gigantic muscles in order to run at high speeds. Overall, *Tyrannosaurus* would have been a formidable and terrifying predator.

That theropod dinosaurs gave rise to birds is now nearly universally accepted, and it was the discovery of a wolf-sized biped called *Deinonychus* that finally pointed the way (Fig. 18.11 and Plate 23). *Deinonychus* was discovered by Yale paleontologist John Ostrom in Lower Cretaceous rocks of Montana in the 1960s, and the chassis of this theropod challenged the then-accepted model of dinosaurs as sluggish, lizard-like evolutionary dead ends (Wilford, 1985). In some ways, *Deinonychus*, which measured over 3 meters from head to tail, has the typical theropod chassis already discussed. However, three skeletal features stand out as distinct from other theropods we have discussed: (1) possession of a semilunate carpal (wrist) bone; (2) the presence of an enlarged, recurved, and sharp toe claw held aloft on digit 2 of the foot; and (3) a tail stiffened distally by elongated bony struts (Fastovsky and Weishampel, 2012; Holtz, 2012) (Fig. 18.11). You may recall from chapter 16 that birds possess a semilunate carpal, and that this bone allows the hand to pivot against the forearm when the wing is folded against the body. The presence of this wrist element in *Deinonychus* shows that this theropod could pivot its hand in a similar way to modern birds, but in this case, the hands are long and bear grappling hook–like claws. Studies of forelimb movements in *Deinonychus* by one of my friends, Phil Senter, has shown convincingly that the birdlike movements of this predator would snatch and deliver prey to the animal's chest in a rather violent bear hug (Senter, 2006). The enlarged second toe claw could be held aloft, somewhat reminiscent of a cat retracting its claws. The retracted claws of cats stay sharp because they don't touch the ground and come out only when it is time for prey (or, unfortunately for most cat owners, time to play) (Page, 2008). This large and recurved nail likely served several prey-subduing purposes, including disemboweling prey or holding it securely in place while the *Deinonychus*'s head ripped flesh from bone (Fowler et al., 2011). Ostrom also recognized that the long, distally stiffened tail of *Deinonychus* would have served as a stabilizer and rudder, allowing the animal to make hairpin turns in pursuit of its prey (Paul, 1988; Fastovsky and Weishampel, 2012).

Not only was *Deinonychus* an apparently dynamic predator, but Ostrom began to realize there were striking skeletal similarities between his newly discovered predator and one of the most famous fossils ever discovered: *Archaeopteryx*, the so-called first bird (Wilford, 1985) (compare Plate 22 and Plate 23). When the first fossil of *Archaeopteryx* was uncovered (currently 11 body fossils are known to science [Foth et al., 2014]) in the Jurassic (~150 Ma; Card 97) Solnhofen limestone quarry of Germany in the 1860s, it was immediately clear that this animal was something special. At first glance, it appeared to be a small crow-sized bipedal reptile having a delicate, toothy skull, long arms with clawed hands, and an elongate tail (Fastovsky and Weishampel, 2012) (Fig. 18.12). In fact, the famous evolutionary biologist Thomas Henry Huxley supposed it might be a small dinosaur (Fastovsky and Weishampel, 2012). But what made *Archaeopteryx* special was that nearly every body fossil discovered in the fine-grained limestone of the Solnhofen quarry preserved detailed impressions of feathers. Not just any filamentous trace, either, but long, asymmetrical feathers nearly identical to the flight feathers of modern birds, which attached deeply into its ulna and radiated into a wing from its arms. Moreover, impressions of feathers were discovered along the tail and much of the body, and a newly described specimen even shows feathers on the hind limbs (Foth et al., 2014). For nearly 100 years, however, the origins of *Archaeopteryx* and birds were disputed. *Archaeopteryx* was certainly recognized as an archosaur because it had an antorbital fenestra in its skull and a hinge-like ankle joint. Many aspects of the skeleton of *Archaeopteryx* looked dinosaurian, but it was unclear whether the similarities were due to convergent evolution or common descent. Without a lot of small, well-documented dinosaur fossils over this time period (1860–1960), it was difficult to know if the theropod-like structure of the *Archaeopteryx* skeleton was not simply a case of the archosaur chassis converging on a bipedal posture (Wilford, 1985).

So when Ostrom went on to compare his *Deinonychus* with the available specimens of *Archaeopteryx*, it was a revelation. If you ignored the feathers, you had in *Archaeopteryx* what was essentially a smaller version of *Deinonychus*! *Deinonychus* and *Archaeopteryx* shared all these characteristics: a fully erect stance made possible by a deep, perforated acetabulum; a special, hinge-like ankle joint with roller-like surfaces; hollow bones; a furcula; a highly flexible neck; a semilunate carpal; an ulna that bows laterally; and a retroverted pubis (Figs. 18.11, 18.12). If

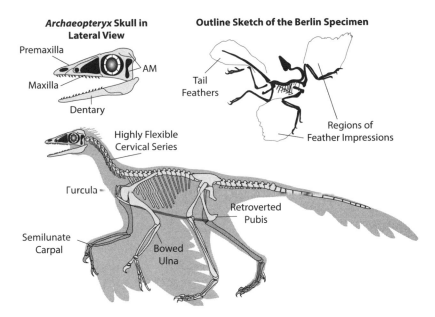

Archaeopteryx Skull in Lateral View

Premaxilla
Maxilla
Dentary
AM

Outline Sketch of the Berlin Specimen

Tail Feathers
Regions of Feather Impressions

Highly Flexible Cervical Series
Furcula
Retroverted Pubis
Semilunate Carpal
Bowed Ulna

18.12. The 'first bird,' *Archaeopteryx*. Note the very theropod-like chassis of *Archaeopteryx*, especially when compared to *Deinonychus* (Fig. 18.11). The famous Berlin specimen is shown as an outline drawing, indicating where the feather impressions were found along the arms and tail of *Archaeopteryx*. Note the elongate arms and clawed hands. AM indicates the region of the jaw-closing muscles. Illustrations based on figures in Fastovsky and Weishampel (2012).

you will recall from chapter 16, this particular combination of skeletal hallmarks is found only in birds. The chances that the skeletal chassis of *Archaeopteryx* and *Deinonychus* were this similar to one another due simply to convergent evolution appeared very remote. The simplest answer was instead that *Archaeopteryx* and *Deinonychus* shared all of this specialized anatomy because they shared a close, common ancestor. If this was the case, then it followed that birds were direct descendants of theropod dinosaurs!

The hypothesis that birds are dinosaurs has been repeatedly tested and bolstered by numerous studies and a plethora of new fossils. Beginning in the 1980s, with the increased use of desktop computers, it was possible to more easily test and scan trait state data to determine the pedigrees of numerous vertebrate groups. When Jacques Gauthier ran the first robust trait state analysis of Saurischian dinosaurs, birds were shown to have branched off theropod dinosaurs (Gauthier, 1986). It quickly became apparent that theropods such as the coyote-sized cousin of *Deinonychus* from Mongolia, *Velociraptor*, and other so-called raptor predators, were within striking distance of being birds themselves. It should be noted that Michael Crichton's 'Velociraptor' in *Jurassic Park* was actually an enlarged *Deinonychus*; Crichton followed the suggestion by Paul (1988) that the genus *Deinonychus* was a junior synonym of *Velociraptor*. It should also be noted that watching any dinosaur movie with a paleontologist is the surest way to kill what laypeople call 'fun.' Moreover, beginning in the 1990s and continuing right up to the present, dinosaur fossil after dinosaur fossil has been found with traces, imprints, and other unquestionable signs of feathers, even in animals that would have been too big to fly (Fastovsky and Weishampel, 2012).

Perhaps most significant have been the embryonic studies that have recently illuminated the development and evolutionary origin of feathers.

Classically, it was hypothesized that feathers were derived from long scales that become more and more finely segmented. Thanks to work by a number of avian biologists, we now know that this is patently false. Feathers are indeed a modified form of keratin-based tissue like hair and scales, and they arise like hair from a follicle, a small, specialized ball-shaped cluster of cells with a hollow center. The importance of a follicle is that this group of cells can actively grow and replace hair and feathers, allowing shedding and molting to occur (Prum and Dyck, 2003; Prum, 2005). It has been shown through careful study of numerous bird embryos that feathers, unlike hair, arise from a follicle as a cylindrical tube (Prum and Dyck, 2003; Prum, 2005). In fact, Richard Prum has described feather development and structure as "totally tubular" (Prum, 2005:571). It is the tubular nature of the feather that gives it its diversity of structure and function. If you played with clay as a child, then you probably remember an extruding device where you fed the clay in one end and it was pushed through a series of closely spaced pores on the other. The end result was the conversion of a large, cylindrical mass of clay into multiple clay 'worms' that radiated out from the center. Although feathers are not clay, they begin as a simple tube (rather than solid) but then divide in a circular pattern along their main axis (the rachis), giving rise to multiple branches or barbs (Prum and Dyck, 2003; Prum, 2005). The barbs, themselves tubular, can divide again, producing the barbules that help 'Velcro' the feather vane together (Prum and Dyck, 2003; Prum, 2005) (see chapter 16 for feather structure).

From the work of Prum and Dyck (2003), Prum (2005), and others, we now also know that feather development proceeds in three developmental stages. In stage 1, there is the development of the hollow tubelike feather from a flat, scalelike precursor, and this change occurs with the modified expression of particular developmental genes. In stage 2, the tubular feather divides into multiple branches (barbs), and in stage 3, there is the development of flight feathers by the 'encouragement' of barb branching on the ventral feather axis and its 'extinction' on the dorsal axis via developmental genes (Prum, 2005). In essence, the asymmetrical flight feather we have discussed in chapter 16 is formed. What is astounding about this particular sequence of developmental events is that feather evolution has been shown to proceed in the same way! The most primitive theropod fossils with feathers, for example, have only hollow filaments, whereas theropods closer and closer to the bird line have either examples of the simple branching barbs or full-on flight feathers (Witmer, 2009). More mind-boggling is that fossils of ornithischian dinosaurs, the line not directly ancestral to birds, also shows evidence of hollow filaments and simple branching barbs, but none shows full-on modern flight feathers (Witmer, 2009). Having the ability to produce feathers must have originated in the common ancestor of all dinosaurs, and perhaps the ability to produce such structures has an even deeper origin, considering that pterosaurs produce those lovely pycnofibers (Witmer, 2009). If you come away from this discussion with nothing else, I hope to have convinced you that the best available data from biology, paleontology, and embryology point like a giant, neon arrow toward the modern chassis of birds over which large, bold letters spell out D I N O S A U R.

The early radiations of birds included the common ancestors of the neornithines we discussed in chapter 16, as well as several lineages that went extinct by the end of the Cretaceous period 66 Ma (Card 99). Trends witnessed during the evolution of birds in the Mesozoic include the reduction of the tail to the pygostyle (see chapter 16) as well as an increasing diversity of body types adapted for water and land as well as the air (Naish, 2012). Alongside the neorinthines were the enantiornithines, a lineage that produced numerous bird species of varying shapes and sizes, but one that sadly went extinct at the close of the Mesozoic (Naish, 2012). In the end, for reasons still not quite clear, only the neornithine birds managed to survive past the end of the Cretaceous (Naish, 2012).

We end this chapter by all too briefly summarizing trends in the other major groups of nonavian dinosaurs that no longer share our planet. Among the saurischian dinosaurs, one lineage, the Sauropodomorpha, gave rise to the largest terrestrial herbivores that ever existed. Early bipedal members of this lineage were already fairly large (~5–10 meters in length) and possibly had an edge over other herbivorous competitors due to their elongate necks, which may have allowed them to reach vegetation out of the reach of their contemporaries (Yates, 2012). As my friend Heinrich Mallison has shown with computer models, the mass distribution of early sauropodomorphs would have precluded a quadrupedal posture (Mallison, 2010). Moreover, Phil Senter and I have shown that the geometry of the forelimb bones in some early sauropodomorphs would have physically prevented them from walking on all fours because their hands could not be rotated to face forward (Bonnan and Senter, 2007). I have been privileged to have helped unearth and describe *Aardonyx*, a sauropodomorph dinosaur from the Early Jurassic of South Africa that shows signs of transition from a biped to a quadruped (Yates et al., 2009). Nearly 7 meters long, *Aardonyx* had a wide mouth (presumably for taking in lots of vegetables) and an interlocking radius and ulna found elsewhere only in sauropods and their direct ancestors (Yates et al., 2009). The significance of these findings is that the ancestors of sauropods were becoming larger bulk browsers, which required a steady forelimb to support their increasing mass (Yates et al., 2009).

From rather humble bipedal beginnings arose the truly giant quadrupedal sauropods, the largest of which reached nearly 30 meters from head to tail and weighed in at 10–30 metric tons or more (Fig. 18.13 and Plate 23) (Lacovara et al., 2014; Bates et al., 2015). Nothing like these animals has since aspired to such sizes and strangeness on land. In typical sauropods, the skull was relatively small, comprising less than 5% of the animal's body length. Very little room was available for the brain; in fact sauropods had one of the smallest brains of all dinosaurs, theirs being only 20% the relative size of an alligator's (Fastovsky and Weishampel, 2012). Pencil- or spoon-shaped teeth lacking any denticles or serrations were

18.13. The gigantic sauropod dinosaurs. For such huge animals, sauropods had relatively tiny heads, with peg- or spoon-like teeth that presumably allowed them to rake in extraordinary amounts of vegetation. Their necks were elongate and flexible, and much of their vertebral column was hollow and pneumatic, possibly being filled by air sacs as in birds. The limb bones were solid to support their massive weight. Their hands were typically digitigrade and semitubular, whereas their feet were semiplantigrade and cushioned by a large fat pad. Skulls and skeleton of *Camarasaurus* and *Diplodocus* based on illustrations in McIntosh, Brett-Surman, and Farlow (1999) and Fastovsky and Weishampel (2012).

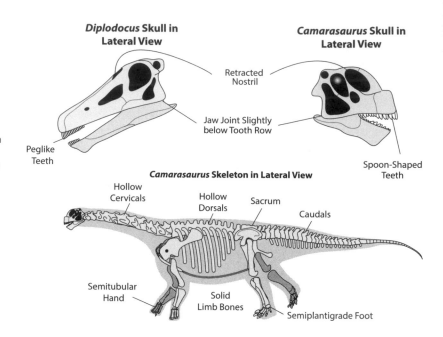

present to snap, tear, and pluck vegetation from tall trees or to sweep over fern prairies, the head gulping down vegetation like a reptilian vacuum cleaner. Recent studies of sauropod teeth even suggest that this ability to efficiently snap off and consume vegetation quickly relied on a steady supply of fresh, unworn teeth, with replacement rates ranging from about one to two months (D'Emic et al., 2013). In all known sauropods, the nostrils were retracted toward the eyes, forming either a large, dome-like opening or smaller cavities flush with the skull. What such retracted nostrils were used for in sauropods remains difficult to determine. However, it has been demonstrated that most dinosaurs, including sauropods, probably had a rostral (that is, forward-directed) fleshy nostril position relative to the bony nasal openings (Witmer, 2001).

The cervical vertebrae numbered from as few as 12 to as many as 19. The cervical vertebrae of most sauropods were incredibly pneumatic, a pattern that matches that of birds (see chapter 16) and suggests the presence of large air-sacs (Yates, Wedel, and Bonnan, 2012). This would have been a major weight-saving feature of the sauropod chassis. The dorsal vertebrae were likewise hollowed-out, but possessed numerous additional joints that inhibited their independent movement, providing sauropods with a light but rigid body core. As we saw for birds (chapter 16), such reduction in dorsal vertebral movement may have further decreased body mass by diminishing the need for large back support muscles. The sacrum of most sauropods was solid and well fused into the pelvic girdle, and an elongate tail composed of dozens of vertebrae contributed to nearly half the length of these giant herbivores (Fig. 18.13).

The long necks and retracted nostrils of sauropods suggested to earlier paleontologists that these behemoths spent their lives armpit deep in stagnant swamp water, moving into deep lakes and snorkeling safely away from predators when the need arose. But the math and physics of this setup just don't work out. If you've ever gone snorkeling, then you know that your air tube is typically short, less than a foot in length. Now, imagine if you wanted to dive deeper and your solution was to increase the length of your snorkel. The problem is, water is dense and the deeper you dive the faster that pressure increases. At the surface of the water, your lungs experience 1 atmosphere of pressure or approximately 101 kPa. But dive beneath the surface and swim down just 10 meters, and you have now doubled the pressure on your lungs. On top of this imagine breathing through a small-bore 10 meter pipe: the air pressure at the surface is half of what it is at 10 meters — this means your lungs would have to work very hard just to pull air down that pipe! This is why SCUBA divers use pressurized tanks — to create a force similar to the water pressure outside the diver's lungs, which enables their lungs to function as if they were on the surface. Now, imagine an animal with a chest nearly 6–10 meters below sea level, and you can see that, unless sauropods had gigantic, unworldly chest muscles, their lungs would collapse (Fastovsky and Weishampel, 2012). Another stake in the heart to swamp-loving sauropods came from computer models by Donald Henderson that showed that with their long, air-filled necks and bodies, sauropods would bob like corks on the surface and be liable to tipping over (Henderson, 2004).

The forelimbs and hind limbs of sauropod dinosaurs show conclusively that these were animals in the slow lane (Fig. 18.13). The humerus and femur are always significantly longer than their distal elements, and the hands and feet of sauropods are relatively short (Upchurch, Barrett, and Dodson, 2004). Based on our pendulum and metronome analogies, this is very slow. In fact, my own research and that of my friend Ray Wilhite on sauropod limbs has shown that, if anything, adult sauropods had the same limb proportions as juveniles (Bonnan, 2004, 2007; Tidwell and Wilhite, 2005). Basically, animals that started with slow proportions as juveniles did nothing to change those as adults. Now, add more weight and you have a formula for even relatively slower adults. The only 'saving grace' might have been that when you have a limb 4.5 meters in length, small movements at the shoulder or hip translate to long strides (Bonnan, 2007). In cross-section, the long bones of sauropods are solid with no hollow spaces: this indicates that resisting compressive forces was of utmost importance to these herbivores (Upchurch, Barrett, and Dodson, 2004). The sauropod hand consisted essentially of metacarpals with tiny phalanges tacked on as an afterthought. In fact, some sauropods such as titanosaurs show trends toward eliminating the need for pesky fingers altogether (Upchurch, Barrett, and Dodson, 2004). In a typical sauropod, the metacarpals were arranged vertically and collectively formed a nearly tubular shape that may have acted to resist literally tons of compression (Bonnan, 2003). Many but not all sauropods possessed a thumb claw that may have served multiple purposes, including traction. The feet were less vertical and the metacarpals were spread over a fleshy heel (we know this from sauropod footprints) presumably to dissipate stress, much as occurs

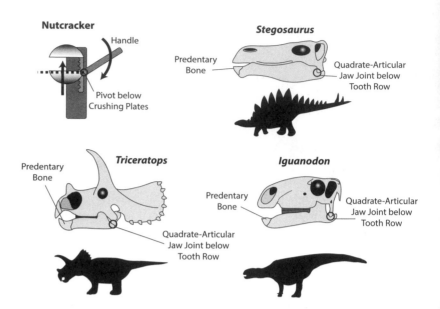

18.14. Selected ornithischian dinosaur skulls and silhouettes. All known ornithischians were herbivorous, and their jaw joint was set below their teeth row, enabling their jaws to act like a nutcracker. Beyond their pelvis, all ornithischians are characterized by having a beak-like predentary bone in the lower jaw. Skull diagrams and silhouettes based on illustrations in Fastovsky and Weishampel (2012).

in the feet of modern elephants (Bonnan, 2005). Unlike elephants, three to four large, dull, but recurved claws projected laterally from the feet, and perhaps aided in traction over variable terrain (Bonnan, 2005).

The ornithischian dinosaurs (Figs. 18.14, 18.15 and Plate 23) were a successful and diverse group of (as far as we know) herbivores that span the entire history of the Dinosauria from 235 to 66 Ma (Cards 95–99). Early forms were small bipeds, but eventually three major lineages arose from these common ancestors, many of which reverted to a quadrupedal posture. One lineage, called the thyreophorans, produced medium to large quadrupedal herbivores with a variety of plates, spikes, and other forms of armor. One of the classic thyreophorans is *Stegosaurus* from the Late Jurassic (160–145 Ma; Card 97) (Fig. 18.14). This iconic dinosaur sported rows of 'plates' (in reality, tall and pointy scutes) that had multiple channels for blood vessels, which may have helped regulate the body temperature of *Stegosaurus* much as it does for alligators (Farlow, Hayashi, and Tattersall, 2010). Notably, the short forelimbs, tall hind limbs, and rearward center of mass in *Stegosaurus* combine to make a slow-moving, unwieldy animal: had the hind limbs moved at the same pace as the forelimbs, this spiky herbivore would have stepped on its own hands (Fastovsky and Weishampel, 2012). Other thyreophorans included the Cretaceous radiations of the Ankylosauria, tanklike dinosaurs covered literally from head to tail with thick armor scutes (Benton, 2005; Fastovsky and Weishampel, 2012).

Another ornithischian lineage, called the Marginocephalians, gave rise to dome-headed bipeds, called pachycephalosaurs, which may have cracked their skulls together like bighorn rams (Benton, 2005; Peterson and Vittore, 2012). Another branch of the Marginocephalia was the diverse array of horned dinosaurs, of which the most familiar is *Triceratops*. The horned dinosaurs, or ceratopsians, ranged in size from dog-sized

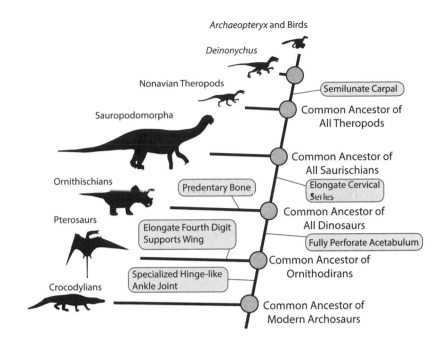

Archaeopteryx and Birds

Deinonychus

Nonavian Theropods

Semilunate Carpal

Common Ancestor of
All Theropods

Sauropodomorpha

Common Ancestor of
All Saurischians

Ornithischians

Predentary Bone

Elongate Cervical
Series

Common Ancestor of
All Dinosaurs

Pterosaurs

Elongate Fourth Digit
Supports Wing

Fully Perforate Acetabulum

Common Ancestor of
Ornithodirans

Specialized Hinge-like
Ankle Joint

Crocodylians

Common Ancestor of
Modern Archosaurs

18.15. Pedigree of relationships among the ornithodiran archosaurs. See text for details.

herbivores to elephant-sized fortresses. Typically, the head of ceratopsians was very large and heavy, with a greatly expanded 'head shield' comprised of the squamosal and other braincase bones that may have served numerous purposes, including jaw muscle expansion and display (Fastovsky and Weishampel, 2012) (Fig. 18.14). A variety of horns typically protruded from the nose and face, which probably served both as protection against predators and for intraspecific combat (Fastovsky and Weishampel, 2012). We know a great deal about the growth and 'herding' behavior of many ceratopsians because they are often found in large skeletal accumulations called bone beds (Brinkman, Eberth, and Currie, 2007).

Yet another diverse ornithischian lineage was the Ornithopoda, a group of mostly bipedal herbivores, some of which are known to the general public as the 'duck-billed' dinosaurs because some of them had flattened snouts (Fig. 18.14). Many ornithopods have been found in large bone beds associated with nests, eggs, and juveniles, suggesting that these dinosaurs gathered in large herds and actively raised their young (Horner and Gorman, 1990), a hypothesis bolstered by abundant ornithopod trackway data (Martin, 2014). Based on their limb proportions, many ornithopods remained functionally bipedal, but others may have favored a quadrupedal posture when moving slowly or standing still (Fastovsky and Weishampel, 2012). Some ornithopods were themselves giants, with a few approaching the size of the smaller sauropods (e.g., Ji et al., 2011).

All the ornithischian dinosaurs, the nonavian saurischian dinosaurs, and the pterosaurs took their last curtain call at the end of the Cretaceous period some 66 Ma (Card 99) (Fastovsky and Weishampel, 2012). Among the archosaurs, only the crocodylians and birds made it through to the present. The end-Cretaceous extinction event was complex, but geological data clearly show that environmental upheaval correlated with a large asteroid impact, extensive global volcanism, and the draining of interior seaways combined to break the long-held dominance of the dinosaurs and other archosaurs (Archibald, 2012). A large hole was left behind in many of the eco-niches once inhabited by dinosaurs and their allies. Although some of these 'empty spaces' were filled by birds and crocodylians, it was primarily another group of vertebrates, the mammals, that radiated into these new job openings. Unlike other vertebrate chassis described previously, the mammal chassis is at once familiar and strange, and it is to our own ancestors and other furry friends that we turn next.

As a coda, during the Cretaceous period (145–66 Ma; Cards, 97–99) the first major wave of flowering plants, the angiosperms, made an appearance. In particular, the common ancestor of all modern flowering plants can probably trace its ancestry back to at least 135 Ma (Card 98) (Magallón et al., 2015) and perhaps even earlier, to the Jurassic (162 Ma; Card 97) (Liu and Wang, 2015). Based on fossils and what we know about primitive living members of the angiosperms, the earliest flowering plants were likely small, sprawling understory shrubs dotted with tiny flowers (*Amborella* Genome Project et al., 2013; Liu and Wang, 2015). Dinosaurs coexisted with flowering plants, and the two groups may have influenced each other's evolution in various ways (Bakker, 1986; Barrett and Willis, 2001; Fastovsky and Weishampel, 2012). We know that modern angiosperms are excellent at using bribes of fruit, nectar, or even intoxicants to entice invertebrates and vertebrates alike to pollinate them (Pollan, 2002). A diversity of pollination strategies may partly explain why, at the close of the Cretaceous (66 Ma; Card 99), the angiosperms had become the dominant terrestrial plants. Their subsequent evolution and diversification during the so-called Age of Mammals (66–0 Ma; Cards 99–100) would continue to influence the evolution of the vertebrates, including the mammals. Later in the book, we will return to the evolution of grass (a flowering plant) and its effect on mammalian herbivores, as well as how the spread of grasslands in eastern Africa may have affected human evolution.

Overcome by Fur: The Mammalian Chassis

7

Glass of milk, sitting in between extinction in the cold and explosive radiating growth.

THEY MIGHT BE GIANTS, *Mammal*

19.1. The male African elephant (*Loxodonta africana*) that nearly took my life . . . sort of.

The Mammalian Chassis: A Primer

THE WHO ONCE SANG, "MEET THE NEW BOSS, SAME AS THE OLD boss," in their classic song, "Won't Get Fooled Again." Unfortunately, it seems that power is a revolving door, and many of those in charge follow in the corrupt footsteps of those before them. That line also resonates throughout vertebrate history: as we have seen, there are only so many ways the skeletal chassis can be adapted for speed, flight, or swimming. Nature abhors a vacuum, and when the nonavian dinosaurs died out, another group rose to take their place: the mammals. Whereas we will see many ways in which the chassis of the mammal skeleton is different from those of other vertebrates we have previously encountered, there are remarkable convergences between mammals and dinosaurs.

I was starkly reminded of this fact during my first visit to South Africa in 2004. My colleague and friend, Adam Yates, and I were finished with five weeks of digging for Early Jurassic dinosaurs in the Free State province, and there were two or three days left before I departed back to the United States. He suggested we visit Pilanesberg National Park in the North West Province of the country. Pilanesberg is one of those game reserves you see on TV where people drive in to a wildlife 'safari': African wildlife abounds out in the open, and you are best advised to remain in your car. Within the first few minutes of entering the park, we encountered some water buffalo several feet from our vehicle knocking heads and pushing each other about, and in the next few miles we saw a group of majestic giraffes cross the road.

Keep in mind we were not on safari with tourists in one of those tanklike jeeps. Being brilliant Ph.D.s, we entered the park in a rusty pink two-door Nissan mostly held together with good feelings and wishes. Eventually, we ended up away from most of the tourists on a dirt road in a more remote corner of the park. And that's when we spied it: a beautiful bull African elephant loping down the road in the distance toward our car. In my excitement, I rolled down the passenger window, leaned out, and snapped several pictures of this incredible sight (Fig. 19.1). I even had a chance to take a short digital movie. But then I felt the car lurch to a stop and begin driving in reverse. Adam was agitated and swearing loudly, all the while increasing our backward speed. I couldn't quite understand what the problem was until I looked back up at the elephant and realized it was gaining on us. The elephant was no longer loping – it was damn near close to running, its long legs hitting the ground in a fast pace and getting uncomfortably close! This giant mammal found us either interesting or annoying, but whatever was going on, it was closing in on us.

I jerked myself back into the passenger seat as Adam tried to find some way to quickly turn the car around, but in our panic our back wheels slipped off the road and we jammed our car in a ditch. We were stuck, and the elephant was coming right at the car. The only thing I could think to do was to roll up the window on my side. My thought process was that this might stop the elephant from reaching in to get me with its trunk, a very laughable action in retrospect, seeing as the elephant had to weigh at least twice as much as our car! And so Adam and I held our collective breaths as this enormous mammal came within mere feet of our car. At that point, all I could see out the windows were the large, gray tree trunk legs that pivoted effortlessly past our tiny car and selves. And then it was all over—at the last minute, it seems, the elephant decided we were just not interesting anymore, and it shifted slightly to maneuver around our car and continue on its way. After the elephant was safely down the road, Adam and I breathed a sigh of relief, our hearts beating in our throats. Then Adam hit me in the arm and said, "Damn it! You should have filmed that! It would have made a fantastic video!"

In that moment with the elephant, I realized a number of things. First, I'm not as incontinent as I thought. Second, there was no sound . . . nothing. Those fat pads on the elephant's feet made it walk very quietly, something I did not in any way expect. (By the way, *Jurassic Park* fans, this means the *Tyrannosaurus* that thundered through the movie would have actually moved silently! Like I said back in chapter 18, we paleontologists do a good job at killing fun.) Third, there was the smell—a rank, zoo- or barn-like smell that wafted through the car as the elephant passed. Fourth, as my wife duly noted, we do not have elephant insurance. But perhaps most of all, fifth, I finally understood the meaning of the word 'awesome.' Here was an animal several metric tons in mass that dwarfed us and our vehicle, and it could have easily killed us both without too much effort. This giant mammal, like me, was a descendant of ancestors far smaller and furrier. Yet, here it strolled on the African plains with little to fear from any predator. An elephant may be the 'new boss,' but it was walking with an upright, parasagittal stance searching for vegetation just as the giant sauropods had millions of years before. Those long, pillar-like legs, like those of the dinosaurs before it, gave that elephant the amazingly long strides it needed to catch up with our car. And those columnar limbs, just like those of dinosaurs, ensured that its limb bones were positioned to best endure the compressive stress of its massive body. "Meet the new boss, same as the old boss."

But we're getting a bit ahead of ourselves here. Most of the mammals that we see every day are small, sleek, and furry. Several we keep as pets. We ourselves are mammals. We seem to recognize that something is a bit different about ourselves and the other mammals around us from other vertebrates, but what is it? I find it instructive in my undergraduate courses to first introduce students to the skeletal chassis and anatomy of modern placental mammals before trying to trace their evolutionary history. As with so much in this book, we will first look to a few modern placental mammals to give us a search image for the fossil record. Having stuck with our story of vertebrate evolution to this point, you may be surprised by what we find.

Mammals, Part 1: An Introduction to the Milk-Giving Fur-Bearers

There are over 5,000 species of mammals in the world today (Wilson and Reeder, 2005), over 90% of which are the so-called placental, or eutherian, mammals. Whole books can be and have been written on mammals, and so it would be impractical for us to detail the rich diversity of our furry family, let alone focus on all eutherian mammals. Here, I will first focus on two common eutherian mammals that have followed humans across the globe and that have both, for good or bad, found their way into our homes: the domestic house cat (*Felis catus*) and the brown rat (*Rattus norvegicus*). Rats and cats are used here for their familiarity, but not because they represent eutherian mammals (or all mammals, for that matter) in their diverse entirety. As in previous chapters, we will use rats and cats as our search image for mammals, and then expand on mammalian evolution and diversity in chapter 20.

Like humans and the elephant that almost crushed me, cats and rats allow their embryos to grow for an extended period (gestation) inside the mother's uterus, kept alive through the life-support device we call the placenta. Many people assume that the placenta is created by the mother alone, but it is actually almost entirely the work of the embryo. You may recall that in chapter 12 we discussed the functioning of the amniotic egg. The embryo of an amniote, whether reptile or mammal, creates its own four life-support membranes: the amnion, chorion, allantois, and yolk sac. In eutherian mammals, the amnion continues its function as a fluid-filled antigravity chamber. The chorion, one of the original gas-exchanging organs of the amniotic embryo, now insinuates itself into the wall of the mother's uterus, anchoring the developing embryo in place (Wilt and Hake, 2003; Gilbert, 2010) (Fig. 19.2). The network of blood vessels that form in the chorion are small, numerous, and close enough to the blood supply of the uterine wall that gas and nutrient exchange is possible between embryo and mother without mixing of the blood (Wilt and Hake, 2003; Gilbert, 2010). If the blood of the embryo (which has a combination of the father's and mother's genes) and mother did mix, her body might well expel the embryo as a foreign invader! The allantois, the other respiratory organ and waste storage sac, has become modified into a long, cord-like structure that runs from the belly of the embryo into the placenta: it has become an umbilical cord (Wilt and Hake, 2003; Freyer and Renfree, 2009; Gilbert, 2010) (Fig. 19.2). Oxygen- and nutrient-rich blood from the chorion pulses up the umbilical cord and flows into the embryo, whereas a descending series of blood vessels removes wastes and carbon dioxide. The combined chorion, allantois, and uterine wall are what

Pedigree The Surviving Synapsid Amniotes

Date of First Appearance 200 Ma (Card 96)

Specialties of the Skeletal Chassis A Dentary-Squamosal Jaw Joint, Three Inner Ear Ossicles Fully Separated from the Jaw

Eco-niche Terrestrial, Aquatic, and Aerial with Tremendous Dietary Diversity

19.2. Stylized eutherian mammal embryo attaching to the uterine wall of its mother. Notice that the same four life-support membranes we saw in other amniotes are still present in mammals. The difference is that they have modified themselves to attach to the uterus rather than adapt to the inside of a shelled egg.

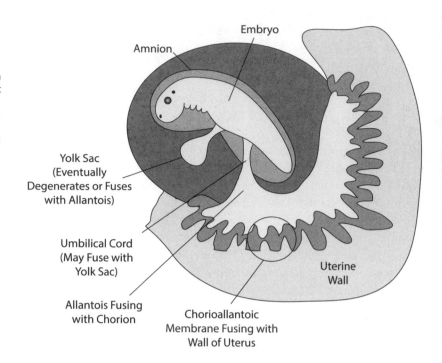

Amnion

Embryo

Yolk Sac
(Eventually
Degenerates or Fuses
with Allantois)

Umbilical Cord
(May Fuse with
Yolk Sac)

Uterine
Wall

Allantois Fusing
with Chorion

Chorioallantoic
Membrane Fusing with
Wall of Uterus

make up the placenta, technically known in eutherian mammals as the chorio-allantoic placenta (Freyer and Renfree, 2009; Gilbert, 2010).

The yolk sac still develops, but depending on the mammal, it will contribute in various degrees to the placental structures (Freyer and Renfree, 2009). For eutherian mammals, yolk is no longer deposited into the egg, but the old yolk sac still plays vital roles in embryonic development that have selected for its retention in mammals. For example, the yolk sac is a center for the development of much of the circulatory system, including the blood vessels and the raw materials (the stem cells) that create the blood itself. Experimentally, when mouse embryos are deprived of their yolk sac, they die because they can no longer appropriately generate a majority of their circulatory system (Wilt and Hake, 2003; Freyer and Renfree, 2009). Thus, the eutherian embryo is essentially a parasite that pulls nutrients and gases from the mother in exchange for its wastes and carbon dioxide. It is not too much of an exaggeration to say that this internal parasite becomes an external parasite after birth because it must suckle the mother's milk, yet another energy drain. Based on their development, milk glands appear to be derived from glands that produce sweat (Blackburn, Hayssen, and Murphy, 1989; Oftedal, 2002a, 2002b). Think on this the next time you pour yourself a cold glass of milk! When milk first appeared and under what evolutionary circumstances this occurred are tackled in the last chapter.

No eutherian mammals lay eggs, although the weirdly primitive monotremes do, as we will see in chapter 21. Given that the embryos develop within the protective uterine environment of the mother, most eutherian mammals invest more resources into each offspring compared

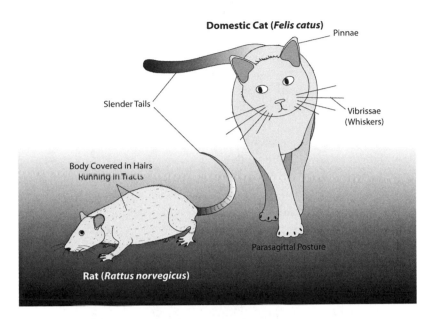

Domestic Cat (*Felis catus*)

Pinnae

Slender Tails

Vibrissae
(Whiskers)

Body Covered in Hairs
Running in Tracts

Parasagittal Posture

Rat (*Rattus norvegicus*)

19.3. The rat (*Rattus norvegicus*) and the domestic cat (*Felis catus*) as our stand-ins for eutherian mammals. Remember that these animals are used to simplify our discussion of mammal characteristics and do not in any way represent the true breadth and diversity of mammalian evolution. Vibrissae, hair running in tracts, and pinnae are some of the key external markers for most mammals.

to a reptile, and therefore produce far fewer descendants with each generation. Certainly, rats and cats are known for their prolific breeding abilities, but far more rat litters and kittens would enter the world if these mammals put fewer resources into their offspring and simply laid lots of eggs. Given the rat population problems that plague cities like New York (Sullivan, 2004), the thought of egg-laying rats is truly terrifying. For example, a single female rat may give birth to several litters per year, with each litter containing 5–20 rat pups (Sullivan, 2004). Thankfully, rats and cats both produce far fewer offspring than could be laid in eggs, and these babies must also be suckled until they become independent. These two factors, resource investment and suckling, curb mammal population growth (somewhat).

Rats and cats, like most mammals, are covered in a coat of thin, keratin-based outgrowths we call hair or fur (Liem et al., 2001; Kardong, 2012) (Fig. 19.3). Fur serves numerous functions in mammals. It can be fluffed up to trap and warm a layer of air next to skin. It can protect the skin from the sun's ultraviolet radiation. Its oils can repel water, and many mammals have relatively loose skin and use a stereotypical side-to-side shake in which they quickly fling water off hair on their bodies (Dickerson, Mills, and Hu, 2012). Hair typically grows in tracts, and has a 'grain,' a direction of slant (which explains why stroking your cat tail to head, opposite the 'grain,' ruffles their fur and feelings) (Kardong, 2012). Specialized facial hairs, technically called vibrissae but known commonly as whiskers, adorn the faces of rats, cats, and many other mammals (Fig. 19.3). Whiskers are stiffened hairs connected to sensitive nerve endings that fire signals to the brain to give a rat or cat the 'feel' of the landscape around them (Liem et al., 2001; Sullivan, 2004; Page, 2008; De Iuliis and Pulerà, 2011). As both rats and cats are nocturnal, whiskers allow them to detect the size of openings in the dark so that in a split second they can decide whether or not they will fit. In fact, as a general rule the width of a cat's whiskers is usually proportional to the girth of its body; thus, fat cats typically have long whiskers (Page, 2008). Whiskers are ultimately touch receptors that allow many nocturnal mammals to feel their way about the world (Pough, Janis, and Heiser, 2002; Kardong, 2012). Similar whisker-like hairs are also present near the back of the wrist in cats (De Iuliis and Puleràà, 2011).

Rats, cats, and other mammals are also endothermic, burning energy at a high, constant rate. Being this active requires a large, double-pump heart as we discussed for birds. Back in chapter 3, we saw generally how body size and volume have an effect on various physiological processes in vertebrates. I often ask my students an odd question I have borrowed from Robert Bakker (1986): if it was a cold night, you were locked in the zoo, and you were naked, would you keep warmer sleeping with the elephants or with the rabbits? The answer, perhaps not surprisingly, is that it would be best to sleep with the rabbits. Rabbits, being small mammals, shed more body heat to the environment than elephants, and before you were arrested for public obscenity, you would do well to sleep with the bunnies. As with feathers, hair probably evolved in mammals for a variety of physiological needs (Kardong, 2012), but retention of body warmth may have been one of them. Burning as much fuel as they do, small mammals lose a tremendous amount of heat to the environment. Hair in many mammals serves as an insulator, slowing down this heat loss (Liem et al., 2001; Pough, Janis, and Heiser, 2002; Kardong, 2012), and it should not be too surprising that, in general, smaller mammals are more likely to be furry than larger ones. Those of us with pet cats have often enjoyed the comfort, on cold nights, of a warm, furry companion that sheds heat from its highly stoked internal furnace.

However, milk and hair are typically not preserved in the fossil record. How, then, can we trace the evolution of mammals if their two well-known traits don't show up reliably in fossils? This was a question I posed in high school to some of my teachers out of curiosity, and I was sorely disappointed when they didn't know. Of course, in retrospect, that was a pretty unrealistic expectation. But my older self now knows the answer – besides their hair and milk, mammals have skeletal traits unique to and shared by only them. If you know what to look for, you can be confident that you have a mammal skeletal chassis when you behold one.

First and foremost, mammals have more solidly constructed skulls than the other amniotes we have focused on up to this point (see Plate 24). In particular, the lower jaw is now comprised of a single bone: our good friend the dentary (Liem et al., 2001; Pough, Janis, and Heiser, 2002; Benton, 2005; Kardong, 2012) (Fig. 19.4). No trace of the postdentary bones is visible, including the articular bone that formed the lower portion of the jaw joint in all other jawed vertebrates. The quadrate, too, has seemingly vanished. Instead of a quadrate-articular jaw joint, mammals have a jaw joint made up of the squamosal and dentary bones (Liem et al., 2001;

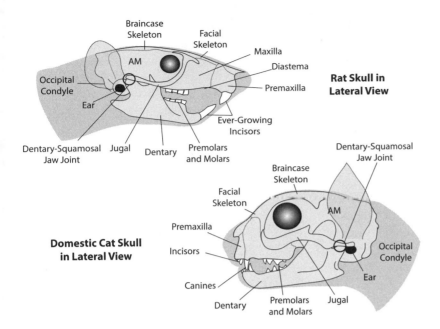

19.4. The skulls of the rat and the domestic cat. Mammalian skulls are solid and akinetic compared with those of reptiles. Note the enlarged braincase skeleton, the 'new' dentary-squamosal jaw joint, and the single lower jaw bone (the dentary). AM, adductor mandibulae muscles (now the temporalis and masseter in mammals). Skull drawings based on skeletal specimens in my lab and artwork in Rohensky (1986) and De Iuliis and Pulerà (2011).

Benton, 2005) (Fig. 19.4). Secondly, the premaxilla, maxilla, and dentary all continue to bear teeth, but mammalian teeth are different from those of reptiles. Instead of similar teeth (homodont dentition) of varying sizes distributed throughout the jaws, mammals have specialized teeth in different regions of the mouth (heterodont dentition) (Liem et al., 2001; Pough, Janis, and Heiser, 2002; Kardong, 2012; Kielan-Jaworowska, 2013) (Fig. 19.4). As a mammal yourself, you can open your mouth and examine these differing regions in your bathroom mirror. Cranially, you have incisors whose chisel-like shapes are good for biting off chewable bits of food. Caudal to the incisors are canines, teeth that are elongated into stabbing fangs in many mammals, but which in us are smaller with blunted crowns. Caudal to these are the premolars and molars, wider teeth with (in us) rounded cusps that provide large surfaces over which food can be smashed, ground, and pulped before being swallowed. This basic pattern (incisors, canines, premolars, molars) is repeated over and over again in mammals, with modifications in number, the occasional loss of certain tooth types, and a vast array of cusp and ridge shapes on the molars (Radinsky, 1987; Kardong, 2012; Kielan-Jaworowska, 2013).

Unlike the teeth of other vertebrates we have examined, the premolars and molars have cusps and bumps that contact each other. Specifically, these cusps or bumps on the upper molars fit into valleys and pits on the lower molars, allowing the teeth to smash and grind food between them (Carroll, 1988; Benton, 2005; Kielan-Jaworowska, 2013). Such precise tooth-to-tooth contact, called occlusion, would be difficult to manage if the teeth of mammals were constantly shed and replaced as they are in many reptiles. Human dentistry is rife with stories and examples of what happens when adult teeth fall out or are misaligned: usually nothing good! For example, uneven tooth pressure on different sides of the jaw can lead to temporomandibular joint (TMJ) syndrome, where the cartilaginous joint between the squamosal and dentary bones is damaged. TMJ syndrome commonly causes clicking sounds in the jaws and resistance to jaw opening on one side of the mouth (Dawson, 2006). Most significantly, if the teeth don't occlude in particular ways, the efficiency of chewing can be diminished. Thus, there has been selective pressure among most mammals to limit the number of times teeth are replaced. For many eutherian mammals, a deciduous set of (baby or 'milk') teeth first emerges, and these are later replaced one by one with a single set of nonrenewable adult teeth (Liem et al., 2001; Pough, Janis, and Heiser, 2002; Kardong, 2012). I swear that this form of tooth replacement was evolution's way of creating the dental profession.

The distinct shape of mammalian teeth is a testament to their eating style—these are champion chewers. Unlike most of the vertebrates we have encountered, mammals have the best table manners: they typically thoroughly process and chew up their meals before swallowing. The incisors nip off tasty bits of a meal, the canines stab and tenderize the food, and the premolars and molars grind the food into small bits that combine with saliva to make a swallow-ready packet of food called a bolus (Liem et al., 2001; Pough, Janis, and Heiser, 2002). As in alligators and other crocodylians we discussed in chapter 16, the premaxilla, maxilla, and other upper jaw bones of mammals form a secondary bony palate that separates the nasal passages from the mouth (Fig. 19.5 and Plate 24). Whereas in crocodylians this separation of nose and mouth was useful for breathing while submerged, this arrangement in mammals allows chewing and breathing to happen simultaneously (Liem et al., 2001; Kardong, 2012). In another parallel with crocodylians, the choanae are shifted caudally to direct air taken in through the nose directly back to the glottis of the trachea (Fig. 19.5). This division of the nasal passages from the mouth also serves most mammals well in their earliest moments of life when they must suckle (Liem et al., 2001; Kardong, 2012). When infant mammals latch onto their mother's teats and drink milk, they can continue to breathe without breaking the gasket-like seal of their lips. The secondary bony palate of mammals also appears to act as a strut that resists the torqueing forces generated by the chewing cycle (Liem et al., 2001; Benton, 2005), much as the bony palate in crocodylians resists the twisting forces of its struggling prey. A series of facial muscles unique to mammals and responsible for both the actions of suckling and facial expression are present. The origin of these muscles will be covered in chapter 21.

Chewing as a form of specialized feeding in most mammals has led to the modification of the jaw-opening and -closing apparatus. As we have discussed throughout this book, increasing the mechanical advantage of a lever usually requires a short out-lever that rotates about a fulcrum close to the in-force (see chapter 6). The new squamosal-dentary jaw joint and the loss of the postdentary bones in mammals shifts the jaw joint (fulcrum) cranially while shortening the out-lever (dentary), increasing the overall biting power (Liem et al., 2001; Benton, 2005; Kardong, 2012)

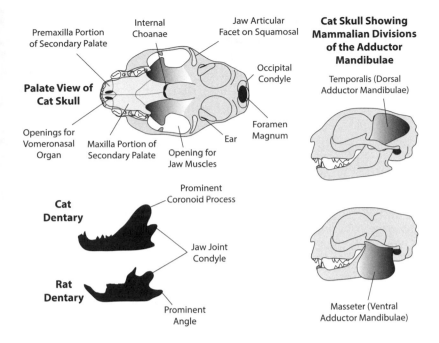

Palate View of Cat Skull

Premaxilla Portion of Secondary Palate

Internal Choanae

Jaw Articular Facet on Squamosal

Occipital Condyle

Foramen Magnum

Ear

Opening for Jaw Muscles

Maxilla Portion of Secondary Palate

Openings for Vomeronasal Organ

Cat Dentary

Prominent Coronoid Process

Jaw Joint Condyle

Rat Dentary

Prominent Angle

Cat Skull Showing Mammalian Divisions of the Adductor Mandibulae

Temporalis (Dorsal Adductor Mandibulae)

Masseter (Ventral Adductor Mandibulae)

19.5. Aspects of mammal skull anatomy. In most mammals, a bony secondary palate has evolved that separates the nasal passages from the mouth. The evolution of this feature, paralleled in crocodylians, seems to have resulted from the competing selective pressures for suckling while breathing and resistance to the repetitive forces associated with chewing. The adductor mandibulae has been subdivided into a dorsal biting muscle called the temporalis and a ventral grinding muscle called the masseter. On the dentary bone, a coronoid process is associated with the temporalis, and the angle is associated with the insertion point of the masseter. Thus, in predatory mammals, the temporalis is enlarged compared to the masseter, and this is reflected in an enlarged coronoid and small angle. The reverse is the case for herbivorous and many omnivorous mammals, where grinding food is most important. Here, the masseter is larger than the temporalis, and as a consequence the angle is large and the coronoid process is small. These differences can be noted in the silhouettes of the cat and rat dentaries. Cat and rat skull details adapted from Bohensky (1986) and De Iuliis and Pulerà (2011).

(Fig. 19.5). The only moving piece of the jaws is now the lone dentary bone. Much like the difference between a solid ceramic bar and one that has been fractured into several pieces and then glued together, a single lower jaw bone, rather than multiple elements, may best resist the repetitive strain of chewing. To accommodate the large muscles necessary for chewing, a single, postorbital fenestra is present. In most mammals, this fenestra has expanded laterally and ventrally, providing an avenue through which thick jaw muscles can pass from the skull to the dentary bone (Romer, 1962; Liem et al., 2001). The jugal bone forms the cheek of this outwardly expanded fenestra, and along its lower border other jaw muscles anchor (Fig. 19.5). This expanded arch formed by portions of the jugal and squamosal bones is called the zygomatic arch.

In mammals, the old adductor mandibulae muscles have become subdivided into two complementary mechanical units. Much like the gears on a bike, these divisions of the old adductor mandibulae have distinct mechanical advantages under different circumstances. The dorsal portion of the adductor mandibulae has become the temporalis muscle (Fig. 19.5). This muscle anchors to the squamosal bone and inserts itself into the dentary bone on a specialized projection called the coronoid process (Liem et al., 2001; Pough, Janis, and Heiser, 2002). The coronoid process itself is located cranial and dorsal to the jaw joint, meaning that the temporalis pulls on this jaw region just in front of where it rotates (Fig. 19.5). This all has the effect of quickly rotating the dentary upward to bring its teeth against those of the upper jaw (Liem et al., 2001; Pough, Janis, and Heiser, 2002). In essence, the temporalis portion of the adductor mandibulae in mammals is for rapid biting and snapping. The other jaw-closing muscle segment, the masseter, anchors along the cheek region of the skull (the jugal) and attaches to a portion of the dentary called the angle or angular process (Fig. 19.5). Essentially, the masseter acts like a powerful sling that encapsulates the jaw joint (Romer, 1962; Liem et al., 2001; Pough, Janis, and Heiser, 2002). Given its close proximity to the jaw joint, the mechanical advantage of the masseter provides substantial grinding forces along the premolars and molars. The pterygoid muscles make an appearance as well, completing the muscular sling on the medial side of the dentary bone and aiding (mostly) in jaw closing (see below).

As you might anticipate, these specialized components of the adductor mandibulae vary among mammals depending on their diet. In most predatory mammals, the temporalis is enlarged compared to the masseter to ensure rapid and powerful bites to disable and kill prey (Hildebrand and Goslow, 1995) (Fig. 19.5). In contrast, most herbivorous and omnivorous mammals have a large masseter, which greatly enhances the grinding and pulping of tough vegetation on the molars (Hildebrand and Goslow, 1995) (Fig. 19.5). These differences in muscle power are also reflected in the size and shape of the coronoid and angular processes on the dentary bone. In predatory mammals such as the house cat, the coronoid process is commonly enlarged and triangular, whereas the angular process is small (Fig. 19.5). In contrast, herbivorous and omnivorous mammals, such as the rat, typically have an expanded angular process or region, with a tiny coronoid process (Fig. 19.5). Moreover, just as we have discussed since chapter 2, the jaw joint of carnivorous mammals is usually in line with the teeth, allowing them to close in scissors-like fashion to cut meat, whereas the jaw joint of many herbivores is situated above the tooth row to enhance the teeth's ability to contact each other simultaneously like a nut cracker. Thus, the paleontologist can use these differences in skull and jaw shape to infer the relative jaw power and approach to feeding in fossil mammals.

You may well wonder, how is the jaw opened in mammals? The old depressor mandibulae has 'disappeared' along with the quadrate and articular jaw joint in mammals (not really: see chapter 20). Instead an old lower jaw elevator and portion of the depressor mandibulae have fused into a single muscle called the digastric, a name that means 'two bellies' for the fleshy parts of a muscle. The digastric pulls on the dentary to lower it; why this should have occurred will be tackled in chapter 20. However, a portion of the pterygoid muscles has become situated across the jaw joint such that its contraction assists the digastric by causing the dentary to rotate (Moore and Dalley, 1999).

Rats and cats have modified their skulls in various ways along this spectrum of mammal head space. Rat incisors are absolutely huge, chisel-shaped teeth that are unusual among mammals in that they are ever-growing (Bohensky, 1986; Pough, Janis, and Heiser, 2002) (Fig. 19.5). More intriguing is that whereas the front of the incisor and its crown are covered in diamond-hard enamel, softer dentine lines the back of the tooth. With a hard front end and softer back, as rats gnaw their food, their incisors are naturally sharpened as the soft dentine is ground down, exposing more of the chisel-like tooth crown (Bohensky, 1986; Pough, Janis, and Heiser, 2002). This is somewhat akin to dragging a blade over a sharpening stone after each use to retain its thin cutting edge. In fact, rats can click their jaws into two different positions: one position juts the lower jaw forward so that the upper and lower incisors contact each other end to end to nip through a variety of tough materials (Hildebrand and Goslow, 2001). Having been bitten by a rat, I can attest that their incisors are indeed sharp, powerful, and painful! The second position of the jaws allows the premolars and molars to come together without interference from the incisors (Hildebrand and Goslow, 2001). Canines are absent in rats, and instead a space called the diastema separates the incisors from the premolars and molars, which have become modified into hard, flat shelves studded with cusps for grinding down the most stubborn of food. As you might predict, the masseter is huge in rats, and it pulls on a large angular process on the dentary with great mechanical power (Fig. 19.5). In rats, the postorbital fenestra is so large and expanded to make room for the powerful jaw closing muscles that it overtakes the orbit. When you look at a rat skull, there is no clear distinction between where the eyes end and the jaw muscles begin.

In cats, the incisors are small but pointed, the canines are long and sharp for stabbing prey, and the premolars and molars have been modified into sharp, blade-like teeth called carnassials, which slide past one another as the jaw closes like a pair of scissor blades (see chapter 2) (Page, 2008; De Iuliis and Pulerà, 2011). In cats, it is the temporalis muscle that has become enlarged, allowing them to quickly bite and clamp down on prey (or crunchy cat food), and this muscle inserts into a prominent coronoid process on the dentary (Hildebrand and Goslow, 2001; Liem et al., 2001; Kardong, 2012) (Fig. 19.5). The orbit and the postorbital fenestra are much more distinct in cats, although the orbit is not completely walled off from the space for the jaw muscles. Those of you with pet cats may have noticed an odd behavior when they sit in front of a window and birds or small mammals come to your yard. The cat will stare intently at the birds or mammals, and then begin rapidly chattering its teeth, sometimes while emitting a warbling mew. It turns out that your cat is fantasizing about having that little bird or chipmunk in its jaws, because this tooth chattering behavior mimics jaw movements that enable the carnassials to quickly slice through the spinal cord of their prey (Morris, 1993).

We should briefly consider the inside of the mammalian nose as well. In most mammals, delicate, scroll-like bones called turbinates fill the nasal passages. These turbinate bones appear to serve multiple functions, two of which we will highlight here. First, they increase the surface area of the nasal epithelium, the soft tissues that cover the nasal passages. The olfactory nerve endings are located within these tissues, so the surface area over which odors can be detected is expanded (Kardong, 2012). Second, the pockets and openings that the turbinate bones create inside the nasal passages modify the speed of incoming and outgoing air. As air is inhaled through the nose, it is warmed and moisturized before proceeding on to the respiratory system. During exhalation, the turbinates slow and cool the air passing through the nose, causing a significant amount of moisture to remain behind (Kardong, 2012). The turbinate bones are strongly associated with endothermic metabolism, and it should be noted that many birds show these as well (Hillenius and Ruben, 2004). However, endothermy is not always associated with the presence of turbinate bones. For example, some birds have no turbinate bones, and newly emerging data on dinosaur noses suggest cartilaginous tissues within the nostrils fulfilled an analogous function (Bourke et al., 2014).

It is once again time to peer into the skull, this time with our eye on the brains of rats and cats. One of the first things you notice when you see a mammal brain is that the cerebrum is an enlarged sheet that covers much of the rest of the brain regions. In both rats and cats, the cerebrum is large. In rats, the cerebrum has a smooth surface, but in cats this brain region is a wrinkled surface composed of hills and valleys called gyri and sulci (Liem et al., 2001; Kardong, 2012) (Fig. 19.6). Unlike in all other vertebrates, the cerebrum of mammals is constructed as a six-layered sandwich of neurons and their associated tissues (Liem et al., 2001; Butler, Reiner, and Karten, 2011) (Fig. 19.6). However, for the cerebrum to become larger, it cannot simply thicken or swell like the DVR region of birds because this would disrupt the communications between the nerve cells in this six-layered cake of information (Liem et al., 2001; Mota and Herculano-Houzel, 2012). Instead of the volume being increased through thickening, the cerebrum of many mammals, like cats, is increased by wrinkling its surface area, thereby expanding the area for neural networks without compromising its neuronal connections (Mota and Herculano-Houzel, 2012). The cerebrum of many mammals is so large that the braincase skeleton has inflated tremendously. In fact, one of the reasons why the postorbital fenestra of mammals is oriented lateroventrally is that the braincase has ballooned outward, displacing the typically more vertical orientation of these openings in most other amniotes.

To say that many mammals are smart is a bit of an understatement. Rats are known for their cleverness, and their persistence in our homes and cities despite our Herculean efforts to destroy them is a testament to their smarts (Sullivan, 2004). House cats are also intelligent problem solvers (Morris, 1993; Page, 2008). In what I can only describe as the mother of all lazy moves, one of our cats once figured out an ingenious and slothful way to eat. Ash, as he was known, found that he could lay supine with his hairy belly aloft and kick his food dish with one of his feet in such a way that crunchy food bits catapulted gracefully onto his chest, just within reach of his mouth. From there, he could lick up the food bits and proceed to kick the food dish again for a refill!

Of course, many birds and reptiles are smart and clever as well (Emery, 2006), so I in no way want to create the impression that mammals have intelligence cornered. But mammal brains are certainly wired differently. For example, most mammals, like rats and cats, have enlarged

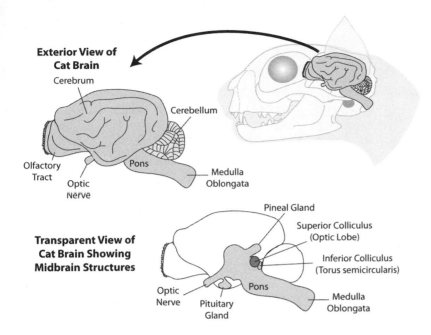

Exterior View of Cat Brain

Cerebrum

Cerebellum

Olfactory Tract

Optic Nerve

Pons

Medulla Oblongata

Transparent View of Cat Brain Showing Midbrain Structures

Pineal Gland

Superior Colliculus (Optic Lobe)

Inferior Colliculus (Torus semicircularis)

Optic Nerve

Pituitary Gland

Pons

Medulla Oblongata

19.6. Domestic cat (*Felis catus*) brain. As in most eutherian mammals, the cerebrum, cerebellum, and olfactory tract are enlarged. Because of its six-layered cortex, the mammalian cerebrum expands not by thickening of nerve clusters (as in reptiles and birds) but by inflation and wrinkling, producing various sulci and gyri. The midbrain and much of the hindbrain are hidden beneath the expanded cerebrum, and the typically enlarged optic lobes and torus semicircularis for sight and hearing are greatly diminished, now transformed into reflexive centers called the superior and inferior colliculi. Based on specimens in my lab and illustrations in De Iuliis and Pulerà (2011).

olfactory tracts and bulbs that protrude from beneath the cerebrum (Fig. 19.6). We humans have tiny and poorly developed olfactory senses, but most mammals understand complex aspects of their world through the detection of odors we never notice. Pheromones, those odorless, species-specific chemicals we talked about in chapter 11, form a related part of mammal senses. Rats, cats, and many other mammals have openings in the roof of their mouth around their premaxilla that give pheromones a direct track to a vomeronasal organ (Liem et al., 2001; Pough, Janis, and Heiser, 2002; Kardong, 2012) (Fig. 19.5). Many mammalian pheromones provide cues for mating or individual identification, and these molecules may be concentrated and directed toward the vomeronasal organ through an odd-looking behavior called the Flehmen response (Liem et al., 2001). Typically, the mammal lifts its upper lips to expose its teeth, sticks out its tongue, closes its nostrils, and inhales through its mouth for a few seconds. Dog owners may have noticed this behavior when their pet picks up a ball or other object and holds it against the roof of the mouth while panting. In this case, pheromones or scents from another dog or animal may be on the ball, and the Flehmen response concentrates and circulates these molecules so the dog's vomeronasal organ can 'decode' this chemical information (Liem et al., 2001).

Mammals are also wired differently when it comes to discerning and interpreting sound and sights. As you may recall, in most vertebrates, the midbrain processes and interprets most of the audio and visual information relayed to it from the ears and eyes. However, in mammals the midbrain region has become relatively small, and the regions that normally process sight (the optic lobe) and sound (the torus semicircularis) have been reduced to tiny lumps called the superior and inferior colliculi, respectively (Fig. 19.6). Thus, the midbrain of mammals mainly handles only auditory or visual reflexes and orienting (Liem et al., 2001; Pough, Janis, and Heiser, 2002; Butler, Reiner, and Karten, 2011; Kardong, 2012). For example, the superior colliculi function to direct the eyes toward unexpected movement and to keep objects of interest 'localized' in the visual field. In other words, if you detect a sudden movement out of the corner of your eye or you are tracking a moving object (such as a batter timing when to hit a moving baseball), your superior colliculi are behind those reflexes. The inferior colliculi do similar things with audio stimuli. For example, turning your head toward sudden noises or the classic behavior of crouching and closing one's eyes when a loud, explosive sound is heard are handled by the inferior colliculi. Television producers long ago caught on to these reflex centers: the fast changing video clips in a music video repeatedly grab the attention of your superior colliculi (you can't look away!), and commercials are often noticeably louder than the program you are watching, drawing the attention of your inferior colliculi.

Strangely enough, most of the audio and visual interpretation in mammals now happens within the cerebrum itself. In some mammals, multiple regions within the cerebrum process and interpret visual and audio signals, forming a complex and three-dimensional impression of environmental sights and sounds (Liem et al., 2001; Kardong, 2012). It is often said that eyewitness accounts of crimes are unreliable, and the biological part of this reason seems to reside in our mammalian brains – much of what we see is channeled, massaged, manipulated, and transmogrified by visual centers in our cerebrum such that we often don't see reality as it exists but 'reality' as we perceive it to be. The same goes for much of our hearing, which is why there are so many tastes in music, for example. Speaking of which, three little auditory bones (ossicles) transmit sound from the tympanic membrane to the inner ear in mammals, unlike the single stapes bone of other tetrapods. Why mammalian brains should be so oddly organized and why there should be three inner ear bones are intriguing evolutionary questions we will reserve for the final chapters.

In terms of the eye, many mammals are naturally crepuscular, or nocturnal, and usually have diminished color vision compared to other vertebrates. The mammal retina is typically dominated by the rod cells, which detect movement, light, and darkness, whereas many fewer cone cells are available to interpret the color spectrum (Pough, Janis, and Heiser, 2002). Rods, as you may recall, also form a somewhat blurry image. In fact, your house cat, even it wanted to, does not have the fine resolution to read the text you are reading now (Page, 2008). However, there is the false conception that most mammals see in black and white like that picture on an old television. Whereas this is not usually the case, rats, cats, and many other mammals detect fewer colors and fewer differences in the shades of color they can see compared to humans. In essence, most mammals see in what is called blue-green dichromatic vision: they see blues and greens, but either can't see or have trouble seeing yellows and reds (Peichl, 2005). Many mammals have a reflective layer associated with their retina called a tapetum lucidum. This layer acts like a mirror that bounces, reflects, and concentrates light, especially under low-light conditions, onto the retina, improving a mammal's night vision (Liem et al., 2001; Pough, Janis, and Heiser, 2002; Kardong, 2012). This is what is behind the phenomenon of 'eye shine,' where the eyes of your cat or dog or a wild mammal glow brightly when you use a flashlight in the dark. For completeness sake, it should be noted that other vertebrates from fishes to reptiles can also have a tapetum lucidum, creating the same 'eye shine' effect. In fact, I have experienced reptile 'eye shine' in the Everglades—on a weekend camping trip, we shone our flashlight out onto the water at night only to be greeted by the reflections of many alligator eyes! Primates, the group of mammals we belong to, are weirdos among mammals, having a good degree of color vision that appears to have evolved from a nocturnal ancestor with dichromatic blue-green vision (Heesy and Hall, 2010; Hall, Kamilar, and Kirk, 2012; Moritz et al., 2013). Why this should be so is again the subject of the final chapter.

Once beyond the skull, the chassis of a rat, cat, or typical mammal is still a bit weird (Figs. 19.7 and 19.8). As you will recall, the bony joint that connects the skull to the atlas vertebra in the neck is called the occipital condyle (Fig. 19.8). In most amniotes, the occipital condyle is a ball-like projection beneath the foramen magnum (the exit for the spinal cord) that allows some rotational and nodding movements of the skull. In mammals, the occipital condyle is divided into two convex wedges on either side of the foramen magnum, which articulate in winglike scoops on the cranial end of the atlas vertebra (Liem et al., 2001; Kardong, 2012). This connection allows quite a range of head nodding at this juncture while restricting rotational movements (Fig. 19.8). The subsequent axis vertebra has a process called the dens into which the atlas articulates and rotates about, somewhat like a vinyl record rotating on the spindle of a turntable (Fig. 19.8). So in mammals, the 'yes' movements of the head occur between the occipital condyle and atlas, whereas shaking the head 'no' is accomplished by the joint between the atlas and axis (Liem et al., 2001; Kardong, 2012) (Fig. 19.8). With few exceptions, most mammals have only seven cervical vertebrae, including the atlas and axis. This is amazingly weird when you realize that giraffes have such long necks and whales and dolphins have short, immobile necks. Despite these vast anatomical differences in length and function, simply changing the shapes and dimensions of the seven cervical vertebrae creates necks long, short, mobile, or stiff. Embryonic and genetic studies have shown that this restraint on the number of cervical vertebrae in mammals is tied into something called pleiotropy, or the ability of one gene to affect several traits simultaneously (Muller et al., 2010; Varela-Lasheras et al., 2011). In this case, it seems that for whatever reason, cervical vertebral number has become enmeshed with cancer regulation. In most mammals, mutations that cause the number of cervical vertebrae to vary from the typical number of seven increase the likelihood that the animals will develop cancers as youngsters (Varela-Lasheras et al., 2011).

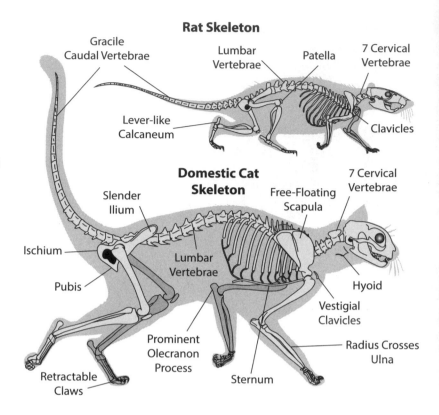

19.7. Rat (*Rattus norvegicus*) and domestic cat (*Felis catus*) skeletal diagrams (not to scale). Among other quintessentially mammalian features represented here are the presence of seven cervical vertebrae; specialized thoracic and lumbar vertebrae; gracile caudal vertebrae; an enlarged, free-floating scapula; reduced or vestigial clavicles; an upright forelimb with crossed radius and ulna bones; prominent, lever-like olecranon processes and calcanea; and the development of the patella sesamoid in the knee. Diagrams based on skeletons in my lab and illustrations in Bohensky (1986) and De Iuliis and Pulerà (2011).

Rat Skeleton

Gracile
Caudal Vertebrae

Lumbar
Vertebrae

Patella

7 Cervical
Vertebrae

Lever-like
Calcaneum

Clavicles

**Domestic Cat
Skeleton**

Slender
Ilium

Free-Floating
Scapula

7 Cervical
Vertebrae

Ischium

Lumbar
Vertebrae

Pubis

Hyoid

Vestigial
Clavicles

Radius Crosses
Ulna

Retractable
Claws

Prominent
Olecranon
Process

Sternum

The dorsal vertebrae of rats, cats, and other mammal skeletal chassis are specialized and subdivided into a cranial rib-bearing thoracic series and a ribless caudal lumbar series. Unlike in many other tetrapods, the ends of the centra (vertebral bodies) are flat or acoelous and surrounded by thickened fibrocartilaginous disks (Fig. 19.8). This arrangement allows dorsoventral sliding of the vertebrae in relation to one another, but reduces the lateral side-to-side movements commonly available to most reptiles (Hildebrand and Goslow, 2001; Liem et al., 2001). Collectively, this allows many mammals, especially those like the rat and cat, to have a dorsal vertebral column that can flex craniocaudally and move dorsoventrally. The rib-lacking lumbar vertebrae further enhance this accordion- and springlike movement, allowing the well-muscled region between the sternum and pelvis to flex and extend across a wide range of movement (Hildebrand and Goslow, 2001; Liem et al., 2001). Cats in particular have especially thick intervertebral cartilages that allow them to twist their backs when falling upside down so they land right side up (Page, 2008). A narrow and segmented sternum that accepts the cartilaginous portions of the ribs is present in most mammals (Hildebrand and Goslow, 2001) and probably contributes to the suppleness of their trunk.

In some senses, the vertebral chassis of a rat or cat acts like an archer's bow (as discussed for chordate notochords back in chapter 3), with the muscles acting as the bowstring and the vertebral column acting as the bow itself (Liem et al., 2001). When the abdominal muscles contract, they pull the lumbar and thoracic vertebrae toward one another, creating

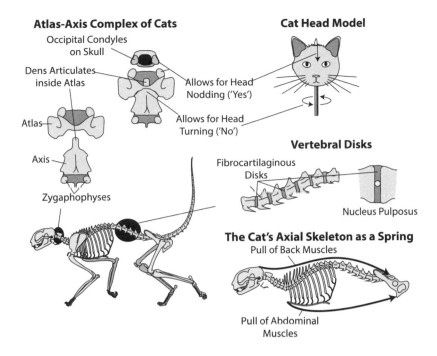

Atlas-Axis Complex of Cats

Occipital Condyles on Skull

Dens Articulates inside Atlas

Atlas

Axis

Zygaphophyses

Cat Head Model

Allows for Head Nodding ('Yes')

Allows for Head Turning ('No')

Vertebral Disks

Fibrocartilaginous Disks

Nucleus Pulposus

The Cat's Axial Skeleton as a Spring

Pull of Back Muscles

Pull of Abdominal Muscles

19.8. Features of the axial skeleton in the domestic cat (*Felis catus*). The atlas of cats, like most mammals, allows for head nodding ('yes') by allowing dorsoventral rocking between the occipital condyles of the skull and the winglike facets on the atlas. The articulation of the axis with the atlas through the dens allows the laterally swiveling movements ('no') that allow cats and other mammals to turn their heads side to side. The dorsal vertebrae of mammals have centra that are acoelous and connected by fibrocartilaginous disks containing a remnant of the notochord called the nucleus pulposus. In cats, these disks are especially thick and contribute to the amazingly flexible spine of these mammals. Due to the morphological differences between the thoracic and lumbar vertebral regions, the dorsal and ventral body muscles of cats and other mammals can act as opposing springs that stretch or compress the body during locomotion. Diagrams based on lab specimens and illustrations in Liem et al. (2001), De Iuliis and Pulerà (2011), and Kardong (2012).

stress and strain and storing elastic energy (Fig. 19.8). When the abdominal muscles relax, the stored energy in the vertebral column is released, allowing it to spring back to a more outstretched posture (Fig. 19.8). Unlike what happens in an archer's bow, muscles throughout the back that extend the vertebral column also assist in stretching it out. One can think of the mammalian dorsal vertebrae as a series of tough segments attached dorsally and ventrally by separate springs. Mechanically, a spring is a simple machine that can store and release energy by changing its shape: in other words, it is elastic (Bloomfield, 2006). When the dorsal spring contracts, the vertebral segments are extended, and the ventral spring is put under tension; when the dorsal spring relaxes, the ventral spring pulls the vertebral segments closer to one another. Replace the springs with dorsal and ventral trunk and abdominal muscles and you have the dorsoventral accordion-like body movements you see in running cats, rats, and other mammals (Liem et al., 2001). This arrangement also increases leaping ability, especially in cats.

As it turns out, this distinction between thoracic and lumbar vertebrae in mammals is correlated with the diaphragm. Mammals inflate their lungs using this specialized muscle that divides the pleural cavity (where the lungs and heart reside) from the peritoneal cavity (where the guts reside) (Fig. 19.9). In essence, the diaphragm roughly marks the transition between the thoracic and lumbar vertebrae. Therefore, when a paleontologist looks at the skeleton of a mammalian relative, if thoracic and lumbar vertebrae are present it is probably a good bet that a diaphragm was present as well. The diaphragm is under both voluntary and involuntary control, and likely evolved from a fusion of skeletal muscles and the involuntary smooth muscles more characteristic of the digestive

19.9. The diaphragm respiratory muscle in mammals. The diaphragm is situated between the thoracic and lumbar vertebrae, and it acts like a plunger that sucks air into and pushes it out of the lungs. When a mammal like a cat inhales, the diaphragm contracts and pulls on a central tendon that pulls it back like a toilet plunger being pulled from a drain. This creates negative pressure, which sucks fresh air into the lungs. When the diaphragm relaxes and springs back to its resting position, it creates positive pressure, which forces air from the lungs.

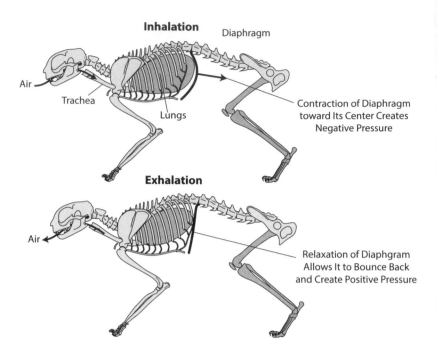

Inhalation

Diaphragm

Air

Trachea

Lungs

Contraction of Diaphragm toward Its Center Creates Negative Pressure

Exhalation

Air

Relaxation of Diaphgram Allows It to Bounce Back and Create Positive Pressure

tract (Pickering and Jones, 2002; Perry et al., 2010). Whatever its exact origin, this is the main muscle of respiration in mammals.

Unlike in other tetrapods, where rib and sternal movements help suck and pump air into and out of the lungs, the mammalian diaphragm is the main driver of respiration, acting somewhat like a giant plunger. When your toilet becomes clogged, you first push the rubber plunger into the drain creating positive pressure, and then you pull back on the plunger, creating negative pressure. The negative pressure sucks back whatever unmentionable things are clogging your toilet drain, loosening them so that water can flow freely again to the sewer. Likewise, the diaphragm muscle is shaped like a bowl-shaped toilet plunger and acts to push and pull air. When the diaphragm muscle contracts, its bowl-like shape expands downward toward the guts, creating negative pressure that pulls air into the lungs. When the diaphragm muscle relaxes, it springs back into the lungs, creating positive pressure that allows you to exhale (Liem et al., 2001) (Fig. 19.9).

The location and function of the diaphragm also seem to have played an important role in mammals' ability to run. When a lizard runs, its vertebral column flexes side-to-side rapidly, squeezing and collapsing the right then left lung repeatedly in such a way that it becomes impossible for the animal to run and breathe at the same time (Pough, Janis, and Heiser, 2002) (Fig. 19.10). In mammals, the ribless lumbar vertebrae allow the vertebral column to flex and extend dorsoventrally, which aids in increasing the stride length while running. You have almost certainly observed this when watching dogs or cats run at full tilt. With the diaphragm situated on the border between the thoracic and lumbar vertebrae, something intriguing occurs: breathing becomes linked to body

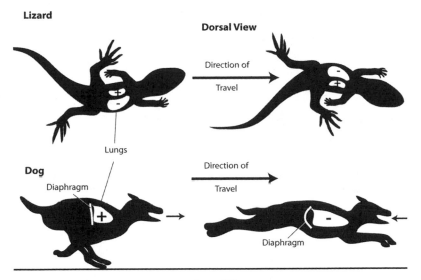

Lizard

Dorsal View

Direction of Travel

Lungs

Dog

Diaphragm

Direction of Travel

Diaphragm

Lateral View

19.10. The difference between a reptile and a mammal in relation to running and breathing. In a typical lizard, the lateral bending of the vertebral column alternately compresses the right and left lungs, preventing the lizard from efficiently breathing and running at the same time. In contrast, the dorsoventral movement of the vertebral column works in synch with the diaphragm, allowing breathing and running to occur simultaneously. Diagrams based on work by Bramble and Carrier (1983) and Deban and Carrier (2002), as well as illustrations in Liem et al. (2001), Pough, Janis, and Heiser (2002), and Kardong (2012).

movements. When a mammal's vertebral column flexes, the diaphragm is physically pushed into the lungs, aiding in exhalation (Bramble and Carrier, 1983). In contrast, when the mammal's vertebral column extends, the diaphragm is physically moved caudally, aiding the drop in pressure that sucks air into the lungs (Bramble and Carrier, 1983) (Fig. 19.10). Thus, mammals can run and breathe at the same time, improving their stamina for traveling at high speeds over longer distances than a lizard of similar mass (Bramble and Carrier, 1983; Pough, Janis, and Heiser, 2002).

The sacrum of most mammals is relatively small and triangular in dorsal profile, being widest where the ilia of the pelvic girdle hug it cranially. In many mammals, a caudal series of various lengths and robustness follows the sacrum. In the rat, the tail is long, flexible, scaly, and relatively naked, whereas that of the cat is furry and often held aloft. In both rats and cats, the tail can act as a stabilizer, its often rotational movements acting to counterbalance a sleek and flexible body on the move (Hickman, 1979; Walker, Vierck, and Ritz, 1998). In cats and dogs the tail can, of course, also act as a behavioral signal (Hickman, 1979). The wagging tail of a friendly dog is a familiar sight, and an anxious or indecisive cat will flick its tail side to side until it makes up its mind or claws the annoying human that bothered it in the first place (Morris, 1993; Page, 2008). Unlike the tails of many reptiles, those of mammals are commonly narrow and not overly muscled, and no large tail muscles such as the caudofemoralis longus of archosaurs and lizards play a significant role in hind limb retraction. Therefore, unlike the back-heavy chassis of archosaurs, the center of mass in mammals is situated much further cranially, usually between the forelimbs and hind limbs (Alexander, 1989; Walter and Carrier, 2011).

Moving onto the limb girdles, there are further modifications here as well. In the rat, cat, and most mammals, the scapula has become the dominant bone of the pectoral girdle (Jenkins, 1974; Liem et al., 2001;

Pough, Janis, and Heiser, 2002; Kardong, 2012) (Fig. 19.7). In rats, a clavicle is still present, but in cats the clavicle has been reduced to a tiny splint. In other mammals such as dogs, horses, and elephants, the clavicle has disappeared altogether (Jenkins, 1974; Pough, Janis, and Heiser, 2002). In fact, with the strange exceptions of the duck-billed platypus and spiny echidna, there is no direct, bony connection to the sternum through a coracoid bone (or bones) as in other amniotes. Instead, the scapula literally floats in space along the ribcage, suspended solely by muscles and soft tissues (Liem et al., 2001). We will return to the functional significance of this arrangement after we explore the rest of the forelimb. The pelvic girdle is still composed of the ilium, ischium, and pubis, but these bones are now usually fused and united in most mammals into a single unit sometimes called the os coxae or the innominate bone (literally, the bone with no name!). The ilia are typically cranially projecting, spoon-shaped bones that articulate with the sacral vertebrae (Fig. 19.7). In most mammals, the upright, parasagittal posture of their hind limbs requires far less adductor muscle mass, and so the pubis and ischium are narrow as we saw for dinosaurs, but are relatively much smaller in comparison.

Returning to the forelimb, the upright posture attained by many mammals has resulted in a number of changes (Fig. 19.7). First, without a prominent coracoid, the scapula does not articulate directly with the sternum. Instead, the remnant of the coracoid (the metacoracoid we touched on back in chapter 12) and the majority of the scapula form the shallow, cup-shaped glenoid socket into which the humerus articulates (Vickaryous and Hall, 2006). The humerus has a semispherical head that, combined with a shallow shoulder socket, allows a large range of flexion, extension, abduction, and adduction, which combine to create sophisticated rotational movements at the shoulder (Liem et al., 2001; Pough, Janis, and Heiser, 2002; Kardong, 2012). Distally, the humerus has two distinct articular surfaces for the radius and ulna called the capitulum and trochlea, respectively (Fig. 19.11). In rats, cats, and most other mammals, the capitulum is raised and convex, which, in combination with the shallowly cupped end of the head of the radius, allows a combination of flexion, extension, and some degree of long-axis rotation. The trochlea has a saddle-shaped, spool-like articular surface, and the proximal end of the ulna has a C-shaped surface that grabs onto this region like a wrench (Jenkins, 1973; Jenkins and Parrington, 1976). In fact, mostly hinge-like flexion and extension movements are available to the ulna, and the ulna's caudally directed olecranon process, into which the large triceps muscles insert, is a distinct and caudally pointing lever through which these antigravity muscles gain mechanical advantage. Combined, this elbow anatomy allows a rat, a cat, and other mammals to assume an upright, parasagittal posture (Jenkins, 1973; Hildebrand and Goslow, 2001; Liem et al., 2001) (Fig. 19.7).

To facilitate an upright forelimb in which the hand was kept pronated, palm-side down on the ground, the radius could not remain parallel to the ulna as it is in many reptiles. Instead, the shaft of the radius is curved outward, allowing its shaft to cross over the ulna (Fig. 19.11).

Human Forelimb Demonstrating Supination and Pronation

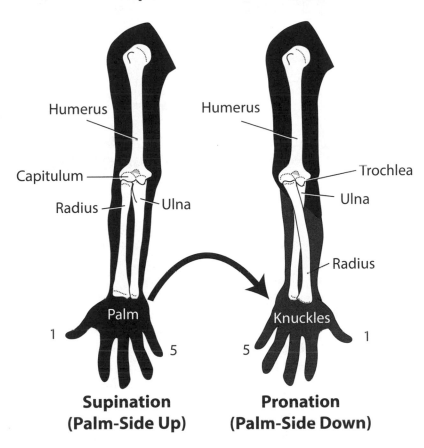

Humerus

Capitulum

Radius — Ulna

Palm

1

5

**Supination
(Palm-Side Up)**

Humerus

Trochlea

Ulna

Radius

Knuckles

5

1

**Pronation
(Palm-Side Down)**

19.11. Pronation and supination in mammals as exemplified by humans. In most mammals, the forelimb has become erect, and the radius crosses the ulna to pronate the hand so that it flexes and extends in concert with the foot. In humans, we have a remarkably well developed ability to actively pronate and supinate our hands with muscles that rotate the radius about the ulna. Some mammals have varying degrees of active pronation and supination, whereas others have minimal ability to do this or are permanently pronated with no ability to 'uncross' the radius and ulna. Diagram based on figures in Moore and Dalley (1999).

This ensures that its distal end is cranial and medial in relation to the distal end of the ulna (VanBuren and Bonnan, 2013) when the forelimb is upright. Most reptiles can pronate the hand only by sticking their elbows out to the side, which has the effect of rotating the hand palm-side down (although radius and ulna movements augment pronation; see chapter 13). Furthermore, the head of the radius becomes more rounded in those mammals that can actively pronate and supinate their hands, with humans and other primates having the most 'enhanced' ability to do so (VanBuren and Bonnan, 2013) (Fig. 19.11). Surprisingly, among the vast Reptilia it is only the chameleons that cross the radius over the ulna to pronate their hands on the narrow branches where they hunt insect prey (Fischer, Krause, and Lilje, 2010; VanBuren and Bonnan, 2013). With the exceptions of the sprawling duck-billed platypus and spiny echidna from the Australian outback, the next time you visit a natural history museum, take pause and have a look at the forearms of the land mammals on display. All of these magnificent mammals will have a radius that curves gracefully over the ulna to plant their hands firmly on the ground in pronation in an upright stance.

19.12. The bounding gait of many mammals, exemplified by the cat. The lumbar vertebrae of most mammals allow the body to stretch and compress during locomotion, enhancing the stride length. A mobile scapula, supported by a muscular sling and powerfully rotated by the serratus ventralis, further enhances stride length while providing shock absorption for the forelimbs as they hit the ground. In the hind limb, the large biceps femoris both flexes the knee and rotates the femur caudally, whereas the gastrocnemius pulls on the lever-like calcaneum to plantarflex the foot against the ground for effective push-off. Running cat based on the classic photographic series by Eadweard Muybridge (Muybridge, 2007).

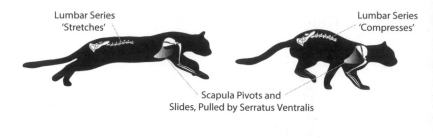

Lumbar Series 'Stretches'

Lumbar Series 'Compresses'

Scapula Pivots and Slides, Pulled by Serratus Ventralis

Biceps Femoris Both Retracts Femur and Flexes the Knee

Biceps Femoris

Gastrocnemius

In rats, cats, and other mammals, seven wrist bones form a series of interconnecting joints that enhance the flexion, extension, and rotational movements of the mammalian hand (Hildebrand and Goslow, 2001; Liem et al., 2001). In rats and cats, five fingers are present, with well-developed joints that allow for an extraordinary range of flexion and extension. Rats have a well-developed ability to reach, grab, and manipulate objects—so much so that their manipulative abilities are used as a model to treat and understand human subjects with forelimb deficits (Klein et al., 2012). In cats, the claw unguals (the claw-bearing distal finger bones) can be extended so they are held off the ground, keeping the claw nails very sharp for when they are engaged (Page, 2008) (Fig. 19.7). House cats can supinate and pronate their paws to a great degree, manipulating and grabbing objects with a surprising and sometimes annoying dexterity (Morris, 1993; Page, 2008).

Returning to the shoulder girdle, one thing you may have noticed in your pet cats is that their scapula is surprisingly mobile. If you've ever watched a cat jump from a table or couch and land on the floor, you have probably noticed that their shoulders seem to move almost independently of their body. If you don't have house cats but have watched a wildlife special on big cats, you have probably viewed scenes of lions or cheetahs stalking prey. In these scenes, you can almost always see the outline of the scapula moving dorsally, ventrally, and craniocaudally. This is no illusion—the scapula of cats and many other eutherian mammals is free of any bony attachments to the sternum and is suspended entirely by muscles (Hildebrand and Goslow, 2001; Liem et al., 2001; Kardong, 2012) (Fig. 19.12). This means that in many eutherian mammals, forelimb posture is entirely modulated and supported by the muscles of the shoulder and chest (Hildebrand and Goslow, 2001; Liem et al., 2001; Kardong, 2012). These muscles arise from the back, flanks, and chest, and collectively they act like a muscular sling that can both support the body and move the forelimb (Carrier, 2006). In cats,

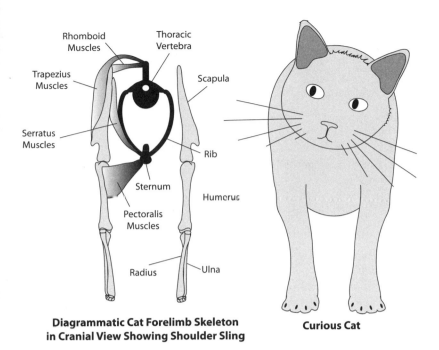

Diagrammatic Cat Forelimb Skeleton in Cranial View Showing Shoulder Sling

Rhomboid Muscles

Thoracic Vertebra

Trapezius Muscles

Scapula

Serratus Muscles

Rib

Sternum

Humerus

Pectoralis Muscles

Radius

Ulna

Curious Cat

19.13. The muscular shoulder-sling of mammals. In cats and many other mammals, the scapula is nearly or completely detached from the sternum and is suspended entirely by muscles and other soft tissues. A sling of muscles consisting of the dorsally placed rhomboids and trapezius and the ventrally placed serratus ventralis and pectoralis support, move, and provide shock absorption to the forelimb. A large forelimb retractor, the latissimus dorsi, also forms part of the muscular sling but is not shown. Diagram of muscular sling based on illustrations in Gray (1968), Liem et al. (2001), Pough, Janis, and Heiser (2002), Carrier (2006), and Kardong (2012).

horses, elephants, and other eutherian mammals where the clavicles have either diminished to splints or vanished, the scapula has become, in effect, an additional limb segment (Hildebrand and Goslow, 2001; Liem et al., 2001; Pough, Janis, and Heiser, 2002). Given the upright, pendulum-like motion of many eutherian mammals, the scapula now acts as a higher pivot point that, as it is swung craniocaudally by powerful muscles, imparts a greater range of motion to the entire forelimb (Liem et al., 2001) (Fig. 19.12).

The sling-like arrangement of the scapula-supporting muscles also acts collectively like a shock absorber on a car. For example, when a cat places its forelimb on the ground as it is walking, the ground-reaction force (the 'shock' or sudden deceleration of the forelimb as it hits the ground) travels up the forelimb. If the scapula were attached rigidly to the ribcage, such a force would cause a jolt that might very well snap bones. If you ever used a pogo stick as a child, then you intuitively know that if it were not for the spring-like action of the toy, many a child would have suffered broken bones and bruises from jumping forcefully on a rigid post. Instead, the spring inside the pogo stick, much like the shock absorber on a car, more gradually decelerated the stick and also allowed movement to dissipate the stress. In a cat, the soft and pliable sling of shoulder-supporting muscles diffuses and damps the ground-reaction force in ways similar to a pogo stick, and the movement of the scapula in particular is crucial in spreading out the impact (Liem et al., 2001; Pough, Janis, and Heiser, 2002; Kardong, 2012) (Fig. 19.12). So, the next time you watch your house cat jump from a high perch and land on the floor, take a moment to admire its scapular movements and the sling of muscles that take the shock (Figs. 19.12–19.13).

The hind limb of rats, cats, and many other eutherian mammals resembles the upright and pendulum-like limbs we have already discussed concerning dinosaurs (see chapters 17 and 18). However, the acetabulum of the pelvis is solid, concave, and somewhat ventrally facing whereas the head of the femur is deflected medially (as we saw for dinosaurs) but has a rounded, semispherical articular surface. In layperson's terms, we have a good old-fashioned ball-and-socket joint (Fig. 19.7). Unlike the socket-wrench mechanics we saw for dinosaurs and their feathery descendants, the ball-and-socket joint of the mammalian hip allows craniocaudal pendulum-like swinging as well as rotational movements and a fair degree of abduction. This combination of movements is controlled and guided by a number of pelvic muscles, including the gluteal muscles that make up our buttocks. The femur of mammals now has an enlarged greater and lesser trochanter, prominent insertion sites for the gluteal muscles. Gluteal muscles are in general abductors that pull the hind limb away from the body wall when it is lifted off the ground. However, when a mammal is standing or placing weight on a limb, the abductors serve as powerful stabilizers that prevent the limb from collapsing inward, especially when transferring weight from side to side as the mammal walks. It appears that gluteal muscles became more prominent in mammals, in part, because of the loss of a large, muscular tail and the reduction or loss of associated muscles such as caudofemorales, which, as you will recall, are the major hind limb retractors in reptiles. Without a large counterbalancing tail, the originally small abductors would have taken on a more significant support and balancing role in the ancestors of mammals.

In many quadrupedal mammals, and especially in cats, one of the large hamstring muscles, called the biceps femoris, is absolutely huge and crosses both the hip and knee joints (Liem et al., 2001; Harris and Steudel, 2002; De Iuliis and Pulerà, 2011) (Fig. 19.12). So large is the biceps femoris that the caudal profile of the hind limb in many mammalian quadrupeds is defined and shaped by this single muscle (Fig. 19.12). By crossing caudal to the knee joint, the biceps femoris acts as a powerful knee flexor. However, because it also crosses caudal to the hip joint, it contributes to caudal rotation of the femur at the hip. Both of these actions combine during walking or running to forcibly push the foot off the ground, giving an extra 'oomph' that propels eutherian mammals toward prey or away from predators (Harris and Steudel, 2002). In cats, the biceps femoris also provides the last part of 'take-off' when the cat jumps, simultaneously pushing the foot back against the ground and flexing the knee to ready the cat for its landing (Harris and Steudel, 2002).

As we discussed for lizards back in chapter 13, many mammals develop a number of small bones around their joints, especially in the hind limbs, called sesamoids (Hildebrand and Goslow, 2001; Liem et al., 2001; Kardong, 2012). You may recall that sesamoids act in concert with soft tissues like an anatomical pulley system, increasing the mechanical advantage of muscles whose tendons bend around joints (Liem et al., 2001; Kardong, 2012) (chapter 13). The largest sesamoid in eutherian mammals is the patella or knee cap, which acts with the tendon of the large knee-extending muscles to increase their mechanical advantage during activities such as jumping and walking (Fig. 19.7).

The tibia and fibula of mammals tend to be 'weird' as well. In mammals, the knee joint occurs only between the distal end of the femur and the proximal end of the tibia. The fibula is excluded from this arrangement, and lies 'below' the articular surface of the tibia (Liem et al., 2001; Kardong, 2012). The proximal end of the tibia has two distinct facets that articulate with the femoral condyles, and a prominent ridge of bone or eminence fully separates the facets from each other. As you may recall from chapter 9, this bony eminence gives a general indication of where the cruciate ligaments (which passively resist overrotation) reside in mammal knees (Haines, 1942). In many non-mammal tetrapods, both the tibia and fibula are capable of some degree of independent long-axis rotation, which controls foot orientation distally (Haines, 1942). In contrast, in most mammals, with the knee joint solely between the femur and tibia, long-axis rotation of the tibia orients the foot, with the fibula going along for the ride (Haines, 1942). This difference seems to have occurred with the development of the mammalian ankle joint that forms, uniquely, between the tibia and astragalus (although see chapter 21).

We take for granted that our style of walking is somehow 'normal,' but our heel-to-toe foot motion is anything but. You may not realize it, but your heel bone is the calcaneum, one of the major amniote ankle bones. Recall that in non-mammals, the calcaneum is usually alongside the astragalus, forming the proximal portion of the relatively hinge-like ankle joint. In contrast, eutherian mammals have an ankle in which the pivoting movement of the foot occurs between the distal end of the tibia and the convex articular surface of the astragalus (Liem et al., 2001; Kardong, 2012). In most mammals, the calcaneum has shifted its position to lie somewhat inferior to the astragalus, where it can act as a lever through which the calf muscles (gastrocnemius in particular) rotate the foot back against the ground (Liem et al., 2001; Pough, Janis, and Heiser, 2002) (Fig. 19.14). The distal end of the fibula is lower and 'cradles' the lateral side of the astragalus in most mammals, helping to form a complete bony perimeter around the ankle with the tibia. This may be why the independent rotational movements observed in some non-mammal tetrapods are restricted or absent in most mammals (though there are exceptions—see chapter 21): any independent movements of the fibula could interfere with the ankle or result in bone breakage (Haines, 1942). It should be noted that many mammalogists and human anatomists refer to the calcaneum as the calcaneus, and the astragalus as the talus. To keep the homologies clear and terminology consistent, I am using standard tetrapod anatomical terminology.

In rats and cats, the ankle is held clear of the ground, which positions the metatarsals (the sole) into a vertically inclined orientation (Fig. 19.14). This means that rats, cats, and many eutherian mammals are digitigrade, standing only on their toes with their calcaneum 'heels' sticking up in

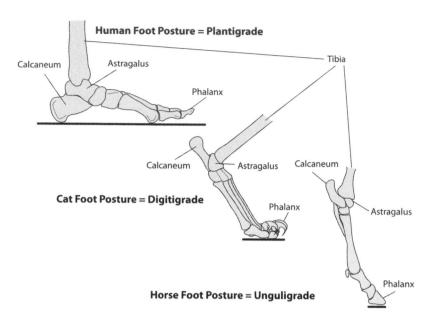

19.14. The foot posture of mammals. As we saw for dinosaurs, many mammals, such as the cat, have digitigrade feet with which they stand on their toes with their ankles off the ground. Humans and some other mammals have a plantigrade posture in which the calcaneum forms the heel at the back of the foot. Stride length is increased in digitigrade mammals for functional reasons similar to those we encountered for dinosaurs: such a posture lengthens the hind limb. The digitigrade posture is taken to extremes in ungulate mammals that stand only on their nails (unguals). Illustrations based on specimens in my lab and illustrations in Moore and Dalley (1999), Liem et al. (2001), and De Iuliis and Pulerà (2011).

the air (Hildebrand and Goslow, 2001; Liem et al., 2001). Some eutherian mammals have gone one step further, literally, by standing on toe nail! These would be the ungulates such as horses and cattle, so-called because they stand only on their unguals, the anatomical word for the finger- and toenails (Fig. 19.14). The digitigrade foot posture is reminiscent of dinosaurs and their birdy descendants, and it mechanically accomplishes the same thing: it lengthens the hind limb by adding an additional segment (the heel and sole) above the foot (see chapters 16 and 17). Humans and some other mammals (such as bears) have adopted a plantigrade foot posture where our calcaneum and metatarsals contact the ground with each footstep (Liem et al., 2001; Pough, Janis, and Heiser, 2002) (Fig. 19.14). With every step we take, our calf muscles pull on our lever-like calcaneum, lifting the metatarsals off the ground and driving the toes down and back against the ground, propelling us forward with a process called plantarflexion. Since the ankle and sole rise to pivot across the toes, we develop the familiar human lope as we walk along, rising and falling with our calcaneum.

At this point, we have developed a search image of a eutherian mammal, which is a good stand-in for modern mammals generally: most living mammal species are eutherians. We know mammals by their hair and milk glands, but if we're looking for fossils then it is the skeletal chassis that holds clues to mammalian ancestry. Many skeletal characteristics of mammals are pretty unusual compared to other vertebrates: solid skulls with a single, lower jaw bone; a new dentary-squamosal jaw joint; an enlarged braincase; typically only seven cervical vertebrae; a springy backbone with thoracic, rib-bearing body vertebrae and ribless lumbar vertebrae; a slender tail; a pectoral girdle dominated by the scapula, itself suspended entirely or almost entirely by muscles; an upright forelimb with caudally pointing elbow and crossed radius and ulna; a prominent patella; and an enlarged, lever-like calcaneum. We also know that rats, cats, and other eutherians can be fairly intelligent animals that commonly have a well-developed sense of smell but reduced color vision, and that three inner ear bones (ossicles) transmit sounds from the tympanic membrane to the brain. Where did this unusual skeletal chassis originate? Under what selective circumstances did the jaw joint and inner ear ossicles change? Why do so many mammals lack good color vision? What driving forces modified the cerebrum of the mammalian brain to become larger and to take over much of the tasks formerly handled by other systems? The origin of mammals and the evolution of the unique mammal chassis are explored in the final two chapters.

20.1. The early synapsid, *Dimetrodon*. Note that *Dimetrodon* has a single, postorbital fenestra, a canine tooth, and a reflected lamina on its lower jaw, all characteristics of synapsids later inherited and modified in the common ancestor of mammals. The elongate neural spines form a 'sail' that may have served multiple purposes, including thermoregulation and display. AM indicates the region of the jaw-closing adductor mandibulae muscles. Skull and skeletal diagrams modified after Carroll (1988) and Benton (2005).

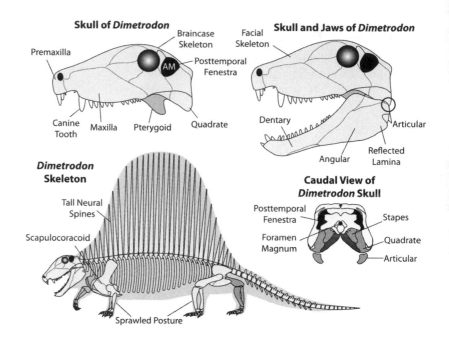

Skull of *Dimetrodon*

Premaxilla
Braincase Skeleton
AM
Posttemporal Fenestra
Canine Tooth
Maxilla
Pterygoid
Quadrate

Skull and Jaws of *Dimetrodon*

Facial Skeleton
Dentary
Articular
Angular
Reflected Lamina

Dimetrodon Skeleton

Tall Neural Spines
Scapulocoracoid
Sprawled Posture

Caudal View of *Dimetrodon* Skull

Posttemporal Fenestra
Foramen Magnum
Stapes
Quadrate
Articular

The Evolution of the Mammal Chassis 20

TO MY CHILDREN, THE THOUGHT OF A WORLD WITHOUT THE INTER-
net, on-demand television, and cell phones with cameras in them is difficult to conceive. Admittedly, from our narrow perspective it is difficult if
not impossible to conceptualize that there was a time without the familiar
technological tools we now take for granted. Having grown up during the
beginning of the microcomputer age, I played the first primitive arcade
games, had an Atari console, and cut my teeth on BASIC programming
and the old DOS system using type commands. I saw the evolution and
development of Windows, and went from using a typewriter to a word
processor during my late high school and early college years. So, I have
some perspective on the development of these technologies, but I certainly could not imagine a world without television or radio, or without
records and cassette tapes. And yet such a world existed, and those people
that lived from the beginning of the 1900s to the cusp of the twenty-first
century went from a world of horse-drawn carriages to supersonic jets and
rockets that took people to the moon.

By the same token, when we watch a cat frolic in sunbeams, or
pet a warm dog curled up next to us, it seems almost unfathomable to
think that modern mammals shared ancestors with reptiles and birds,
let alone that there could have been a world without our furry friends.
And yet, such a world existed. The mammal chassis, like those of reptiles, descended from a common ancestor similar to our very old friend
Paleothyris, a small, insectivorous lizard-like amniote less than 30 cm
long (see chapter 12). Just as it would be difficult to explain to someone
familiar only with the telegraph that such technology would ultimately
result in mobile communications devices that captured moving pictures,
so it is difficult to imagine how the specialized mammalian chassis could
have originated in a small, lizard-like frame. But of course, we now have
a search image—we know what features to be on the lookout for that act
as tells of the mammalian lineage. To effectively trace the evolution of
the mammalian chassis, we have to jump back in time to the end of the
Carboniferous and beginning of the Permian periods (~304–271 Ma;
Card 94).

Mammals are the last remaining branch of the Synapsida, and as
such, they have (as we have seen) numerous derived trait states. Moreover, a number of fossils closely approach the mammal condition but fall
just short. Therefore, to avoid confusion, we will divide the Synapsida
into the following artificial groups: non-mammal synapsids; non-mammal
cynodonts; mammaliaforms; and true mammals. Again, I emphasize
that Synapsida is a natural group that includes the ancestors of modern

mammals and all their descendant branches, but for convenience we divide the family tree up to make our story of mammal chassis evolution less confusing. As before, we are only just scratching the surface of synapsid and mammal diversity in this chapter. It is my hope that this simplified picture of mammalian evolution best conveys how our unique skeletal chassis arose from reptile-like ancestors.

From their beginnings, the Synapsida were characterized by the development of a single, lateral posttemporal fenestra that, like those of diapsids, was associated with the adductor mandibulae muscles (Carroll, 1988; Pough, Janis, and Heiser, 2002; Benton, 2005) (see Fig. 12.2). This characteristic of the skull is ubiquitous across all synapsids, including modern mammals, and is therefore a key sign paleontologists look for to confirm their identification. In addition, the possession of one or more canine teeth in the maxilla is another hallmark of the Synapsida (Pough, Janis, and Heiser, 2002, 2013; Benton, 2005). At a quick glance, the earliest synapsids could well be mistaken for small to medium-sized lizards, and their sprawled limb posture and low-slung bodies are again inferred to have allowed crank-like limb mechanics coupled with a laterally undulating torso. Synapsids quickly radiated into terrestrial environments in the Early Permian, and were soon the most numerous and dominant terrestrial vertebrates of their time (Benton, 2005; Pough, Janis, and Heiser, 2013).

Non-mammal Synapsida – Distant, Early Ancestors of the Mammals

Pedigree Amniotes Closer to Mammals Than to Reptiles

Date of First Appearance 304 Ma (Card 94)

Specialties of the Skeletal Chassis Single Posttemporal Skull Opening and Canine Teeth

Eco-niche Small to Large Herbivores, Omnivores, and Carnivores in Terrestrial Environments

Among the several early radiations in the Late Carboniferous and Early Permian (304–271 Ma; Card 94) probably no non-mammal synapsid is better known than *Dimetrodon* (Fig. 20.1 and Plate 25). *Dimetrodon* was a large predator for its day, stretching nearly 2.5 meters from head to tail. *Dimetrodon* is most readily recognized by the elongate neural spines, arranged like a picket fence, that projected from its dorsal vertebrae. However, before we tackle the functional significance of the 'sail,' let's first explore a few other aspects of *Dimetrodon* anatomy. It had a large, hatchet-shaped skull that housed impressive jaws studded with pointed teeth (including an enlarged canine), robust limb girdles, sprawling but somewhat gracile limbs, and a long tail. In the skull, the snout is long, the braincase is short, and prominent pterygoid bones project well below the level of the tooth row (Fig. 20.1). As you may recall from our discussions about crocodylians in chapter 16, such a feature is strongly associated with enlarged, deep pterygoid muscles that pull the jaws closed in synergy with the adductor mandibulae. In combination with this feature, a large coronoid process is present on the lower jaw in *Dimetrodon*, a feature correlated with enlarged adductor mandibulae muscles (see, for example, chapters 13 and 19) (Carroll, 1988; Benton, 2005). Together, the expanded coronoid and pterygoid regions of the lower jaw and skull suggest strong jaw-closing muscles were present to drive this synapsid's teeth into its prey.

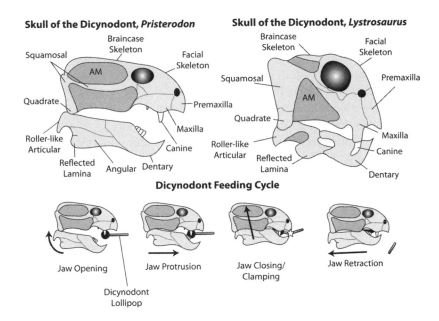

Skull of the Dicynodont, *Pristerodon*

Braincase
Skeleton

Squamosal

AM

Facial
Skeleton

Quadrate

Premaxilla

Roller-like
Articular

Maxilla

Canine

Reflected
Lamina

Angular Dentary

Skull of the Dicynodont, *Lystrosaurus*

Braincase
Skeleton

Facial
Skeleton

Squamosal

Premaxilla

AM

Quadrate

Maxilla

Roller-like
Articular

Canine

Reflected
Lamina

Dentary

Dicynodont Feeding Cycle

Jaw Opening

Jaw Protrusion

Jaw Closing/
Clamping

Jaw Retraction

Dicynodont
Lollipop

20.2. Dicynodont skull variation and mechanics. In dicynodonts, the adductor mandibulae were greatly enlarged, as indicated by widening of the postorbital fenestra and expansion of the squamosal bone. Keratinous beaks probably adorned the premaxilla and dentary bones, and the large canines may have served a multitude of tusklike functions. Jaw studies of *Pristerodon* suggest that a typical feeding cycle involved the protrusion and retraction of the dentary and other lower jaw bones. Figures based on Benton (2005).

A closer look at the jaw joint itself reveals a few interesting anatomical quirks. First, whereas there is still the typical articular-quadrate jaw joint, the quadrate is butted up against a large stapes bone (Fig. 20.1). In other words, the articular, quadrate, and stapes bones are connected through joints in a chain. What does this mean? Well, one could imagine that such a chain, hypothetically, might pass vibrations from the jaw into the stapes, allowing *Dimetrodon* and other early synapsids to hear using their jaws (McLoughlin, 1980). But is this just conjecture? We will return to this odd but significant part of the synapsid chassis later on.

As we have seen for many of the early amniotes and several of the reptilian lineages, the pectoral girdle is comprised of a large scapulocoracoid (containing a distinct procoracoid and metacoracoid) that articulated directly with the sternum, along with other elements such as clavicles (Fig. 20.1). A large, saddle-shaped glenoid faces laterally off the scapulocoracoid, providing a wide area for rotational movements of the humerus. The pelvic girdle is bowl-shaped in cross-section, the acetabulum is shallow and wide, and the ischium and pubis are especially robust, indicating large adductor muscles were present to hold and stabilize the hind limbs in a sprawled orientation (Carroll, 1988; Benton, 2005; Angielczyk, 2009; Pough, Janis, and Heiser, 2013). As in many reptiles, the humerus and femur are robust and expanded mediolaterally, giving them enlarged joint regions that allowed for rotation, abduction, and adduction at their respective girdles. The radius and ulna of *Dimetrodon* parallel one another and, as in many reptiles, would have been held perpendicular relative to the more horizontally directed humerus. The wrist and ankle consist of several small, interlocking bones, and five large claw-bearing digits adorn the hand and foot (Fig. 20.1).

Returning to the vertebral column, it has long been suspected that a 'sail' membrane of skin covered the dorsal vertebral spines (McLoughlin,

1980). Such a 'sail' could have been used for numerous and interrelated functions including thermoregulation, display, and even enhancement of locomotion (Angielczyk, 2009; Huttenlocker, Mazierski, and Reisz, 2011; Rega et al., 2012). Certain *Dimetrodon* specimens preserve neural spines that were broken and rehealed (as evidenced by a bone callus like those seen in injured human skeletons) at their tips (Rega et al., 2012). It has been noted that if the neural spines were not united in some way into a skin-covered membrane, the tips of these broken bones should be missing (Carroll, 1988; Benton, 2005). In other words, the distal end of the neural spine, with no soft tissue to hold it in place, would be expected to fall off. The fact that multiple *Dimetrodon* specimens show broken and healed neural spine tips has been taken as evidence of their being united into a single, membranous unit. Since this membranous unit was alive and required a blood supply, it is likely that is was invested with radiating vessels and capillaries that could deliver cold blood from the body core for warming in the sun. The warmed blood could then return to the body, heating the animal's core more rapidly than if it waited for warmth to pass through its bulky mass. The process could also work in reverse, with the large but thin sail radiating heat to the environment much like the fins in a water radiator rapidly dump heat into a home during the winter. Heating and cooling of physical models of a related, sail-backed herbivorous synapsid, *Edaphosaurus*, further support this hypothesis (Bennett, 1996).

Certainly, there are other functional possibilities beyond thermoregulation for the 'sail' on *Dimetrodon*. With such a large 'billboard' on their backs, it is possible (and not mutually exclusive) that *Dimetrodon* and kin sported patterns and colors on their sail-like backs to establish dominance and attract mates (Bakker, 1986; Benton, 2005). It is also possible that, as in some actinopterygian fish, elongate spines could have contributed to improved lateral balance and enhanced locomotor prowess (Rega et al., 2012). It should also be noted that many non-mammalian synapsids related to *Dimetrodon* do not possess elongate neural spines. Thus, the condition in *Dimetrodon* and related forms like *Edaphosaurus* was an exception rather than the rule, suggesting that, if a thermoregulatory function was present, this may have evolved after the fact rather than under direct selective pressure (Benton, 2005). Whatever the function or functions of the 'sail,' for its day *Dimetrodon* must have been a ferocious predator. One can almost imagine this proud beast sunning itself on a river bed, its sharp teeth gleaming in the sun, its large sail shining brightly with colors, waiting for the right moment to scramble forth and pounce on unsuspecting prey.

The synapsids were the first terrestrial vertebrates to establish communities of multiple carnivores, omnivores, and, most significantly, herbivores (Benton, 2005; Pough, Janis, and Heiser, 2013). There had, of course, existed the occasional herbivorous amniote (recall *Diadectes* from chapter 12), but a diverse variety of specialized herbivores radiated from the synapsids very soon after their origin. For example, some pioneering herbivores called the caseids had log-like bodies (some species grew over 2.5 meters long) supported on stumpy limbs. The broad ribcages of caseids suggest that their bodies housed massive guts for digesting nothing but plant matter (McLoughlin, 1980; Benton, 2005). Somewhat like the sauropod dinosaurs that would later evolve in the Mesozoic, caseid synapsids had tiny heads that must have vacuumed up plants without the need for chewing. More importantly, the radiation of non-mammal synapsids called the therapsids successfully capitalized on herbivory in several lineages.

One of the more significant lineages of herbivorous non-mammal therapsid synapsids was the dicynodonts, so named for their two prominent doglike canines that formed tusks in many species (Fig. 20.2). These strange herbivores ranged in mass from chipmunk-like burrowers to hippo-sized behemoths. Shortly after their evolutionary appearance nearly 270 Ma (Card 95), dicynodonts remained a significant part of the biota, surviving into the latest Triassic (200 Ma; Card 96) despite suffering a large die out during the end-Permian extinction event (Benton, 2005; Dzik, Sulej, and Niedźwiedzki, 2008). In fact, one species of dicynodont, *Lystrosaurus*, was not only widespread throughout the earliest Triassic (~245 Ma; Card 95) but represented 95% of all terrestrial vertebrates in some locales (Benton, 2005)! All dicynodonts had somewhat dorsoventrally flattened beaked skulls with large canines but otherwise typically small teeth or, in some species, no other teeth at all (Fig. 20.2). Multiple skeletons and skulls of many dicynodonts have shed some light on the 'purpose' of the tusklike canines. In one species, *Diictodon*, animals with otherwise identical postcranial skeletons bear one of two distinct skull types. In one form, the skull is robust with enlarged canines and a boss-like growth on the skull (Sullivan, Reisz, and Smith, 2003). This has suggested that there was sexual dimorphism among some dicynodonts, with the males being larger and more robust than the females. If this is the case, perhaps some dicynodont males were using their canines for intimidation of other competitors, just as canine displays of aggression occur in modern mammals (Sullivan, Reisz, and Smith, 2003).

As we have seen for other herbivores, in many dicynodonts the jaw joint is located below the tooth row, which would bring the teeth and beak together simultaneously to crush and grind plant matter. The dicynodont jaw joint was peculiar in that the articular bone possessed a long, tracklike groove which allowed it to slide craniocaudally against the quadrate (Carroll, 1988; Benton, 2005) (Fig. 20.2). Based on both manipulations of the actual skull material and more modern computer analyses, it appears that the lower jaw of dicynodonts operated by clamping onto plant material and then sliding caudally (Benton, 2005; Jasinoski, Rayfield, and Chinsamy, 2009). Many smaller dicynodonts had an elongate head and jaw well suited for this kind of chewing stroke, whereas some later, larger forms had tall, deep skulls that may have been capable of rapidly and powerfully snapping off tough pieces of vegetation (Benton, 2005; Jasinoski, Rayfield, and Chinsamy, 2009) (Fig. 20.2). Not surprisingly, the squamosal region of the skull in dicynodonts is expanded, presumably for

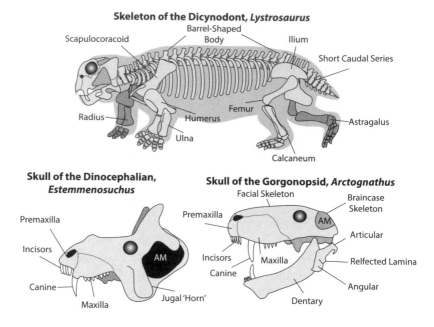

20.3. Dicynodont skeleton and therapsid skulls. *Lystrosaurus* was a pig-sized dicynodont that shows the basic form of these synapsids, including robust, sprawled forelimbs, an enlarged pelvis, and a diminutive tail. The skulls of the dinocephalian *Estemmenosuchus* and the gorgonopsid *Arctognathus* show some of the diversity present in non-cynodont therapsids. *Lystrosaurus* skeletal diagram based on Williston (1914); skull of *Estemmenosuchus* based on illustrations by Chudinov (1965); skull of *Arctognathus* based on Benton (2005).

Skeleton of the Dicynodont, *Lystrosaurus*

Scapulocoracoid — Barrel-Shaped Body — Ilium — Short Caudal Series — Femur — Astragalus — Calcaneum — Radius — Humerus — Ulna

Skull of the Dinocephalian, *Estemmenosuchus*

Premaxilla — Incisors — Canine — Maxilla — Jugal 'Horn' — AM

Skull of the Gorgonopsid, *Arctognathus*

Facial Skeleton — Braincase Skeleton — Premaxilla — AM — Articular — Incisors — Maxilla — Relfected Lamina — Canine — Angular — Dentary

expansive adductor mandibulae musculature, but curiously the expansion occurs so as to create a mammal-like zygomatic arch (Carroll, 1988; Benton, 2005). As you will recall from the previous chapter, the zygomatic arch of mammals is associated with the specialized masseter muscle of the adductor mandibulae, which plays a significant role in grinding food. However, it is doubtful whether such a specialized division of labor in the adductor mandibulae group had occurred in dicynodonts because they lack the hallmark of the masseter on their dentary: the masseteric fossa. It is also difficult to know what role the deep pterygoid muscles played, but they may have helped to draw the lower jaw forward as the depressor mandibulae pulled on the large retroarticular process caudal to the jaw pivot (Benton, 2005).

Dicynodont skeletons themselves were rather short and stocky, even in the smaller species. A typical dicynodont had a short neck, a robust, barrel-shaped trunk with expansive ribs, a tall and deep pelvis, and a short tail (Carroll, 1988; Benton, 2005; Pough, Janis, and Heiser, 2013) (Fig. 20.3). In smaller forms, the scapulocoracoid could be somewhat gracile, but in the largest dicynodonts the scapulocoracoid was a large, robust element associated with prominent clavicles and an interclavicle. In the pelvis, the ilium was large and expanded, and it would have supported large knee extensor and hip abductor muscles. The forelimbs and hind limbs were short, sprawled, and, based on their shapes, probably crank-like in their movements (just as we saw for so many amphibians and amniotes). Moreover, the wide regions of articulation on their respective girdles leave little doubt that these herbivores were not capable of assuming an upright, parasagittal posture (Benton, 2005) (Fig. 20.3).

Dicynodonts are also the first synapsids for which we have evidence of burrowing. Throughout South Africa, in the Permian-aged rocks of the

Karoo Basin (270–245 Ma; Card 95), there are many well-preserved cork-screw-like burrows with terminal chambers at their base (Damiani et al., 2003; Bordy et al., 2011). Contained within many of these burrows are the skeletal remains of one or more dicynodonts (Benton, 2005). In modern ecosystems, corkscrew-shaped burrows are created by mammals to insulate the terminal burrow chamber from environmental conditions above. In essence, the corkscrew shape prevents cold or hot air from directly blowing into the burrow, allowing the mammal in question to maintain a stable, equable burrow microclimate (Benton, 2005). The dicynodont burrows are preserved in mudstones that were laid down near riverbanks. This suggests that many of these synapsids were living near meandering river systems that at times, like modern rivers, experienced flash floods. Such events could have buried and trapped the hapless dicynodonts in their underground homes (Benton, 2005; Bordy et al., 2011).

Other radiations of therapsid synapsids included the dinocephalians, a lineage of herbivores and carnivores that were characterized by enlarged heads commonly adorned with bony ornamentation in many species (McLoughlin, 1980; Carroll, 1988; Benton, 2005; Antón, 2013; Pough, Janis, and Heiser, 2013) (Fig. 20.3). Another group, the gorgonopsids, were chiefly predatory animals, and several species had enlarged, saber-toothed canines (McLoughlin, 1980; Carroll, 1988; Benton, 2005; Antón, 2013; Pough, Janis, and Heiser, 2013) (Fig. 20.3). As successful as the dinocephalians, gorgonopsids, and other lineages of non-mammal therapsid synapsids were, nearly all of them went extinct or were severely devastated by the end-Permian extinction event (Benton, 2005). As you will recall from chapter 13, the coalescing of the continents into the supercontinent Pangea appears to have caused a cascading of environmental events that led to huge, inhospitable deserts, reduced water access, and extreme climate change. One can begin to appreciate why so many synapsids, such as the dicynodonts, became avid burrowers: it may have been a way to escape an otherwise extreme environment (Bordy et al., 2011).

Non-mammal Cynodonts—Close Ancestors of the Mammals

It is against the background of environmental devastation wrought by a changing Earth at the end of the Permian that the synapsid lineage to which modern mammals directly trace their ancestry arose and diversified. Called cynodonts, these were generally smaller-bodied synapsids compared with other groups we have examined. Before the end-Permian extinction, cynodonts were a smaller part of a larger, more diverse synapsid-dominated ecosystem in which dinocephalians, gorgonopsids, dicynodonts, and other relatives flourished. After the Permian, dicynodonts and some other non-mammal therapsid synapsids hung on, but by the dawn of the Jurassic period (~200 Ma; Card 96), only the cynodonts were left to carry on the synapsid lineage (Carroll, 1988; Benton, 2005; Pough, Janis, and Heiser, 2013). Some fragmentary evidence from Australia suggests that at least one species of dicynodont made it to the Cretaceous

Pedigree Synapsids Closer to Mammals Than to other Non-mammal Synapsids

Date of First Appearance 260 Ma (Card 95)

Specialties of the Skeletal Chassis Expanded Zygomatic Arch, Interlocking Postcanine Teeth

Eco-niche Small to Large Herbivores, Omnivores, and Carnivores in Terrestrial Environments

period (~110 Ma; Card 98) (Thulborn and Turner, 2003), but otherwise only the cynodonts carried on after the Jurassic. These derived cynodonts, called mammals, were the ancestral population that would later spawn the diverse lineages of furry friends we know today. Therefore, because mammals are cynodonts, we refer to the animals discussed here as non-mammal cynodonts.

One of the best known non-mammal cynodonts to emerge at the beginning of the Triassic period (~245 Ma; Card 95) was *Thrinaxodon*, a little animal 30–50 cm in overall length (McLoughlin, 1980; Carroll, 1988; Benton, 2005; Pough, Janis, and Heiser, 2013) (Fig. 20.4). It is in *Thrinaxodon* that we begin to see the emergence of features we already know well for the mammal chassis. Beginning with the skull, the premaxilla, maxilla, and dentary bones bear heterodont teeth; that is, we see in *Thrinaxodon* incisors, large canines, and molar-like teeth. Moreover, the premaxilla and maxilla have expanded inward, creating a rudimentary secondary bony palate that separated the nasal passages from the mouth as in modern mammals (Benton, 2005; Pough, Janis, and Heiser, 2013). There is the potential that *Thrinaxodon* possessed primitive whiskers! Foramina, pores that typically transmit blood vessels and nerves, are present in the bones of its facial skeleton. Such pores are noted in many different mammals, especially in those where the whiskers are extremely important organs of touch, such as in a walrus (Reep et al., 2011). If *Thrinaxodon* and other cynodonts had whiskers, it is also possible they had developed sensory hairs or even primitive fur, although no direct evidence for these traits is currently known (McLoughlin, 1980; Benton, 2005; Pough, Janis, and Heiser, 2013). We will return to the subject of mammal endothermy in the last chapter. The braincase has enlarged, and the postorbital opening has expanded in concert with a well-developed zygomatic arch (Fig. 20.4). The postorbital opening has also expanded dorsally through the development of a sagittal crest–a raised ridge of bone down the middle of the braincase bones that provides additional attachment surface for the adductor mandibulae (Benton, 2005). Such a trend is observed today in many mammals from big cats to gorillas, and it is always associated with the expansion of the temporalis muscle. These skull features in *Thrinaxodon* strongly suggest that the adductor mandibulae was becoming specialized into the temporalis and masseter groups we see in modern mammals (Benton, 2005; Pough, Janis, and Heiser, 2013).

In the lower jaw, the dentary has become enlarged to the point where it is the dominant bone. A large coronoid process has developed on the dentary, reminiscent of many predatory mammals, and this suggests that a strong, mechanically effective temporalis-like musculature was present that pulled the jaws shut (Carroll, 1988; Benton, 2005) (Fig. 20.4). In fact, as in modern mammals, the coronoid process on the dentary bone extends well up inside the zygomatic arch. The dentary bone also has a depression along its caudal border, a feature expanded in mammals into the masseteric fossa where the masseter inserts (Fig. 20.4). This feature in *Thrinaxodon* leaves little doubt that a well-developed mammal-style

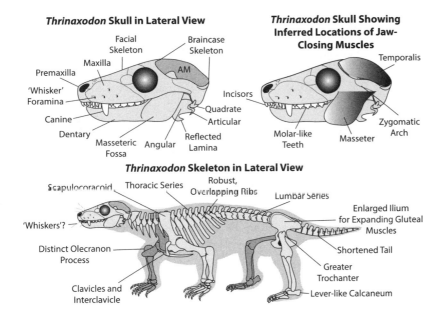

Thrinaxodon Skull in Lateral View

Facial Skeleton
Braincase Skeleton
Maxilla
AM
Premaxilla
'Whisker' Foramina
Incisors
Canine
Quadrate
Dentary
Articular
Masseteric Fossa
Angular
Reflected Lamina

Thrinaxodon Skull Showing Inferred Locations of Jaw-Closing Muscles

Temporalis
Molar-like Teeth
Masseter
Zygomatic Arch

Thrinaxodon Skeleton in Lateral View

Scapulocoracoid
Thoracic Series
Robust, Overlapping Ribs
Lumbar Series
Enlarged Ilium for Expanding Gluteal Muscles
'Whiskers'?
Shortened Tail
Distinct Olecranon Process
Greater Trochanter
Clavicles and Interclavicle
Lever-like Calcaneum

20.4. The cynodont therapsid, *Thrinaxodon*. Note how mammal-like the skull and skeleton appear, especially the large dentary bone in the jaw, and the lumbar vertebral series. Note as well that in *Thrinaxodon* the zygomatic arch has developed to the point where it is likely that a mammal-like masseter muscle was present. In the jaws, the dentary bone has become so large that the postdentary bones are rather small and displaced in comparison. Skull and skeletal diagrams based on illustrations in Carroll (1988), Benton (2005), and Kielan-Jaworowska (2013).

masseter muscle was already present in cynodonts (Benton, 2005; Pough, Janis, and Heiser, 2013). The postdentary bones of *Thrinaxodon* are small, crowded toward the ear, and loosely associated with one another. In fact, these postdentary bones are almost swallowed up by the dentary bone so that they sit in a channel on its medial side. The old quadrate-articular jaw joint was still present and functioning, but based on the greatly enlarged dentary and expanded zygomatic arch, most of the jaw musculature was now firmly associated with the dentary. Thus, the pivot of the jaw and the muscles that powered its closing were now dissociated from one another (Benton, 2005; Pough, Janis, and Heiser, 2013). This change in jaw muscle attachments would have significant impacts on both chewing and hearing in the living cynodont descendants, the mammals (Luo, 2011). We will return to this surprisingly evolutionary transition later in the chapter.

Beyond the skull, other features of the skeletal chassis of *Thrinaxodon* are also mammal-like. As in almost all modern mammals, there were now only seven cervical vertebrae, and the atlas and axis had shapes reminiscent of those we previously discussed in chapter 19 (Benton, 2005; Pough, Janis, and Heiser, 2013). In fact, the occipital condyle surrounding the foramen magnum is divided into the two convex wedges we have seen for mammals, and these articulated with facets on the atlas vertebra. In addition, a primitive dens-like process projects from the axis vertebra to articulate with the atlas, just as it does in mammals. Although small bony elements (the proatlas) are still present, they are reduced, and this suggests that *Thrinaxodon* and other cynodonts had the beginnings of the head nodding and side-to-side shaking movements typical of mammals (Pough, Janis, and Heiser, 2013). The dorsal vertebrae are distinctly divided in *Thrinaxodon* into a thoracic set with prominent ribs and a series of seven lumbar vertebrae with extremely reduced ribs (Benton, 2005; Pough, Janis, and Heiser, 2013) (Fig. 20.4). This arrangement suggests that

Thrinaxodon, like modern mammals, may have possessed a diaphragm muscle that could have powered its breathing. Moreover, the stage was now set for developing a skeletal chassis that could flex dorsoventrally. However, given that the ribs were quite robust and overlapping, the trunk of *Thrinaxodon* was likely quite rigid and not capable of much if any dorsoventral flexion. The tail is shortened compared with primitive synapsids and most reptiles, and it is also slender like those of many mammals (Benton, 2005; Pough, Janis, and Heiser, 2013).

The pectoral girdle is more slender than those of other non-mammal synapsids, but *Thrinaxodon* and other non-mammal cynodonts still have a scapulocoracoid in which both the procoracoid and metacoracoid were distinct and which articulated directly with the sternum, as well as prominent clavicles and an interclavicle (Carroll, 1988; Benton, 2005; Pough, Janis, and Heiser, 2013) (Fig. 20.4). The glenoid faces ventrally and laterally as it does in many sprawling amniotes, and this suggests that *Thrinaxodon* and other cynodonts did not have a mammal-like, upright forelimb posture. The pelvic girdle has a well-developed and craniocaudally expanded ilium, which likely anchored powerful gluteal and knee-extending muscles, and the pubis and ischium are much reduced in size compared with earlier synapsids. Moreover, the acetabulum is rotated significantly ventrally. These features suggest, in contradistinction to the forelimb, that the pelvis was capable of holding the hind limb in a more erect, somewhat parasagittal posture, analogous to that of archosaurs like the crocodylians (Fig. 20.4).

The postures suggested by the pectoral and pelvic girdles are bolstered by the shapes of the forelimb and hind limb bones. In the forelimb, the humerus is expanded mediolaterally at both its proximal and distal ends, and a large deltopectoral crest for receiving forelimb adductors is present (Carroll, 1988; Benton, 2005). The radius and ulna are straight bones that could not cross each other and that would only have held the hand in a pronated position with the elbow directed laterally, with the humerus in a sprawled orientation as is typical of many reptiles (see chapters 13 and 19) (Fig. 20.4). Numerous small wrist bones comprise the wrist, and the hand sports five well-developed digits with small claws. For the hind limb, the femur has a well-developed ball-like head and a prominent greater trochanter (Carroll, 1988; Benton, 2005). These two features of the femur are very mammalian, and suggest, along with the shape of the ilium, that enlarged gluteal muscles for holding the hind limb erect and preventing its inward collapse during locomotion were present (Fig. 20.4). The distal end of the femur is rather narrow and articulates almost exclusively with the tibia, a condition close to that of modern mammals, where the knee joint consists solely of the femur and tibia (Benton, 2005). The fibula is less robust than those of primitive synapsids, and has shifted somewhat caudally in relation to the tibia, articulating in a more posterior position both on the tibia and where it does contact the femur. This shift in fibula position seems to be correlated with a well-developed and caudally shifted calcaneum in *Thrinaxodon*. In fact, the calcaneum has a large, lever-like projection that almost undoubtedly anchored the powerful Achilles' tendon from the gastrocnemius and synergistic muscles that collectively rotated the foot into plantarflexion (Carroll, 1988; Benton, 2005). Five long, clawed toes adorn the foot (Fig. 20.4).

Fossils of *Thrinaxodon* and several closely related cynodonts have been preserved in ways that have shed some light on their potential lifestyle. If the interpretations are correct, *Thrinaxodon* and other close kin did not just have a mammal-like skeletal chassis—they had incipient mammalian behaviors. As with many dicynodonts (Groenewald, 1991), *Thrinaxodon* and other cynodont fossils from South Africa are often found associated with burrows (Groenewald, Welman, and MacEachern, 2001; Damiani et al., 2003; Benton, 2005). Whereas this in and of itself seems to go along with an ancient synapsid behavior (i.e., digging yourself a home) (Groenewald, Welman, and MacEachern, 2001), it is how cynodonts are found in these burrows that is truly amazing. For one, it is clear that many of these cynodonts were making a home in the burrow and were not simply washed in there by a flood. For example, their skeletons are often well preserved and curled up nose to tail (Damiani et al., 2003), not a typical position assumed by a small animal violently washed into a hole in the ground. Even more fascinating, in some specimens teeth have fallen from their sockets and lay with the skull, a kind of forensic clue suggesting that the animal died where it lived and that the teeth later fell out of the decaying creature's mouth (Damiani et al., 2003). Whereas these are typically lone specimens, a closely related cynodont species, *Trirachodon*, has been found in or near burrow complexes where up to 20 specimens ranging from juvenile to adult are associated (Groenewald, Welman, and MacEachern, 2001). Again, these specimens are often curled up, and perhaps they were the victims of a flash flood that caught their family group (or groups) by surprise (Groenewald, Welman, and MacEachern, 2001; Damiani et al., 2003). This close association of juveniles and adults near a complex burrow system strongly suggests these cynodonts were living together communally much as many mammals do now.

The non-mammal cynodonts continued to diversify and become more diminutive throughout the Triassic period, becoming small-bodied vertebrates dashing beneath the feet of the ever-enlarging dinosaurs. A quick glance during the Triassic at some of these non-mammal cynodonts might remind us of rats: in particular, some were sleek little animals with large gnawing teeth and robust jaws and maybe, just maybe, fur. Given that the dinosaurs were relatively large and diurnal predators with reasonably good eyesight, direct competition with these archosaurs would have spelled disaster for the cynodonts. Instead, cynodonts became smaller and probably led a more nocturnal existence to avoid the gaze of these large and hungry hunters. Against this backdrop, sometime during the close of the Triassic or the opening of the Jurassic, the very first cynodonts to become mammaliaforms made their appearance.

Pedigree The Common Cynodont Ancestor of All Living Mammals

Date of First Appearance 200 Ma (Card 96)

Specialties of the Skeletal Chassis A Dentary-Squamosal Jaw Joint, Three Inner Ear Ossicles Still Associated with Lower Jaw

Eco-niche Insectivores

The mammaliaforms were mammalian ancestors that had a skeletal chassis just north of the non-mammal cynodont line, meaning there were a number of features we would consider a bit odd. One of the best known and reasonably complete mammaliaforms is *Morganucodon*, a tiny animal (<100 grams: the size of a shrew) from Wales approximately 200–190 Ma (Card 96) (Benton, 2005; Kielan-Jaworowska, 2013). *Morganucodon* was small even when compared with the smallest non-mammal cynodonts (~500 grams) (Benton, 2005; Kielan-Jaworowska, 2013; Pough, Janis, and Heiser, 2013). Close inspection of the skull would reveal the following weirdness. The dentary bone was so large that it covered the now tiny postdentary bones, which were still held in a trough on its medial side as they were in non-mammal cynodonts like *Thrinaxodon*. Here we see a fully formed dentary-squamosal jaw joint as we see in all mammals today, and the dentary bears a condyle just as it does in you and me. But, surprisingly, the quadrate-articular jaw joint was still present and . . . drum roll . . . still functional (Carroll, 1988; Benton, 2005; Luo, 2011; Pough, Janis, and Heiser, 2013) (Fig. 20.5). In essence, the old vertebrate jaw joint we have followed for most of the length of this book was now very tiny but still present, sitting just behind the 'new' dentary-squamosal joint in *Morganucodon*. You have probably never thought of your jaw joint being 'double-jointed,' and yet, here in this tiny mammaliaform there were two jaw joints side by side! *Morganucodon*, its close relatives, and some of the last non-mammal cynodonts all show this pattern (Luo, 2011) . . . but what does it all mean?

We now have all the skeletal pieces in place, so to speak, to rewind back to *Dimetrodon* and the early synapsids to flesh out this amazing evolutionary story. Back in chapter 9, we discussed the basic mechanics and evolution of the tetrapod ear, but left out one important detail – the ear of tetrapods has apparently evolved at least three different times in different tetrapod lineages (Manley, 2010)! What is mind-blowing is that most of the basic bony components have remained the same – for example, the ear is always situated near the jaw, and the quadrate bone in the skull is always very close to the stapes, which in all tetrapods transmits airborne vibrations into the inner ear. In amphibians and again in the Reptilia there was the independent acquisition of a tympanic membrane in a notch or depression near the jaw (on the quadrate bone), whose vibrations were passed onto the stapes to be interpreted as sound in the inner ear and brain (Liem et al., 2001; Manley, 2010; Kardong, 2012; Pough, Janis, and Heiser, 2013). But we can't look to these tetrapods as the starting point for hearing in mammals. Why? Because for reasons not altogether clear, the earliest synapsids held their stapes fast to their quadrate . . . and this means that hearing in mammals was acquired independently of amphibians and reptilians (Luo, 2011; Pough, Janis, and Heiser, 2013).

Thus, non-mammal synapsids like *Dimetrodon* could not have listened to airborne sounds with a tympanic membrane attached to the

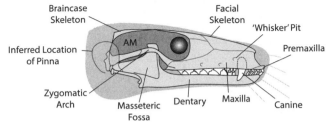

Skull of the Mammaliaform *Morganucodon*

Braincase Skeleton

Facial Skeleton

'Whisker' Pit

Premaxilla

AM

Inferred Location of Pinna

Zygomatic Arch

Masseteric Fossa

Dentary

Maxilla

Canine

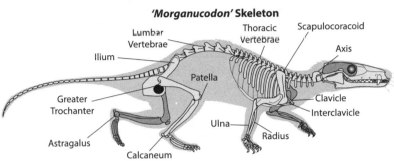

'*Morganucodon*' Skeleton

Lumbar Vertebrae

Thoracic Vertebrae

Scapulocoracoid

Ilium

Axis

Patella

Greater Trochanter

Clavicle

Interclavicle

Ulna

Radius

Astragalus

Calcaneum

20.5. One of the best known mammaliaforms, *Morganucodon*. Notice the further development of the dentary and its masseteric fossa and zygomatic arch. Also note that the postorbital fenestra has expanded to the point that it encroaches on the orbit. The postcranial skeleton is based in part on skeletons of the closely related *Megazostrodon*. Diagrams based on illustrations in Jenkins and Parrington (1976), Carroll (1988), Benton (2005), and Kielan-Jaworowska (2013).

stapes—there was no conceivable way for this to happen (Allin, 1975; Crompton and Parker, 1978; Benton, 2005; Pough, Janis, and Heiser, 2013). In fact, looking at the skull of *Dimetrodon*, one is hard pressed to see anywhere that a tympanic membrane would have found a home (Fig. 20.1). It could very well be that the earliest synapsids were hard of hearing and, like snakes, detected low-frequency vibrations when their jaws were pressed to the ground (Crompton and Parker, 1978). And curiously, the skull and jaws of *Dimetrodon* may point the way. The angular bone is primitively a part of the amniote lower jaw, and it connects to the articular bone, which, as we are now very familiar with, forms the lower part of the jaw joint. The angular bone of *Dimetrodon*, and in fact all synapsids from their evolutionary inception, has a part called the reflected lamina that curls ever so slightly outward (Allin, 1975; Luo, 2011; Pough, Janis, and Heiser, 2013) (Fig. 20.1). If the lower jaw was placed on the ground, low-frequency vibrations could pass up through the reflected lamina of the angular, onward through the articular, upward into the quadrate, and finally vibrate the stapes against the inner ear (Allin, 1975; McLoughlin, 1980; Pough, Janis, and Heiser, 2013).

If this hypothesis has merit, then we would predict that the angular bone and its reflected lamina should continue to play a part in mammal hearing today . . . and we would be correct in that prediction! As it turns out, the tympanic membrane of all mammals, including you, is stretched out and supported by a tiny, semilunate bone called the ectotympanic, which itself sits near the inner ear bone called the malleus (Manley, 2010; Luo, 2011) (Fig. 20.6). The ectotympanic bone, as it turns out, is the old angular bone and the malleus is the old articular (Liem et al., 2001; Manley, 2010; Luo, 2011; Pough, Janis, and Heiser, 2013) (Fig. 20.6). How do we know that? Before we delve into fossils: mammal embryos point the way

20.6. Evolutionary changes in the jaw joint and middle ear bones from non-mammal cynodonts to therian mammals. The amniote quadrate-articular jaw joint is indicated by QA, whereas the new mammalian squamosal-dentary jaw joint is indicated by SQD. Notice that in *Morganucodon,* there are two functional jaw joints operating synchronously. Also note that in *Thrinaxodon* and *Morganucodon* there is a groove on the medial (lingual) side of the dentary, which indicates the presence of a Meckel's cartilage uniting the small postdentary bones into a functional chain. Diagrams based on illustrations in Luo (2011). See text for further details.

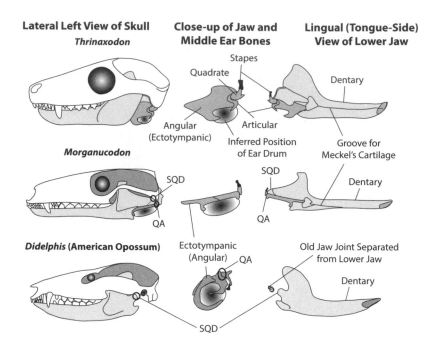

Lateral Left View of Skull — *Thrinaxodon*

Close-up of Jaw and Middle Ear Bones

Lingual (Tongue-Side) View of Lower Jaw

Stapes
Quadrate
Dentary
Angular (Ectotympanic)
Articular
Inferred Position of Ear Drum
Groove for Meckel's Cartilage

Morganucodon
SQD
SQD
Dentary
QA
QA

Didelphis (American Opossum)
Ectotympanic (Angular) QA
Old Jaw Joint Separated from Lower Jaw
Dentary
SQD

during their development. In the small domestic or laboratory opossum (*Monodelphis domestica*), the development of the jaw and other aspects of anatomy have been researched in detail because these mammals are born very underdeveloped and complete most of their development outside attached to their mother's teats (Fadem et al., 1982; Sanchez-Villagra et al., 2002). In other words, domestic opossums, being marsupials, have done us an extraordinary favor—most of the developmental events that would occur inside the womb in placental mammals occur for all to see externally. This has made domestic opossums a convenient study subject for mammal development.

When a newborn, embryo-like domestic opossum emerges into the world and attaches to its mother's teat, its middle ear has not yet finished developing (Sanchez-Villagra et al., 2002; Luo, 2011). A look inside the ears of these newborn and developing opossums shows something remarkable—they, like *Morganucodon* before them, have a double jaw joint (Sanchez-Villagra et al., 2002; Luo, 2011; Kielan-Jaworowska, 2013) (Fig. 20.7). Not only is the dentary-squamosal jaw joint of mammals present, but the articular is still part of the lower jaw, connecting to the quadrate. And here we have to reach way back to chapter 5, where we introduced the undergirding of the lower jaw, to our old friend the Meckel's cartilage. In all bony vertebrates, the Meckel's cartilage develops embryonically but, as you may recall, becomes covered and replaced to varying degrees by dermal bones that form the dentary and other elements of the lower jaw. In amniotes like reptiles, the Meckel's cartilage persists and holds all of the postdentary bones together. The Meckel's cartilage sits in a groove inside the dentary (Luo, 2011), and this is the same groove we encountered in *Thrinaxodon* where we

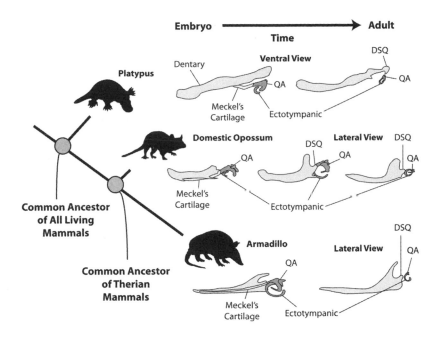

20.7. Embryonic series of lower jaw and ear development in exemplar mammals from each of the living groups. In primitive mammals (duck-billed platypus) and therian mammals (domestic opossum and armadillo), the postdentary bones (ectotympanic, quadrate, and articular) are initially bound to the dentary through the Meckel's cartilage. As development proceeds, the Meckel's cartilage dissolves and the connection between the postdentary bones and the dentary is lost, freeing the quadrate and articular (QA) to vibrate in a chain with the stapes in the middle ear. The old angular bone, now called the ectotympanic, becomes incorporated into the ear as a bony suspender of the tympanic membrane (ear drum). Diagrams based on illustrations in Luo (2011).

found the angular, articular, and quadrate all crowded together (Figs. 20.4, 20.6).

In *Monodelphis* babies, we observe the same situation as seen in *Thrinaxodon*, but notice something very important – the angular, with its reflected lamina, is holding tissue that will become the tympanum or ear drum (Sanchez-Villagra et al., 2002; Pough, Janis, and Heiser, 2013). As development continues, the braincase expands, the dentary enlarges, and the Meckel's cartilage begins to break down and release its hold on the angular, articular, and quadrate (Anthwal, Joshi, and Tucker, 2013) (Fig. 20.7). Once the connection of these three minute bones with the lower jaw is released, they are pushed away from the dentary-squamosal joint, while the quadrate still hangs on to its connection with the stapes (Anthwal, Joshi, and Tucker, 2013) (Fig. 20.7). In essence, what we see played out during *Monodelphis* development is a transition from a double jaw joint anatomy like we observe in *Morganucodon* to the typical mammalian dentary-squamosal joint and three inner-ear ossicles (Luo, 2011; Anthwal, Joshi, and Tucker, 2013; Pough, Janis, and Heiser, 2013) (Fig. 20.8). We also see how the angular bone provides a semicircular 'notch' that holds taut the tympanic membrane. These bones are now renamed in mammals as the ectotympanic (angular), the malleus (the articular), and the incus (the quadrate) (Fig. 20.8). The old stapes bone is still there doing what it has done for most lineages of tetrapods for over 350 million years – vibrating against the inner ear, acting like a sound transducer. The difference for mammals is that our ancestors incorporated the jaw into hearing, and as a consequence we have two other inner ear ossicles in addition to the stapes. This had the fortuitous side effect of creating three small levers (the malleus, incus, and stapes) that could amplify and heighten awareness of high-frequency sounds (Luo, 2011).

20.8. Fossil and embryonic development sequences compared side by side. On the left, the fossil sequence of jaw and middle ear evolution from a non-mammal cynodont (*Thrinaxodon*) to a modern therian mammal (*Didelphis*); on the right, the developmental sequence of the jaw and middle ear in the laboratory opossum, *Monodelphis*. Note that the basic sequence of events both in the fossil record and in modern mammal development shows fairly clearly how a new jaw joint and a rearrangement of the middle ear evolved. Illustrations based on figures in Luo (2011) and Kielan-Jaworowska (2013).

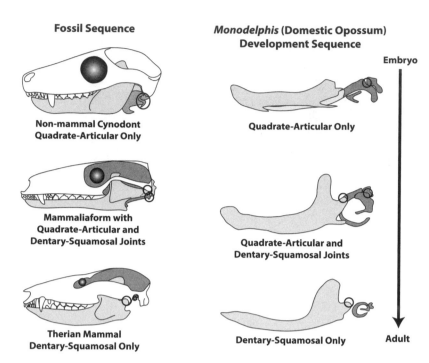

Fossil Sequence

Non-mammal Cynodont
Quadrate-Articular Only

Mammaliaform with
Quadrate-Articular and
Dentary-Squamosal Joints

Therian Mammal
Dentary-Squamosal Only

***Monodelphis* (Domestic Opossum)
Development Sequence**

Embryo

Quadrate-Articular Only

Quadrate-Articular and
Dentary-Squamosal Joints

Dentary-Squamosal Only   Adult

In fact, mammals on the whole can hear much higher frequencies than most other vertebrates because of this unique transformation, and echolocating mammals that rely on high-frequency vocal pulses (bats and whales) would not exist if not for this remarkable evolutionary occurrence (Manley, 2010).

Returning to the fossil record with the eyes of an embryologist, we can now clearly see that the angular bone becomes flatter and smaller and that the reflected lamina becomes more prominent, as we move from primitive non-mammal synapsids to mammaliaforms (Allin, 1975; Luo, 2011) (Fig. 20.8). Whether non-mammal synapsids like *Dimetrodon* had a tympanic membrane is unclear. However, given the shape of the angular bone and its proximity to the articular, even thinner skin in this region of the lower jaw may have allowed some higher-frequency airborne sounds to be detected (Allin, 1975; Pough, Janis, and Heiser, 2013). It is clearer in non-mammal cynodonts that the angular held a tympanic membrane because its shape is eerily similar to that of the ectotympanic of mammals (Figs. 20.6, 20.8). And whereas the quadrate and articular bones in non-mammal cynodonts like *Thrinaxodon* were still part of the jaw joint, their small size and their incorporation into the groove on the dentary occupied by the Meckel's cartilage leave little doubt that they were capable of transmitting higher-frequency sounds to the stapes and into the inner ear (Manley, 2010; Luo, 2011; Anthwal, Joshi, and Tucker, 2013) (see also *Probelesodon* in Plate 25). By the time we get to *Morganucodon* and into early true mammals, whereas the quadrate and articular are still attached loosely to the jaw, the ectotympanic (the angular) has a shape reminiscent of our opossum friends (Fig. 20.8). This strongly suggests that the ectotympanic of *Morganucodon* would have held a tiny, delicate tympanic membrane ideal for detecting and transferring high-frequency sounds through a chain of three tiny lever-like ossicles into the inner ear (Allin, 1975; Luo, 2011; Kielan-Jaworowska, 2013). One final thought on all of this—the next time you snack on potato chips or eat cereal and hear the crunching and crackling of the food deep in your ear, remember that this is because you are hearing with the old quadrate-articular jaw joint of our deep amniote and synapsid ancestors (McLoughlin, 1980; Pough, Janis, and Heiser, 2013).

What also goes along with three inner ear ossicles and higher-frequency hearing? Pinnae, the soft, cartilaginous projections encased in skin around the ear opening (auditory meatus), appear to go hand-in-hand (sound-in-ear?) with this development. Essentially, the pinnae of mammals act like radar or satellite dishes, collecting and funneling sounds into the middle ear (McLoughlin, 1980; Manley, 2010) (Fig. 20.5). Moreover, specialized facial muscles attach to and move the pinnae in many mammals, providing them with the ability to gather sounds stereoscopically, which helps fine tune and detect where particular high-frequency sounds in the environment are emanating from (McLoughlin, 1980; Manley, 2010) (Fig. 20.9). Not all mammals have well-developed pinnae (some marine mammals lack them altogether), and not all mammals have active control over those pinnae. Nonetheless, it is conceivable that, given the evolution of ear elements that we have discussed, mammaliaforms like *Morganucodon* may have sported little pinnae.

With the dentary-squamosal now the major jaw joint, and with the dentary being the major bone into which the jaw closing and opening muscles inserted, *Morganucodon* and early true mammals were probably chewing more or less like their modern descendants. The masseter would act as a powerful sling that in addition to providing impressive grinding forces would rotate the jaw side-to-side in a pattern stereotypical for most mammals alive today (Gorniak, 1985; Liem et al., 2001; Kardong, 2012; Pough, Janis, and Heiser, 2013) (Fig. 20.9). Since the lower jaw was narrower than the upper jaw, the teeth could only be brought into occlusion one side at a time, but this motion would allow the food to be rolled around and smashed repeatedly until it was pulped into a bolus ready for swallowing (Kielan-Jaworowska, 2013; Pough, Janis, and Heiser, 2013). The enlarged coronoid process on the dentary of *Morganucodon* and early true mammals would provide the ideal anchoring point for the temporalis. And then, of course, there were the teeth themselves. Like modern mammals, *Morganucodon* had well-developed premolars and molars with ridges and valleys perfect for quickly pulping food into manageable bites (Fig. 20.5). Given its tiny size and small, pointed teeth, *Morganucodon* and early true mammals must have primarily been insectivores, somewhat like tiny modern shrews with their pointy teeth and voracious appetites. But most tellingly, *Morganucodon* and early true mammals now had deciduous teeth—that is, they had a set of milk teeth that were later shed to make way for the adult teeth, a condition

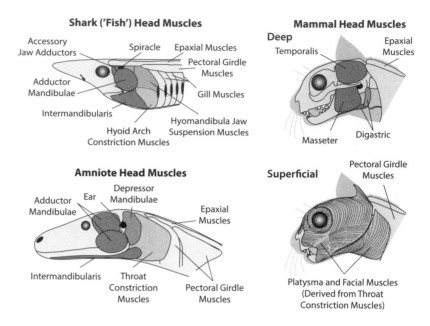

20.9. Evolution of head muscles in mammals. In mammals, the original jaw-closing adductor mandibulae muscles have become subdivided into a dorsal temporalis and a more ventral, sling-like masseter. With the rearrangement of the jaw joint, the old depressor mandibulae system became inadequate for opening the jaws. As a consequence, the fusion of muscles descended from the old intermandibularis and portions of the depressor mandibulae became a new muscle, the digastric, which lowers the jaw in modern mammals. Sequence of muscle evolution based on diagrams and information in Liem et al. (2001) and Diogo and Abdala (2010).

called diphyodonty (Benton, 2005; Pough, Janis, and Heiser, 2013). As you will recall from chapter 19, having permanent adult teeth ensures a stable and unchanging set of occluding surfaces for properly chewing and grinding food.

However, we are leaving out one important detail here: how did *Morganucodon* and early true mammals open their jaws? This is a good question, since the old depressor mandibulae was associated with opening the jaws at the quadrate-articular jaw joint—a joint now transformed into a miniature auditory chain within the middle ear. In fact, the gradual miniaturization or loss of the postdentary bones in mammals caused some of the original jaw-opening and -closing muscles to lose their moorings. Three key muscle changes occurred in mammals related to jaw opening as a consequence (Fig. 20.9).

First, the depressor mandibulae, while still remaining attached to the skull, had a complex history in which portions of it grabbed onto different bones as the quadrate-articular jaw joint miniaturized (Diogo and Abdala, 2010). Some of the original portions of the depressor mandibulae apparently migrated into the middle ear with the old jaw joint. For example, a tiny muscle associated with the stapes in mammals (called stapedius) is derived from part of the muscle bundles that give rise to the depressor mandibulae in other vertebrates (Diogo and Abdala, 2010). However, second, a significant slip of the old depressor mandibulae seems to have remained in its original position, and eventually reached out to insert into tissues around the hyoid bone (Diogo and Abdala, 2010). Probably simultaneously, the intermandibularis, an old deep jaw elevator nestled in between the dentaries, also reached out for a new insertion point around the hyoid. The consequence of these two muscles finding new places to insert was the development of a unique mammal muscle, the digastric

(Liem et al., 2001; Diogo and Abdala, 2010; Pough, Janis, and Heiser, 2013). This sequence of events is supported by embryology and nerve studies. Embryonically, these two muscles grow from different places to find a common 'middle ground' on the hyoid. Moreover, the anterior digastric (the old intermandibularis) is innervated by a different cranial nerve (5, the trigeminal) than the posterior digastric (the old depressor mandibulae) (7, the facial), and this further supports that the mammal digastric is a muscle 'fused' from two others (Diogo and Abdala, 2010). Collectively, when the anterior and posterior 'bellies' of the digastric contract, they act to lift the hyoid bone, which, in turn, assists in swallowing. However, when the hyoid is immobilized by other muscles, contraction of the digastric pulls through the dentary, which depresses the lower jaw (Fig. 20.9). In addition, third, the lateral portion of the pterygoid muscle (an original deep jaw closer as you will recall) still inserts on the dentary, but now things have changed. The line of action for this muscle passes right next to the new dentary-squamosal jaw joint. As a consequence, as mentioned in chapter 19, when the lateral pterygoid is contracted, it causes the condyloid process of the dentary to rotate cranially, which lowers the jaw (Moore and Dalley, 1999; Liem et al., 2001). In essence, the decoupling of the old quadrate-articular jaw joint in mammals 'forced' the evolution of a new muscle arrangement to lower the jaws and open the mouth.

Beyond the ear, jaws, and teeth, *Morganucodon* and early true mammals had other mammalian adaptations as well. As in many modern mammals, the posttemporal opening was greatly enlarged and the orbit merged with this opening (McLoughlin, 1980; Benton, 2005; Kielan-Jaworowska, 2013) (Fig. 20.5). The braincase itself was greatly expanded compared with non-mammal cynodonts like *Thrinaxodon*, presaging the beginnings of mammalian brain expansion (Carroll, 1988; Benton, 2005; Kielan-Jaworowska, 2013). Examining the roof of the mouth, one can see that there are incisive foramina (Benton, 2005; Kielan-Jaworowska, 2013), which, as you will recall, signal the presence of a well-developed vomeronasal organ for pheromone detection. Additionally, the premaxilla and maxilla help form a bony secondary palate that separated the nasal cavity from the mouth (Carroll, 1988; Benton, 2005; Kielan-Jaworowska, 2013).

The skeleton of *Morganucodon* itself is rather fragmentary, but a closely related form known as *Megazostrodon* gives a good approximation of what the postcranial skeleton of these tiny mammaliaforms was like (Benton, 2005; Kielan-Jaworowska, 2013; Pough, Janis, and Heiser, 2013). To avoid confusion and mixing of different taxonomic names, we will simply refer to the postcranial skeleton of *Megazostrodon* as 'Morganucodon' throughout the text. The axial skeleton of these mammaliaforms resembles in many ways what we have already encountered in rats. There are seven cervical vertebrae with a specialized atlas and axis that would have allowed the nodding and side-to-side movements of the head we typically see in mammals. There is a distinct thoracic vertebral series with ribs, followed by a rib-lacking lumbar series, a tiny sacrum, and a gracile, ratlike caudal series (Jenkins and Parrington, 1976). The clear separation between the thoracic and lumbar vertebral series leaves little doubt that 'Morganucodon' would have possessed a diaphragm and the ability to walk (but not quite bound) with dorsoventral flexion (Fig. 20.5).

The pectoral girdle of 'Morganucodon' resembles in many ways that of non-mammal cynodonts and, to some extent, even lizards. There is a T-shaped interclavicle that unites with the clavicles, all of which articulate with a narrow sternum, suggesting that, like in lizards (see chapter 13) these elements acted as spokes that held and guided the forelimbs as the animal walked (Gray, 1968; Jenkins, 1974; Jenkins and Parrington, 1976). There is still a coracoid bone (the metacoracoid) that articulates with the sternum, a long but relatively narrow scapula, and a wide and mostly laterally directed glenoid, all traits found in the crank-like forelimbs of many reptiles and primitive living mammals like the duck-billed platypus (Jenkins and Parrington, 1976) (Fig. 20.5). The pelvis of 'Morganucodon' is very mammalian, consisting of a long, cranially projecting ilium coupled with a short pubis and ischium. Moreover, the acetabulum is a deep socket that, as in most mammals, faces lateroventrally (Jenkins and Parrington, 1976; Kielan-Jaworowska, 2013; Pough, Janis, and Heiser, 2013) (Fig. 20.5). These are all hallmarks of a switch to the gluteal muscle–driven parasagittal posture common to mammals (Benton, 2005).

As we saw for dinosaurs, the earliest mammals had forelimbs with a more sprawled posture than the hind limbs (Kielan-Jaworowska, 2013). In 'Morganucodon,' the humerus has some very mammalian characteristics, including a smooth, spherical head and a relatively narrow shaft (Fig. 20.5). However, in other respects the humerus is still somewhat like what we have encountered in many reptiles. The deltopectoral crest, while not as large and medially deflected as in many reptiles, is still relatively prominent, and distally the epicondyles that act as anchoring regions for the extensor and flexor muscles of the wrist and hand are expanded as they commonly are in many reptiles. Moreover, whereas there are articular surfaces for the radius and ulna, these surfaces form low, rounded condyles rather than the distinctly convex capitulum and groove-like trochlea seen in modern mammals (Jenkins and Parrington, 1976) (Fig. 20.5). The radius and ulna look surprisingly like those of small, modern mammals such as rats (Fig. 20.5). The ulna itself has the beginnings of the wrench-like shape we saw for eutherian mammals, and even a distinct olecranon process to enhance the insertion of the triceps muscles (Jenkins and Parrington, 1976; Carroll, 1988; Benton, 2005; Kielan-Jaworowska, 2013). The radius has an oval rather than a rounded head, which suggests it could not actively turn against the ulna to enable much active pronation and supination. However, we see that the radius shaft in 'Morganucodon' crosses cranially to the ulna (Jenkins and Parrington, 1976), a harbinger of this more extensive trend in therian mammals. As we discussed in chapter 19, this would help in at least partially pronating the hand. As we saw for many reptiles, if this anatomy was combined with a more sprawled forelimb in which the elbow was directly laterally, this would, in conjunction with the partially crossed radius, have effectively pronated the hand in full. A tiny clustering of wrist bones similar to those we saw for the rat is present, and a delicate five-fingered hand is present (Fig. 20.5).

The hind limb is more mammalian in 'Morganucodon.' The femur has a prominent medially oriented semispherical head supported on a short but distinct 'neck.' As we saw for the rat and cat, the femur bears greater and lesser trochanters (Jenkins and Parrington, 1976), which are related to the insertion of the gluteal muscles associated with a more parasagittal posture in mammals (Benton, 2005) (Fig. 20.5). Distally, the femur has distinct condyles for articulation with the tibia, and there is evidence of a patellar groove (Jenkins and Parrington, 1976). Although no patella is known from 'Morganucodon,' the presence of a clear patellar groove is good indirect evidence that this sesamoid bone was present and assisted the quadriceps muscles in extending the leg. What is known of the tibia shows that, like we saw for eutherian mammals, the proximal end of the tibia is divided into two distinct facets for articulation with the femur separated by an eminence. However, the fibula has a craniocaudally broad proximal end that would have articulated with the lateral corner of the femur (Jenkins and Parrington, 1976) much like what we have seen for non-mammal tetrapods (Fig. 20.5). The fibula, therefore, may have been capable of some independent rotational movements in relation to the tibia, which suggests that whereas the hind limb was more erect than the forelimb, it could not have assumed the condition we see in certain eutherian mammals like the cat. The astragalus and calcaneum are distinct but not as developed as we are familiar with from eutherian mammals (Fig. 20.5). The calcaneum does bear a small but distinct tuberosity (Jenkins and Parrington, 1976), which would have provided the gastrocnemius and other plantarflexors with a point of lever-like insertion. However, both the astragalus and calcaneum articulate proximally with the tibia and fibula, and this suggests, along with what we observed in the knee, that 'Morganucodon' had an ankle that operated in a more reptilian fashion, allowing some rotational movements to occur between the tarsus and tibia/fibula. The foot is rather long with five spindly digits (Fig. 20.5).

Overall, the chassis of 'Morganucodon' and early true mammals is that of a tiny, crouched insectivorous machine that was relatively agile while still hanging on to more primitive non-mammal synapsid or even reptilian-like features. What we see among the synapsids is an overall trend toward smaller body size, a reorganization of the jaws and their musculature, the development of deciduous teeth, the remarkable development of three ossicles in the middle ear, and the harbingers of a skeleton capable of flexing dorsoventrally. But other anatomical changes were occurring and being enhanced in mammals throughout the Mesozoic Era (245–66 Ma; Cards 95–99), including an ever-enlarging brain and the appearance of hair and milk. One can think of the Mesozoic Era as a funnel through which mammals passed, squeezed by various natural selection pressures, until they exited at the end of Cretaceous Period (66 Ma; Card 99) with highly modified skeletal chassis. In our final chapter, we turn to other changes that occurred in mammals and their amazing evolutionary diversification.

21.1. Comparison of 'primitive' tetrapod long bone growth (as exemplified by an alligator) with that of mammals. In many reptiles (but not lizards), adult size can vary depending on numerous physiological and environmental factors. Reptiles such as alligators do not form epiphyseal bones. Therefore, cessation of growth at adult size in these reptiles is not correlated with the fusion of epiphyseal bones onto the diaphysis. In contrast, mammals (as well as lizards) develop epiphyseal bones that eventually fuse with their shafts at a region called the metaphysis, and this process is correlated with the halting of growth at adult size. Diagrams after Carter and Beaupré (2001), Chinsamy-Turan (2005), and Holliday et al. (2010).

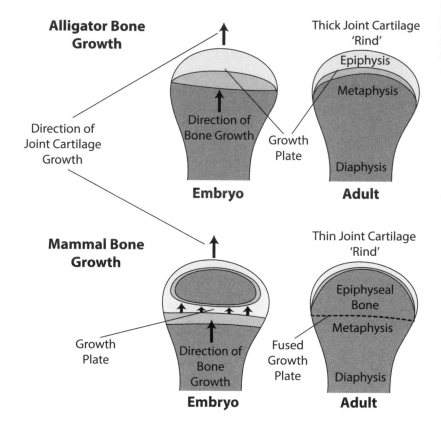

SOMETIMES INVENTIONS SIT IN THE BACKGROUND UNTIL THE TIME is right: a change in culture, perhaps, or the advent of complementary technologies. Many times, the original purpose of the invention or device, while useful in its original context, becomes explosively successful in a new era. During the Mesozoic Era, the mammal chassis was evolving alongside those of the dinosaurs, their archosaur brethren, and other tetrapods. As we have already seen, the mammalian chassis had a number of quirks that, while perfectly capable of keeping our long lost relatives alive, did not appear to place them in direct competition with dinosaurs. In fact, many of the adaptations were those that kept our mammalian ancestors out of contact with the dinosaurs so as to ensure our survival.

The evidence is pretty clear that 66 Ma (Card 99) an asteroid or comet about 10 kilometers across smashed into what is the now the Gulf of Mexico near the Yucatan Peninsula, leaving as evidence a crater 180 kilometers wide (Powell, 1998; Archibald, 2012). The energy released was significant enough to generate mega-tsunamis, to disrupt the molecular structure of quartz where it struck, and to produce a global dust and ash cloud that darkened the Earth for months or more (Powell, 1998; Schulte et al., 2010; Archibald, 2012). Whether this impact was the primary cause of the great extinction at the end of the Cretaceous or whether this event compounded other environmental stresses (Archibald, 2012), the non-avian dinosaurs and many other organisms disappeared in a geologically short period of time. Arising from this great die-off, the ancestors of modern mammals emerged, blinking into the future. Much like opening a soda bottle releases pressure and allows the trapped carbon dioxide gases to escape, so the natural selection pressure of the nonavian dinosaurs was released from mammals.

Mammals brought with them the interesting traits we have discussed in previous chapters, and we are now in a position to appreciate the probable origin of these features more fully. As you will recall, most mammals have limited color vision but excellent sound and odor detection. What pressures would have selected for this combination of features? During the Mesozoic Era, the earliest mammals had been moderately successful but lived under several stressors. First, many dinosaurs were keen-eyed and diurnal, and would have snapped up tiny mammals in the daylight hours. To survive this onslaught, the earliest mammals were most likely crepuscular, or nocturnal in their habits. Being nocturnal was not a total fail-safe, however, because some dinosaurs were also nocturnal or crepuscular, based on the size of their sclerotic rings (Schmitz and Motani, 2011; see also chapter 15 for a discussion on inferred eyeball size from sclerotic rings in ichthyosaurs). Second, turn out the lights and eyesight, particularly color vision, becomes next to useless. However, if even some light is available, any ability to capture it and form images and detect movement is better than nothing. As you will recall from chapter 19, mammal retinas typically have numerous rods (those cells in the retina that detect light and dark) and fewer cones (for detecting color), and many have the reflective tapetum lucidum that creates 'eye shine' and amplifies any light that falls on the eye. Thus, the limited color detection of most of our furry friends may trace itself back to a time when living in the darkness was necessary for survival (McLoughlin, 1980; Heesy and Hall, 2010; Rowe, Macrini, and Luo, 2011).

If you are nearly blind, then odor and auditory detection become that much more crucial. Just watch the family dog or cat go about its daily routine, and you begin to notice a lot of sniffing and the turning of the pinnae around the ears. If you've kept pet rodents, or observed rats or mice, you know that it is common to find them with their noses tilted up, their nostrils twitching rapidly to take in all the odors the environment has to offer. Many of the earliest mammals and the mammaliaforms, such as *Morganucodon*, were small and likely insectivorous. One can imagine our small ancestors emerging after dark, straining to see any sign of movement while their noses and ears worked overtime creating an odor and auditory map in their minds to track down delicious, crunchy insects (McLoughlin, 1980; Heesy and Hall, 2010; Rowe, Macrini, and Luo, 2011). This lifestyle may explain yet another unusual feature of mammalian anatomy: the reduced optic and auditory regions of the midbrain (the superior and inferior colliculi) and the expansion of visual and sound processing areas of the cerebrum. We do know that in modern mammals the cerebral hemispheres have taken over most of the routing and interpretation of sight and sound, whereas the inferior and superior colliculi mostly control visual and auditory reflexes. If you compare the endocast of a non-mammal cynodont such as *Thrinaxodon* to that of a mammaliaform, such as *Morganucodon*, you will note that the cerebrum of the latter is greatly expanded over the reduced optic and auditory centers of the midbrain (Rowe, Macrini, and Luo, 2011). These observations would support the hypothesis that living a nocturnal existence, probably to avoid direct competition with the dinosaurs, created selective pressures that resulted in a change of brain architecture in Mesozoic mammals.

When hair as we know it for mammals first appeared is still a bit uncertain. The earliest fossils of mammals with distinct hair impressions come to us from the Jurassic period, in forms such as the beaver-like *Castorocauda*, the rat-like *Megaconodon*, and the earliest known eutherian mammal, *Juramaia* (Luo et al., 2011; Rowe, Macrini, and Luo, 2011; Kielan-Jaworowska, 2013; Zhou et al., 2013). However, hair in some form may have made an appearance back to 245 Ma (Card 95) or even earlier. If you recall, distinct nutrient foramina on the snout of *Thrinaxodon* may be indicative of vibrissae (whiskers?) and perhaps a hairlike pelage (Maderson, 2003). Where did hair come from and why?

One of the key features in the skin of tetrapods is the enhancement of lamellar bodies that secrete a combination of fats and oils that prevent dehydration of the skin (Feingold, 2007). These lamellar bodies prevent what is clinically known as cutaneous water loss (Maderson, 2003). Although some early tetrapods and ancient amphibians had scales (see chapters 10 and 11), recall that the earliest amniotes evolved under the warm and moist environments of the ancient coal forests (see chapter 12) nearly 310 Ma (Card 94). Only after the formation of the supercontinent Pangea 300 Ma (Card 94) did the environment begin drying out (see chapter 13). It is possible that a smooth and glandular skin somewhat like a toad's was present in the common ancestor of amniotes (Maderson, 2003). In fact, a smooth and glandular skin may represent the original amniote skin condition because not only is it present in mammals, but birds have developed it as well under their feathers (Maderson, 2003). Moreover, somewhat anecdotal evidence suggests that a skin impression from the therapsid synapsid *Estemmenosuchus* shows a smooth skin with evidence for glands (Kardong, 2012). However, in mammals that retain scale-like structures, such as the tail of a rat, hairs grow in regular patterns between the scales (Kardong, 2012) which might suggest, alternatively, that our ancestors had a lightly scaled skin with patches of hairs. Whatever was initially present, it seems clear that the scaly skin we associate with modern reptilians is apparently a derived adaptation to truly dry environments.

Although we await better fossil and developmental evidence to be more certain, it seems most likely that mammalian hair first appeared in patches as so-called mechanoreceptors: long, tough filaments intimately connected with nerve bundles that, when disturbed, provide additional perception of the physical environment (Maderson, 2003; Kardong, 2012). Support for this hypothesis comes from the presence of such structures in most mammals, the vibrissae (whiskers) we discussed in chapter 19. Moreover, mammalian hairs are well innervated (Maderson, 2003; Vaughan, Ryan, and Czaplewski, 2011; Kardong, 2012). It is therefore hypothesized that the initial function of mammalian hair was to expand the mechanical perception of our synapsid ancestors (Maderson, 2003; Kardong, 2012). So-called protohairs probably first originated around the face, later migrating across the body (Maderson, 2003). These sensory hairs would have also formed a physical barrier between the skin and the outside world, and so may have functioned collectively as a sensitive skin barrier (Maderson, 2003; Rowe, Macrini, and Luo, 2011). In the earliest mammal fossils preserved with hair traces, it is clear that by about 200 Ma (Card 96) vibrissae, an outer coat of tough guard hairs (the part on a dog or cat that you pet), and a shorter, softer layer close to the skin called the underfur were all present.

One of the successful strategies in mammals has been the development of milk. Milk is an amazing substance that provides in easily digestible form a liquid rich in calories, nutrients, and antibiotics that in many mammals feeds multiple offspring simultaneously (Oftedal, 2012).

What is often underappreciated about mammals is that milk has freed the adult mammal to specialize in feeding on numerous foodstuffs that would be indigestible to their young. In other words, by suckling, young gain needed nutrition so they can grow to a size and develop complex dentitions that later allow them to exploit specialized diets in their adult lives that would otherwise be impossible (Oftedal, 2012). But when did milk first come onto the scene, and how? Again, it is difficult to say with certainty, but there are a few hints in the modern biology of mammals. As you will recall, milk is generated in mammary glands, which happen to share a lot in common with the plumbing of sweat (apocrine) glands (Blackburn, Hayssen, and Murphy, 1989; Vaughan, Ryan, and Czaplewski, 2011; Oftedal, 2012). Moreover, most mammals have sweat glands distributed across the body in varying patterns, and these are often associated with hair follicles (Oftedal, 2002a, 2002b, 2012). If hair goes back to animals like *Thrinaxodon*, it is possible that the early forms of sweat glands, and thus milk glands, do as well (Oftedal, 2002a, 2002b, 2012). But how could we know when milk glands and suckling first appeared? The skeletal chassis once again points the way.

As we discussed in chapter 19, most living mammals start life with a set of deciduous teeth that are eventually replaced with permanent adult teeth. Compared to most other vertebrates, the number of teeth replaced is reduced, and the timing of that replacement is greatly slowed, in mammals. In chapter 19, it was suggested that proper tooth-to-tooth occlusion was the selective pressure that led to fewer shed teeth because gaps in the tooth arcade can lead to serious problems with chewing and jaw health. Yet that scenario is not necessarily the whole story. Two other issues were at play: rapid growth to a predetermined size and the nonnecessity of functional adult teeth in infancy because suckling milk provides the needed nutrition.

In many reptiles, but not birds, growth to adult size is slower or less time-constrained. As many reptiles grow along these variable trajectories, the size of their jaws continues to change, and so older, smaller teeth are shed to make way for newer, larger teeth that better fit the mouth (Sanchez-Villagra, 2010). However, with some exceptions, growth is rapid and ceases at a determined time in modern mammals: the juvenile grows quickly to adult size and then growth ceases at a specific age. Telltale signs of this growth characteristic show up in limb bones. First, unlike most reptiles, mammals develop secondary bone centers within their joint cartilage that fuse with the growing bone shaft as they reach sexual maturity (Carter and Beaupré, 2001; Chinsamy-Turan, 2005) (Fig. 21.1). This physically stops the bone from growing and provides the paleontologist with a means of determining who is full-grown and who is still growing: juveniles have unfused ends on their limb bones. Second, thin sections of limb bones can be ground into fine slices that are viewed under a microscope. This area of scientific specialization is called bone histology. In this way, the texture of the limb bone and the arrangement of its cells can be observed and quantified.

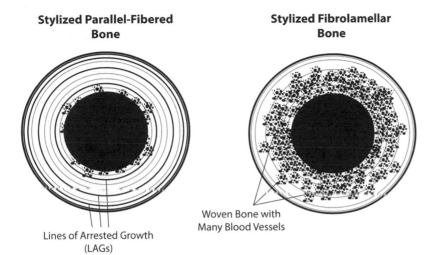

Stylized Parallel-Fibered Bone

Stylized Fibrolamellar Bone

Lines of Arrested Growth (LAGs)

Woven Bone with Many Blood Vessels

21.2. Stylized diagrams of parallel-fibered and fibrolamellar bone. Typically, ectothermic amniotes tend to have parallel-fibered bone, whereas endotherms such as birds and mammals show a more fibrolamellar pattern of bone deposition. See the text for more details. Diagrams after Currey (2002) and Chinsamy-Turan (2005).

To understand what a bone histologist examines under the microscope, imagine if you were a contractor who built small brick sheds. One of your customers gives you plenty of time and resources, and you are able to lay the bricks and mortar neatly. However, another customer demands you finish the shed in record time, and so you hurriedly construct theirs as best you can, but the bricklaying ends up being more chaotic and less neat as a consequence. Further, imagine that for yet another customer you began building their shed in the fall but were forced to stop for the winter. When you resumed in the spring, the bricks available to you are a slightly different color. Therefore, when completed, the bottom of the shed is a different hue than the top, creating a line demarcating where you stopped and resumed.

In the same way, when limb bones grow slowly, just like the slowly built brick shed, the tissues in their shafts have time to form themselves into well-organized concentric rings (parallel-fibered bone) (Currey, 2002; Chinsamy-Turan, 2005, 2011) (Fig. 21.2). On the other hand, much like your shed built under duress, limb bones that grow fast and continuously appear 'disorganized' histologically: the arrangement of the bone tissues appears as a sequence of random and sponge-like layers (woven bone) interspersed with thicker layers (fibrolamellar bone) (Currey, 2002; Chinsamy-Turan, 2005, 2011) (Fig. 21.2). As with your project that required you to stop and then resume building, when limb bones grow in a series of starts and stops, so-called lines of arrested growth (LAGs) are created (Chinsamy-Turan, 2005, 2011) (Fig. 21.2). This occurs as the bone tissues halt and then resume their building, usually in relation to changing seasons or water and food availability. Thus, the trained histologist can examine thin sections of recent and fossil limb bones and decipher how fast or slowly the animal was growing. Although this is a great oversimplification of how bone histology is studied and deciphered, you can at least appreciate the basics of how a paleontologist would infer growth rate in a fossil vertebrate.

For example, when the bone histology of recent adult mammals and birds is examined, we tend to find fibrolamellar bone tissues (indicating rapid, continuous growth) followed by a thin layer of parallel-fibered bone (indicating growth has slowed and ceased) (Chinsamy-Turan, 2005, 2011). This pattern makes sense given that they are endothermic vertebrates that grow fast but stop growing upon reaching maturity. In contrast, the histology of most recent reptile limb bones shows a more leisurely, parallel-fibered pattern of bone growth, typically with several LAGs, as would befit ectothermic vertebrates growing more slowly in starts and stops (Chinsamy-Turan, 2005, 2011). One of the remarkable discoveries of the past several decades has been the realization that dinosaurs and pterosaurs among the archosaurs typically show fibrolamellar bone histology, indicative of rapid growth and some form of endothermy (Horner, de Ricqlès, and Padian, 1999; Horner, Padian, and de Ricqlès, 2001; Padian, Horner, and de Ricqlès, 2004; Padian, 2012; Erickson, 2014). It should also be noted that LAGs indicate only starts and stops in bone growth, not metabolic rate. Recent histological work on modern ungulate mammals shows that LAGs are formed in these fast-growing herbivores during times of dietary stress brought about by changes in seasonal rainfall (Köhler et al., 2012; Padian, 2012). At this point, we should also dispel the notion that reptiles have 'indeterminate' growth: new bone histology research shows that specialized, dense layering of bone tissues, called an external fundamental system (EFS), occurs in most tetrapods as they reach adult size and cease growth (Woodward, Horner, and Farlow, 2011). That we see an EFS in dinosaurs, crocodylians, and pseudosuchians, for example, argues strongly against the notion that these reptiles had unlimited growth (Woodward, Horner, and Farlow, 2011).

Bone histology studies of non-mammal synapsids and early mammals show an intriguing evolutionary pattern. Early synapsids such as *Dimetrodon* show bone growth patterns akin to those typical of many reptiles (Sanchez-Villagra, 2010; Chinsamy-Turan, 2011). However, non-mammal cynodonts such as *Thrinaxodon* show fibrolamellar patterns that suggest fast and determinate growth (Botha and Chinsamy, 2005), and the limb bones of mammaliaforms such as *Morganucodon* show mammal-like histology (Chinsamy and Hurum, 2006). Moreover, it is in the limb bones of mammaliaforms such as '*Morganucodon*' that we see the beginnings of secondary bony centers (Sues and Jenkins, 2006) that fuse with the shaft, halting growth.

Complementing histological studies, studies of the dentition across synapsid evolution show that the sequence and generations of replacing teeth were changing. It is possible with physically broken or sectioned skulls or, more commonly now, with X-ray technology, to peer into the jaws of long dead synapsids and see how many sets of teeth are hiding under the exposed crowns. The results of such studies are astonishing. Early synapsids, such as *Dimetrodon*, had multiple layers of developing teeth in different stages of eruption, suggesting a pattern of constant and rapid tooth replacement like those of most modern reptiles (Sanchez-Villagra,

2010). Later non-mammal cynodonts such as *Thrinaxodon* have fewer layers of teeth and a more orderly pattern of replacement along the jaws, which suggests that tooth replacement was slowed (Abdala, Jasinoski, and Fernandez, 2013). By the time we get to our little friend *Morganucodon*, we have true diphyodont dentition: there are only two sets of teeth, and their eruption and replacement were slow, like those in our own mouths (Sanchez-Villagra, 2010).

So we come full circle back to our initial question: how could we know when milk glands and suckling first appeared? To review, we know that the evolutionary line of synapsids leading to mammals was becoming smaller, was developing a soft, glandular skin covered in sensory hairs, and was growing rapidly to adult size. Concurrently, tooth replacement rates were dropping while a secondary bony palate was forming that separated the nasal passages from the mouth. It is informative at this point to consider the most primitive living mammals, the monotremes, represented by the duck-billed platypus and the spiny anteater known as the echidna. In these mammals, teats are absent and milk is secreted from glands scattered across abdominal skin patches onto hairs, which young-sters subsequently lap up (Vaughan, Ryan, and Czaplewski, 2011). These mammals lay very small eggs with parchment-like shells that are kept moist in a pouch on the mother's underbelly (echidnas) or in a burrow (platypus) (Vaughan, Ryan, and Czaplewski, 2011). Milk is a skin secre-tion, and one of its major benefits is the transfer of antimicrobial products and immune-boosting substances from mother to offspring. Given that glandular skin may go back to non-mammal synapsids such *Thrinaxodon*, which were likely laying small eggs with parchment-like shells, it seems plausible that milk's initial role was moisturizing and protecting such eggs during incubation (Oftedal, 2002a, 2012).

As the ancestors of mammals were becoming smaller and smaller, so were their eggs. There would reach a point where young could not be born as fully functional (precocial), independent offspring because the tiny eggs would not contain enough yolk for such a developmental jump-start. This is nothing new, because many of the small birds we are familiar with have the same issue: they lay small eggs with just enough yolk to hatch very immature (altricial), dependent chicks. As you are well aware, these underdeveloped offspring cry constantly for food, and the parents provide sustenance in the form of regurgitated insects and, in some cases, milk-like secretions from their esophagus and guts (Oftedal, 2012). If our mammalian ancestors already inherited glandular skin that was secreting a milky substance over these tiny eggs, this liquid would be a ready source of nutrition for underdeveloped but fast-growing juve-niles. Such altricial young, hatched with a still-developing jaw, would not have had functional adult teeth (Sanchez-Villagra, 2010). It is this lack of functional adult teeth during infancy–diphyodonty–that corre-lates with lactation. Therefore, the presence of diphyodont dentition in *Morganucodon* and other early mammals strongly suggests that lactation was already well developed in their ancestors (Oftedal, 2012). In terms of jaws, two sets of functional teeth – a small set of teeth that fit in juvenile jaws, and a permanent set of interlocking and specialized occluding adult teeth – are possible in mammals because the growth stages are predictable and determined (Sanchez-Villagra, 2010).

Suckling itself requires muscles that can control the lips and cheeks so that the appropriate suction can be generated by the infant to gain milk from its mother, and this requires facial muscles (Vaughan, Ryan, and Czaplewski, 2011; Kardong, 2012). However, no other vertebrates have facial muscles, so where did these come from in mammals? The evolution of muscles is traceable because they have conservative innervation (Diogo and Abdala, 2010). Imagine if you had a series of color-coded electrical cords that you used to supply electrical power to various tools and machines in a workshop. Now, imagine that to improve your workflow, you needed to move one of those tools from where it used to sit in the middle of the room to a spot near the door. Although the tool was moved, the electrical cord that powered it remained the same and was simply dragged to this new location. Similarly, throughout all vertebrates, the same major groups of muscles are innervated by the same power supplies, the nerves, even when they are shuffled around (Liem et al., 2001; Kardong, 2012). By following the embryonic development of head and neck muscles in mammals, following their paths and the nerves that innervate them, and comparing those data to similar stages of development in reptiles and birds, we know that the facial muscles of mammals are in fact old hyoid muscles repurposed for suckling (Diogo and Abdala, 2010). In particular, the old hyoid constrictors that typically assist in swallowing and pharyngeal compression in other vertebrates stretch across the face of mammals during their development to form our facial muscles (Diogo and Abdala, 2010). Though many groups of mammals, including humans, now use the facial muscles for communication and emotional expression, they were probably first commandeered to assist the infants of our distant ancestors with suckling. Therefore, if *Morganucodon* and earlier ancestors were producing milk, they must also have had the earliest forms of muscles that now allow us to make goofy smiles in selfies.

We are finally in a position to return to the oddities of the mammalian brain (beyond politics, anyway). You are probably familiar with computer crashes in which your machine locks up due to information overload. To counteract this problem, memory and processing power must increase to keep up with a greater number of peripheral devices. Likewise, each hair on a mammal is part of a follicle that is innervated, and that means a lot of data on location and movement is constantly streaming to the brain. To keep up with this biological version of Internet broadband, the mammalian cerebrum apparently expanded in tandem with the evolution of hair (Rowe, Macrini, and Luo, 2011). How do we know this? A significant change in the region of the braincase that houses the cerebrum occurs in *Morganucodon* and a close relative, *Hadrocodium*, some of the earliest to possess furry coats (Rowe, Macrini, and Luo, 2011). Although *Morganucodon* and *Hadrocodium* fossils do not directly preserve traces of hair, we can be almost certain they were fuzzy. In most living mammals, a specialized gland, the Harderian gland, surrounds the orbit and drains an oily fluid down to the nose (Ruben, 2000). You have probably noted in your pet cats or dogs that their noses are wet, and this is a direct result of the Harderian gland's oily secretions. Hair grooming is common among most mammals, and the oily secretions of this gland are rubbed from the nose onto the paws, which subsequently spread these oils to the hairs. Harderian gland secretions act as a hair conditioner that enhances the insulation properties of the fur (Ruben, 2000). The skull of *Morganucodon* is known well enough that the tiny canal that runs from the Harderian gland is preserved, thus leaving little doubt that this mammaliaform and its close kin had and maintained a furry coat (Ruben, 2000). If one compares the braincase of *Morganucodon* and *Hadrocodium* to those of non-mammal cynodonts such as *Thrinaxodon*, the differences are remarkable. Where *Thrinaxodon* has a modest, reptile-like brain, the brains of *Morganucodon* and *Hadrocodium* have an inflated cerebrum that presages what is common for all living mammals (Rowe, Macrini, and Luo, 2011).

In tandem with the sensory overload from hairs, the reliance of underdeveloped but fast-growing young on mother's milk likely encouraged further enlargement of the cerebrum. All vertebrates possess an integrated series of nerve clusters called the limbic system in the forebrain that tie together olfaction, memory, emotion, visceral physiological drives (hunger, thirst, sex), and ordering of tasks (Liem et al., 2001; Butler and Hodos, 2005; Kardong, 2012). In mammals, the portions that are especially well developed are those that closely interlink olfaction, emotion, and memory (Liem et al., 2001; Butler and Hodos, 2005; Butler, Reiner, and Karten, 2011). Consider now that lactation is an expensive calorie drain on the mother mammal and that the youngsters draining this life energy from her are the equivalent of parasites! However, mother mammals who abandoned their young to save their calories are now lost to the mists of time because selection for the enhancement of the olfactory-emotional-memory link in the limbic system seems to have gone into overdrive. In essence, there has evolved in mammals a strong mother-child bond. As soon as infant mammals come into the world, both mother and children come to associate the smell of one another with comfort and food-sharing. The 'love' chemical, oxytocin, is an ancient vertebrate hormone exuded by the brain that primitively rewards behavior that propagates the species (e.g., reproduction) (Knobloch and Grinevich, 2014). In mammals, oxytocin has been implicated in feelings of romantic attachment, sexual arousal, food-sharing, childbirth, lactation, and parent-child bonding. Perhaps not surprisingly, the amount and distribution of oxytocin in mammals is expanded compared with most vertebrates, having enhanced effects on the forebrain and limbic system (Knobloch and Grinevich, 2014). To be cheeky, one could go so far as to say milk and hair formed the cornerstone upon which mammals built love.

Finally, tie all this back into living a nocturnal existence and avoiding diurnal dinosaurian predators, and you have all the pieces of the mammal brain puzzle. Here were vertebrates whose brain wiring had to change to decipher a dangerous landscape by relying on odors, pheromones, and sounds rather than vision. To overlay the mechano-sensory input from their furry pelage onto their other senses, they also required larger somato-sensory processing regions in the cerebrum. With small, underdeveloped but fast-growing young that depended on skin secretions from the mother to survive infancy, strong emotional bonds were selected for. As the ecological collapse that marked the end of the Cretaceous period subsided, mammals emerged into the world with a sensitive and insulating body covering, an unusual means of reproduction and infant-rearing, and a brain geared toward pair-bonding and strong emotional links between parents and their offspring.

Mammals, Part 2: Final Return to the Milk-Giving Fur-Bearers

Pedigree The Surviving Synapsid Amniotes

Date of First Appearance 200 Ma (Card 96)

Specialties of the Skeletal Chassis Dentary-Squamosal Jaw Joint, Three Inner Ear Ossicles Fully Separated from the Lower Jaw

Eco-niche Terrestrial, Aquatic, and Aerial with Tremendous Dietary Diversity

The evolutionary radiations that led to the modern diversity of mammals are complex (Fig. 21.3). Although we've learned much more about mammaliaforms and early true mammals in the past few decades, nearly two-thirds of their history is poorly known and pieced together from teeth and fragmentary remains throughout the Mesozoic Era (Vaughan, Ryan, and Czaplewski, 2011; Kielan-Jaworowska, 2013). After early mammaliaforms such as *Morganucodon*, the mammal family tree split several times. One branch, the Monotremata, is represented today by the egg-laying platypus and spiny echidna (Fig. 21.4) (Plate 26). These odd mammals sport a mosaic of primitive and specialized traits, and the surviving members of this lineage call Australia their home. The platypus, *Ornithorhynchus anatinus*, weighs at most 1.4 kilograms and may stretch to a little over 45 cm in length. This mammal is covered in dense and insulating hair, has an expanded region of skin (the rhinarium or 'bill') that is invested with electrosensitive nerve endings, bears a venomous spur on its hind limbs, and has a dorsoventrally flattened, beaver-like tail (Vaughan, Ryan, and Czaplewski, 2011). Regarding the venomous spur, it is intriguing that a number of mammaliaforms and early mammals have just such a structure on their ankles as well, which suggests this is an ancient trait among our common ancestors (Kielan-Jaworowska, 2013). The skull, though mammalian, is absolutely weird. The facial bones form a cranially open-ended bill-like platform upon which the electrosensitive rhinarium rests. Although juveniles sport some deciduous teeth, adults lose all traces of them and instead have a thickened and rough palate that, in combination with tongue movements, allows them to smash up the small invertebrates that serve as their prey (Vaughan, Ryan, and Czaplewski, 2011). The limb posture is sprawled, especially in the forelimb, and the platypus retains a large coracoid bone that connects the scapula to the sternum. This mammal dives in streams and can remain submerged for up to a minute in search of mollusks and other small invertebrates, using its 'bill' to turn over rocks, stir up river sediments, and sense the weak electric discharges of its prey. Intriguingly, this diving body plan

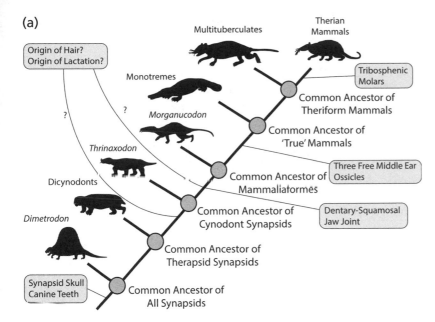

(a)

Origin of Hair?
Origin of Lactation?

Multituberculates

Therian
Mammals

Monotremes

Tribosphenic
Molars

Common Ancestor of
Theriform Mammals

Morganucodon

Common Ancestor of
'True' Mammals

Thrinaxodon

Three Free Middle Ear
Ossicles

Common Ancestor of
Mammaliaformes

Dicynodonts

Common Ancestor of
Cynodont Synapsids

Dentary-Squamosal
Jaw Joint

Dimetrodon

Common Ancestor of
Therapsid Synapsids

Synapsid Skull
Canine Teeth

Common Ancestor of
All Synapsids

21.3. Family tree of mammalian relationships. (a) The major relationships among the synapsids, including mammals. (b) The major relationships among the mammaliaforms and mammals. See the text for more details.

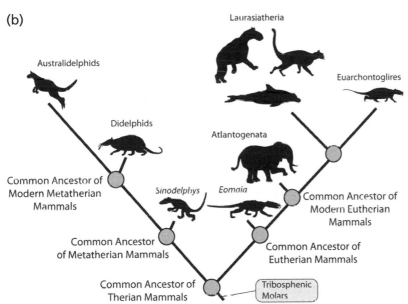

(b)

Laurasiatheria

Australidelphids

Euarchontoglires

Didelphids

Atlantogenata

Common Ancestor of
Modern Metatherian
Mammals

Sinodelphys

Eomaia

Common Ancestor of
Modern Eutherian
Mammals

Common Ancestor
of Metatherian Mammals

Common Ancestor of
Eutherian Mammals

Common Ancestor of
Therian Mammals

Tribosphenic
Molars

was presaged by the beaver-like mammaliaform *Castorocauda* from the Jurassic (Ji et al., 2006)! The spiny echidna is represented by two genera, *Tachyglossus* and *Zaglossus*, the largest individuals of which can stretch just over a 45 cm in length. As their common name suggests, spiny echidnas are covered in coarse, spinelike hairs. Their rhinarium is also invested with electrosensitive nerve endings, but not to the same extent as in the platypus. In going after ants, they plug the opening of an anthill with their rhinarium, which then engorges with blood. A long and sticky tongue laps up the hapless insects (Vaughan, Ryan, and Czaplewski, 2011). As with the platypus, the echidna is low-slung with a sprawling gait,

21.4. Monotreme mammal skeletal diagrams. Note that despite the possession of typical mammalian traits, the platypus has a sprawled posture and a venomous spur. Note as well that monotremes have an epipubic bone, as do multituberculates and metatherians.

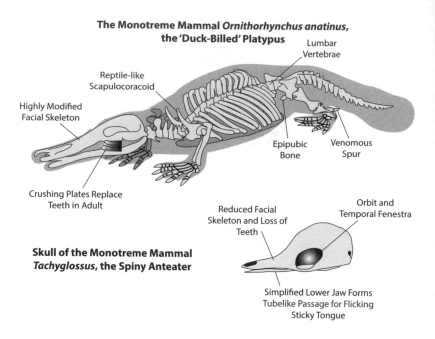

The Monotreme Mammal *Ornithorhynchus anatinus*, the 'Duck-Billed' Platypus

Lumbar Vertebrae

Reptile-like Scapulocoracoid

Highly Modified Facial Skeleton

Epipubic Bone

Venomous Spur

Crushing Plates Replace Teeth in Adult

Reduced Facial Skeleton and Loss of Teeth

Orbit and Temporal Fenestra

Skull of the Monotreme Mammal *Tachyglossus*, the Spiny Anteater

Simplified Lower Jaw Forms Tubelike Passage for Flicking Sticky Tongue

although it can walk in a surprisingly upright manner at times (Vaughan, Ryan, and Czaplewski, 2011).

The mammalian line leading to the Theria (the marsupials and placentals) contains a number of so-called theriimorph mammals. We have only recently begun to appreciate one of these branches, the Haramiyida, due to the very recent discovery of new body fossils (Bi et al., 2014). Here, we will focus one of the most successful theriimorph branches, the multituberculates (Fig. 21.5). These theriimorph mammals were diverse and survived the end-Cretaceous extinction, but sadly died out by about 50 Ma (Card 99) (Benton, 2005). Multituberculates have a lot in common with rodents, and in many cases were rodent and squirrel analogs in their lifestyles. The skulls of the earliest known multituberculate, *Rugosodon* from the Jurassic (160 Ma; Card 97), as well as later forms like *Ptilodus* from the Paleogene (50 Ma; Card 99), have enlarged, evergrowing incisors and well-developed premolars and molars (Krause, 1982; Yuan et al., 2013). The lower premolars of multituberculates are unusual in that they are laterally compressed with ridge-like cusps that would have sliced into food held against the upper premolars (Krause, 1982). Unlike the pattern of tooth movements we will discuss shortly for therian mammals, the architecture of the teeth and jaws in multituberculates suggests they had a two-stage chewing cycle. Food was first held in the jaws and sliced by their odd premolars, after which the jaw would retract, dragging the bottom teeth against the upper ones, shearing animal and vegetable matter (Krause, 1982).

In the postcranial skeleton, perhaps the most intriguing adaptations occur in the ankle. In opossums and several other arboreal mammals, the foot can pivot so that the toes face caudally. This unusual movement allows an opossum to descend tree trunks head-first: with the feet turned

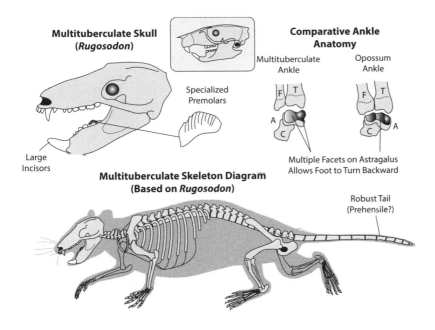

Multituberculate Skull
(*Rugosodon*)

Specialized Premolars

Large Incisors

Multituberculate Skeleton Diagram
(Based on *Rugosodon*)

Comparative Ankle Anatomy

Multituberculate Ankle

Opossum Ankle

F T

A

C

F T

A

C

Multiple Facets on Astragalus Allows Foot to Turn Backward

Robust Tail
(Prehensile?)

21.5. The multituberculate skeletal chassis. Multituberculate mammals, as exemplified here by *Rugosodon,* were very rodent-like in their overall body form and skull shape, including the enlarged incisors. Multituberculates had specialized premolars with numerous ridges. The ankle anatomy of these mammals is very similar to those of opossums and other tree-dwelling mammals that descend trees head-first, requiring their feet to be rotated backward. The robust tails of many multituberculates may have been used, as in opossums and some primates, as a prehensile appendage. Inset is a skull of a rat for comparison. The skull and skeleton of *Rugosodon* are modified from Yuan et al. (2013); the ankle diagrams are modified from Jenkins and McClearn (1984).

back, they grip the bark like a pair of 'hands,' controlling the mammal's descent (Jenkins and Krause, 1983). As we discussed in chapter 19, the main ankle joint of most mammals lies between the cup-shaped distal end of the tibia and the convex articular surface of the astragalus. However, primitively in mammaliaforms the astragalus sported facets for articulation with both the fibula and tibia, and these independent articulations are retained in modified form both in monotremes and in the opossum (Jenkins and McClearn, 1984) (Fig. 21.5). This configuration allows both plantarflexion-dorsiflexion movements as well as abduction-adduction motions that collectively rotate the foot into a wide range of positions (Jenkins and McClearn, 1984). In multituberculates, ankle facets nearly identical to those observed in opossums must have also allowed the astragalus to pivot the foot into a wide range of orientations, including ones allowing the descent of trees head-first (Jenkins and Krause, 1983). The robust caudal vertebrae of multituberculates are reminiscent of opossums and monkeys with prehensile tails, and further support the hypothesis that these theriimorph mammals were arboreal in their habits (Jenkins and Krause, 1983; Benton, 2005). As successful as they were, the multituberculates occupied an eco-niche that came to be shared with the eutherian mammals we call the glires, among which we find the rodents. There is no indication that the dental diversity of multituberculates was somehow 'inferior' to the teeth of rodents (Wilson et al., 2012). Instead, given how prolifically rodents reproduce, the fate of multituberculates may have been sealed not by inferior anatomy, but by fewer numbers (Krause, 1986).

The most successful branch of the mammal family tree is the Theria, represented today by the Metatheria (marsupials) and Eutheria (placentals) (Plate 26). One of the key traits of therian mammals is the possession

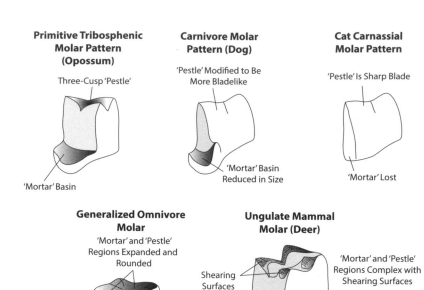

21.6. Basic patterns of the tribosphenic molars of therian mammals. From a generalized form as still found today in the Virginia opossum, the tribosphenic molar has been modified in various therian mammal lineages to enhance slicing, crushing, or grinding. Diagrams modified from Radinsky (1987).

of what are called tribosphenic molars (Benton, 2005; Vaughan, Ryan, and Czaplewski, 2011) (Fig. 21.6). The significance of these teeth is best understood as a mortar and pestle. In your kitchen, there is no better tool if you wish to grind up cloves, garlic, or other spices. You place the object you wish to grind into the bowl-like mortar, and then use the club-like pestle to grind in a rotational pattern. In a nutshell, the tribosphenic molars are the mortar and pestle of the mammalian world. Three cusps on the upper molars press and grind food in a so-called talonid basin on the lower molars. Therian mammals move their jaws in a rotational pattern, enhancing the grinding action of their tribosphenic molars, and also pass the food from side-to-side. The next time you chew gum, notice that you normally don't chew on both sides of your mouth at the same time. Instead, you chew on one side, and then switch to the other. This side-to-side processing enhances the mortar and pestle action of the tribosphenic molars, allowing them to both compress and shear food particles. The possession of tribosphenic molars was likely one of the key evolutionary adaptations that allowed therian mammals to exploit a large and diverse variety of food resources.

It is in the therian mammal chassis that we see the complete reduction of the already shrinking coracoid bone. Losing the direct connection between the scapula and the sternum freed the pectoral girdle to become more supple in therians, as we have already discussed in some detail in chapter 19. However, this loss of the coracoid also freed the supracoracoideus muscle, which in most other tetrapods functions as a synergist with the pectoralis muscles, or in birds as the wing elevator (see chapter 16) (Romer, 1962; Diogo and Abdala, 2010). Muscles typically do not simply disappear—they may fuse with other muscles, become reduced, or, as was the case with the supracoracoideus, they move to new locations. In the chassis of therian mammals, this former adductor of the forelimb kept its

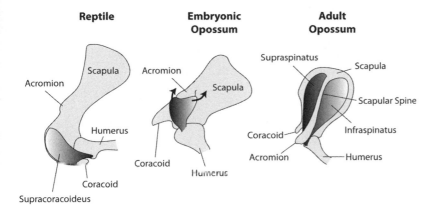

Reptile **Embryonic Opossum** **Adult Opossum**

Acromion

Scapula

Acromion

Humerus

Coracoid

Supracoracoideus

Acromion

Scapula

Coracoid

Humerus

Supraspinatus

Scapula

Scapular Spine

Coracoid

Infraspinatus

Acromion

Humerus

21.7. The evolutionary transformation of the supracoracoideus into the infraspinatus and supraspinatus muscles in therian mammals. In reptiles and the ancestors of therian mammals, the coracoid was a large element of the shoulder girdle to which the supracoracoideus muscle anchored. As revealed by therian mammal embryology and muscle innervation studies, with the reduction of the coracoid bone, the supracoracoideus gained a new anchorage on the scapula and transformed into the supraspinatus and infraspinatus muscles of the rotator cuff. Diagram after Romer (1962).

insertion on the deltopectoral crest, but switched its anchoring point to the scapula, where it now lies on either side of the scapular spine. If you know your human anatomy, you might be surprised to learn that the old supracoracoideus of our synapsid ancestors has become the infraspinatus and supraspinatus muscles, two of the four muscles that wrap around the proximal end of our humerus: the so-called rotator cuff muscles (Romer, 1962; Diogo and Abdala, 2010) (Fig. 21.7). Thus, not only was the scapula freed from a direct, bony connection with the sternum, but the supracoracoideus muscle found a new function in the supple pectoral girdle of our ancestors.

We have already detailed much of the skeletal chassis and soft anatomy of eutherians (see chapter 19), but not those of metatherians. Modern metatherians include many familiar forms such as kangaroos, wallabies, opossums, koalas, and wombats (Plate 26). Their common name, marsupials, refers to the marsupium, or pouch, that many of them possess and in which their young are first reared. Therefore, first and foremost, metatherians are distinguished from eutherians by their reproductive anatomy. In metatherians, the embryos do attach to the mother's uterus and form a placenta, but this is a fleeting moment in their development. Instead, they are born very early and underdeveloped compared with eutherians. What occurs next is remarkable: the tiny embryonic newborn crawls from the mother's pelvis to her pouch (if she has one) where it attaches to a teat and begins to suckle (Vaughan, Ryan, and Czaplewski, 2011). Intriguingly, the forelimb skeleton of metatherians develops faster than much of the rest of its anatomy, providing the tiny newborn with the means to make its trek from vulva to teat (Keyte and Smith, 2010). In opossums and other metatherians that lack a pouch, the tiny youngsters crawl to one of the numerous teats on the mother and hang on until they are mature enough to begin exploring the world on their own (Vaughan, Ryan, and Czaplewski, 2011).

Although metatherians have had a diverse and successful evolutionary history, they have almost always lost out to eutherian mammals when these two groups have come into close contact. For example, a diverse population of metatherians occupied South America for nearly

50 million years while it was an island continent, separated from North America by the ocean. However, approximately 2 Ma, the land bridge that now connects North and South America formed, allowing mammals north and south to travel to new lands and interact with one another. The outcome was terrible for the metatherians: diverse forms, including saber-toothed predators, met their demise at the hands of the eutherian invaders. One victory for the metatherians was the successful migration and invasion of North America by the opossums. Today, the only continent with a substantial metatherian population is Australia, but even here, various species continue to be decimated by eutherian competitors introduced by both aboriginal peoples (e.g., the dingo about 5,000 years ago) (Savolainen et al., 2004) and later European colonizers (everything from rabbits to sheep to camels) (Vaughan, Ryan, and Czaplewski, 2011).

Part of this sad record may be tied to metatherian reproduction, which is actually an advantage in the extreme climates many of them inhabit. Typically, metatherians do not produce offspring in the numbers that eutherians can, and their tiny newborns are dependent on the mother for longer than most eutherians. Therefore, in the same amount of time it takes for eutherians to produce many independent offspring, far fewer metatherians are produced (Vaughan, Ryan, and Czaplewski, 2011). However, in the unforgiving heat and aridity of the Australian outback, the metatherian reproductive strategy is advantageous for the survival of the mother. First, an extended pregnancy is less desirable in extreme environments because it unduly taxes the mother's energy and puts her at heightened risk of death. By having a short pregnancy, this energy-intensive time is greatly diminished (Vaughan, Ryan, and Czaplewski, 2011). Second, because less energy and resources have been invested in each offspring, their loss, especially early on, is not as catastrophic to the survival of the species. In essence, because less energy is invested in each offspring up front, if environmental extremes cause the mother to shut down her lactation to save her own life, she will still be around to reproduce when the climate becomes favorable. Overall, though, it has been the opossums and their relatives, with their rapid reproductive output, that have been able to hold their own against the eutherians (Benton, 2005; Vaughan, Ryan, and Czaplewski, 2011).

However, it should be noted that because of the history of their evolutionary migrations (see below), metatherians ended up for most of their history on the southern continents (Sánchez-Villagra, 2013). Currently, the northern continents have more accessible landmass, and this is where a great deal of eutherian evolution occurred. This is important to consider because Antarctica, which spans nearly 14 million square kilometers, became 'unavailable' for colonization by about 30 Ma, becoming locked away and frozen under ice (Benton, 2005). Therefore, part of the reduction in metatherian numbers could well be tied to their evolution on the southern continents, with much less landmass and available resources (Sánchez-Villagra, 2013).

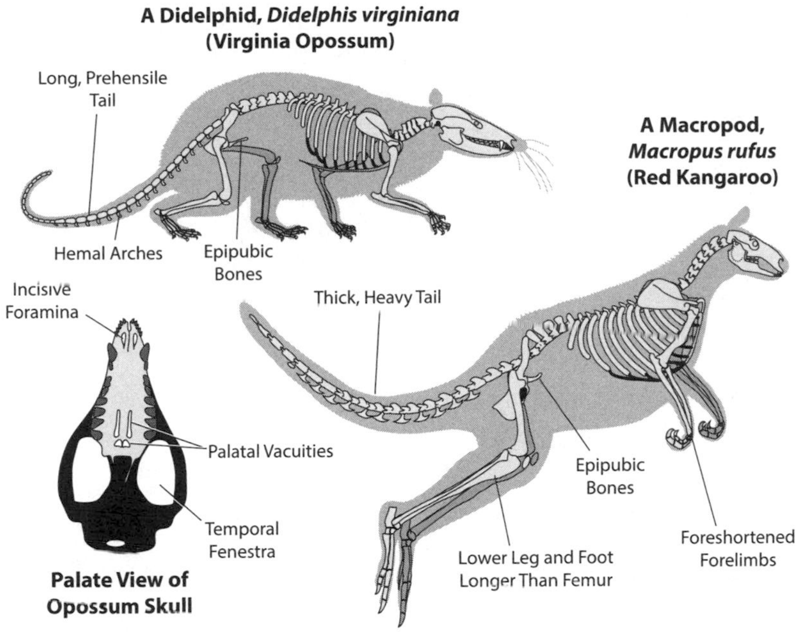

**A Didelphid, *Didelphis virginiana*
(Virginia Opossum)**

Long, Prehensile Tail

Hemal Arches

Epipubic Bones

Incisive Foramina

Palatal Vacuities

Temporal Fenestra

Palate View of Opossum Skull

**A Macropod,
Macropus rufus
(Red Kangaroo)**

Thick, Heavy Tail

Epipubic Bones

Foreshortened Forelimbs

Lower Leg and Foot Longer Than Femur

21.8. Modern examples of metatherian (marsupial) mammals. Metatherians are identified skeletally by palatal vacuities in their bony palates and by an unusual tooth eruption pattern where only certain teeth are replaced. As with monotremes and multituberculates, metatherian mammals have epipubic bones.

The skeletons of metatherians are very similar to those of eutherians, but there are a few differences (Fig. 21.8). Tooth replacement differs from eutherians in that only one tooth in each jaw replaces a deciduous tooth (Cifelli and de Muizon, 1998; Vaughan, Ryan, and Czaplewski, 2011). Apparently, during development the deciduous incisors and canines begin to form but then are reabsorbed and replaced by the secondary, adult teeth (Cifelli and de Muizon, 1998; Vaughan, Ryan, and Czaplewski, 2011). The first two premolars (top and bottom) are deciduous but are never replaced. Only the third premolars are actually erupted first as deciduous teeth and then replaced with their adult counterparts. As with the first two premolars, there is only ever one set of molars that slowly erupt to complete the adult tooth arcade when growth ceases (Cifelli and de Muizon, 1998; Vaughan, Ryan, and Czaplewski, 2011). Evidence for this type of tooth replacement is present in several fossil metatherians (Cifelli and de Muizon, 1998).

There are typically large openings or vacuities in the bony secondary palate in metatherians, and for most of these mammals there is never an equal number of upper and lower incisors (Vaughan, Ryan, and Czaplewski, 2011). The forelimb skeletons of most metatherians are quite conservative because of their early rapid development, and this may be one of the factors that has limited these mammals from developing hooves, flippers, and wings (Cooper and Steppan, 2010). The hind limbs and especially the feet can be diverse, however, given the various locomotor habits of metatherians, including hopping and tree-climbing (Vaughan, Ryan, and Czaplewski, 2011).

Many metatherians possess so-called epipubic bones: small, strut-like bones that point cranially from their pubis. Traditionally, it was

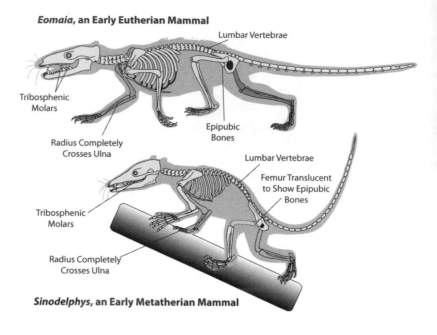

21.9. Some of the earliest known therian mammals. *Sinodelphys* is the earliest known metatherian, whereas *Eomaia* is one of the earliest known eutherians, both from the Cretaceous period of China. *Eomaia* is preceded only by *Juramaia* from the Jurassic of China. Note that even the earliest eutherian mammals had epipubic bones. Skeletal diagrams after Ji et al. (2002) and Luo et al. (2003).

Eomaia, an Early Eutherian Mammal

Lumbar Vertebrae

Tribosphenic Molars

Epipubic Bones

Radius Completely Crosses Ulna

Lumbar Vertebrae

Femur Translucent to Show Epipubic Bones

Tribosphenic Molars

Radius Completely Crosses Ulna

Sinodelphys, an Early Metatherian Mammal

hypothesized that the epipubic bones supported the marsupium or pouch. However, we now know that epipubic bones are not associated with the pouch for several reasons. First, both male and female metatherians possess epipbuic bones, but only females have a pouch. Second, epipubic bones were a common anatomical feature of many early mammals, including the earliest eutherian mammals, none of which possess a marsupium (Kielan-Jaworowska, 2013). Third, recent anatomical studies have shown that the epipubic bones provided mechanical leverage to abdominal muscles that stabilize the body core in metatherians during locomotion (Reilly et al., 2010). Modern eutherian mammals have apparently lost these bones because of a shift in which muscles were recruited to stabilize their body core, no longer requiring the leverage provided by the epipubic bones (Reilly et al., 2010).

Metatherians and eutherians split as separate lineages by at least the middle of the Jurassic period some 160 Ma (Card 97). We know this because the earliest definitive eutherian, *Juramaia*, a small, insectivorous shrew-like mammal, comes to us from the middle Jurassic rocks of China (Luo et al., 2011). Although the first definitive metatherian, *Sinodelphys*, is known from Early Cretaceous rocks (120 Ma; Card 98) in China (Luo et al., 2003), the presence of a eutherian in the Middle Jurassic shows that we are simply missing metatherian fossils from this time. *Juramaia* is known only from its skull and the cranial part of its skeleton (everything from the pelvis back is missing), but a similar eutherian, *Eomaia*, from the Early Cretaceous of China is similar in overall form and complete head to tail (Ji et al., 2002). Therefore, the skeletons of *Sinodelphys* and *Eomaia* make for an informative comparison (Fig. 21.9).

To begin with, both mammals are tiny (both are about 10 cm in length) and were likely insectivorous given their size and sharp, pointed teeth. Both were also found with carbonaceous imprints of hair (Ji et al., 2002; Luo et al., 2003), showing that they had the full complement of guard hairs and underfur common to all mammals and their close ancestors. Much of their skeletal anatomy is similar to that of small arboreal or scansorial (those that climb trees but also are at home on the ground) mammals like opossums and tree shrews (Ji et al., 2002; Luo et al., 2003). The radius and ulna of *Sinodelphys* and *Eomaia* are similar to what we have encountered in rodents: the former bone crosses the latter, allowing the hand to be actively pronated. Such an adaptation is critical for arboreal and scansorial mammals that must grip branches and navigate narrow perches. The hands and feet of *Sinodelphys* and *Eomaia* are perhaps the most telling regarding arboreal and scansorial habits. In both mammals, the hands and feet are naturally curved when their bones are articulated, such that finger flexion was more accentuated than extension (Ji et al., 2002; Luo et al., 2003), a trend particularly well developed in modern primates (Benton, 2005; Vaughan, Ryan, and Czaplewski, 2011). For example, relax your hand and notice that your fingers naturally curl toward your palm: this is a trait of mammals with an arboreal or scansorial ancestor. Finally, the elongate tails of *Sinodelphys* and *Eomaia* suggest they were utilized for balance and might have been prehensile to some degree, just like many modern arboreal and scansorial mammal species.

It seems clear that the earliest therian mammals originated in the Northern continents (Laurasia), and then migrated west and south. From ancestral metatherians like *Sinodelphys*, the marsupial lineage appears to have first spread to what is now North America and then migrated into the southern continents (Gondwana) via South America and then on to Antarctica and Australia, which were all interconnected during the Cretaceous (144–66 Ma; Cards 97–99) (Nilsson et al., 2010). Among the metatherians, the radiations that remained in the Americas mostly comprise the didelphids, of which the opossums are the most well recognized and successful (Fig. 21.3b). The metatherians that traversed across the southern continents into Australia diversified in the absence of direct competition (until recently) with eutherians. These became the kangaroos and wombats, the koalas and bandicoots (Fig. 21.3b). Kangaroos are among the few large mammals that have become bipedal (humans being another rare exception). Their hopping locomotor style is very efficient at high speeds and relies, in part, on a reciprocal pattern of storing elastic energy in their long leg tendons and then releasing it (Alexander and Vernon, 1975). However, the kangaroo style of locomotion is apparently so rare because large, hopping mammals can effectively cover long distances only in shrub land and desert (which characterizes much of Australia) but not in jungles and forests (Vaughan, Ryan, and Czaplewski, 2011).

In contrast, the eutherians descending from mammals like *Eomaia* apparently had a somewhat more complex dispersal (Fig. 21.3b). Initially, the common ancestral populations of eutherians were divided into the Northern and Southern hemispheres by the separating continents during Cretaceous (~100 Ma; Card 98) (Springer et al., 2004, 2005; Wildman

21.10. Examples of eutherians from the Atlantogenata radiation. The Afrotheria, one of the most primitive radiations, includes elephants, manatees, and aardvarks. The Xenarthra is an oddball group composed of armadillos, sloths, and anteaters. Diagrams based, in part, on Carroll (1988), Shoshani (1996), Benton (2005), and Vaughan, Ryan, and Czaplewski (2011).

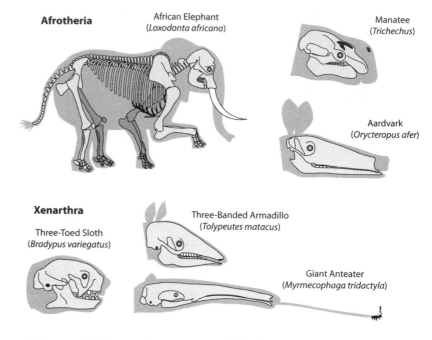

Afrotheria

African Elephant
(*Loxodonta africana*)

Manatee
(*Trichechus*)

Aardvark
(*Orycteropus afer*)

Xenarthra

Three-Banded Armadillo
(*Tolypeutes matacus*)

Three-Toed Sloth
(*Bradypus variegatus*)

Giant Anteater
(*Myrmecophaga tridactyla*)

et al., 2007). The southern group, called the Atlantogenata, were then later separated into two different groups as Africa and South America drifted apart (~100 Ma; Card 98) (Wildman et al., 2007). The remaining northern landmass eutherians, the Boreoeutheria, migrated across North America and Eurasia, but eventually found their ways back south into Africa and South America (Wildman et al., 2007).

Among the Atlantogenata were the common ancestors of the eutherian lineages called Afrotheria and Xenarthra. Afrotherians include among their ranks elephants, sirenians (manatees and dugongs), hyraxes, aardvarks, and a number of small mole- and shrew-like mammals native to the African continent. Elephants and sirenians share a close, common ancestor and form the group Tethytheria (Fig. 21.10). As odd as it sounds, the common ancestor of elephants and sirenians was a semiaquatic, hippo-like herbivore with cranially placed orbits and tusks (Benton, 2005; Pough, Janis, and Heiser, 2013). On the line to sirenians, the bones became dense, acting as diving weights to hold these herbivores underwater, where they munch on sea vegetation; the hind limbs were lost; and the forelimbs were modified into paddle-like flippers with long fingers capped by elephant-like hooves (Vaughan, Ryan, and Czaplewski, 2011). On the line to elephants, their lineage moved from semiaquatic environments to a purely terrestrial existence (Benton, 2005). However, elephant anatomy still bears telltale signs of their previous semiaquatic ancestry. As in most marine mammals, the testes of male elephants are internal and do not descend into a scrotum (Gaeth, Short, and Renfree, 1999), and all elephants lack the pleural membranes that surround the lungs of most terrestrial mammals (West et al., 2003). Such membranes are lost in marine mammals to prevent nitrogen bubbles from forming around their lungs after deep dives. Like the sauropod dinosaurs before them, elephants have heavy, columnar limb bones that are solid through and through to support their massive weight (McGowan, 1999), and complex fat pads with gel-like chambers cushion their footfalls (Weissengruber et al., 2006).

The Xenarthra are an odd bunch of eutherians that include anteaters, sloths, and armadillos (Springer et al., 2004, 2005) (Fig. 21.10). This grouping gets their name from the odd extra joints found along their vertebral column, which, together with the zygapophyses, apparently brace the lumbar dorsal vertebrae (Vaughan, Ryan, and Czaplewski, 2011). The incisors, if not absent altogether, are usually very reduced; deciduous teeth do not erupt at all; and premolars and molars, when present (they are absent in anteaters, for example), lack enamel (Vaughan, Ryan, and Czaplewski, 2011). Xenarthrans are almost entirely a South American group, and giant armadillo and sloth relatives used to roam that continent until very recently (some as recently as 10,000 years ago) (Benton, 2005). Most xenarthrans have large, spade-like claws on their hands and feet, and many are burrowers and diggers. In particular, the robust claws on the hands of anteaters cause these mammals to walk on the knuckles of their hands (Orr, 2005) and provide them with powerful tools for extracting ants and termites from their nests (Vaughan, Ryan, and Czaplewski, 2011). Sloths are odd among the odd in that they typically have fewer than seven cervical vertebrae, extremely low metabolic rates, and such a leisurely lifestyle that algae grow on their fur (Vaughan, Ryan, and Czaplewski, 2011)!

As the Boreoeutheria radiated across the northern continents, they split into two major, multisyllabic groups, the Laurasiatheria and the Euarchontoglires (Springer et al., 2004, 2005; Wildman et al., 2007). The diversity of the Laurasiatheria can fill and has filled books, and represents the most speciose of eutherian mammal groups. Basal members of this group include the bats or Chiroptera (Fig. 21.11). The evolutionary origins of bats remain obscure, and the first good bat fossils, *Icaronycteris* and *Onychonycteris* from the Early Eocene (~55 Ma; Card 99), are nearly indistinguishable from modern bats (Benton, 2005; Vaughan, Ryan, and Czaplewski, 2011). As with the pterosaurs, bats fly using a patagium of skin that connects from the forelimbs to the hind limbs, and round out their airfoil-like profile with a propatagium that fills the space cranial to the elbow between the shoulder and wrist. However, instead of having a single, elongate fourth digit to support the wing as in pterosaurs, bats incorporate the entirety of the hand skeleton into their wing membrane. As in birds and pterosaurs, much of the vertebral column is fused or braced to prevent extraneous movements during flight, and the sternum, clavicle, and scapula provide robust platforms for the pectoralis and other muscles that power flight. As with birds, the muscles of the forearm and hand are reduced and interconnected in such a way as to automate the wing stroke (Vaughan, Ryan, and Czaplewski, 2011). In fact, some forearm muscles have been reduced to tough connective tissue–like bands that act as weight-saving cables that transfer wing movement. Bat evolution

21.11. Some examples of eutherians from the laurasiatherian radiation. Chiroptera, or bats, modified the eutherian skeleton for flight. The Perissodactyla, the odd-toed ungulates, include the horses, tapirs, and rhinos, as well as extinct forms such as brontotheres and the unusual bipedal chalicotheres. Note the major changes in the teeth and jaws of one of the earliest horses (*Eohippus*) and modern horses (*Equus*). Diagrams based, in part, on Carroll (1988), Benton (2005), and Vaughan, Ryan, and Czaplewski (2011).

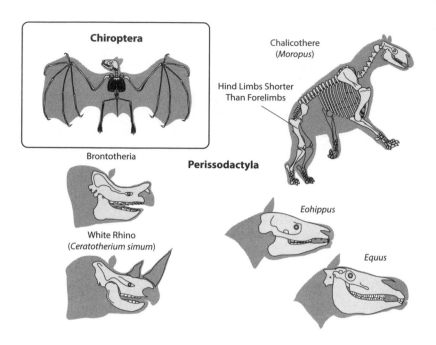

has resulted in two major groups, the larger fruit bats and the smaller insectivorous microbats. It is among the microbats that echolocation has evolved: these flying mammals bounce high-frequency sound waves off the surroundings to detect moving insect prey, using their enlarged pinnae acting like a sonar dish (Vaughan, Ryan, and Czaplewski, 2011). Along with bats, the hedgehogs, shrews, and other tiny insectivores comprise the remaining branches of the basal Laurasiatheria radiations (Springer et al., 2004, 2005; Wildman et al., 2007).

A variety of hoofed mammals, the so-called ungulates, derive from two major laurasiatherian groups, the perissodactyls and the cetartiodactyls. Ungulates have an unguligrade posture in which these mammals stand on the ends of their fingernails and toenails, the unguals, thereby increasing their relative stride length (see chapter 19). The perissodactyls comprise the extant and extinct odd-toed ungulates, and include among their ranks the horses, tapirs, and rhinos, as well as extinct horse relations such as the massive horned brontotheres and the strangely bipedal chalicotheres (Prothero and Schoch, 2002) (Fig. 21.11). Among the rhinos are the indricotheres, the largest land mammals of all time, which stood over over 4 meters tall and may have weighed nearly 15 metric tons (Prothero, 2013)! The earliest horses, such as *Eohippus*, were no larger than small dogs, but the spread of modern grasslands about 17–15 Ma (Card 100) appears to have selected for increasing body and skull size in this group (Benton, 2005). Modern grasses take up silica from the soil and stud their leaves with this hard mineral as a means of defense against grazing. That's right: there is glass in grass! Given that mammals have only one set of adult teeth, such a defense tactic by grass would grind their molars down at a phenomenal rate, rendering their dentition useless in record time. In the evolutionary line of horses, the lower jaw becomes large and

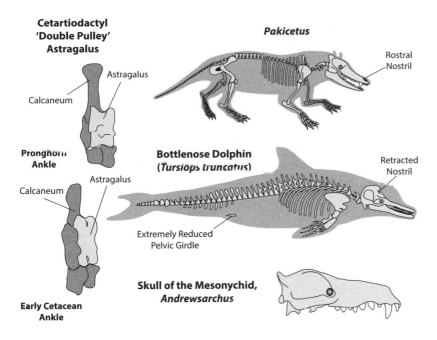

Cetartiodactyl 'Double Pulley' Astragalus

Calcaneum
Astragalus

Pronghorn Ankle

Calcaneum
Astragalus

Early Cetacean Ankle

Pakicetus

Rostral Nostril

Bottlenose Dolphin (*Tursiops truncatus*)

Retracted Nostril

Extremely Reduced Pelvic Girdle

Skull of the Mesonychid, *Andrewsarchus*

21.12. Examples of eutherians from the laurasiatherian lineage called Cetartiodactyla. Cetartiodactyls share a 'double-pulley' astragalus as a trait that links together seemingly odd bedfellows such as camels, hippos, dolphins, and whales. One of the earliest members of the cetacean lineage, *Pakicetus,* was a terrestrial predator about the size of a raccoon. The modern cetacean skeleton, as exemplified here by the bottlenose dolphin, has been modified wholesale to adapt the chassis to the water. An odd, ungulate-like group, the mesonychids, may be related distantly to cetartiodactyls and the carnivorous mammals. Cetartiodactyl ankle diagrams after Gingerich (2012). Skeletal diagram of *Pakicetus* after Thewissen et al. (2009); dolphin after Gatesy et al. (2013); *Andrewsarchus* skull after Osborn (1924).

deep, the angle where the masseter muscles attach becomes huge, and the teeth become long and deeply rooted (Benton, 2005; Vaughan, Ryan, and Czaplewski, 2011). Beyond longer teeth, each molar becomes a folded sandwich of alternating layers of hard enamel and softer dentine, topped with a coat of mineral tissue called cementum. When horses chew grass, the grinding action of their molars and the abrasive silica in the leaves wears down the cementum over the enamel and dentine sandwich in such a way that it produces high, sharp grinding ridges and lower, mortar-like valleys (Benton, 2005; Vaughan, Ryan, and Czaplewski, 2011). In essence, during the evolution of horses, their teeth adapted to the silica content of grass and used this defense against the plant by constantly sculpting sharper teeth with each bite.

Similar adaptations occurred in the jaws and teeth of the other major even-toed ungulate group, the cetartiodactyls, in a similar 'war' with grasses. These laurasiatherians are characterized by a specialized 'double-pulley' astragalus that enhances flexion and extension movements between it and the tibia proximally and the navicular (an ankle bone) distally (Fig. 21.12). Among the cetartiodactyls we find the pigs and peccaries, sheep and goats, cows and deer, camels and llamas (some with hats), and hippos. Many of these mammals are ruminants: they have multichambered stomachs that process grasses and other plant material to break down the cellulose (plant cell walls) in cooperation with symbiotic protists and bacteria. Many ruminants, such as cows, regularly regurgitate partially digested plant material (the cud) to rechew it before passing it on for further digestion. Surprisingly, cetaceans, the dolphins and whales, are also members of this group, and share a close common ancestor with hippos! Although this may sound odd, among the cetartiodactyls, only hippos and whales are capable of giving birth and suckling underwater, and both groups lack sweat glands (Gatesy et al., 2013). On top of these striking similarities, numerous studies of cetartiodactyl DNA consistently place hippos next to cetaceans (Gatesy et al., 2013).

The earliest cetaceans were in fact terrestrial carnivores, and *Pakicetus*, a little raccoon-sized mammal, is one of the very first of its kind (Thewissen et al., 2007, 2009; Gingerich, 2012) (Fig. 21.12). We know that these diminutive mammals are some of the earliest cetaceans because their ectotympanic bones (recall that this is the old angular that now holds the tympanic membrane) form what is called an auditory bulla (Gatesy et al., 2013). Many other mammal lineages have an auditory bulla, and it functions as a sort of 'echo chamber' that amplifies the vibrations of the tympanic membrane. However, in most mammals the auditory bulla is derived from portions of the squamosal, not the ectotympanic (Vaughan, Ryan, and Czaplewski, 2011). This is a significant difference. Cetaceans, being descendants of terrestrial mammals, had ancestors with a tympanic membrane that was sensitive to vibrations in air. Stick your head underwater, and water fills your outer ears, pressing against your tympanic membrane and muffling the sound. Moreover, because your skull is about the same density as the water surrounding it, sounds pass from all directions into your inner ears, and you can't tell where they're coming from. Modern cetaceans have sequestered their middle ears away from the braincase through an expansion of the ectotympanic bone and by filling the gap between the middle ear and skull with air sinuses (Vaughan, Ryan, and Czaplewski, 2011). This separation of the middle ear from the braincase insulates the ears, much like placing each of them in its own soundproof box, so that sounds can once again be detected as coming from the right or left (Vaughan, Ryan, and Czaplewski, 2011). Therefore, the beginning of this expansion of the ectotympanic in *Pakicetus* shows it is on the path to what would become the Cetacea (Gatesy et al., 2013).

During the evolution of cetaceans, the skeleton was modified wholesale (Fig. 21.12). The external nostrils have retracted to the top of the head to form a blowhole, and this transition from a rostral nostril is well documented in fossils (Gatesy et al., 2013). In essence, during cetacean evolution the premaxilla became more and more elongated, as we saw for ichthyosaurs (see chapter 15). Since the nasal bones always lie caudal and just dorsal to the premaxilla, the more the latter lengthens, the farther back the external nostrils are pushed until they end up in their present position in modern cetaceans (Gilbert, 2010; Gatesy et al., 2013). Seven cervical vertebrae are still present but are greatly shortened craniocaudally, precluding a mobile neck. As with all mammals, ribless lumbar vertebrae allow dorsoventral flexion, and therefore the major swimming movements are up-and-down rather than side-to-side. A fleshy dorsal fin is present in many cetaceans, and soft tissues on the end of the tail form a large, dorsoventrally compressed fluke that works in concert with the up-and-down swimming movements, aiding in propulsion. The clavicles are lost, and the forelimb is modified into a flipper wherein the radius, ulna, and wrist bones are compressed dorsoventrally, enhancing the winglike profile of this appendage. The only significant movements in the flippers occur at the glenoid joint, muscles distal to the elbow are either absent or greatly reduced, and the fingers are lengthened into the distal end of the flipper by the addition of more phalanges (Thewissen et al., 2009). All that remains of the hind limb is a small vestigial floating pelvis. Modern cetaceans are divided into two major lineages, the odontocetes, or toothed whales, and the mysticetes, those cetaceans with the large, hairlike keratinous filter-feeding plates in their mouths called baleen (Thewissen et al., 2009; Gatesy et al., 2013).

The carnivorous mammals, the Carnivora, are also found among the laurasiatherians, and these furry meat eaters have diversified into two major lineages. One line gave rise to the pinnipeds (seals, sea lions, walruses), dogs, bears, mustelids (weasels, skunks, and their kin) and raccoons, whereas the other lineage led to the mongooses, hyenas, and cats, large and small (Benton, 2005; Vaughan, Ryan, and Czaplewski, 2011). Close relatives of the carnivores may be the pangolins, odd ant-eating mammals with a scale-like body covering formed from modified hair keratin (Vaughan, Ryan, and Czaplewski, 2011). It remains unclear whether the carnivorous mammals are cousins to all ungulate mammals, or whether they are more closely related to the perissodactyls (Benton, 2005; Vaughan, Ryan, and Czaplewski, 2011). Adding to this confusion are early predatory mammals called mesonychids that, while once hypothesized to be whale ancestors, share many primitive features with the cetartiodactyl ungulates (Fig. 21.12). Strange as it sounds, the earliest large predatory mammals were not members of the Carnivora, but rather were flesh-eating ungulates (Benton, 2005)! It should also be noted that an entirely different branch of ungulates, the so-called notoungulates, populated South America for nearly 60 million years, but sadly left no modern descendants. Until recently, it was unclear whether the notoungulates were related to other ungulate groups, or whether they were a member of the Atlantogenata mammals that convergently evolved a similar suite of traits (Benton, 2005). New data from bone proteins and collagen suggest that the notoungulates were 'true' ungulates and members of the perissodactyls (Buckley, 2015; Welker et al., 2015). As with so much of our understanding of the fossil record, more fossils and more data will (hopefully) clarify this picture.

Finally, we turn to the euarchontoglires, which comprise the lineages that led to the rodents and rabbits (the so-called Glires) and the primates and their close relatives, tree shrews and the colugo, or 'flying lemur,' a strange, gliding mammal from Asia (Benton, 2005; Springer et al., 2005; Vaughan, Ryan, and Czaplewski, 2011). The close relationship between primates and Glires has not always been appreciated, but one look at the skulls and skeletons of the early rodent *Paramys* and the archaic primate relative *Plesiadapis* leaves little doubt of our shared kinship (Fig. 21.13). In both early rodents and primate ancestors, large incisors protrude from the front of the jaws, the forelimbs and hind limbs have hooked claws ideal for climbing, and long and presumably prehensile

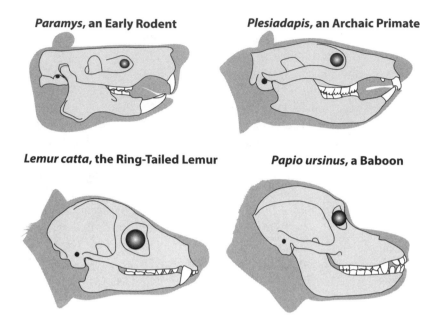

21.13. Comparative skull anatomy of the Euarchontoglires eutherian lineage. Note the striking similarities in skull and tooth form in the early rodent, *Paramys,* and the early, archaic primate, *Plesiadapis.* The ring-tailed lemur skull is a stand-in for the so-called Prosimian primates, whereas a baboon skull represents a monkey. Skull diagrams modified, in part, after Radinsky (1987), Benton (2005), and Vaughan, Ryan, and Czaplewski (2011).

Paramys, an Early Rodent

Plesiadapis, an Archaic Primate

Lemur catta, the Ring-Tailed Lemur

Papio ursinus, a Baboon

tails are present (Benton, 2005). Primates themselves have a number of telling traits. The radius becomes significantly bowed away from the ulna, greatly enhancing pronation and supination, and the development of flat nails allows the hands and feet to grip branches, actions that would be difficult with large claws (Benton, 2005; Vaughan, Ryan, and Czaplewski, 2011). The eyes have become forward facing, improving depth perception, and primates are one of the few mammal groups with enhanced color vision. On the other hand, the sense of smell is not always well developed in some lineages, including humans (Vaughan, Ryan, and Czaplewski, 2011). All of these features combine to make adept and dexterous arboreal mammals that can pick out colorful and tasty fruits and vegetables among a sea of green leaves in the tree canopy. On the whole, primates have brain masses much larger than that predicted for their body size, and the complexities of primate societies are well known. Moreover, primates usually have an extended and unhurried childhood where enough time is given for the young to learn the 'dos and don'ts' of their community (Benton, 2005; Vaughan, Ryan, and Czaplewski, 2011).

Among the living primate groups are a diverse bunch including the smaller, catlike prosimians (the lemurs, lorises, and perhaps the big-eyed tarsiers) and the anthropoids (monkeys and apes) (Fig. 21.13, Plate 26). Monkeys radiated across Eurasia and Africa as well as into South America, whereas apes have predominantly been restricted to Africa and Asia. Unlike monkeys, apes lack tails, have five cusps on their lower molars arranged in a 'Y' pattern, and have modified elbows that lock when extended to 180°. This last skeletal trait is related to the ancestral mode of locomotion among apes called brachiation, using the forelimbs to swing from tree to tree. An overwhelming amount of data from anatomy, the fossil record, and DNA shows that humans are apes and that we share

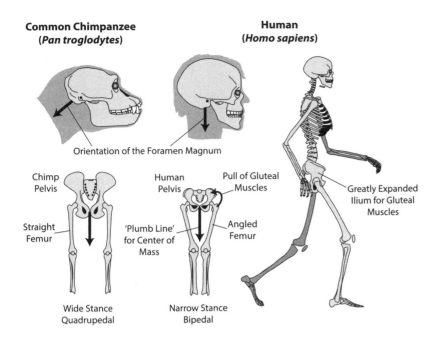

Common Chimpanzee
(*Pan troglodytes*)

Human
(*Homo sapiens*)

Orientation of the Foramen Magnum

Chimp Pelvis

Human Pelvis

Pull of Gluteal Muscles

Greatly Expanded Ilium for Gluteal Muscles

Straight Femur

'Plumb Line' for Center of Mass

Angled Femur

Wide Stance Quadrupedal

Narrow Stance Bipedal

21.14. Being human. Humans are apes, but two of the major traits that set us apart from our closest relatives, the chimpanzees, are an absolutely huge braincase with a ventrally pointing foramen magnum and a bipedal posture. The angle of the human femur, the so-called valgus angle, positions our feet directly below our center of mass, whereas the straight femora of chimps places them to either side of their center of mass. Diagrams based, in part, on Benton (2005).

our closest common ancestors with chimpanzees. Human evolution has been covered in glorious detail in many sources, and the intricacies of that history go well beyond the scope of this book.

From the perspective of the skeletal chassis, two traits for humans stand out more than any others: a bipedal posture and a huge braincase (Fig. 21.14). The femur of most apes, like those of many quadrupedal mammals, has a straight shaft such that when standing or walking, the knee and foot are aligned. This has the effect of placing the feet to either side of the body's midline, which makes for a stable quadrupedal stance. However, although chimps are capable of walking on their hind limbs from time to time, their straight-shafted femur causes them to have a less-efficient, Frankenstein-like walk. In contrast, humans have a femur where the shaft curves medially, so that when standing upright we are knock-kneed. Given that the foot is directly under the knee, this change in femur shape places the feet closer to the center of mass, so that when we walk, we place one foot in front of the other (Benton, 2005). Think of human footprints on a beach – they are typically placed one in front of the other, not side by side. As we discussed in chapter 17 on archosaur locomotion, this has the effect of giving humans dynamic stability in a bipedal posture.

However, walking bipedally with a narrow gait would, without compensation, lead to the legs collapsing inward. This is actually a functional problem for all tetrapods, especially those with upright limbs, and this is 'solved' by gluteal muscles and other abductors that contract when a limb is supporting the body while walking or running. Although contracting an abductor when the foot is firmly planted on the ground will not cause much movement, it provides a force that counteracts the tendency of the limb to collapse inward. Given that humans stand bipedally with a vertical vertebral column and without the help of a counterbalancing tail, these abductors, especially the gluteals, have become the primary means by which our posture is maintained. In fact, the superficial gluteal muscle, called the gluteus maximus in humans, has become huge, and its anchoring region, the ilium, has expanded in tandem (Lovejoy, 1988). The shape of human buttocks, unique among the vertebrates, goes part and parcel with our upright bipedal posture and locomotion.

The huge braincase of humans houses the mind that makes us what we are. So large is our braincase that our face is flattened, our temporal musculature is reduced, and the process of childbirth continues to pose a serious risk for mother and child (Zimmer, 2001; Benton, 2005). Having such an unusually large head makes us top heavy and prone to falling! To adequately balance our huge brains on top of our vertically oriented bodies, the foramen magnum of humans points almost directly ventrally, allowing the spinal cord to exit the braincase vertically and centering the head above our center of mass (Zimmer, 2001; Benton, 2005) (Fig. 21.14).

What selective factors spurred the evolution of our bipedal posture and the concomitant changes to our skeletal chassis remains debated, but these changes do correlate with the loss of forested habitat and expanding grasslands in eastern Africa (Benton, 2005). As with kangaroos, a large, upright, bipedal ape is ungainly in a forest but perhaps more efficient at covering open spaces. Whatever the ultimate factors were that selected for our bipedal posture and large brains, the world has never been same. The great migrations of modern humans across the globe have occurred in the geological blink of an eye, in the last 100,000–200,000 years or so, on the thin edge of our last card of time (Zimmer, 2001). Civilization and agriculture as we know them are an even more recent invention of our species, occurring a mere 10,000 years ago (Zimmer, 2001; Diamond, 2011). With agriculture, the Industrial Revolution, and improved access to water, food, and medicine, we have been able to produce far more humans than would typically survive, having recently passed the 7-billion-person goalpost.

Along with our successes have come great challenges. Beyond the complexities of managing our resources and maintaining our large populations, our technological genius has often resulted in unintended but terrible consequences. Currently, one of many challenges we face is the recognition that our use of energy from fossil fuels has increased atmospheric carbon dioxide to levels higher than they have been in more than 1 million years (Honisch et al., 2009). Essentially, we are burning the Carboniferous coal forests and the concentrated remains of marine invertebrates to fuel our society, millions of years of stored sunlight energy, and releasing all of this energy nearly spontaneously from a geological perspective. Human-aided climate change is already having real impacts on our lives and the planet (Rosenzweig et al., 2008). Difficulties stemming from vested interests, ignorance, and fear, as well as technical challenges in the creation of more sustainable technologies, must be overcome if we are to survive.

Although it would be easy to become cynical and despair in the face of such challenges, the evolution of the human skeletal chassis brought with it a huge and complex brain combined with the power to use our hands for creative purposes, and that gives reason for hope. Why? Because beyond being intelligent, there are two things that, among all of the organisms that have ever lived, only humans can do. First, we can anticipate the future and, most importantly, act on it to alter the course of events. Second, we can use our imagination to bring into concrete existence the tools to make the necessary changes. You and I, and everyone else alive, have the power to change our world, for better or worse, here and now. Our skeletal chassis bears the hallmarks of over 500 million years of the struggle for existence in a world full of predators and environmental upheaval. The brain in our head is descended from a long line of survivors, but unlike every other vertebrate on the planet, only we can contemplate and understand our place on this grand family tree. Our diverse vertebrate family members, and the members of our human family tree, are yours and mine to understand, protect, and admire. Just like some mornings, the keys are there to drive us into the future; we just have to find them first.

Appendix: The Cards of Time

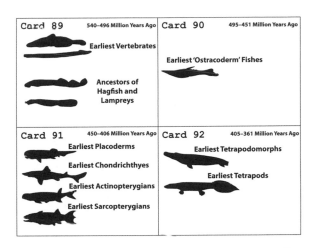

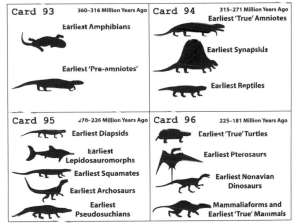

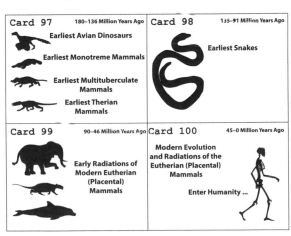

This appendix provides a visual compilation of the cards of time we have referred to throughout the book. Each card depicts the silhouettes of the earliest examples of particular vertebrate groups.

Card 89 — 540–496 Million Years Ago
Earliest Vertebrates
Ancestors of Hagfish and Lampreys

Card 90 — 495–451 Million Years Ago
Earliest 'Ostracoderm' Fishes

Card 91 — 450–406 Million Years Ago
Earliest Placoderms
Earliest Chondrichthyes
Earliest Actinopterygians
Earliest Sarcopterygians

Card 92 — 405–361 Million Years Ago
Earliest Tetrapodomorphs
Earliest Tetrapods

Card 93 — 360–316 Million Years Ago
Earliest Amphibians
Earliest 'Pre-amniotes'

Card 94 — 315–271 Million Years Ago
Earliest 'True' Amniotes
Earliest Synapsids
Earliest Reptiles

Card 95 — 270–226 Million Years Ago
Earliest Diapsids
Earliest Lepidosauromorphs
Earliest Squamates
Earliest Archosaurs
Earliest Pseudosuchians

Card 96 — 225–181 Million Years Ago
Earliest 'True' Turtles
Earliest Pterosaurs
Earliest Nonavian Dinosaurs
Mammaliaforms and Earliest 'True' Mammals

Card 97 — 180–136 Million Years Ago
Earliest Avian Dinosaurs
Earliest Monotreme Mammals
Earliest Multituberculate Mammals
Earliest Therian Mammals

Card 98 — 135–91 Million Years Ago
Earliest Snakes

Card 99 — 90–46 Million Years Ago
Early Radiations of Modern Eutherian (Placental) Mammals

Card 100 — 45–0 Million Years Ago
Modern Evolution and Radiations of the Eutherian (Placental) Mammals
Enter Humanity ...

Abdala, F., J. C. Cisneros, and R. M. H. Smith. 2006. Faunal aggregation in the Early Triassic Karoo Basin: earliest evidence of shelter-sharing behavior among tetrapods? PALAIOS 21:507–512.

Abdala, F., S. C. Jasinoski, and V. Fernandez. 2013. Ontogeny of the Early Triassic cynodont Thrinaxodon liorhinus (Therapsida): dental morphology and replacement. Journal of Vertebrate Paleontology 33:1408–1431.

Abourachid, A., R. Hackert, M. Herbin, P. A. Libourel, F. Lambert, H. Gioanni, P. Provini, P. Blazevic, and V. Hugel. 2011. Bird terrestrial locomotion as revealed by 3D kinematics. Zoology 114:360–368.

Adnet, S., and R. A. Martin. 2007. Increase of body size in sixgill sharks with change in diet as a possible background of their evolution. Historical Biology 19:279–289.

Aerts, P., R. Van Damme, K. D'Août, and B. Van Hooydonck. 2003. Bipedalism in lizards: whole–body modelling reveals a possible spandrel. Philosophical Transactions of the Royal Society of London. Series B: Biological Sciences 358:1525–1533.

Ahlberg, P. E., J. A. Clack, and H. Blom. 2005. The axial skeleton of the Devonian tetrapod Ichthyostega. Nature 437:137–140.

Alexander, R. M. 1965. The lift produced by heterocercal tails of Selachii. Journal of Experimental Biology 43:131–138.

Alexander, R. M. 1989. Dynamics of Dinosaurs and Other Extinct Giants. Columbia University Press, New York, 167 pp.

Alexander, R. M., and A. Vernon. 1975. The mechanics of hopping by kangaroos (Macropodidae). Journal of Zoology 177:265–303.

Allen, R. C. 2010. The Anatomy and Systematics of Terrestrisuchus gracilis (Archosauria, Crocodylomorpha). Northern Illinois University, DeKalb, Illinois, 186 pp.

Allen, V., J. Molnar, W. Parker, A. Pollard, G. Nolan, and J. R. Hutchinson. 2014. Comparative architectural properties of limb muscles in Crocodylidae and Alligatoridae and their relevance to divergent use of asymmetrical gaits in extant Crocodylia. Journal of Anatomy 225:569–582.

Allin, E. F. 1975. Evolution of the mammalian middle ear. Journal of Morphology 147:403–438.

Amborella Genome Project, V. A. Albert, W. B. Barbazuk, C. W. dePamphilis, J. P. Der, J. Leebens-Mack, H. Ma, J. D. Palmer, S. Rounsley, D. Sankoff, S. C. Schuster, D. E. Soltis, P. S. Soltis, S. R. Wessler, R. A. Wing, V. A. Albert, J. S. S. Ammiraju, W. B. Barbazuk, S. Chamala, A. S. Chanderbali, C. W. dePamphilis, J. P. Der, R. Determann, J. Leebens-Mack, H. Ma, P. Ralph, S. Rounsley, S. C. Schuster, D. E. Soltis, P. S. Soltis, J. Talag, L. Tomsho, B. Walts, S. Wanke, R. A. Wing, V. A. Albert, W. B. Barbazuk, S. Chamala, A. S. Chanderbali, T.-H. Chang, R. Determann, T. Lan, D. E. Soltis, P. S. Soltis, S. Arikit, M. J. Axtell, S. Ayyampalayam, W. B. Barbazuk, J. M. Burnette, S. Chamala, E. De Paoli, C. W. dePamphilis, J. P. Der, J. C. Estill, N. P. Farrell, A. Harkess, Y. Jiao, J. Leebens-Mack, K. Liu, W. Mei, B. C. Meyers, S. Shahid, E. Wafula, B. Walts, S. R. Wessler, J. Zhai, X. Zhang, V. A. Albert, L. Carretero-Paulet, C. W. dePamphilis, J. P. Der, Y. Jiao, J. Leebens-Mack, E. Lyons, D. Sankoff, H. Tang, E. Wafula, C. Zheng, V. A. Albert, N. S. Altman, W. B. Barbazuk, L. Carretero-Paulet, C. W. dePamphilis, J. P. Der, J. C. Estill, Y. Jiao, J. Leebens-Mack, K. Liu, W. Mei, E. Wafula, N. S. Altman, S. Arikit, M. J. Axtell, S. Chamala, A. S. Chanderbali, F. Chen, J.-Q. Chen, V. Chiang, E. De Paoli, C. W. dePamphilis, J. P. Der, R. Determann, B. Fogliani, C. Guo, J. Harholt, A. Harkess, C. Job, D. Job, S. Kim, H. Kong, J. Leebens-Mack, G. Li, L. Li, J. Liu, H. Ma, B. C. Meyers, J. Park, X. Qi, L. Rajjou, V. Burtet-Sarramegna, R. Sederoff, S. Shahid, D. E. Soltis, P. S. Soltis, Y.-H. Sun, P. Ulvskov, M. Villegente, J.-Y. Xue, T.-F. Yeh, X. Yu, J. Zhai, J. J. Acosta, V. A. Albert, W. B. Barbazuk, R. A. Bruenn, S. Chamala, A. de Kochko, C. W. dePamphilis, J. P. Der, L. R. Herrera-Estrella, E. Ibarra-Laclette, M. Kirst, J. Leebens-Mack, S. P. Pissis, V. Poncet, S. C. Schuster, D. E. Soltis, P. S. Soltis, and L. Tomsho. 2013. The Amborella genome and the evolution of flowering plants. Science 342:1241089–1241089.

Anderson, J. S. 2008. Focal review: the origin(s) of modern amphibians. Evolutionary Biology 35:231–247.

Anderson, J. S., R. R. Reisz, D. Scott, N. B. Fröbisch, and S. S. Sumida. 2008. A stem batrachian from the Early Permian of Texas and the origin of frogs and salamanders. Nature 453:515–518.

Anderson, P. S. L. 2009. Biomechanics, functional patterns, and disparity in Late Devonian arthrodires. Paleobiology 35:321–342.

Anderson, P. S. L., and M. W. Westneat. 2009. A biomechanical model of feeding kinematics for Dukelosteus terrelli (Arthrodira, Placodermi). Paleobiology 35:251–269.

Angielczyk, K. D. 2009. Dimetrodon is not a dinosaur: using tree thinking to understand the ancient relatives of mammals and their evolution. Evolution: Education and Outreach 2:257–271.

Antón, M. 2013. Sabertooth. Indiana University Press, Bloomington, Indiana, 243 pp.

Anthwal, N., L. Joshi, and A. S. Tucker. 2013. Evolution of the mammalian middle ear and jaw: adaptations and novel structures: the mammalian middle ear and jaw. Journal of Anatomy 222:147–160.

Archibald, J. D. 2012. Dinosaur extinction: past and present perceptions; pp. 1027–1038 in M. K. Brett-Surman, T. R. Holtz, and J. O. Farlow (eds.), The Complete Dinosaur, 2nd ed. Indiana University Press, Bloomington, Indiana.

Attenborough, D. 1998. The Life of Birds. BBC Books, London, 320 pp.

Atz, J. W. 1952. Narial breathing in fishes and the evolution of internal nares. The Quarterly Review of Biology 27:366–377.

Autumn, K., S. T. Hsieh, D. M. Dudek, J. Chen, C. Chitaphan, and R. J. Full. 2006. Dynamics of geckos running vertically. Journal of Experimental Biology 209:260–272.

Autumn, K., Y. A. Liang, S. T. Hsieh, W. Zesch, W. P. Chan, T. W. Kenny, R. Fearing, and R. J. Full. 2000. Adhesive force of a single gecko foot-hair. Nature 405:681–685.

Baier, D. B., and S. M. Gatesy. 2013. Three-dimensional skeletal kinematics of the shoulder girdle and forelimb in walking *Alligator*. Journal of Anatomy 223:462–473.

Baier, D. B., S. M. Gatesy, and F. A. Jenkins. 2007. A critical ligamentous mechanism in the evolution of avian flight. Nature 445:307–10.

Bakker, R. T. 1986. The Dinosaur Heresies: New Theories Unlocking the Mystery of the Dinosaurs and Their Extinction. William Morrow, New York, 165 pp.

Barrett, P. M., and K. J. Willis. 2001. Did dinosaurs invent flowers? Dinosaur–angiosperm coevolution revisited. Biological Reviews of the Cambridge Philosophical Society 76:411–447.

Bateman, P. W., and P. A. Fleming. 2009. To cut a long tail short: a review of lizard caudal autotomy studies carried out over the last 20 years. Journal of Zoology 277:1–14.

Bates, K. T., and P. L. Falkingham. 2012. Estimating maximum bite performance in *Tyrannosaurus rex* using multi-body dynamics. Biology Letters 8:660–664.

Bates, K. T., P. L. Falkingham, S. Macaulay, C. Brassey, and S. C. R. Maidment. 2015. Downsizing a giant: re-evaluating Dreadnoughtus body mass. Biology Letters 11:20150215.

Bemis, W. E., and L. Grande. 1992. Early development of the actinopterygian head. I. External development and staging of the paddlefish *Polyodon spathula*. Journal of Morphology 213:47–83.

Bemis, W. E., and B. Kynard. 2002. Sturgeon rivers: an introduction to acipenseriform biogeography and life history. Environmental Biology of Fishes 48:167–183.

Bemis, W. E., and G. V. Lauder. 1986. Morphology and function of the feeding apparatus of the lungfish, *Lepidosiren paradoxa* (Dipnoi). Journal of Morphology 187:81–108.

Bemis, W., E. Findeis, and L. Grande. 2002. An overview of Acipenseriformes. Sturgeon Biodiversity and Conservation 25–71.

Bennett, S. C. 1996. Aerodynamics and thermoregulatory function of the dorsal sail of *Edaphosaurus*. Paleobiology 22:496–506.

Bennett, S. C. 2001. The evolution of the pterosaurian pectoral girdle. Strata 11:15–17.

Bennett, S. C. 2003. Morphological evolution of the pectoral girdle of pterosaurs: myology and function; pp. 191–215 in Evolution and Palaeobiology of Pterosaurs. Special Publications Geological Society, London.

Bennett, S. C. 2012. New information on body size and cranial display structures of *Pterodactylus antiquus*, with a revision of the genus. Paläontologische Zeitschrift 87:269–289.

Benton, M. J. 2005. Vertebrate Palaeontology, 3rd ed. Blackwell Publishing, Oxford, UK, 455 pp.

Benton, M. J. 2012. Origin and early evolution of dinosaurs; pp. 320–345 in M. K. Brett-Surman, T. R. Holtz, and J. O. Farlow (eds.), The Complete Dinosaur, 2nd ed. Indiana University Press, Bloomington, Indiana.

Berman, D. S., and P. J. Regal. 1967. The loss of the ophidian middle ear. Evolution 21:641–643.

Berman, D. S., A. C. Henrici, S. S. Sumida, and T. Martens. 2000. Redescription of *Seymouria sanjuanensis* (Seymouriamorpha) from the Lower Permian of Germany based on complete, mature specimens with a discussion of paleoecology of the Bromacker locality assemblage. Journal of Vertebrate Paleontology 20:253–268.

Berman, D. S., R. R. Reisz, D. Scott, A. C. Henrici, S. S. Sumida, and T. Martens. 2000. Early Permian bipedal reptile. Science 290:969–972.

Bernal, D., J. K. Carlson, K. J. Goldman, and C. G. Lowe. 2012. Energetics, metabolism, and endothermy in sharks and rays; pp. 211–237 in J. C. Carrier, J. A. Musick, and M. R. Heithaus (eds.), Biology of Sharks and Their Relatives. CRC Press, Boca Raton, Florida.

Bertmar, G. 1981. Evolution of vomeronasal organs in vertebrates. Evolution 35: 359–366.

Bever, G. S., T. R. Lyson, D. J. Field, and B.-A. S. Bhullar. 2015. Evolutionary origin of the turtle skull. Nature. doi:10.1038/nature14900.

Beznosov, P. 2009. A redescription of the Early Carboniferous acanthodian *Acanthodes lopatini* Rohon, 1889. Acta Zoologica 90:183–193.

Bi, S., Y. Wang, J. Guan, X. Sheng, and J. Meng. 2014. Three new Jurassic euharamiyidan species reinforce early divergence of mammals. Nature 514:579–584.

Biewener, A. A. 1989. Mammalian terrestrial locomotion and size. Bioscience 39:776–783.

Biewener, A. A. 2005. Biomechanical consequences of scaling. Journal of Experimental Biology 208:1665–76.

Birks, S. M. 1997. Paternity in the Australian brush-turkey, *Alectura lathami*, a megapode bird with uniparental male care. Behavioral Ecology 8:560–568.

Blackburn, D. G., V. Hayssen, and C. J. Murphy. 1989. The origins of lactation and the evolution of milk: a review with new hypotheses. Mammal Review 19:1–26.

Blakey, R. 2010. Global Paleogeography. Global Paleogeography. Available at http://jan.ucc.nau.edu/~rcb7/globaltext2.html. Accessed November 2010.

Blakey, R., and W. Ranney. 2008. Ancient Landscapes of the Colorado Plateau. Grand Canyon Association, Grand Canyon, Arizona, 156 pp.

Bloomfield, L. A. 2006. How Things Work: The Physics of Everyday Life. John Wiley and Sons, Inc., New Jersey, 561 pp.

Boggs, S. 2011. Principles of Sedimentology and Stratigraphy, 5th ed. Prentice Hall, 600 pp.

Bohensky, F. 1986. Photo Manual and Dissection Guide of the Rat. Avery Publishing Group, Inc., New Jersey, 154 pp.

Boisvert, C. A. 2009. The humerus of *Panderichthys* in three dimensions and its significance in the context of the fish–tetrapod transition. Acta Zoologica 90:297–305.

Bonnan, M. F. 2003. The evolution of manus shape in sauropod dinosaurs: implications for functional morphology, forelimb orientation, and phylogeny. Journal of Vertebrate Paleontology 23:595–613.

Bonnan, M. F. 2004. Morphometric analysis of humerus and femur shape in Morrison sauropods: implications for functional morphology and paleobiology. Paleobiology 30:444–470.

Bonnan, M. F. 2005. Pes anatomy in sauropod dinosaurs: implications for functional morphology, evolution, and phylogeny; pp. 346–380 in V. Tidwell and K. Carpenter (eds.), Thunder Lizards: the Sauropodomorph Dinosaurs. Indiana University Press, Bloomington, Indiana.

Bonnan, M. F. 2007. Linear and geometric morphometric analysis of long bone scaling patterns in Jurassic neosauropod dinosaurs. their functional and paleobiological implications. Anatomical Record 290:1089–111.

Bordy, E. M., O. Sztanó, B. S. Rubidge, and A. Bumby. 2011. Early Triassic vertebrate burrows from the Katberg Formation of the south-western Karoo Basin, South Africa. Lethaia 44:33–45.

Botha, J , and A. Chinsamy. 2005. Growth patterns of *Thrinaxodon liorhinus*, a non-mammalian cynodont from the Lower Triassic of South Africa. Palaeontology 48:385–394.

Bourke, J. M., W. Ruger Porter, R. C. Ridgely, T. R. Lyson, E. R. Schachner, P. R. Bell, and L. M. Witmer. 2014. Breathing life into dinosaurs: tackling challenges of soft-tissue restoration and nasal airflow in extinct species. dinosaur nasal airflow. Anatomical Record 297:2148–2186.

Brainerd, E. L., K. Liem, and C. T. Samper. 1989. Air ventilation by recoil aspiration in polypterid fishes. Science 246:1593–1595.

Bramble, D. M., and D. R. Carrier. 1983. Running and breathing in mammals. Science 219:251–256.

Brett-Surman, M. K., Holtz, T. R., Jr., and Farlow, J. O. (eds.). 2012. The Complete Dinosaur, (2nd ed.). Indiana University Press, Bloomington, Indiana, 1112 pp.

Brazeau, M. D. 2009. The braincase and jaws of a Devonian 'acanthodian' and modern gnathostome origins. Nature 457:305–308.

Brazeau, M. D., and P. E. Ahlberg. 2006. Tetrapod-like middle ear architecture in a Devonian fish. Nature 439:318–321.

Brinkman, D., D. A. Eberth, and P. J. Currie. 2007. From bonebeds to paleobiology: applications of bonebed data; pp. 221–264 in R. R. Rogers, D. A. Eberth, and A. R. Fiorillo (eds.), Bonebeds: Genesis, Analysis, and Paleobiological Significance. University of Chicago Press, Chicago, Illinois.

Brinkmann, H., A. Denk, J. Zitzler, J. J. Joss, and A. Meyer. 2004. Complete mitochondrial genome sequences of the South American and the Australian lungfish: testing of the phylogenetic performance of mitochondrial data sets for phylogenetic problems in tetrapod relationships. Journal of Molecular Evolution 59:834–848.

Brochu, C. A. 2000. A digitally-rendered endocast for *Tyrannosaurus rex*. Journal of Vertebrate Paleontology 20:1–6.

Brochu, C. A. 2003. Phylogenetic approaches toward crocodylian history. Annual Review of Earth and Planetary Sciences 31:357–397.

Brochu, C. A., J. R. Wagner, S. Jouve, C. D. Sumrall, and L. D. Densmore. 2009. A correction corrected: consensus over the meaning of Crocodylia and why it matters. Systematic Biology 58:537–543.

Brown, S. J. 2004. Scurvy: How a Surgeon, a Mariner, and a Gentleman Solved the Greatest Medical Mystery of the Age of Sail. Thomas Dunne Books, New York, 256 pp.

Brusatte, S. L., G. Niedzwiedzki, and R. J. Butler. 2010. Footprints pull origin and diversification of dinosaur stem lineage deep into Early Triassic. Proceedings of the Royal Society of London. Series B: Biological Sciences 278:1107–1113.

Brusatte, S. L., M. J. Benton, J. B. Desojo, and M. C. Langer. 2010. The higher-level phylogeny of Archosauria (Tetrapoda: Diapsida). Journal of Systematic Palaeontology 8:3–47.

Buckley, M. 2015. Ancient collagen reveals evolutionary history of the endemic South American "ungulates." Proceedings of the Royal Society B: Biological Sciences 282:20142671–20142671.

Burggren, W. W. 1982. "Air gulping" improves blood oxygen transport during aquatic hypoxia in the goldfish *Carassius auratus*. Physiological Zoology 55:327–334.

Burggren, W. W., and K. Johansen. 1986. Circulation and respiration in lungfishes (Dipnoi). Journal of Morphology Supplement 1:217–236.

Burrow, C. 2003. Response to comment on "Separate evolutionary origins of teeth from evidence in fossil jawed vertebrates." Science 300:1661c–1661.

Butler, A. B., and W. Hodos. 2005. Comparative Vertebrate Neuroanatomy: Evolution and Adaptation. Wiley-Liss, New York, 600 pp.

Butler, A. B., A. Reiner, and H. J. Karten. 2011. Evolution of the amniote pallium and the origins of mammalian neocortex. Annals of the New York Academy of Sciences 1225:14–27.

Caldwell, M. W. 2003. "Without a leg to stand on": on the evolution and development of axial elongation and limblessness in tetrapods. Canadian Journal of Earth Sciences 40:573–588.

Caprette, C. L., M. S. Y. Lee, R. Shine, A. Mokany, and J. F. Downhower. 2004. The origin of snakes (Serpentes) as seen through eye anatomy. Biological Journal of the Linnean Society 81:469–482.

Carrier, D. R. 2006. Locomotor function of the pectoral girdle "muscular sling" in trotting dogs. Journal of Experimental Biology 209:2224–2237.

Carrier, D. R., and C. G. Farmer. 2000. The evolution of pelvic aspiration in archosaurs. Paleobiology 26:271–293.

Carroll, A. M., and P. C. Wainwright. 2003. Functional morphology of prey capture in the sturgeon, *Scaphirhynchus albus*. Journal of Morphology 256:270–284.

Carroll, R. L. 1969. A Middle Pennsylvanian captorhinomorph, and the interrelationships of primitive reptiles. Journal of Paleontology 43:151–170.

Carroll, R. L. 1988. Vertebrate Paleontology and Evolution. W. H. Freeman and Company, New York, 698 pp.

Carroll, R. L. 2009. The Rise of the Amphibians: 365 Million Years of Evolution. Johns Hopkins University Press, Baltimore, Maryland, 360 pp.

Carroll, R. L., and D. Baird. 1968. The Carboniferous amphibian *Tuditanus* (*Eosauravus*) and the distinction between microsaurs and reptiles. American Museum novitates; no. 2337.

Carter, D. R., and G. S. Beaupré. 2001. Skeletal Function and Form : Mechanobiology of Skeletal Development, Aging, and Regeneration. Cambridge University Press, Cambridge; New York, 318 pp.

Case, E. C. 1910. New or little known reptiles and amphibians from the Permian (?) of Texas. Bulletin of the American Museum of Natural History 28:163–181.

Cerny, R., P. Lwigale, R. Ericsson, D. Meulemans, H.-H. Epperlein, and M. Bronner-Fraser. 2004. Developmental origins and evolution of jaws: new interpretation of "maxillary" and "mandibular." Developmental Biology 276:225–236.

Chadwick, K. P., S. Regnault, V. Allen, and J. R. Hutchinson. 2014. Three-dimensional anatomy of the ostrich (*Struthio camelus*) knee joint. PeerJ 2:e706.

Charvet, C. J., T. Owerkowicz, and G. F. Striedter. 2009. Phylogeny of the telencephalic subventricular zone in sauropsids: evidence for the sequential evolution of pallial and subpallial subventricular zones. Brain, Behavior and Evolution 73:285–294.

Chew, S. F., N. K. Y. Chan, A. M. Loong, K. C. Hiong, W. L. Tam, and Y. K. Ip. 2004. Nitrogen metabolism in the African lungfish (*Protopterus dolloi*) aestivating in a mucus cocoon on land. Journal of Experimental Biology 207:777–786.

Chinsamy, A., and J. Hurum. 2006. Bone microstructure and growth patterns of early mammals. Acta Palaeontologica Polonica 51:325–338.

Chinsamy-Turan, A. 2005. The Microstructure of Dinosaur Bone–Deciphering Biology with Fine-scale Techniques. Johns Hopkins University Press, Baltimore, Maryland, 195 pp.

Chinsamy-Turan, A. (ed.). 2011. Forerunners of Mammals: Radiation, Histology, Biology. Indiana University Press, Bloomington, Indiana, 352 pp.

Choo, B., J. A. Long, and K. Trinajstic. 2009. A new genus and species of basal actinopterygian fish from the Upper Devonian Gogo Formation of Western Australia. Acta Zoologica 90:194–210.

Chudinov, P. K. 1965. New facts about the fauna of the Upper Permian of the U.S.S.R. Journal of Geology 73:117–130.

Cieri, R. L., B. A. Craven, E. R. Schachner, and C. G. Farmer. 2014. New insight into the evolution of the vertebrate respiratory system and the discovery of unidirectional airflow in iguana lungs. Proceedings of the National Academy of Sciences 201405088.

Cifelli, R. L., and C. de Muizon. 1998. Tooth eruption and replacement pattern in early marsupials. Comptes Rendus de l'Academie Des Sciences 326:215–220.

Cisneros, J. C., and M. Ruta. 2010. Morphological diversity and biogeography of procolophonids (Amniota: Parareptilia). Journal of Systematic Palaeontology 8:607–625.

Clack, J. A. 2002. Gaining Ground: The Origin and Evolution of Tetrapods, 1st ed. Indiana University Press, Bloomington, Indiana, 369 pp.

Clack, J. A. 2007. Devonian climate change, breathing, and the origin of the tetrapod stem group. Integrative and Comparative Biology 47:510–523.

Clack, J. A. 2009a. The fish–tetrapod transition: new fossils and interpretations. Evolution: Education and Outreach 2:213–223.

Clack, J. A. 2009b. The fin to limb transition: new data, interpretations, and hypotheses from paleontology and developmental biology. Annual Review of Earth and Planetary Sciences 37:163–179.

Clack, J. A., and S. M. Finney. 2005. *Pederpes finneyae,* an articulated tetrapod from the Tournaisian of Western Scotland. Journal of Systematic Palaeontology 2:311–346.

Clack, J. A., P. E. Ahlberg, S. M. Finney, P. D. Alonso, J. Robinson, and R. A. Ketcham. 2003. A uniquely specialized ear in a very early tetrapod. Nature 425:65–69.

Claessens, L. P. A. M. 2004. Archosaurian respiration and the pelvic girdle aspiration breathing of crocodyliforms. Proceedings of the Royal Society of London. Series B: Biological Sciences 271:1461–1465.

Claessens, L. P. A. M. 2009. The skeletal kinematics of lung ventilation in three basal bird taxa (emu, tinamou, and guinea fowl). Journal of Experimental Zoology Part A: Ecological Genetics and Physiology 311:586–599.

Claessens, L. P. A. M., P. M. O'Connor, and D. M. Unwin. 2009. Respiratory evolution facilitated the origin of pterosaur flight and aerial gigantism. PLoS ONE 4:e4497.

Cohn, M. J. 2002. Evolutionary biology: lamprey Hox genes and the origin of jaws. Nature 416:386–387.

Cohn, M. J., and C. Tickle. 1999. Developmental basis of limblessness and axial patterning in snakes. Nature 399:474–478.

Colbert, E. H., and C. M. Mook. 1951. The ancestral crocodilian *Protosuchus.* Bulletin of the American Museum of Natural History 97:143–182.

Cooper, W. J., and S. J. Steppan. 2010. Developmental constraint on the evolution of marsupial forelimb morphology. Australian Journal of Zoology 58:1.

Crawford, N. G., B. C. Faircloth, J. E. McCormack, R. T. Brumfield, K. Winker, and T. C. Glenn. 2012. More than 1000 ultraconserved elements provide evidence that turtles are the sister group of archosaurs. Biology Letters 8:783–786.

Crawford, N. G., J. F. Parham, A. B. Sellas, B. C. Faircloth, T. C. Glenn, T. J. Papenfuss, J. B. Henderson, M. H. Hansen, and W. B. Simison. 2015. A phylogenomic analysis of turtles. Molecular Phylogenetics and Evolution 83:250–257.

Crompton, A. W., and P. Parker. 1978. Evolution of the mammalian masticatory apparatus: the fossil record shows how mammals evolved both complex chewing mechanisms and an effective middle ear, two structures that distinguish them from reptiles. American Scientist 66:192–201.

Cucherousset, J., S. Boulêtreau, F. Azémar, A. Compin, M. Guillaume, and F. Santoul. 2012. "Freshwater killer whales": beaching behavior of an alien fish to hunt land birds. PLoS ONE 7:e50840.

Cupello, C., P. M. Brito, M. Herbin, F. J. Meunier, P. Janvier, H. Dutel, and G. Clément. 2015. Allometric growth in the extant coelacanth lung during ontogenetic development. Nature Communications 6:8222.

Currey, J. D. 2002. Bones: Structure and Mechanics. Princeton University Press, Princeton, New Jersey, 436 pp.

D'Emic, M. D., J. A. Whitlock, K. M. Smith, D. C. Fisher, and J. A. Wilson. 2013. Evolution of high tooth replacement rates in sauropod dinosaurs. PLoS ONE 8:e69235.

Daeschler, E. B., N. H. Shubin, and F. A. Jenkins. 2006. A Devonian tetrapod-like fish and the evolution of the tetrapod body plan. Nature 440:757–763.

Damiani, R., S. Modesto, A. Yates, and J. Neveling. 2003. Earliest evidence of cynodont burrowing. Proceedings of the Royal Society of London. Series B: Biological Sciences 270:1747–1751.

Darwin, C. 1859. On the Origin of Species. John Murray, Albemarle Street, London, 502 pp.

David, P. A. 1985. Clio and the economics of QWERTY. American Economic Review 77:332–337.

Davies, W. L., J. A. Cowing, J. K. Bowmaker, L. S. Carvalho, D. J. Gower, and D. M. Hunt. 2009. Shedding light on serpent sight: the visual pigments of henophidian snakes. Journal of Neuroscience 29:7519–7525.

Dawson, M. M., K. A. Metzger, D. B. Baier, and E. L. Brainerd. 2011.

Kinematics of the quadrate bone during feeding in mallard ducks. Journal of Experimental Biology 214:2036–2046.

Dawson, P. E. 2006. Functional Occlusion: From TMJ to Smile Design. Mosby, St. Louis, Missouri, 648 pp.

De Iuliis, G., and D. Pulerà. 2011. The Dissection of Vertebrates, 2nd ed. Elsevier, Academic Press, New York, 332 pp.

De Monchaux, N. 2011. Spacesuit: Fashioning Apollo. MIT Press, Cambridge, Massachusetts, 380 pp.

Dean, B., and B. Bhushan. 2010. Shark-skin surfaces for fluid-drag reduction in turbulent flow: a review. Philosophical Transactions of the Royal Society of London. Series A: Mathematical, Physical and Engineering Sciences 368:4775–4806.

Deban, S. M., and D. R. Carrier. 2002. Hypaxial muscle activity during running and breathing in dogs. Journal of Experimental Biology 205:1953–1967.

Deban, S. M., and S. B. Marks. 2002. Metamorphosis and evolution of feeding behaviour in salamanders of the family Plethodontidae. Zoological Journal of the Linnean Society 134:375–400.

Deban, S. M., J. C. O'Reilly, U. Dicke, and J. L. van Leeuwen. 2007. Extremely high-power tongue projection in plethodontid salamanders. Journal of Experimental Biology 210:655–667.

DeBraga, M. 2003. The postcranial skeleton, phylogenetic position, and probable lifestyle of the Early Triassic reptile Procolophon trigoniceps. Canadian Journal of Earth Sciences 40:527–556.

DeBraga, M., and O. Rieppel. 1997. Reptile phylogeny and the interrelationships of turtles. Zoological Journal of the Linnean Society 120:281–354.

Degrange, F. J., C. P. Tambussi, K. Moreno, L. M. Witmer, and S. Wroe. 2010. Mechanical analysis of feeding behavior in the extinct "terror bird" Andalgalornis steulleti (Gruiformes: Phorusrhacidae). PLoS ONE 5:e11856.

Degrange, F. J., C. P. Tambussi, M. L. Taglioretti, A. Dondas, and F. Scaglia. 2015. A new Mesembriornithinae (Aves, Phorusrhacidae) provides new insights into the phylogeny and sensory capabilities of terror birds. Journal of Vertebrate Paleontology 35:e912656.

Diamond, J. M. 2011. Collapse: How Societies Choose to Fail or Succeed, revised ed. Penguin Books, New York, 600 pp.

Dickerson, A. K., Z. G. Mills, and D. L. Hu. 2012. Wet mammals shake at tuned frequencies to dry. Journal of the Royal Society Interface 9:3208–3218.

Dickson, J. D. 1992. The Wild Turkey: Biology and Management. Stackpole Books, Mechanicsburg, Pennsylvania, 480 pp.

Dilkes, D. W., J. R. Hutchinson, C. M. Holliday, and L. M. Witmer. 2012. Reconstructing the musculature of dinosaurs; pp. 151–190 in M. K. Brett-Surman, T. R. Holtz, and J. O. Farlow (eds.), The Complete Dinosaur, 2nd ed. Indiana University Press, Bloomington, Indiana.

DiMichelle, W. A., and R. W. Hook. 1992. Paleozoic terrestrial ecosystems; pp. 205–325 in A. K. Behrensmeyer, W. A. DiMichelle, R. Potts, and H. D. Sues (eds.), Terrestrial Ecosystems through Time. University of Chicago Press, Chicago, Illinois.

Diogo, R., and V. Abdala. 2010. Muscles of Vertebrates: Comparative Anatomy, Evolution, Homologies and Development. Science Publishers, New York, 500 pp.

Dodson, P. 1975. Functional and ecological significance of relative growth in Alligator. Journal of Zoology 175:315–355.

Dodson, P. 2003. Allure of El Lagarto—why do dinosaur paleontologists love alligators, crocodiles, and their kin? Anatomical Record. Part A: Discoveries in Molecular, Cellular, and Evolutionary Biology 274:887–890.

Donoghue, P. C. J., and I. J. Sansom. 2002. Origin and early evolution of vertebrate skeletonization. Microscopy Research and Technique 59:352–372.

Donoghue, P. C. J., and M. Rücklin. 2014. The ins and outs of the evolutionary origin of teeth: evolutionary origin of teeth. Evolution & Development. doi: 10.1111/ede.12099.

Downs, J. P., E. B. Daeschler, F. A. Jenkins, and N. H. Shubin. 2008. The cranial endoskeleton of Tiktaalik roseae. Nature 455:925–929.

Drucker, E. G., and G. V. Lauder. 2003. Function of pectoral fins in rainbow trout: behavioral repertoire and hydrodynamic forces. Journal of Experimental Biology 206:813–826.

Dzik, J., T. Sulej, and G. Niedźwiedzki. 2008. A dicynodont-theropod

association in the Latest Triassic of Poland. Acta Palaeontologica Polonica 53:733–738.

Ebinger, P., and M. Röhrs. 1995. Volumetric analysis of brain structures, especially of the visual system in wild and domestic turkeys (Meleagris gallopavo). Journal Für Hirnforschung 36:219–228.

Elias, J. A., L. D. McBrayer, and S. M. Reilly. 2000. Prey transport kinematics in Tupinambis teguixin and Varanus exanthematicus: conservation of feeding behavior in "chemosensory-tongued" lizards. Journal of Experimental Biology 203:791–801.

Emerson, S. B. 1983. Functional analysis of frog pectoral girdles: the epicoracoid cartilages. Journal of Zoology (London) 201:293–308.

Emery, N. J. 2006. Cognitive ornithology: the evolution of avian intelligence. Philosophical Transactions of the Royal Society of London. Series B: Biological Sciences 361:23–43.

Erickson, G. M. 2014. On dinosaur growth. Annual Review of Earth and Planetary Sciences 42:675–697.

Erickson, G. M., A. K. Lappin, and K. A. Vliet. 2003. The ontogeny of bite-force performance in American alligator (Alligator mississippiensis). Journal of Zoology 260:317–327.

Erickson, G. M., P. J. Makovicky, P. J. Currie, M. A. Norell, S. A. Yerby, and C. A. Brochu. 2004. Gigantism and comparative life-history parameters of tyrannosaurid dinosaurs. Nature 430:772–775.

Essner, R. L., D. J. Suffian, P. J. Bishop, and S. M. Reilly. 2010. Landing in basal frogs: evidence of saltational patterns in the evolution of anuran locomotion. Naturwissenschaften 97:935–939.

Evans, S. E. 1994. A new anguimorph lizard from the Jurassic and Lower Cretaceous of England. Palaeontology 37:33–49.

Evans, S. E. 2003. At the feet of the dinosaurs: the early history and radiation of lizards. Biological Reviews 78:513–551.

Everhart, M. J. 2005. Oceans of Kansas: A Natural History of the Western Interior Sea. Indiana University Press, Bloomington, Indiana, 322 pp.

Everhart, M. J. 2007. Sea Monsters: Prehistoric Creatures of the Deep. National Geographic, Washington, D.C., 191 pp.

Ewer, R. F. 1965. The anatomy of the thecodont reptile Euparkeria capensis Broom. Philosophical Transactions of the Royal Society of London. Series B: Biological Sciences 248:379–435.

Eyal, S., E. Blitz, Y. Shwartz, H. Akiyama, S. Ronen, and E. Zelzer. 2015. On the development of the patella. Development 142:1–9.

Fadem, B. H., G. L. Trupin, E. Maliniak, J. L. VandeBerg, and V. Hayssen. 1982. Care and breeding of the gray, short-tailed opossum (Monodelphis domestica). Laboratory Animal Science 32:405–409.

Falcón, J., L. Besseau, M. Fuentès, S. Sauzet, E. Magnanou, and G. Boeuf. 2009. Structural and functional evolution of the pineal melatonin system in vertebrates. Annals of the New York Academy of Sciences 1163:101–111.

Falcon-Lang, H. J., M. J. Benton, and M. Stimson. 2007. Ecology of earliest reptiles inferred from basal Pennsylvanian trackways. Journal of the Geological Society 164:1113–1118.

Farlow, J. O., S. Hayashi, and G. J. Tattersall. 2010. Internal vascularity of the dermal plates of Stegosaurus (Ornithischia, Thyreophora). Swiss Journal of Geosciences 103:173–185.

Farlow, J. O., S. M. Gatesy, T. R. Holtz, J. R. Hutchinson, and J. M. Robinson. 2000. Theropod locomotion. American Zoologist 40:640–663.

Farmer, C. G., and K. Sanders. 2010. Unidirectional airflow in the lungs of alligators. Science 327:338–340.

Fastovsky, D. E., and D. B. Weishampel. 2012. Dinosaurs: A Concise Natural History, 2nd ed. Cambridge University Press, Cambridge; New York, 423 pp.

Feingold, K. R. 2007. Thematic review series: skin lipids. The role of epidermal lipids in cutaneous permeability barrier homeostasis. Journal of Lipid Research 48:2531–2546.

Field, D. J., J. A. Gauthier, B. L. King, D. Pisani, T. R. Lyson, and K. J. Peterson. 2014. Toward consilience in reptile phylogeny: miRNAs support an archosaur, not lepidosaur, affinity for turtles: reptile phylogeny from miRNAs. Evolution & Development 16:189–196.

Fischer, M. S., C. Krause, and K. E. Lilje. 2010. Evolution of chameleon locomotion, or how to become arboreal as a reptile. Zoology 113:67–74.

Flammang, B. E. 2009. Functional morphology of the radialis muscle in shark tails. Journal of Morphology 271:340–352.

Flammang, B. E., S. Alben, P. G. A. Madden, and G. V. Lauder. 2013. Functional morphology of the fin rays of teleost fishes: functional morphology of fish fin rays. Journal of Morphology 274:1044–1059.

Flammang, B. E., G. V. Lauder, D. R. Troolin, and T. Strand. 2011. Volumetric imaging of shark tail hydrodynamics reveals a three-dimensional dual-ring vortex wake structure. Proceedings of the Royal Society of London. Series B: Biological Sciences 278:3670–3678.

Flanagan, J. P. 2015. Chelonians (Turtles, Tortoises); pp. 27–38 in R. E. Miller and M. E. Fowler (eds.), Fowler's Zoo and Wild Animal Medicine, Volume 8. Elsevier, Saunders: St. Louis.

Fleishman, L. J., E. R. Loew, and M. J. Whiting. 2011. High sensitivity to short wavelengths in a lizard and implications for understanding the evolution of visual systems in lizards. Proceedings of the Royal Society of London. Series B: Biological Sciences 278:2891–2899.

Foster, J. R. 2014. Cambrian Ocean World: Ancient Sea Life of North America. Indiana University Press, Bloomington, Indiana, 416 pp.

Foth, C., H. Tischlinger, and O. W. M. Rauhut. 2014. New specimen of Archaeopteryx provides insights into the evolution of pennaceous feathers. Nature 511:79–82.

Foureaux, G., M. I. Egami, C. Jared, M. M. Antoniazzi, R. C. Gutierre, and R. L. Smith. 2010. Rudimentary eyes of squamate fossorial reptiles (Amphisbaenia and Serpentes). Anatomical Record: Advances in Integrative Anatomy and Evolutionary Biology 293:351–357.

Fowler, D. W., E. A. Freedman, J. B. Scannella, and R. E. Kambic. 2011. The predatory ecology of Deinonychus and the origin of flapping in birds. PLoS ONE 6:e28964.

Freeman, S., and J. C. Herron. 2007. Evolutionary Analysis, 4th ed. Benjamin Cummings, New York, 800 pp.

Freyer, C., and M. B. Renfree. 2009. The mammalian yolk sac placenta. Journal of Experimental Zoology Part B: Molecular and Developmental Evolution 312B:545–554.

Fricke, H., and K. Hissmann. 2000. Feeding ecology and evolutionary survival of the living coelacanth Latimeria chalumnae. Marine Biology 136:379–386.

Friedman, M., and M. D. Brazeau. 2010. A reappraisal of the origin and basal radiation of the Osteichthyes. Journal of Vertebrate Paleontology 30:36–56.

Friedman, M., and M. I. Coates. 2006. A newly recognized fossil coelacanth highlights the early morphological diversification of the clade. Proceedings of the Royal Society of London. Series B: Biological Sciences 273:245–250.

Fritz, J., E. Kienzle, M. Hummel, O. Wings, W. J. Streich, and M. Clauss. 2011. Gizzard vs. teeth, it's a tie: food-processing efficiency in herbivorous birds and mammals and implications for dinosaur feeding strategies. Paleobiology 37:577–586.

Fröbisch, N. B., and R. R. Schoch. 2009. Testing the impact of miniaturization on phylogeny: Paleozoic dissorophoid amphibians. Systematic Biology 58:312–327.

Fröbisch, N. B., J. C. Olori, R. R. Schoch, and F. Witzmann. 2010. Amphibian development in the fossil record. Seminars in Cell & Developmental Biology 21:424–431.

Frobisch, N. B., J. Frobisch, P. M. Sander, L. Schmitz, and O. Rieppel. 2013. Macropredatory ichthyosaur from the Middle Triassic and the origin of modern trophic networks. Proceedings of the National Academy of Sciences. 110:1393–1397.

Fry, B. G., N. Vidal, L. van der Weerd, E. Kochva, and C. Renjifo. 2009. Evolution and diversification of the Toxicofera reptile venom system. Journal of Proteomics 72:127–136.

Fry, B. G., N. Vidal, J. A. Norman, F. J. Vonk, H. Scheib, S. F. R. Ramjan, S. Kuruppu, K. Fung, S. Blair Hedges, M. K. Richardson, W. C. Hodgson, V. Ignjatovic, R. Summerhayes, and E. Kochva. 2005. Early evolution of the venom system in lizards and snakes. Nature 439:584–588.

Gaeth, A. P., R. V. Short, and M. B. Renfree. 1999. The developing renal, reproductive, and respiratory systems of the African elephant suggest an aquatic ancestry. Proceedings of the National Academy of Sciences 96:5555–5558.

Gaffney, E. S. 1990. The comparative osteology of the Triassic turtle Proganochelys. Bulletin of the AMNH; no. 194.

Gaffney, E. S., and L. J. Meeker. 1983. Skull morphology of the oldest turtles: a preliminary description of Proganochelys quenstedti. Journal of Vertebrate Paleontology 3:25–28.

Gans, C., and G. M. Hughes. 1967. The mechanism of lung ventilation in the tortoise *Testudo graeca* Linne. Journal of Experimental Biology 47:1–20.

Gans, C., and R. G. Northcutt. 1983. Neural crest and the origin of vertebrates: a new head. Science 220:268–273.

Gao, K.-Q., and N. H. Shubin. 2001. Late Jurassic salamanders from northern China. Nature 410:574–577.

Gao, K.-Q., and N. H. Shubin. 2003. Earliest known crown-group salamanders. Nature 422:424–428.

Gardner, J. 2001. Monophyly and affinities of albanerpetontid amphibians (Temnospondyli; Lissamphibia). Zoological Journal of the Linnean Society 131:309–352.

Gatesy, J., J. H. Geisler, J. Chang, C. Buell, A. Berta, R. W. Meredith, M. S. Springer, and M. R. McGowen. 2013. A phylogenetic blueprint for a modern whale. Molecular Phylogenetics and Evolution 66:479–506.

Gatesy, S. M. 1990. Caudofemoral musculature and the evolution of theropod locomotion. Paleobiology 16:170–186.

Gatesy, S. M., R. E. Kambic, and T. Roberts. 2010. Beyond hinges: 3-D joint function in erect bipeds. Journal of Vertebrate Paleontology 30:94A.

Gauthier, J. A. 1986. Saurischian monophyly and the origin of birds.; pp. 1–56 in K. Padian (ed.), The Origin of Birds and the Evolution of Flight. California Academy of Sciences Memoir. University of California Press, Berkeley, California.

Gauthier, J. A., S. J. Nesbitt, E. R. Schachner, G. S. Bever, and W. G. Joyce. 2011. The bipedal stem crocodilian *Poposaurus gracilis*: inferring function in fossils and innovation in archosaur locomotion. Bulletin of the Peabody Museum of Natural History 52:107–126.

Gilbert, S. F. 2010. Developmental Biology, 9th ed. Sinauer Associates, Inc., Sunderland, Massachusetts, 711 pp.

Gilbert, S. F., and A. M. Raunio. 1997. Embryology: Constructing the Organism. Sinauer Associates, Inc., Sunderland, Massachusetts, 537 pp.

Giles, S., M. Friedman, and M. D. Brazeau. 2015. Osteichthyan-like cranial conditions in an Early Devonian stem gnathostome. Nature. doi:10.1038/nature14065.

Gill, F. H. 2006. Ornithology, 3rd ed. W. H. Freeman and Company, New York, 720 pp.

Gingerich, P. D. 2012. Evolution of whales from land to sea. Proceedings of the American Philosophical Society 156:309–323.

Gorniak, G. C. 1985. Trends in the actions of mammalian masticatory muscles. American Zoologist 25:331–338.

Gould, S. J. 1989. Wonderful Life: The Burgess Shale and the Nature of History. W. W. Norton & Company, New York, 347 pp.

Gould, S. J. 1991. The panda's thumb of technology; pp. 59–75 in Bully for Brontosaurus: Reflections in Natural History. W. W. Norton & Company, New York.

Gower, D. J., and M. Wilkinson. 2009. Caecilians (Gymnophiona); pp. 369–372 in S. Blair Hedges and Sudhir Kumar (eds.), The Timetree of Life. Oxford University Press, Oxford, U.K.

Gould, S. J., and E. S. Vrba. 1982. Exaptation—a missing term in the science of form. Paleobiology 8:4–15.

Graham, J. B., N. C. Wegner, L. A. Miller, C. J. Jew, N. C. Lai, R. M. Berquist, L. R. Frank, and J. A. Long. 2014. Spiracular air breathing in polypterid fishes and its implications for aerial respiration in stem tetrapods. Nature Communications 5:1–6.

Grande, L., and W. E. Bemis. 1998. A comprehensive phylogenetic study of amiid fishes (Amiidae) based on comparative skeletal anatomy: an empirical search for interconnected patterns of natural history. Journal of Vertebrate Paleontology 18:1–696.

Gray, J. 1968. Animal Locomotion. William Clowes & Sons, Ltd., London, 479 pp.

Groenewald, G. H. 1991. Burrow casts from the Lystrosaurus-Procolophon Assemblage-zone, Karoo Sequence, South Africa. Koedoe 34:13–22.

Groenewald, G. H., J. Welman, and J. A. MacEachern. 2001. Vertebrate burrow complexes from the Early Triassic Cynognathus Zone (Driekoppen Formation, Beaufort Group) of the Karoo Basin, South Africa. PALAIOS 16:148–160.

Grossi, B., J. Iriarte-Díaz, O. Larach, M. Canals, and R. A. Vásquez. 2014. Walking like dinosaurs: chickens with artificial tails provide clues about non-avian theropod locomotion. PLoS ONE 9:e88458.

Habib, M. B. 2008. Comparative evidence for quadrupedal launch in pterosaurs. Zitteliana 28:159–166.

Haines, R. W. 1942a. The evolution of epiphyses and of endochondral bone. Biological Reviews 17:267–292.

Haines, R. W. 1942b. The tetrapod knee joint. Journal of Anatomy 76:270–301.

Haines, R. W. 1969. Epiphyses and sesamoids; pp. 81–115 in A. d'A. Bellairs and T. S. Parsons (eds.), Biology of the Reptilia, Morphology A vol. 1. Academic Press, New York.

Hall, M. I., J. M. Kamilar, and E. C. Kirk. 2012. Eye shape and the nocturnal bottleneck of mammals. Proceedings of the Royal Society of London. Series B: Biological Sciences 279:4962–4968.

Harris, M. A., and K. Steudel. 2002. The relationship between maximum jumping performance and hind limb morphology/physiology in domestic cats (*Felis silvestris catus*). Journal of Experimental Biology 205:3877–3889.

Hasiotis, S. T., R. W. Wellner, A. J. Martin, and T. M. Demko. 2004. Vertebrate burrows from Triassic and Jurassic continental deposits of North America and Antarctica: their paleoenvironmental and paleoecological significance. Ichnos 11:103–124.

Head, J. J., J. I. Bloch, A. K. Hastings, J. R. Bourque, E. A. Cadena, F. A. Herrera, P. D. Polly, and C. A. Jaramillo. 2009. Giant boid snake from the Palaeocene neotropics reveals hotter past equatorial temperatures. Nature 457:715–717.

Hedges, S. B., and N. Vidal. 2009. Lizards, snakes, and amphisbaenians (Squamata); pp. 383–389 in S. B. Hedges and S. Kumar (eds.), The Timetree of Life. Oxford University Press, Oxford, U.K.

Heesy, C. P., and M. I. Hall. 2010. The nocturnal bottleneck and the evolution of mammalian vision. Brain, Behavior and Evolution 75:195–203.

Henderson, D. M. 2004. Tipsy punters: sauropod dinosaur pneumaticity, buoyancy and aquatic habits. Proceedings of the Royal Society of London. Series B: Biological Sciences 271:S180–S183.

Henderson, D. M. 2010. Pterosaur body mass estimates from three-dimensional mathematical slicing. Journal of Vertebrate Paleontology 30:768–785.

Herrel, A., and G. J. Measey. 2010. The kinematics of locomotion in caecilians:

effects of substrate and body shape. Journal of Experimental Zoology Part A: Ecological Genetics and Physiology 313A:301–309.

Herrel, A., V. Schaerlaeken, J. J. Meyers, K. A. Metzger, and C. F. Ross. 2007. The evolution of cranial design and performance in squamates: consequences of skull-bone reduction on feeding behavior. Integrative and Comparative Biology 47:107–117.

Hickman, G. C. 1979. The mammalian tail: a review of functions. Mammal Review 9:143–157.

Hildebrand, M., and G. E. Goslow. 1995. Analysis of Vertebrate Structure. John Wiley and Sons, Inc., New York, 635 pp.

Hildebrand, M., and G. E. Goslow. 2001. Analysis of Vertebrate Structure, 3rd ed. John Wiley and Sons, 635 pp.

Hill, R. 2005. Integration of morphological data sets for phylogenetic analysis of Amniota: the importance of integumentary characters and increased taxonomic sampling. Systematic Biology 54:530–547.

Hillenius, W. J., and J. A. Ruben. 2004. The evolution of endothermy in terrestrial vertebrates: who? when? why? Physiological and Biochemical Zoology 76:1019–1042.

Holliday, C. M., and L. M. Witmer. 2007. Archosaur adductor chamber evolution: integration of musculoskeletal and topological criteria in jaw muscle homology. Journal of Morphology 268:457–484.

Holliday, C. M., and L. M. Witmer. 2008. Cranial kinesis in dinosaurs: intracranial joints, protractor muscles, and their significance for cranial evolution and function in diapsids. Journal of Vertebrate Paleontology 28:1073–1088.

Holliday, C. M., R. C. Ridgely, J. C. Sedlmayr, and L. M. Witmer. 2010. Cartilaginous epiphyses in extant archosaurs and their implications for reconstructing limb function in dinosaurs. PLoS ONE 5:e13120.

Holliday, C. M., H. P. Tsai, R. J. Skiljan, I. D. George, and S. Pathan. 2013. A 3D interactive model and atlas of the jaw musculature of *Alligator mississippiensis*. PLoS ONE 8:e62806.

Holmes, R. B., R. L. Carroll, and R. R. Reisz. 1998. The first articulated skeleton of *Dendrerpeton acadianum* (Temnospondyli, Dendrerpetontidae) from the Lower Pennsylvanian locality of Joggins, Nova Scotia, and a

review of its relationships. Journal of Vertebrate Paleontology 18:64–79.

Holtz, T. R. 2012. Theropods; pp. 346–377 in M. K. Brett-Surman, T. R. Holtz, and J. O. Farlow (eds.), The Complete Dinosaur, 2nd ed. Indiana University Press, Bloomington, Indiana.

Hone, D. W. E. 2012. Pterosaur research: recent advances and a future revolution. Acta Geologica Sinica—English Edition 86:1366–1376.

Honisch, B., N. G. Hemming, D. Archer, M. Siddall, and J. F. McManus. 2009. Atmospheric carbon dioxide concentration across the mid-Pleistocene transition. Science 324:1551–1554.

Horner, J. R., and J. Gorman. 1990. Digging Dinosaurs: The Search That Unraveled the Mystery of Baby Dinosaurs. Harper Perennial, New York, 208 pp.

Horner, J. R., A. de Ricqlès, and K. Padian. 1999. Variation in dinosaur skeletochronology indicators: implications for age assessment and physiology. Paleobiology 25:295–304.

Horner, J. R., K. Padian, and A. de Ricqlès. 2001. Comparative osteohistology of some embryonic and perinatal archosaurs: developmental and behavioral implications for dinosaurs. Paleobiology 27:39–58.

Hossain, T., and J. Morgan. 2009. The quest for QWERTY. American Economic Review 99:435–440.

Hsieh, S. T. 2003. Three-dimensional hindlimb kinematics of water running in the plumed basilisk lizard (*Basiliscus plumifrons*). Journal of Experimental Biology 206:4363–4377.

Hsieh, S. T., and G. V. Lauder. 2004. Running on water: three-dimensional force generation by basilisk lizards. Proceedings of the National Academy of Sciences 101:16784–16788.

Hu, D. L., J. Nirody, T. Scott, and M. J. Shelley. 2009. The mechanics of slithering locomotion. Proceedings of the National Academy of Sciences 106:10081–10085.

Hu, Y., J. Meng, Y. Wang, and C. Li. 2005. Large Mesozoic mammals fed on young dinosaurs. Nature 433:149–152.

Hutchinson, J. R., and M. Garcia. 2002. *Tyrannosaurus* was not a fast runner. Nature 415:1018–1021.

Hutchinson, J. R., and S. M. Gatesy. 2000. Adductors, abductors, and the evolution of archosaur locomotion. Paleobiology 26:734–751.

Huttenlocker, A. K., D. Mazierski, and R. R. Reisz. 2011. Comparative osteohistology of hyperelongate neural spines in the Edaphosauridae (Amniota: Synapsida). Palaeontology 54:573–590.

Ibrahim, N., P. C. Sereno, C. Dal Sasso, S. Maganuco, M. Fabbri, D. M. Martill, S. Zouhri, N. Myhrvold, and D. A. Iurino. 2014. Semiaquatic adaptations in a giant predatory dinosaur. Science 345:1613–1616.

Iordansky, N. N. 2011. Cranial kinesis in lizards (Lacertilia): origin, biomechanics, and evolution. Biology Bulletin 38:868–877.

Irschick, D. J., and B. C. Jayne. 1999. Comparative three-dimensional kinematics of the hindlimb for high-speed bipedal and quadrupedal locomotion of lizards. Journal of Experimental Biology 202:1047–1065.

Isaak, M. 2007. The Counter-Creationism Handbook. University of California Press, Berkeley, California, 362 pp.

Iwabe, N., Y. Hara, Y. Kumazawa, K. Shibamoto, Y. Saito, T. Miyata, and K. Katoh. 2005. Sister group relationship of turtles to the bird-crocodilian clade revealed by nuclear DNA–coded proteins. Molecular Biology and Evolution 22:810–813.

Iwasaki, S. 2002. Evolution of the structure and function of the vertebrate tongue. Journal of Anatomy 201:1–13.

Jacoby, D. M. P., D. S. Busawon, and D. W. Sims. 2010. Sex and social networking: the influence of male presence on social structure of female shark groups. Behavioral Ecology 21:808–818.

Janvier, P. 1996. Early Vertebrates. Claredon Press, Oxford, U.K., 393 pp.

Janvier, P. 2008a. Early jawless vertebrates and cyclostome origins. Zoological Science 25:1045–1056.

Janvier, P. 2008b. The brain in early fossil vertebrates: evolution information from an empty nutshell. Brain Research Bulletin 75:314–318.

Jarvis, E. D., S. Mirarab, A. J. Aberer, B. Li, P. Houde, C. Li, S. Y. W. Ho, B. C. Faircloth, B. Nabholz, J. T. Howard, A. Suh, C. C. Weber, R. R. da Fonseca, J. Li, F. Zhang, H. Li, L. Zhou, N. Narula, L. Liu, G. Ganapathy, B. Boussau, M. S. Bayzid, V. Zavidovych, S. Subramanian, T. Gabaldon, S. Capella-Gutierrez, J. Huerta-Cepas, B. Rekepalli, K. Munch, M. Schierup, B. Lindow, W. C. Warren, D. Ray, R. E. Green, M. W. Bruford, X. Zhan, A. Dixon,

S. Li, N. Li, Y. Huang, E. P. Derryberry, M. F. Bertelsen, F. H. Sheldon, R. T. Brumfield, C. V. Mello, P. V. Lovell, M. Wirthlin, M. P. C. Schneider, F. Prosdocimi, J. A. Samaniego, A. M. V. Velazquez, A. Alfaro-Nunez, P. F. Campos, B. Petersen, T. Sicheritz-Ponten, A. Pas, T. Bailey, P. Scofield, M. Bunce, D. M. Lambert, Q. Zhou, P. Perelman, A. C. Driskell, R. Shapiro, Z. Xiong, Y. Zeng, S. Liu, Z. Li, B. Liu, K. Wu, J. Xiao, X. Yinqi, Q. Zheng, Y. Zhang, H. Yang, J. Wang, L. Smeds, F. E. Rheindt, M. Braun, J. Fjeldsa, L. Orlando, F. K. Barker, K. A. Jonsson, W. Johnson, K.-P. Koepfli, S. O'Brien, D. Haussler, O. A. Ryder, C. Rahbek, E. Willerslev, G. R. Graves, T. C. Glenn, J. McCormack, D. Burt, H. Ellegren, P. Alstrom, S. V. Edwards, A. Stamatakis, D. P. Mindell, J. Cracraft, E. L. Braun, T. Warnow, W. Jun, M. T. P. Gilbert, and G. Zhang. 2014. Whole-genome analyses resolve early branches in the tree of life of modern birds. Science 346:1320–1331.

Jasinoski, S. C., E. J. Rayfield, and A. Chinsamy. 2009. Comparative feeding biomechanics of *Lystrosaurus* and the generalized dicynodont *Oudenodon*. Anatomical Record: Advances in Integrative Anatomy and Evolutionary Biology 292:862–874.

Jayne, B. C. 1988. Muscular mechanisms of snake locomotion: an electromyographic study of the sidewinding and concertina modes of *Crotalus cerastes, Nerodia fasciata* and *Elaphe obsoleta*. Journal of Experimental Biology 140:1–33.

Jenkins, F. 1973. The functional anatomy and evolution of the mammalian humero-ulnar articulation. American Journal of Anatomy 137:281–297.

Jenkins, F. A. 1974. The movement of the shoulder in claviculate and aclaviculate mammals. Journal of Morphology 144:71–84.

Jenkins, F. A., and G. E. Goslow. 1983. The functional anatomy of the shoulder of the savannah monitor lizard (*Varanus exanthematicus*). Journal of Morphology 175:195–216.

Jenkins, F. A., and D. W. Krause. 1983. Adaptations for climbing in North American multituberculates (Mammalia). Science 220:712–715.

Jenkins, F. A., and D. McClearn. 1984. Mechanisms of hind foot reversal in climbing mammals. Journal of Morphology 182:197–219.

Jenkins, F. A., and F. R. Parrington. 1976. The postcranial skeletons of the Triassic mammals *Eozostrodon, Megazostrodon* and *Erythrotherium*. Philosophical Transactions of the Royal Society of London. Series B: Biological Sciences 387–431.

Jenkins, F. A., and N. H. Shubin. 1998. *Prosalirus bitis* and the anuran caudopelvic mechanism. Journal of Vertebrate Paleontology 18:495–510.

Jenkins, F. A., K. P. Dial, and G. E. Goslow. 1988. A cineradiographic analysis of bird flight: the wishbone in starlings is a spring. Science 241:1495–1498.

Jenkins, F. A., D. M. Walsh, and R. L. Carroll. 2007. Anatomy of *Eocaecilia micropodia*, a limbed caecilian of the Early Jurassic. Bulletin of the Museum of Comparative Zoology 158:285–366.

Ji, Q., Z.-X. Luo, C.-X. Yuan, and A. R. Tabrum. 2006. A swimming mammaliaform from the Middle Jurassic and ecomorphological diversification of early mammals. Science 311:1123–1127.

Ji, Y., X. Wang, Y. Liu, and Q. Ji. 2011. Systematics, behavior and living environment of *Shantungosaurus giganteus* (Dinosauria: Hadrosauridae). Acta Geologica Sinica—English Edition 85:58–65.

Ji, Q., Z.-X. Luo, C.-X. Yuan, J. R. Wible, J.-P. Zhang, and J. A. Georgi. 2002. The earliest known eutherian mammal. Nature 416:816–822.

Johansen, K., and C. Lenfant. 1968. Respiration in the African lungfish *Protopterus aethiopicus*. II. Control of breathing. Journal of Experimental Biology 49:453–468.

Jones, M. E., N. Curtis, M. J. Fagan, P. O'Higgins, and S. E. Evans. 2011. Hard tissue anatomy of the cranial joints in *Sphenodon* (Rhynchocephalia): sutures, kinesis, and skull mechanics. Palaeontologia Electronica 14:17A.

Jones, M. E. H., P. O'Higgins, M. J. Fagan, S. E. Evans, and N. Curtis. 2012. Shearing mechanics and the influence of a flexible symphysis during oral food processing in *Sphenodon* (Lepidosauria: Rhynchocephalia). Anatomical Record: Advances in Integrative Anatomy and Evolutionary Biology 295:1075–1091.

Jones, N., and R. Jones. 1982. The structure of the male genital system of the Port Jackson shark, *Heterodontus portjacksoni,* with particular reference to the genital ducts. Australian Journal of Zoology 30:523.

Jones, S. 1999. Darwin's Ghost: The Origin of Species Updated. Doubleday, London, 377 pp.

Jorgensen, J. M., J. P. Lomholt, R. E. Weber, and H. Malte. 1998. The Biology of Hagfishes. Thomson Science, London, 579 pp.

Joyce, W. G., and J. A. Gauthier. 2004. Palaeoecology of Triassic stem turtles sheds new light on turtle origins. Proceedings of the Royal Society of London. Series B: Biological Sciences 271:1–5.

Joyce, W. G., S. G. Lucas, T. M. Scheyer, A. B. Heckert, and A. P. Hunt. 2009. A thin-shelled reptile from the Late Triassic of North America and the origin of the turtle shell. Proceedings of the Royal Society of London. Series B: Biological Sciences 276:507–513.

Kambic, R. E., T. J. Roberts, and S. M. Gatesy. 2014. Long-axis rotation: a missing degree of freedom in avian bipedal locomotion. Journal of Experimental Biology 217:2770–2782.

Kammerer, C. F., L. Grande, and M. W. Westneat. 2006. Comparative and developmental functional morphology of the jaws of living and fossil gars (Actinopterygii: Lepisosteidae). Journal of Morphology 267:1017–1031.

Kardong, K. V. 2012. Vertebrates: Comparative Anatomy, Function, Evolution, 6th ed. McGraw Hill Company, New York, 794 pp.

Kawasaki, K., and K. M. Weiss. 2006. Evolutionary genetics of vertebrate tissue mineralization: the origin and evolution of the secretory calcium-binding phosphoprotein family. Journal of Experimental Zoology Part B: Molecular and Developmental Evolution 306B:295–316.

Kellner, A. W. A. 2004. Chapter 2: the ankle structure of two pterodactyloid pterosaurs from the Santana Formation (Lower Cretaceous), Brazil. Bulletin of the American Museum of Natural History 285:25–35.

Kellner, A. W. A., X. Wang, H. Tischlinger, D. de Almeida Campos, D. W. E. Hone, and X. Meng. 2009. The soft tissue of *Jeholopterus* (Pterosauria, Anurognathidae, Batrachognathinae) and the structure of the pterosaur wing membrane. Proceedings of the Royal Society of London. Series B: Biological Sciences 277:321–329.

Kelly, A., and M. Kelly. 2009. Darwin: For the Love of Science. Bristol Cultural Development Parnership (BCDP), Bristol, U.K., 280 pp.

Keyte, A. L., and K. K. Smith. 2010. Developmental origins of precocial forelimbs in marsupial neonates. Development 137:4283–4294.

Kielan-Jaworowska, Z. 2013. In Pursuit of Early Mammals. Indiana University Press, Bloomington, Indiana, 253 pp.

Kimmel, C. B., C. T. Miller, and R. J. Keynes. 2001. Neural crest patterning and the evolution of the jaw. Journal of Anatomy 199:105–119.

Kirkland, J. I. 1994. Predation of dinosaur nests by terrestrial crocodilians; pp. 124–134 in K. Carpenter, K. F. Hirsch, and J. R. Horner (eds.), Dinosaurs Eggs and Babies. Cambridge University Press, New York.

Kis, A., L. Huber, and A. Wilkinson. 2015. Social learning by imitation in a reptile (*Pogona vitticeps*). Animal Cognition 18:325–331.

Klein, A., L.-A. R. Sacrey, I. Q. Whishaw, and S. B. Dunnett. 2012. The use of rodent skilled reaching as a translational model for investigating brain damage and disease. Neuroscience & Biobehavioral Reviews 36:1030–1042.

Klein, H., S. Voigt, A. Hminna, H. Saber, J. Schneider, and D. Hmich. 2010. Early Triassic archosaur-dominated footprint assemblage from the Argana Basin (Western High Atlas, Morocco). Ichnos 17:215–227.

Klein, W., and J. R. Codd. 2010. Breathing and locomotion: comparative anatomy, morphology and function. Respiratory Physiology & Neurobiology 173:S26–S32.

Kleinteich, T., A. Haas, and A. P. Summers. 2008. Caecilian jaw-closing mechanics: integrating two muscle systems. Journal of the Royal Society Interface 5:1491–1504.

Knell, S. J. 2013. The Great Fossil Enigma: The Search for the Conodont Animal. Indiana University Press, Bloomington, Indiana, 413 pp.

Knobloch, H. S., and V. Grinevich. 2014. Evolution of oxytocin pathways in the brain of vertebrates. Frontiers in Behavioral Neuroscience 8:1–18.

Köhler, M., N. Marín-Moratalla, X. Jordana, and R. Aanes. 2012. Seasonal bone growth and physiology in endotherms shed light on dinosaur physiology. Nature 487:358–361.

Köppl, C. 2011. Birds—same thing, but different? Convergent evolution in the avian and mammalian auditory systems provides informative comparative models. Hearing Research 273:65–71.

Krause, D. W. 1982. Jaw movement, dental function, and diet in the Paleocene multituberculate *Ptilodus*. Paleobiology 8:265–281.

Krause, D. W. 1986. Competitive exclusion and taxonomic displacement in the fossil record; the case of rodents and multituberculates in North America. Rocky Mountain Geology Contributions to Geology, University of Wyoming, Special Paper 3:95–117.

Kubo, T., and M. J. Benton. 2009. Tetrapod postural shift estimated from Permian and Triassic trackways. Palaeontology 52:1029–1037.

Kuhel, S., S. Reinhard, and A. Kupfer. 2010. Evolutionary reproductive morphology of amphibians: an overview. Bonn Zoological Bulletin 57:119–126.

Kupfer, A., H. Müller, M. M. Antoniazzi, C. Jared, H. Greven, R. A. Nussbaum, and M. Wilkinson. 2006. Parental investment by skin feeding in a caecilian amphibian. Nature 440:926–929.

Kuratani, S. 2004. Evolution of the vertebrate jaw: comparative embryology and molecular developmental biology reveal the factors behind evolutionary novelty. Journal of Anatomy 205:335–347.

Kuratani, S., S. Kuraku, and H. Nagashima. 2011. Evolutionary developmental perspective for the origin of turtles: the folding theory for the shell based on the developmental nature of the carapacial ridge. Evolution & Development 13:1–14.

Kuratani, S., Y. Nobusada, N. Horigome, and Y. Shigetani. 2001. Embryology of the lamprey and evolution of the vertebrate jaw: insights from molecular and developmental perspectives. Philosophical Transactions of the Royal Society of London. Series B: Biological Sciences 356:1615–1632.

Lacovara, K. J., M. C. Lamanna, L. M. Ibiricu, J. C. Poole, E. R. Schroeter, P. V. Ullmann, K. K. Voegele, Z. M. Boles, A. M. Carter, E. K. Fowler, V. M. Egerton, A. E. Moyer, C. L. Coughenour, J. P. Schein, J. D. Harris, R. D. Martínez, and F. E. Novas. 2014. A gigantic, exceptionally complete titanosaurian sauropod dinosaur from Southern Patagonia, Argentina. Scientific Reports 4:6196.

Lambert, D. 1998. The Field Guide to Geology, Updated. Facts on File, Inc., New York, 256 pp.

Lance, S. L., T. D. Tuberville, L. Dueck, C. Holz-Schietinger, P. L. Trosclair, R. M. Elsey, and T. C. Glenn. 2009. Multiyear multiple paternity and mate fidelity in the American alligator, *Alligator mississippiensis*. Molecular Ecology 18:4508–4520.

Landsmeer, J. M. F. 1981. Digital morphology in *Varanus* and *Iguana*. Journal of Morphology 168:289–295.

Landsmeer, J. M. F. 1983. The mechanism of forearm rotation in *Varanus exanthematicus*. Journal of Morphology 175:119–130.

Langone, J. 2004. The New How Things Work: Everyday Technology Explained. National Geographic, Washington, D.C., 272 pp.

Lauder, G. V. 1980. Evolution of the feeding mechanism in primitive actinopterygian fishes: a functional anatomical analysis of *Polypterus, Lepisosteus,* and *Amia.* Journal of Morphology 163:283–317.

Lauder, G. V., and P. G. A. Madden. 2006. Learning from fish: kinematics and experimental hydrodynamics for roboticists. International Journal of Automation and Computing 3:325–335.

Lauder, G. V., E. C. Drucker, J. C. Nauen, and C. D. Wilga. 2003. Experimental hydrodynamics and evolution: caudal fin locomotion in fishes; pp. 117–135 in V. J. Bels, J.-P. Gasc, and A. Casinos (eds.), Vertebrate Biomechanics and Evolution. BIOS Scientific Publishers, Ltd., Oxford, U.K.

Laurin, M. 1996. A redescription of the cranial anatomy of *Seymouria baylorensis,* the best known Seymouriamorph (Vertebrata: Seymouriamorpha). PaleoBios 17:1–16.

Laurin, M. 2004. The evolution of body size, Cope's rule and the origin of amniotes. Systematic Biology 53:594–622.

Laurin, M., F. J. Meunier, D. Germain, and M. Lemoine. 2007. A microanatomical and histological study of the paired fin skeleton of the Devonian sarcopterygian *Eusthenopteron foordi.* Journal of Paleontology 81:143–153.

Lee, M. S. Y. 2005. Molecular evidence and marine snake origins. Biology Letters 1:227–230.

Lee, M. S. Y., G. L. Bell, and M. W. Caldwell. 1999. The origin of snake feeding. Nature 400:655–659.

Lee, M. S. Y., C. E. Gow, and J. W. Kitching. 1997. Anatomy and relationships of the pareiasaur

Pareiasuchus nasicornis from the Upper Permian of Zambia. Palaeontology 40:307–335.

Leitch, D. B., and K. C. Catania. 2012. Structure, innervation and response properties of integumentary sensory organs in crocodilians. Journal of Experimental Biology 215:4217–4230.

Leurs, F., Y. P. Ivanenko, A. Bengoetxea, A.-M. Cebolla, B. Dan, F. Lacquaniti, and G. A. Cheron. 2011. Optimal walking speed following changes in limb geometry. Journal of Experimental Biology 214:2276–2282.

Levine, R. P. 2004. Contribution of eye retraction to swallowing performance in the northern leopard frog, *Rana pipiens.* Journal of Experimental Biology 207:1361–1368.

Li, C., X. C. Wu, O. Rieppel, L. T. Wang, and L. J. Zhao. 2008. An ancestral turtle from the Late Triassic of southwestern China. Nature 456:497–501.

Liebowitz, S. J., and S. E. Margolis. 1990. The fable of the keys. Journal of Law and Economics 33:1–25.

Liem, K., W. Bemis, W. Walker, and L. Grande. 2001. Functional Anatomy of the Vertebrates: An Evolutionary Perspective. Thomson Brooks/Cole, Belmont, California, 703 pp.

Lim, J., D. S. Fudge, N. Levy, and J. M. Gosline. 2006. Hagfish slime ecomechanics: testing the gill-clogging hypothesis. Journal of Experimental Biology 209:702–710.

Lindgren, J., J. W. M. Jagt, and M. W. Caldwell. 2007. A fishy mosasaur: the axial skeleton of *Plotosaurus* (Reptilia, Squamata) reassessed. Lethaia 40:153–160.

Lindgren, J., M. J. Polcyn, and B. A. Young. 2011. Landlubbers to leviathans: evolution of swimming in mosasaurine mosasaurs. Paleobiology 37:445–469.

Lindgren, J., C. Alwmark, M. W. Caldwell, and A. R. Fiorillo. 2009. Skin of the Cretaceous mosasaur *Plotosaurus:* implications for aquatic adaptations in giant marine reptiles. Biology Letters 5:528–531.

Lingham-Soliar, T. 1995. Anatomy and functional morphology of the largest marine reptile known, *Mosasaurus hoffmanni* (Mosasauridae, Reptilia) from the Upper Cretaceous, Upper Maastrichtian of the Netherlands. Philosophical Transactions of the Royal Society of London. Series B: Biological Sciences 347:155–172.

Lisney, T. J., and S. P. Collin. 2006. Brain morphology in large pelagic fishes: a comparison between sharks and teleosts. Journal of Fish Biology 68:532–554.

Liu, Z.-J., and X. Wang. 2015. A perfect flower from the Jurassic of China. Historical Biology 1–13.doi: 10.1080/08912963.2015.1020423.

Liu, J., J. C. Aitchison, Y. Y. Sun, Q. Y. Zhang, C. Y. Zhou, and T. Lv. 2011. New mixosaurid ichthyosaur specimen from the Middle Triassic of SW China: further evidence for the diapsid origin of ichthyosaurs. Journal of Paleontology 85:32–36.

Livingston, V. J., M. F. Bonnan, R. M. Elsey, J. L. Sandrik, and D. R. Wilhite. 2009. Differential limb scaling in the American alligator (*Alligator mississippiensis*) and its implications for archosaur locomotor evolution. Anatomical Record: Advances in Integrative Anatomy and Evolutionary Biology 292:787–797.

Loew, E. R., L. J. Fleishman, R. G. Foster, and I. Provencio. 2002. Visual pigments and oil droplets in diurnal lizards: a comparative study of Caribbean anoles. Journal of Experimental Biology 205:927–938.

Lombard, R. E., and J. R. Bolt. 1995. A new primitive tetrapod, *Whatcheeria deltae,* from the Lower Carboniferous of Iowa. Palaeontology 38:471–494.

Long, J. A. 2001. On the relationships of *Psarolepis* and the onychodontiform fishes. Journal of Vertebrate Paleontology 21:815–820.

Long, J. A. 2006. Swimming in Stone: The Amazing Gogo Fossils of the Kimberley. Freemantle Arts Centre Press, Freemantle, Western Australia, 320 pp.

Long, J. A., K. Trinajstic, and Z. Johanson. 2009. Devonian arthrodire embryos and the origin of internal fertilization in vertebrates. Nature 457:1124–1127.

Long, J. A., E. Mark-Kurik, Z. Johanson, M. S. Y. Lee, G. C. Young, Z. Min, P. E. Ahlberg, M. Newman, R. Jones, J. den Blaauwen, B. Choo, and K. Trinajstic. 2014. Copulation in antiarch placoderms and the origin of gnathostome internal fertilization. Nature 517:196–199.

Longrich, N. R., B.-A. S. Bhullar, and J. A. Gauthier. 2012. A transitional snake from the Late Cretaceous period of North America. Nature 488:205–208.

Lovejoy, C. O. 1988. Evolution of human walking. Scientific American 259:118–125.

Lu, J., D. M. Unwin, X. Jin, Y. Liu, and Q. Ji. 2009. Evidence for modular evolution in a long-tailed pterosaur with a pterodactyloid skull. Proceedings of the Royal Society of London. Series B: Biological Sciences 277:383–389.

Lund, R. 1985a. Stethacanthid elasmobranch remains from the Bear Gulch Limestone (Namurian E2b) of Montana. American Museum Noviates 2828:1–24.

Lund, R. 1985b. The morphology of *Falcatus falcatus* (St. John and Worthen), a Mississippian stethacanthid chondrichthyan from the Bear Gulch Limestone of Montana. Journal of Vertebrate Paleontology 5:1–19.

Lund, R., and W. Lund. 1984. New genera and species of coelacanths from the Bear Gulch Limestone (Lower Carboniferous) of Montana (U.S.A.). Geobios 17:237–244.

Luo, Z.-X. 2011. Developmental patterns in Mesozoic evolution of mammal ears. Annual Review of Ecology, Evolution, and Systematics 42:355–380.

Luo, Z.-X., Q. Ji, J. R. Wible, and C.-X. Yuan. 2003. An Early Cretaceous tribosphenic mammal and metatherian evolution. Science 302:1934–1940.

Luo, Z.-X., C.-X. Yuan, Q.-J. Meng, and Q. Ji. 2011. A Jurassic eutherian mammal and divergence of marsupials and placentals. Nature 476:442–445.

Lutterschmidt, D. I., W. I. Lutterschmidt, and V. H. Hutchison. 2003. Melatonin and thermoregulation in ectothermic vertebrates: a review. Canadian Journal of Zoology 81:1–13.

Lyman, R. L. 1994. Vertebrate Taphonomy. Cambridge University Press, New York, 524 pp.

Lyson, T. R., G. S. Bever, B. A. S. Bhullar, W. G. Joyce, and J. A. Gauthier. 2010. Transitional fossils and the origin of turtles. Biology Letters 6:830–833.

Lyson, T. R., G. S. Bever, T. M. Scheyer, A. Y. Hsiang, and J. A. Gauthier. 2013. Evolutionary origin of the turtle shell. Current Biology 23:1113–1119.

Lyson, T. R., E. A. Sperling, A. M. Heimberg, J. A. Gauthier, B. L. King, and K. J. Peterson. 2011. MicroRNAs support a turtle + lizard clade. Biology Letters 8:104–107.

Lyson, T. R., E. R. Schachner, J. Botha-Brink, T. M. Scheyer, M. Lambertz, G. S. Bever, B. S. Rubidge, and K. de Queiroz. 2014. Origin of the unique ventilatory apparatus of turtles. Nature Communications 5:5211.

Macaulay, D. 1988. The Way Things Work. Houghton Mifflin Company, Great Britain, 384 pp.

Maderson, P. F. A. 2003. Mammalian skin evolution: a reevaluation. Experimental Dermatology 12:233–236.

Magallón, S., S. Gómez-Acevedo, L. L. Sánchez-Reyes, and T. Hernández-Hernández. 2015. A metacalibrated time-tree documents the early rise of flowering plant phylogenetic diversity. New Phytologist 207:437–453.

Mahler, D. L., and M. Kearney. 2006. The palatal dentition in squamate reptiles: morphology, development, attachment, and replacement. Fieldiana Zoology 108:1–61.

Maisey, J., R. Miller, and S. Turner. 2009. The braincase of the chondrichthyan *Doliodus* from the Lower Devonian Campbellton Formation of New Brunswick, Canada. Acta Zoologica 90:109–122.

Mallatt, J. 2009. Early vertebrate evolution: pharyngeal structure and the origin of gnathostomes. Journal of Zoology 204:169–183.

Mallison, H. 2010. The digital Plateosaurus I: body mass, mass distribution, and posture assessed using CAD and CAE on a digitally mounted complete skeleton. Palaeontological Electronica 13:1–26.

Manley, G. A. 2010. An evolutionary perspective on middle ears. Hearing Research 263:3–8.

Martill, D. M., H. Tischlinger, and N. R. Longrich. 2015. A four-legged snake from the Early Cretaceous of Gondwana. Science 349:416–419.

Martin, A. J. 2014. Dinosaurs without Bones: Dinosaur Lives Revealed by Their Trace Fossils. Pegasus Books, New York, 460 pp.

Martin, R. B., D. B. Burr, and N. A. Sharkey. 1998. Skeletal Tissue Mechanics. Springer-Verlag, New York, 392 pp.

Mascarelli, A. 2009. Dead whales make for an underwater feast. Audobon Magazine November–December.

Mason, M. 2006. Pathways for sound transmission to the inner ear in amphibians; pp. 147–183 in P. M. Narins, A. S. Feng, R. R. Fay, and A. N. Popper (eds.), Hearing and Sound Communication in Amphibians, Springer Handbook of Auditory Research vol. 28. Springer, Germany.

Mason, M. J., and P. M. Narins. 2002. Vibrometric studies of the middle ear of the bullfrog *Rana catesbeiana*.

II. The operculum. Journal of Experimental Biology 205:3167–3176.

Massare, J. A. 1987. Tooth morphology and prey preference of Mesozoic marine reptiles. Journal of Vertebrate Paleontology 7:121–137.

Massare, J. A. 1988. Capabilities of Mesozoic marine reptiles: implications for method of predation. Paleobiology 14:187–205.

Massare, J. A., W. R. Wahl, M. Ross, and M. V. Connely. 2014. Palaeoecology of the marine reptiles of the Redwater Shale Member of the Sundance Formation (Jurassic) of central Wyoming, USA. Geological Magazine 151:167–182.

Mazin, J.-M., J.-P. Billon-Bruyat, and K. Padian. 2009. First record of a pterosaur landing trackway. Proceedings of the Royal Society of London. Series B: Biological Sciences 276:3881–3886.

Mazin, J.-M., J.-P. Billon-Bruyat, P. Hantzpergue, and G. Lafaurie. 2003. Ichnological evidence for quadrupedal locomotion in pterodactyloid pterosaurs: trackways from the Late Jurassic of Crayssac (southwestern France). Geological Society, London, Special Publications 217:283–296.

McArthur, K. L., and J. D. Dickman. 2011. State-dependent sensorimotor processing: gaze and posture stability during simulated flight in birds. Journal of Neurophysiology 105:1689–1700.

McDiarmid, R. W., and R. Altig. 2009. Morphology of amphibian larvae; pp. 39–53 in C. K. Dodd (ed.), Amphibian Ecology and Conservation: A Handbook of Techniques. Oxford University Press, New York.

McGowan, C. 1999. A Practical Guide to Vertebrate Mechanics. Cambridge University Press, New York, 316 pp.

McHenry, C. R., A. G. Cook, and S. Wroe. 2005. Bottom-feeding plesiosaurs. Science 310:75–75.

McIntosh, J. S., M. K. Brett-Surman, and J. O. Farlow. 1999. Sauropods; pp. 264–290 in J. O. Farlow and M. K. Brett-Surman (eds.), The Complete Dinosaur, 1st ed. Indiana University Press, Bloomington, Indiana.

McLoughlin, J. C. 1980. Synapsida: A New Look into the Origin of Mammals. Viking Press, New York, 148 pp.

McPhee, J. 2000. Annals of the Former World. Farrar, Straus, and Giroux, New York, 696 pp.

Meers, M. B. 2003. Crocodylian forelimb musculature and its relevance to Archosauria. Anatomical Record. Part A: Discoveries in Molecular, Cellular, and Evolutionary Biology 274:891–916.

Metzger, K. A. 2009. Quantitative analysis of the effect of prey properties on feeding kinematics in two species of lizards. Journal of Experimental Biology 212:3751–3761.

Miles, M. P., and P. M. Clarkson. 1994. Exercise-induced muscle pain, soreness, and cramps. Journal of Sports Medicine and Physical Fitness 34:203–216.

Miller, R. F., R. Cloutier, and S. Turner. 2003. The oldest articulated chondrichthyan from the Early Devonian period. Nature 425:501–504.

Milner, A. R. 1980. The temnospondyl amphibian *Dendrerpeton* from the Upper Carboniferous of Ireland. Palaeontology 23:125–141.

Milner, A. R., and S. E. K. Sequeria. 1994. The temnospondyl amphibians from the Visean of East Kirkton, West Lothian, Scotland. Transactions of the Royal Society of Edinburgh: Earth Sciences 84:331–361.

Modesto, S. P., D. M. Scott, and R. R. Reisz. 2009. Arthropod remains in the oral cavities of fossil reptiles support inference of early insectivory. Biology Letters 5:838–840.

Mohun, S. M., W. L. Davies, J. K. Bowmaker, D. Pisani, W. Himstedt, D. J. Gower, D. M. Hunt, and M. Wilkinson. 2010. Identification and characterization of visual pigments in caecilians (Amphibia: Gymnophiona), an order of limbless vertebrates with rudimentary eyes. Journal of Experimental Biology 213:3586–3592.

Montuelle, S. J., A. Herrel, V. Schaerlaeken, K. A. Metzger, A. Mutuyeyezu, and V. L. Bels. 2009. Inertial feeding in the teiid lizard *Tupinambis merianae*: the effect of prey size on the movements of hyolingual apparatus and the cranio-cervical system. Journal of Experimental Biology 212:2501–2510.

Moon, C. 2011. Infrared-sensitive pit organ and trigeminal ganglion in the crotaline snakes. Anatomy & Cell Biology 44:8–13.

Moore, K. L., and A. F. Dalley. 1999. Clinically Oriented Anatomy, 4th ed. Lippincott Williams & Wilkins, Philadelphia, 1164 pp.

Moreno, K., S. Wroe, P. Clausen, C. McHenry, D. C. D'Amore, E. J.

Rayfield, and E. Cunningham. 2008. Cranial performance in the Komodo dragon (*Varanus komodoensis*) as revealed by high-resolution 3-D finite element analysis. Journal of Anatomy 212:736–746.

Moritz, G. L., N. T.-L. Lim, M. Neitz, L. Peichl, and N. J. Dominy. 2013. Expression and evolution of short wavelength sensitive opsins in colugos: a nocturnal lineage that informs debate on primate origins Evolutionary Biology 40:542–553.

Morris, D. 1993. Cat Watching: Why Cats Purr and Other Feline Mysteries Explained. Three Rivers Press, New York, 144 pp.

Morris, S. C., and J.-B. Caron. 2014. A primitive fish from the Cambrian of North America. Nature 512:419–422.

Moss, S. 1977. Feeding mechanisms in sharks. Integrative and Comparative Biology 17:355–364.

Mota, B., and S. Herculano-Houzel. 2012. How the cortex gets its folds: an inside-out, connectivity-driven model for the scaling of mammalian cortical folding. Frontiers in Neuroanatomy 6:1–14.

Motani, R. 2005. Evolution of fish-shaped reptiles (Reptilia: Ichthyopterygia) in their physical environments and constraints. Annual Review of Earth and Planetary Sciences 33:395–420.

Motani, R. 2009. The evolution of marine reptiles. Evolution: Education and Outreach 2:224–235.

Motani, R., B. M. Rothschild, and W. Wahl Jr. 1999. Large eyeballs in diving ichthyosaurs. Nature 402:747–747.

Motani, R., D.-Y. Jiang, G.-B. Chen, A. Tintori, O. Rieppel, C. Ji, and J.-D. Huang. 2014. A basal ichthyosauriform with a short snout from the Lower Triassic of China. Nature. doi:10.1038/nature13866.

Moyle, P. B., and J. J. Cech. 1996. Fishes: An Introduction to Ichthyology, 3rd ed. Prentice Hall, New Jersey, 590 pp.

Muller, J., T. M. Scheyer, J. J. Head, P. M. Barrett, I. Werneburg, P. G. P. Ericson, D. Pol, and M. R. Sanchez-Villagra. 2010. Homeotic effects, somitogenesis and the evolution of vertebral numbers in recent and fossil amniotes. Proceedings of the National Academy of Sciences 107:2118–2123.

Muybridge, E. 2007. Muybridge's Animals in Motion (Dover Electronic Clip Art) (CD-ROM and Book). Dover Publications, New York, 48 pp.

Naish, D. 2012. Birds; pp. 378–423 in M. K. Brett-Surman, T. R. Holtz, and J. O. Farlow (eds.), The Complete Dinosaur, 2nd ed. Indiana University Press, Bloomington, Indiana.

Nesbitt, S. J., P. M. Barrett, S. Werning, C. A. Sidor, and A. J. Charig. 2012. The oldest dinosaur? A Middle Triassic dinosauriform from Tanzania. Biology Letters 9:20120949–20120949.

Nesbitt, S. J., C. A. Sidor, R. B. Irmis, K. D. Angielczyk, R. M. H. Smith, and L. A. Tsuji. 2010. Ecologically distinct dinosaurian sister group shows early diversification of Ornithodira. Nature 464:95–98.

Nesbitt, S. J., N. D. Smith, R. B. Irmis, A. H. Turner, A. Downs, and M. A. Norell. 2009. A complete skeleton of a Late Triassic saurischian and the early evolution of dinosaurs. Science 326:1530–1533.

Newton, T. 1999. How Cars Work: An Illustrated Guide to the 250 Most Important Car Parts. Black Apple Press, Vallejo, California, 96 pp.

Nilsson, M. A., G. Churakov, M. Sommer, N. V. Tran, A. Zemann, J. Brosius, and J. Schmitz. 2010. Tracking marsupial evolution using archaic genomic retroposon insertions. PLoS Biology 8:e1000436.

Northcutt, R. G. 2002. Understanding vertebrate brain evolution. Integrative and Comparative Biology 42:743–756.

Northcutt, R. G. 2011. Evolving large and complex brains. Science 332:926–927.

Novas, F. E., L. Salgado, M. Suarez, F. L. Agnolin, M. D. Ezcurra, N. R. Chimento, R. de la Cruz, M. P. Isasi, A. O. Vargus, and D. Rubilar-Rogers. 2015. An enigmatic plant-eating theropod from the Late Jurassic period of Chile. Nature 522:331–334.

Oftedal, O. T. 2002a. The origin of lactation as a water source for parchment-shelled eggs. Journal of Mammary Gland Biology and Neoplasia 7:253–266.

Oftedal, O. T. 2002b. The mammary gland and its origin during synapsid evolution. Journal of Mammary Gland Biology and Neoplasia 7:225–252.

Oftedal, O. T. 2012. The evolution of milk secretion and its ancient origins. Animal 6:355–368.

O'Keefe, F. R. 2001. Ecomorphology of plesiosaur flipper geometry. Journal of Evolutionary Biology 14:987–991.

O'Keefe, F. R. 2002. The evolution of plesiosaur and pliosaur morphotypes in the Plesiosauria (Reptilia:

Sauropterygia). Paleobiology 28:101–112.

O'Keefe, F. R., and M. T. Carrano. 2005. Correlated trends in the evolution of the plesiosaur locomotor system. Paleobiology 31:656–675.

Olberding, J. P., L. D. McBrayer, and T. E. Higham. 2012. Performance and three-dimensional kinematics of bipedal lizards during obstacle negotiation. Journal of Experimental Biology 215:247–255.

Orr, C. M. 2005. Knuckle-walking anteater: a convergence test of adaptation for purported knuckle-walking features of African Hominidae. American Journal of Physical Anthropology 128:639–658.

Osborn, H. F. 1924. *Andrewsarchus,* giant mesonychid of Mongolia. American Museum Noviates 146:1–5.

Padian, K. 2003. Pterosaur stance and gait and the interpretation of trackways. Ichnos 10:115–126.

Padian, K. 2012. Evolutionary physiology: a bone for all seasons. Nature 487:310–311.

Padian, K., J. R. Horner, and A. de Ricqlès. 2004. Growth in small dinosaurs and pterosaurs: the evolution of archosaurian growth strategies. Journal of Vertebrate Paleontology 24:555–571.

Page, J. 2008. Do Cats Hear with Their Feet? Harper Collins Publishers, New York, 204 pp.

Palmer, C., and G. Dyke. 2011. Constraints on the wing morphology of pterosaurs. Proceedings of the Royal Society of London. Series B: Biological Sciences 279:1218–1224.

Parker, S. 1991. The Practical Paleontologist. Fireside, New York, 160 pp.

Parrish, J. M. 1987. The origin of crocodilian locomotion. Paleobiology 13:396–414.

Parrish, J. M. 2012. Evolution of the archosaurs; pp. 317–329 in M. K. Brett-Surman, T. R. Holtz, and J. O. Farlow (eds.), The Complete Dinosaur, 2nd ed. Indiana University Press, Bloomington, Indiana.

Paul, G. S. 1988. Predatory Dinosaurs of the World. Simon & Schuster, New York, 464 pp.

Pawley, K., and A. Warren. 2006. The appendicular skeleton of *Eryops magacephalus* Cope, 1877 (Temnospondyli: Eryopoidea) from the Lower Permian of North America. Journal of Paleontology 80:561–580.

Payne, S. L., C. M. Holliday, and M. K. Vickaryous. 2011. An osteological and histological investigation of cranial joints in geckos. Anatomical Record: Advances in Integrative Anatomy and Evolutionary Biology 294:399–405.

Peichl, L. 2005. Diversity of mammalian photoreceptor properties: adaptations to habitat and lifestyle? Anatomical Record Part A: Discoveries in Molecular, Cellular, and Evolutionary Biology 287A:1001–1012.

Perkins, S. 2015. Four-legged snake fossil stuns scientists—and ignites controversy. Science. doi: 10.1126/science.aac8899.

Perry, S. F., T. Similowski, W. Klein, and J. R. Codd. 2010. The evolutionary origin of the mammalian diaphragm. Respiratory Physiology & Neurobiology 171:1–16.

Peterson, J. E., and C. P. Vittore. 2012. Cranial pathologies in a specimen of *Pachycephalosaurus*. PLoS ONE 7:e36227.

Pickering, M., and J. F. X. Jones. 2002. The diaphragm: two physiological muscles in one. Journal of Anatomy 201:305–312.

Pierce, S. E., J. A. Clack, and J. R. Hutchinson. 2012. Three-dimensional limb joint mobility in the early tetrapod *Ichthyostega*. Nature 486:523–526.

Pietsch, T. W., and D. B. Grobecker. 1987. Frogfishes of the World: Systematics, Zoogeography, and Behavioral Ecology. Stanford University Press, Stanford, California, 420 pp.

Pimiento, C., and C. F. Clements. 2014. When did *Carcharocles megalodon* become extinct? A new analysis of the fossil record. PLoS ONE 9:e111086.

Polcyn, M. J., L. L. Jacobs, R. Araújo, A. S. Schulp, and O. Mateus. 2014. Physical drivers of mosasaur evolution. Palaeogeography, Palaeoclimatology, Palaeoecology 400:17–27.

Pollan, M. 2002. The Botany of Desire: A Plant's-Eye View of the World, Paperback ed. Random House, New York, NY, 271 pp.

Pollard, J. E. 1968. The gastric contents of an ichthyosaur from the Lower Lias of Lyme Regis, Dorset. Palaeontology 11:376–388.

Popper, K. R. 1959. The Logic of Scientific Discovery. Basic Books, Inc., Oxford, U.K., 480 pp.

Pough, F. H., C. M. Janis, and J. B. Heiser. 2002. Vertebrate Life, 6th ed. Prentice Hall, New Jersey, 699 pp.

Pough, F. H., C. M. Janis, and J. B. Heiser. 2013. Vertebrate Life, 9th ed. Pearson, Boston, 720 pp.

Pough, F. H., R. M. Andrews, J. E. Cadle, M. L. Crump, A. H. Savitzky, and K. D. Wells. 1998. Herpetology, 1st ed. Prentice Hall, New Jersey, 577 pp.

Powell, J. L. 1998. Night Comes to the Cretaceous: Dinosaur Extinction and the Transformation of Modern Geology. W. H. Freeman and Company, New York, 250 pp.

Pradel, A., J. G. Maisey, P. Tafforeau, R. H. Mapes, and J. Mallatt. 2014. A Palaeozoic shark with osteichthyan-like branchial arches. Nature 509:608–611.

Pratt, H. L., and J. C. Carrier. 2001. A review of elasmobranch reproductive behavior with a case study on the nurse shark, *Ginglymostoma cirratum*. Environmental Biology of Fishes 60:157–188.

Prothero, D. R. 2007. Evolution: What the Fossils Say and Why It Matters. Columbia University Press, New York, 408 pp.

Prothero, D. R. 2013. Rhinoceros Giants: The Paleobiology of Indricotheres. Indiana University Press, Bloomington, Indiana, 160 pp.

Prothero, D. R., and R. Schoch. 2002. Horns, Tusks, and Flippers: The Evolution of Hoofed Mammals. Johns Hopkins University Press, Baltimore, Maryland, 315 pp.

Prum, R. O. 2005. Evolution of the morphological innovations of feathers. Journal of Experimental Zoology. Part B: Molecular and Developmental Evolution 304B:570–579.

Prum, R. O., and J. Dyck. 2003. A hierarchical model of plumage: morphology, development, and evolution. Journal of Experimental Zoology. Part B: Molecular and Developmental Evolution 298:73–90.

Putnam, N. H., T. Butts, D. E. K. Ferrier, R. F. Furlong, U. Hellsten, T. Kawashima, M. Robinson-Rechavi, E. Shoguchi, and A. T. J. K. Yu. 2008. The amphioxus genome and the evolution of the chordate karyotype. Nature 453:1064–1071.

Radinsky, L. 1987. The Evolution of Vertebrate Design. University of Chicago Press, Chicago, Illinois, 188 pp.

Rage, J.-C., and F. Escuillié. 2003. The Cenomanian: stage of hindlimbed snakes. Carnets De Géologie 1:1–11.

Rage, J.-C., and Z. Roček. 1999. Redescription of *Triadobatrachus massinoti* (Piveteau, 1936) an anuran amphibian from the Early Triassic. Palaeontographica Abteilung A 206:1–16.

Rayfield, E. J., and A. C. Milner. 2008. Establishing a framework for archosaur cranial mechanics. Paleobiology 34:494–515.

Rayfield, E. J., D. B. Norman, C. C. Horner, J. R. Horner, P. M. Smith, J. J. Thomason, and P. Upchurch. 2001. Cranial design and function in a large theropod dinosaur. Nature 409:1033–1037.

Redfern, M. 2003. The Earth: A Very Short Introduction. Oxford University Press, New York, 160 pp.

Reep, R. L., J. C. Gaspard, D. Sarko, F. L. Rice, D. A. Mann, and G. B. Bauer. 2011. Manatee vibrissae: evidence for a "lateral line" function. Annals of the New York Academy of Sciences 1225:101–109.

Rega, E. A., K. Noriega, S. S. Sumida, A. Huttenlocker, A. Lee, and B. Kennedy. 2012. Healed fractures in the neural spines of an associated skeleton of *Dimetrodon:* implications for dorsal sail morphology and function. Fieldiana Life and Earth Sciences 5:104–111.

Regnault, S., A. A. Pitsillides, and J. R. Hutchinson. 2014. Structure, ontogeny and evolution of the patellar tendon in emus (*Dromaius novaehollandiae*) and other palaeognath birds. PeerJ 2:e711.

Reilly, S. M. 1998. Sprawling locomotion in the lizard *Sceloporus clarkii:* speed modulation of motor patterns in a walking trot. Brain, Behavior and Evolution 52:126–138.

Reilly, S. M., and M. E. Jorgensen. 2011. The evolution of jumping in frogs: morphological evidence for the basal anuran locomotor condition and the radiation of locomotor systems in crown group anurans. Journal of Morphology 272:149–168.

Reilly, S. M., and G. V. Lauder. 1989. Kinetics of tongue projection in *Ambystoma tigrinum:* quantitative kinematics, muscle function, and evolutionary hypotheses. Journal of Morphology 199:223–243.

Reilly, S. M., E. J. McElroy, R. Andrew Odum, and V. A. Hornyak. 2006. Tuataras and salamanders show that walking and running mechanics are ancient features of tetrapod locomotion. Proceedings of the Royal Society of London. Series B: Biological Sciences 273:1563–1568.

Reilly, S. M., E. J. McElroy, T. D. White, A. R. Biknevicius, and M. B. Bennett. 2010. Abdominal muscle and epipubic bone function during locomotion in Australian possums: insights into basal mammalian conditions and eutherian-like tendencies in Trichosurus. Journal of Morphology 271:438–450.

Reiner, A. 2009. Avian evolution: from Darwin's finches to a new way of thinking about avian forebrain organization and behavioural capabilities. Biology Letters 5:122–124.

Reisz, R. R. 1977. The oldest known diapsid reptile. Science 196:1091–1093.

Rewcastle, S. C. 1983. Fundamental adaptations in the Lacertilian hind limb: a partial analysis of the sprawling limb posture and gait. Copeia 1983:476–487.

Rieppel, O. 1993. Studies on skeleton formation in reptiles. IV. The homology of the reptilian (amniote) astragalus revisited. Journal of Vertebrate Paleontology 13:31–47.

Rieppel, O., and R. R. Reisz. 1999. The origin and early evolution of turtles. Annual Review of Ecology and Systematics 1–22.

Rieppel, O., and H. Zaher. 2000. The braincases of mosasaurs and Varanus, and the relationships of snakes. Zoological Journal of the Linnean Society 129:489–514.

Rieppel, O., H. Zaher, E. Tchernov, and M. J. Polcyn. 2003. The anatomy and relationships of Haasiophis terrasanctus, a fossil snake with well-developed hind limbs from the mid-Cretaceous of the Middle East. Journal of Paleontology 77:536–558.

Roberts, R., R. Lind, and S. Chatterjee. 2011. Flight dynamics of a pterosaur-inspired aircraft utilizing a variable-placement vertical tail. Bioinspiration & Biomimetics 6:026010.

Robinson, J., P. E. Ahlberg, and G. Koentges. 2005. The braincase and middle ear region of Dendrerpeton acadianum (Tetrapoda: Temnospondyli). Zoological Journal of the Linnean Society 143:577–597.

Rogers, R. R., D. A. Eberth, and A. R. Fiorillo. 2007. Bonebeds: Genesis, Analysis, and Paleobiological Significance. University of Chicago Press, Chicago, Illinois, 499 pp.

Rogers, S. W. 1998. Exploring dinosaur neuropaleobiology: computed tomography scanning and analysis of an Allosaurus fragilis endocast. Neuron 21:673–679.

Romer, A. S. 1956. Osteology of the Reptiles. University of Chicago Press, Chicago, Illinois, 800 pp.

Romer, A. S. 1962. The Vertebrate Body, 3rd ed. University of Chicago Press, Chicago, Illinois, pp.

Romer, A. S., and R. V. Witter. 1941. The skin of the rhachitomous amphibian Eryops. American Journal of Science 239:822–824.

Rosenzweig, C., D. Karoly, M. Vicarelli, P. Neofotis, Q. Wu, G. Casassa, A. Menzel, T. L. Root, N. Estrella, B. Seguin, P. Tryjanowski, C. Liu, S. Rawlins, and A. Imeson. 2008. Attributing physical and biological impacts to anthropogenic climate change. Nature 453:353–357.

Rowe, T. B., T. E. Macrini, and Z.-X. Luo. 2011. Fossil evidence on origin of the mammalian brain. Science 332:955–957.

Ruben, J. A. 2000. Selective factors associated with the origin of fur and feathers. Integrative and Comparative Biology 40:585–596.

Ruben, J. A., and A. A. Bennett. 1987. The evolution of bone. Evolution 41:1187–1197.

Rücklin, M., P. C. J. Donoghue, Z. Johanson, K. Trinajstic, F. Marone, and M. Stampanoni. 2012. Development of teeth and jaws in the earliest jawed vertebrates. Nature 491:748–752.

Russell, A. P., and V. Bels. 2001. Biomechanics and kinematics of limb-based locomotion in lizards: review, synthesis and prospectus. Comparative Biochemistry and Physiology. Part A: Molecular & Integrative Physiology 131:89–112.

Ruta, M., M. I. Coates, and D. L. J. Quicke. 2003. Early tetrapod relationships revisited. Biological Reviews of the Cambridge Philosophical Society 78:251–345.

Ruta, M., J. E. Jeffery, and M. I. Coates. 2003. A supertree of early tetrapods. Proceedings of the Royal Society of London. Series B: Biological Sciences 270:2507–2516.

Ruta, M., D. Pisani, G. T. Lloyd, and M. J. Benton. 2007. A supertree of Temnospondyli: cladogenetic patterns in the most species-rich group of early tetrapods. Proceedings of the Royal Society of London. Series B: Biological Sciences 274:3087–3095.

Ruta, M., J. C. Cisneros, T. Liebrecht, L. A. Tsuji, and J. Müller. 2011. Amniotes through major biological crises: faunal turnover among Pararepiles and the end-Permian mass extinction. Palaeontology 54:1117–1137.

Sanchez-Villagra, M. R. 2010. Developmental palaeontology in synapsids: the fossil record of ontogeny in mammals and their closest relatives. Proceedings of the Royal Society of London. Series B: Biological Sciences 277:1139–1147.

Sánchez-Villagra, M. R. 2013. Why are there fewer marsupials than placentals? On the relevance of geography and physiology to evolutionary patterns of mammalian diversity and disparity. Journal of Mammalian Evolution 20:279–290.

Sanchez-Villagra, M. R., S. Gemballa, S. Nummela, K. K. Smith, and W. Maier. 2002. Ontogenetic and phylogenetic transformations of the ear ossicles in marsupial mammals. Journal of Morphology 251:219–238.

Sánchez-Villagra, M. R., H. Müller, C. A. Sheil, T. M. Scheyer, H. Nagashima, and S. Kuratani. 2009. Skeletal development in the Chinese soft-shelled turtle Pelodiscus sinensis (Testudines: Trionychidae). Journal of Morphology 270:1381–99.

Savolainen, P., T. Leitner, A. N. Wilton, E. Matisoo-Smith, and J. Lundeberg. 2004. A detailed picture of the origin of the Australian dingo, obtained from the study of mitochondrial DNA. Proceedings of the National Academy of Sciences 101:12387–12390.

Schachner, E. R., J. R. Hutchinson, and C. Farmer. 2013a. Pulmonary anatomy in the Nile crocodile and the evolution of unidirectional airflow in Archosauria. PeerJ 1:e60.

Schachner, E. R., R. L. Cieri, J. P. Butler, and C. G. Farmer. 2013b. Unidirectional pulmonary airflow patterns in the savannah monitor lizard. Nature 506:367–370.

Schachner, E. R., C. G. Farmer, A. T. McDonald, and P. Dodson. 2011. Evolution of the dinosauriform respiratory apparatus: new evidence from the postcranial axial skeleton. Anatomical Record: Advances in Integrative Anatomy and Evolutionary Biology 294:1532–1547.

Schaeffer, B. 1941. The morphological and functional evolution of the tarsus in amphibians and reptiles. Bulletin of the American Museum of Natural History 78:395–472.

Schmidt-Nielsen, K. 1984. Scaling: Why Is Animal Size So Important? Cambridge University Press, New York, 256 pp.

Schmitz, L., and R. Motani. 2011. Nocturnality in dinosaurs inferred

from scleral ring and orbit
morphology. Science 332:705–708.

Schoch, R. R. 2002. The evolution of
metamorphosis in temnospondyls.
Lethaia 35:309–327.

Schoch, R. R. 2003. Early larval ontogeny
of the Permo-Carboniferous
temnospondyl *Sclerocephalus*.
Palaeontology 46:1055–1072.

Schoch, R. R. 2009. Evolution of life
cycles in early amphibians. Annual
Review of Earth and Planetary
Sciences 37:135–162.

Schoch, R., and H.-D. Sues. 2015. A
Middle Triassic stem-turtle and the
evolution of the turtle body plan.
Nature doi:10.1038/nature14472.

Schuh, R. T., and A. V. Z. Brower. 2009.
Biological Systematics: Principles
and Applications, 2nd ed. Comstock
Publishing Associates, USA, 328 pp.

Schulte, P., L. Alegret, I. Arenillas,
J. A. Arz, P. J. Barton, P. R. Bown,
T. J. Bralower, G. L. Christeson,
P. Claeys, C. S. Cockell, G. S. Collins,
A. Deutsch, T. J. Goldin, K. Goto,
J. M. Grajales-Nishimura, R. A. F.
Grieve, S. P. S. Gulick, K. R. Johnson,
W. Kiessling, C. Koeberl, D. A. Kring,
K. G. MacLeod, T. Matsui, J. Melosh,
A. Montanari, J. V. Morgan, C. R.
Neal, D. J. Nichols, R. D. Norris,
E. Pierazzo, G. Ravizza, M. Rebolledo-
Vieyra, W. U. Reimold, E. Robin,
T. Salge, R. P. Speijer, A. R. Sweet,
J. Urrutia-Fucugauchi, V. Vajda, M. T.
Whalen, and P. S. Willumsen. 2010.
The Chicxulub asteroid impact and
mass extinction at the Cretaceous-
Paleogene boundary. Science
327:1214–1218.

Schweitzer, M. H., R. M. Elsey, C. G.
Dacke, J. R. Horner, and E.-T. Lamm.
2007. Do egg-laying crocodilian
(*Alligator mississippiensis*) archosaurs
form medullary bone? Bone
40:1152–1158.

Schwimmer, D. 2002. King of the
Crocodylians: The Paleobiology of
Deinosuchus. Indiana University Press,
Bloomington, Indiana, 220 pp.

Senter, P. 2006. Comparison of forelimb
function between *Deinonychus*
and *Bambiraptor* (Theropoda:
Dromaeosauridae). Journal of
Vertebrate Paleontology 26:897–906.

Sereno, P. C., and A. B. Arcucci.
1994. Dinosaurian precursors from
the Middle Triassic of Argentina:
Marasuchus lilloensis, gen. nov.
Journal of Vertebrate Paleontology
14:53–73.

Setright, L. J. K. 2003. Drive On! A Social
History of the Motor Car. Granta,
London, 406 pp.

Shedlock, A. M., and S. V. Edwards.
2009. Amniotes (Amniota); pp.
375–379 in S. Blair Hedges and Sudhir
Kumar (eds.), The Timetree of Life.
Oxford University Press, Oxford, U.K.

Shoshani, J. 1996. Skeletal and
other basic anatomical features of
elephants; pp. 9–20 in J. Shoshani
and P. Tassy (eds.), The Proboscidea:
Evolution and Palaeoecology of
Elephants and Their Relatives. Oxford
University Press, New York.

Shu, D.-G., S. Conway Morris, X.-L.
Zhang, S. Hu, L. Chen, J. Han, M. Zhu,
Y. Li, and L.-Z. Chen. 1999. Lower
Cambrian vertebrates from south
China. Nature 402:42–46.

Shu, D.-G., S. Conway Morris, J. Han,
Z.-F. Zhang, K. Yasui, P. Janvier,
L. Chen, X.-L. Zhang, J.-N. Liu, Y. Li,
and H.-Q. Liu. 2003. Head and
backbone of the Early Cambrian
vertebrate *Haikouichthys*. Nature
421:526–529.

Shubin, N. H., E. B. Daeschler, and F. A.
Jenkins. 2006. The pectoral fin of
Tiktaalik roseae and the origin of the
tetrapod limb. Nature 440:764–771.

Shubin, N. H., E. B. Daeschler, and F. A.
Jenkins. 2014. Pelvic girdle and fin
of *Tiktaalik roseae*. Proceedings of
the National Academy of Sciences
111:893–899.

Sigurdsen, T., and J. R. Bolt. 2009. The
lissamphibian humerus and elbow
joint, and the origins of modern
amphibians. Journal of Morphology
270:1443–1453.

Sigurdsen, T., and J. R. Bolt. 2010.
The Lower Permian amphibamid
Doleserpeton (Temnospondyli:
Dissorophoidea), the interrelationships
of amphibamids, and the origin
of modern amphibians. Journal
of Vertebrate Paleontology
30:1360–1377.

Sigurdsen, T., and D. M. Green. 2011.
The origin of modern amphibians: a
re-evaluation. Zoological Journal of
the Linnean Society 162:457–469.

Simon, W. H. 1970. Scale effects in
animal joints. I. Articular cartilage
thickness and compressive stress.
Arthritis & Rheumatism 13:244–255.

Sites, J. W., Jr., T. W. Reeder, and J. J.
Wiens. 2011. Phylogenetic insights
on evolutionary novelties in lizards
and snakes: sex, birth, bodies, niches,
and venom. Annual Review of
Ecology, Evolution, and Systematics
42:227–244.

Skutschas, P., and T. Martin. 2011.
Cranial anatomy of the stem
salamander *Kokartus honorarius*
(Amphibia: Caudata) from the Middle
Jurassic of Kyrgyzstan. Zoological
Journal of the Linnean Society
161:816–838.

Slack, J. M. W. 2013. Essential
Developmental Biology, 3rd ed.
Wiley, Chichester, West Sussex, U.K. ;
Hoboken, New Jersey, 479 pp.

Smith, A. S. 2008. Plesiosaurs. Geology
Today 24:71–75.

Smith, H. W. 1931. Observations of
the African lung-fish, *Protopteris
aethiopicus,* and on evolution from
water to land environments. Ecology
12:164–181.

Smith, J. J. B. 1968. Hearing in terrestrial
urodeles: a vibration-sensitive
mechanism in the ear. Journal of
Experimental Biology 48:191–205.

Smith, K. K., and W. L. Hylander.
1985. Strain gauge measurement of
mesokinetic movement in the lizard
Varanus exanthematicus. Journal of
Experimental Biology 114:53–70.

Smith, M. M., and Z. Johanson. 2003.
Separate evolutionary origins of
teeth from evidence in fossil jawed
vertebrates. Science 299:1235–1236.

Smithson, T. R., R. L. Carroll, A. L.
Panchen, and S. M. Andrews.
1994. *Westlothiana lizziae* from
the Visean of East Kirkton, West
Lothian, Scotland, and the amniote
stem. Transactions of the Royal
Society of Edinburgh: Earth Sciences
84:383–412.

Snively, E., J. R. Cotton, R. C. Ridgely,
and L. M. Witmer. 2013. Multibody
dynamics model of head and neck
function in *Allosaurus* (Dinosauria,
Theropoda). Palaeontologica
Electronica 16:1–29.

Soares, M. C., and M. R. de Carvalho.
2013. Comparative myology of the
mandibular and hyoid arches of
sharks of the order Hexanchiformes
and their bearing on its monophyly
and phylogenetic relationships
(Chondrichthyes: Elasmobranchii).
Journal of Morphology 274:203–214.

Soukup, V., H.-H. Epperlein, I. Horácek,
and R. Cerny. 2008. Dual epithelial
origin of vertebrate oral teeth. Nature
455:795–798.

Springer, M. S., W. J. Murphy, E. Eizirik,
and S. J. O'Brien. 2005. Molecular
evidence for major placental clades;
pp. 37–49 in K. D. Rose and J. D.
Archibald (eds.), The Rise of Placental
Mammals: Origins and Relationships
of the Major Extant Clades. Johns

Hopkins University Press, Baltimore, Maryland.

Springer, M. S., M. J. Stanhope, O. Madsen, and W. W. de Jong. 2004. Molecules consolidate the placental mammal tree. Trends in Ecology & Evolution 19:430–438.

Standen, E. M., and H. C. Larsson. 2012. A living analogue to the fin-limb transition: locomotion and fin use of an air breathing fish on land. Journal of Vertebrate Paleontology Program and Abstracts, 2012:178.

Standen, E. M., T. Y. Du, and H. C. E. Larsson. 2014. Developmental plasticity and the origin of tetrapods. Nature 513:54–58.

Stein, W. E., F. Mannolini, L. V. Hernick, E. Landing, and C. M. Berry. 2007. Giant cladoxylopsid trees resolve the enigma of the Earth's earliest forest stumps at Gilboa. Nature 446:904–907.

Sterli, J., S. Marcelo, and D. La Fuente. 2010. Anatomy of *Condorchelys antiqua* Sterli, 2008, and the origin of the modern jaw closure mechanism in turtles. Journal of Vertebrate Paleontology 30:351–366.

Stocker, M. R., and R. J. Butler. 2013. Phytosauria. Geological Society, London, Special Publications 379:91–117.

Sues, H. D., and F. A. Jenkins. 2006. The postcranial skeleton of *Kayentatherium wellesi* from the Lower Jurassic Kayenta Formation of Arizona and the phylogenetic significance of postcranial features in tritylodontid cynodonts; pp. 114–152 in M. T. Carrano (ed.), Amniote Paleobiology: Perspectives on the Evolution of Mammals, Birds, and Reptiles. University of Chicago Press, Chicago, Illinois.

Sues, H. D., and R. R. Reisz. 1998. Origins and early evolution of herbivory in tetrapods. Trends in Ecology & Evolution 13:141–145.

Sues, H. D., S. P. Modesto, D. M. Scott, and R. R. Reisz. 2009. A new pararaptile with temporal fenestration from the Middle Permian of South Africa. Canadian Journal of Earth Sciences 46:9–20.

Sulej, T. 2007. Osteology, variability, and evolution of *Metoposaurus,* a temnospondyl from the Late Triassic of Poland. Palaeontologia Polonica 64:29–139.

Sullivan, C., R. R. Reisz, and R. M. H. Smith. 2003. The Permian mammal-like herbivore *Diictodon,* the oldest known example of sexually dimorphic

armament. Proceedings of the Royal Society of London. Series B: Biological Sciences 270:173–178.

Sullivan, R. 2004. Rats: Observations on the History and Habitat of the City's Most Unwanted Inhabitants. Bloomsbury, New York, 242 pp.

Szidon, J. P., S. Lahiri, M. Lev, and A. P. Fishman. 1969. Heart and circulation of the African lungfish. Circulatory Research 25:23–38.

Tambussi, C. P., R. de Mendoza, F. J. Degrange, and M. B. Picasso. 2012. Flexibility along the neck of the Neogene terror Bird *Andalgalornis steulleti* (Aves phorusrhacidae). PLoS ONE 7:e37701.

Thewissen, J. G. M., L. N. Cooper, J. C. George, and S. Bajpai. 2009. From land to water: the origin of whales, dolphins, and porpoises. Evolution: Education and Outreach 2:272–288.

Thewissen, J. G. M., L. N. Cooper, M. T. Clementz, S. Bajpai, and B. N. Tiwari. 2007. Whales originated from aquatic artiodactyls in the Eocene epoch of India. Nature 450:1190–1194.

Thomas, K. S., and H. J. McMann. 2011. U.S. Spacesuits, 2nd ed. Springer, New York, 507 pp.

Thulborn, T., and S. Turner. 2003. The last dicynodont: an Australian Cretaceous relict. Proceedings of the Royal Society of London. Series B: Biological Sciences 270:985–993.

Tickle, P. G., A. R. Ennos, L. E. Lennox, S. F. Perry, and J. R. Codd. 2007. Functional significance of the uncinate processes in birds. Journal of Experimental Biology 210:3955–3961.

Tidwell, V., and D. Wilhite. 2005. Ontogenetic variation and isometric growth in the forelimb of the Early Cretaceous sauropod *Venenosaurus;* pp. 187–196 in V. Tidwell and K. Carpenter (eds.), Thunder Lizards: The Sauropodomorph Dinosaurs. Indiana University Press, Bloomington, Indiana.

Tsai, H. P., and C. M. Holliday. 2011. Ontogeny of the alligator cartilago transiliens and its significance for sauropsid jaw muscle evolution. PLoS ONE 6:e24935.

Turner, A. H., P. J. Makovicky, and M. A. Norell. 2007. Feather quill knobs in the dinosaur *Velociraptor*. Science 317:1721.

Unwin, D. M. 1999. Pterosaurs: back to the traditional model? Trends in Ecology & Evolution 14:263–268.

Unwin, D. M. 2003. Palaeontology: Smart-winged pterosaurs. Nature 425:910–911.

Upchurch, P., P. Barrett, and P. Dodson. 2004. Sauropoda; pp. 259–322 in The Dinosauria, 2nd ed. University of California Press, Berkeley, California.

VanBuren, C. S., and M. Bonnan. 2013. Forearm posture and mobility in quadrupedal dinosaurs. PLoS ONE 8:e74842.

Varela-Lasheras, I., A. J. Bakker, S. D. van der Mije, J. A. Metz, J. van Alphen, and F. Galis. 2011. Breaking evolutionary and pleiotropic constraints in mammals: on sloths, manatees and homeotic mutations. EvoDevo 2.11.

Vaughan, T. A., J. M. Ryan, and N. J. Czaplewski. 2011. Mammalogy, 5th ed. Jones and Bartlett Publishers, Boston, 750 pp.

Vazquez, R. J. 1994. The automating skeletal and muscular mechanisms of the avian wing (Aves). Zoomorphology 114:59–71.

Vélez-Zuazo, X., and I. Agnarsson. 2011. Shark tales: a molecular species-level phylogeny of sharks (Selachimorpha, Chondrichthyes). Molecular Phylogenetics and Evolution 58:207–217.

Venkatesh, B., A. P. Lee, V. Ravi, A. K. Maurya, M. M. Lian, J. B. Swann, Y. Ohta, M. F. Flajnik, Y. Sutoh, M. Kasahara, S. Hoon, V. Gangu, S. W. Roy, M. Irimia, V. Korzh, I. Kondrychyn, Z. W. Lim, B.-H. Tay, S. Tohari, K. W. Kong, S. Ho, B. Lorente-Galdos, J. Quilez, T. Marques-Bonet, B. J. Raney, P. W. Ingham, A. Tay, L. W. Hillier, P. Minx, T. Boehm, R. K. Wilson, S. Brenner, and W. C. Warren. 2014. Elephant shark genome provides unique insights into gnathostome evolution. Nature 505:174–179.

Vergne, A. L., M. B. Pritz, and N. Mathevon. 2009. Acoustic communication in crocodilians: from behaviour to brain. Biological Reviews 84:391–411.

Vickaryous, M. K., and B. K. Hall. 2006. Homology of the reptilian coracoid and a reappraisal of the evolution and development of the amniote pectoral apparatus. Journal of Anatomy 208:263–285.

Vickaryous, M. K., and J.-Y. Sire. 2009. The integumentary skeleton of tetrapods: origin, evolution, and development. Journal of Anatomy 214:441–464.

Vidal, N., and S. B. Hedges. 2009. The molecular evolutionary tree of lizards, snakes, and amphisbaenians.

Comptes Rendus Biologies 332:129–139.

Vogel, S. 2003. Comparative Biomechanics: Life's Physical World. Princeton University Press, Princeton, New Jersey, 580 pp.

Vonk, F. J., and M. K. Richardson. 2008. Serpent clocks tick faster. Nature 454:282–283.

Wake, D. B., and J. Hanken. 1996. Direct development in the lungless salamanders: what are the consequences for developmental biology, evolution and phylogenesis? Integrative Journal of Developmental Biology 40:859–869.

Walker, C., C. J. Vierck, and L. A. Ritz. 1998. Balance in the cat: role of the tail and effects of sacrocaudal transection. Behavioural Brain Research 91:41–47.

Walker, R., R. Tames, J. Man, and C. Freeman. 1999. Inventions That Changed the World. Reader's Digest Association, Inc., New York, 160 pp.

Walsh, S. A., and A. Milner. 2011. Evolution of the avian brain and senses; pp. 282–305 in G. J. Dyke, G. Kaiser, S. Walsh, and A. Milner (eds.), Living Dinosaurs: The Evolutionary History of Modern Birds. Wiley Publishers, New York.

Walsh, S. A., P. M. Barrett, A. C. Milner, G. Manley, and L. M. Witmer. 2009. Inner ear anatomy is a proxy for deducing auditory capability and behaviour in reptiles and birds. Proceedings of the Royal Society of London. Series B: Biological Sciences 276:1355–1360.

Walter, R. M., and D. R. Carrier. 2011. Effects of fore-aft body mass distribution on acceleration in dogs. Journal of Experimental Biology 214:1763–1772.

Wang, X., and R. H. Tedford. 2008. Dogs: Their Fossil Relatives and Evolutionary History. Columbia University Press, New York, 219 pp.

Warrell, D. A. 2010. Snake bite. Lancet 375:77–88.

Warren, A., and A. M. Yates. 2000. The phylogeny of the "higher" temnospondyls (Vertebrata: Choanata) and its implications for the monophyly and origins of the Stereospondyli. Zoological Journal of the Linnean Society 128:77–121.

Warrick, D. R., M. W. Bundle, and K. P. Dial. 2002. Bird maneuvering flight: blurred bodies, clear heads. Integrative and Comparative Biology 42:141–148.

Webb, G. J. W., and C. Gans. 1982. Galloping in Crocodylus johnstoni—a reflection of terrestrial activity? Records of the Australian Museum 34:607–618.

Wedel, M. J. 2009. Evidence for bird-like air sacs in saurischian dinosaurs. Journal of Experimental Zoology. Part A: Ecological Genetics and Physiology 311A:611–628.

Weissengruber, G. E., G. F. Egger, J. R. Hutchinson, H. B. Groenewald, L. Elsässer, D. Famini, and G. Forstenpointner. 2006. The structure of the cushions in the feet of African elephants (Loxodonta africana). Journal of Anatomy 209:781–792.

Welker, F., M. J. Collins, J. A. Thomas, M. Wadsley, S. Brace, E. Cappellini, S. T. Turvey, M. Reguero, J. N. Gelfo, A. Kramarz, J. Burger, J. Thomas-Oates, D. A. Ashford, P. D. Ashton, K. Rowsell, D. M. Porter, B. Kessler, R. Fischer, C. Baessmann, S. Kaspar, J. V. Olsen, P. Kiley, J. A. Elliott, C. D. Kelstrup, V. Mullin, M. Hofreiter, E. Willerslev, J.-J. Hublin, L. Orlando, I. Barnes, and R. D. E. MacPhee. 2015. Ancient proteins resolve the evolutionary history of Darwin's South American ungulates. Nature doi:10.1038/nature14249.

Wellnhoffer, P. 1991. The Illustrated Encyclopedia of Pterosaurs. Crescent Books, New York, 192 pp.

West, J. B., Z. Fu, A. P. Gaeth, and R. V. Short. 2003. Fetal lung development in the elephant reflects the adaptations required for snorkeling in adult life. Respiratory Physiology & Neurobiology 138:325–333.

Westneat, M. W. 2004. Evolution of levers and linkages in the feeding mechanisms of fishes. Integrative Comparative Biology 44:378–389.

Wildman, D. E., M. Uddin, J. C. Opazo, G. Liu, V. Lefort, S. Guindon, O. Gascuel, L. I. Grossman, R. Romero, and M. Goodman. 2007. Genomics, biogeography, and the diversification of placental mammals. Proceedings of the National Academy of Sciences 104:14395–14400.

Wilford, J. N. 1985. Riddle of the Dinosaur. Alfred A. Knopf, New York, 304 pp.

Wilga, C. D., and G. V. Lauder. 2000. Three-dimensional kinematics and wake structure of the pectoral fins during locomotion in leopard sharks Triakis semifasciata. Journal of Experimental Biology 203:2261–2278.

Wilga, C. D., and G. V. Lauder. 2004. Biomechanics: hydrodynamic function of the shark's tail. Nature 430:850.

Wilkinson, M. T. 2008. Three-dimensional geometry of a pterosaur wing skeleton, and its implications for aerial and terrestrial locomotion. Zoological Journal of the Linnean Society 154:27–69.

Wilkinson, M., M. K. Richardson, D. J. Gower, and O. V. Oommen. 2002. Extended embryo retention, caecilian oviparity and amniote origins. Journal of Natural History 36:2185–2198.

Wilkinson, M., E. Sherratt, F. Starace, and D. J. Gower. 2013. A new species of skin-feeding caecilian and the first report of reproductive mode in Microcaecilia (Amphibia: Gymnophiona: Siphonopidae). PLoS ONE 8:e57756.

Willey, J. S., A. R. Biknevicius, S. M. Reilly, and K. D. Earls. 2004. The tale of the tail: limb function and locomotor mechanics in Alligator mississippiensis. Journal of Experimental Biology 207:553–563.

Williston, S. W. 1914. Water Reptiles of the Past and Present. University of Chicago Press, Chicago, Illinois, 251 pp.

Wilson, D. E., and D. M. Reeder. 2005. Mammal Species of the World: A Taxonomic and Geographic Reference, 3rd ed. Johns Hopkins University Press, Baltimore, Maryland, 2000 pp.

Wilson, G. P., A. R. Evans, I. J. Corfe, P. D. Smits, M. Fortelius, and J. Jernvall. 2012. Adaptive radiation of multituberculate mammals before the extinction of dinosaurs. Nature 483:457–460.

Wilson, J. A., D. M. Mohabey, S. E. Peters, and J. J. Head. 2010. Predation upon hatchling dinosaurs by a new snake from the Late Cretaceous of India. PLoS Biology 8:e1000322.

Wilson, M. V. H., and M. W. Caldwell. 1993. New Silurian and Devonian fork-tailed "thelodonts" are jawless vertebrates with stomachs and deep bodies. Nature 361:442–444.

Wilt, F. H., and S. C. Hake. 2003. Principles of Developmental Biology. W. W. Norton & Company, New York, 430 pp.

Winchester, S. 2009. The Map That Changed the World: William Smith and the Birth of Modern Geology. Harper Perennial, New York, 368 pp.

Wink, C. S., and R. M. Elsey. 1986. Changes in femoral morphology during egg-laying in Alligator *mississippiensis.* Journal of Morphology 189:183–188.

Withers, P. C., G. Morrison, G. T. Hefter, and T.-S. Pang. 1994. Role of urea and methylamines in buoyancy of elasmobranchs. Journal of Experimental Biology 188:175–189.

Witmer, L. M. 1995. The extant phylogenetic bracket and the importance of reconstructing soft tissues in fossils; pp. 19–33 in J. J. Thomason (ed.), Functional Morphology in Vertebrate Paleontology. Cambridge University Press, New York.

Witmer, L. M. 1997. The evolution of the antorbital cavity of archosaurs: a study in soft-tissue reconstruction in the fossil record with an analysis of function of pneumaticity. Journal of Vertebrate Paleontology 17:1–76.

Witmer, L. M. 2001. Nostril position in dinosaurs and other vertebrates and its significance for nasal function. Science 293:850–853.

Witmer, L. M. 2009. Dinosaurs: fuzzy origins for feathers. Nature 458:293–295.

Witmer, L. M., and R. C. Ridgely. 2009. New insights into the brain, braincase, and ear region of tyrannosaurs (Dinosauria, Theropoda), with implications for sensory organization and behavior. Anatomical Record: Advances in Integrative Anatomy and Evolutionary Biology 292:1266–1296.

Witmer, L. M., S. Chatterjee, J. Franzosa, and T. Rowe. 2003. Neuroanatomy of flying reptiles and implications for flight, posture and behaviour. Nature 425:950–953.

Witton, M. P. 2010. *Pteranodon* and beyond: the history of giant pterosaurs from 1870 onwards. Geological Society, London, Special Publications 343:313–323.

Witton, M. P. 2013. Pterosaurs: Natural History, Evolution, Anatomy. Princeton University Press, New Jersey, 291 pp.

Witton, M. P., and M. B. Habib. 2010. On the size and flight diversity of giant pterosaurs, the use of birds as pterosaur analogues and comments on pterosaur flightlessness. PLoS ONE 5:e13982.

Witton, M. P., and D. Naish. 2008. A reappraisal of azhdarchid pterosaur functional morphology and paleoecology. PLoS ONE 3:e2271.

Witzmann, F., H. Scholz, J. Müller, and N. Kardjilov. 2010. Sculpture and vascularization of dermal bones, and the implications for the physiology of basal tetrapods. Zoological Journal of the Linnean Society 160:302–340.

Woltering, J. M., F. J. Vonk, H. Müller, N. Bardine, I. L. Tuduce, M. A. G. de Bakker, W. Knöchel, I. O. Sirbu, A. J. Durston, and M. K. Richardson. 2009. Axial patterning in snakes and caecilians: evidence for an alternative interpretation of the Hox code. Developmental Biology 332:82–89.

Woodward, H. N., J. R. Horner, and J. O. Farlow. 2011. Osteohistological evidence for determinate growth in the American alligator. Journal of Herpetology 45:339–342.

Wroe, S., D. R. Huber, M. Lowry, C. McHenry, K. Moreno, P. Clausen, T. L. Ferrara, E. Cunningham, M. N. Dean, and A. P. Summers. 2008. Three-dimensional computer analysis of white shark jaw mechanics: how hard can a great white bite? Journal of Zoology 276:336–342.

Yates, A. M. 2012. Basal Sauropodomorpha: the "prosauropods"; pp. 424–443 in M. K. Brett-Surman, T. R. Holtz, and J. O. Farlow (eds.), The Complete Dinosaur, 2nd ed. Indiana University Press, Bloomington, Indiana.

Yates, A. M., M. J. Wedel, and M. F. Bonnan. 2012. The early evolution of postcranial skeletal pneumaticity in sauropodomorph dinosaurs. Acta Palaeontologica Polonica 57:85–100.

Yates, A. M., M. F. Bonnan, J. Neveling, A. Chinsamy, and M. G. Blackbeard. 2009. A new transitional sauropodomorph dinosaur from the Early Jurassic of South Africa and the evolution of sauropod feeding and quadrupedalism. Proceedings of the Royal Society of London. Series B: Biological Sciences 277:787–794.

Yokoyama, S. 1997. Molecular genetic basis of adaptive selection: examples from color vision in vertebrates.

Annual Review of Genetics 31:315–336.

Young, G. C. 2003. Did placoderm fish have teeth? Journal of Vertebrate Paleontology 23:987–990.

Young, G. C. 2008. Early evolution of the vertebrate eye—fossil evidence. Evolution: Education and Outreach 1:427–438.

Yuan, C.-X., Q. Ji, Q.-J. Meng, A. R. Tabrum, and Z.-X. Luo. 2013. Earliest evolution of multituberculate mammals revealed by a new Jurassic fossil. Science 341:779–783.

Zaher, H., S. A. N. Apesteguía, and C. A. Scanferla. 2009. The anatomy of the Upper Cretaceous snake *Najash rionegrina* Apesteguía & Zaher, 2006, and the evolution of limblessness in snakes. Zoological Journal of the Linnean Society 156:801–826.

Zani, P. A. 1996. Patterns of caudal-autotomy evolution in lizards. Journal of Zoology 240:201–220.

Zhou, C.-F., S. Wu, T. Martin, and Z.-X. Luo. 2013. A Jurassic mammaliaform and the earliest mammalian evolutionary adaptations. Nature 500:163–167.

Zhu, M., and X. Yu. 2009. Stem sarcopterygians have primitive polybasal fin articulation. Biology Letters 5:372–375.

Zhu, M., W. Zhao, L. Jia, J. Lu, T. Qiao, and Q. Qu. 2009. The oldest articulated osteichthyan reveals mosaic gnathostome characters. Nature 458:469–474.

Zhu, M., X. Yu, P. E. Ahlberg, B. Choo, J. Lu, T. Qiao, Q. Qu, W. Zhao, L. Jia, H. Blom, and Y. Zhu. 2013. A Silurian placoderm with osteichthyan-like marginal jaw bones. Nature 502:188–193.

Zhu, M. 2014. Bone gain and loss: insights from genomes and fossils. National Science Review 1:490–492.

Zimmer, C. 2001. Evolution: The Triumph of an Idea. Harper Collins Publishers, New York, 364 pp.

Zug, G. R. 1974. Crocodilian galloping: an unique gait for reptiles. Copeia 1974:550–552.

Index

*A*ardonyx, 383
abdominal oblique muscles, 169, 203, 279–280
abduction, 148–149
absolute dating, 16–17. *See also* relative dating
acanthodians, 90–91, 103, 115–118, 121
Acanthostega, 185–189, 192
acetabulum: in birds, 334–336; in chondrichthyes, 104; in crocodylians and kin, 321–322, 324, 355; in dinosaurs (non-avian), 373–374, 376, 380, 387; in early archosaurs, 352–353; functional significance, 84; in lepospondyls, 238; in mammals and mammaliaforms, 416, 440; in non-mammal synapsids, 423, 430; in ornithodirans, 358, 360; in pterosaurs, 368, 372; in tetrapodomorphs, 149–150, 189; in tetrapods, 164; in turtles, 281. *See also* ilium; ischium; pubis
actinopterygians: acrodine enamel, 123–124; adaptations, 110, 114–115, 123; bowfin (*Amia calva*), 127–128; chondrostei (sturgeon and paddlefish), 125–126; gar, 126; primitive members and 'paloeoniscoids', 124–125; semionotids, 129; teleosts, 129–134
adduction, 148–149
adductor mandibulae: in actinopterygians, 119–121, 124–125, 127–129; in amniotes, 232; in amphibians, 205, 207, 212, 214, 218–220, 223; in archosaurs, 316, 352; in basal tetrapods, 186; in chondrichthyes, 102; in crocodylians and kin, 317–319, 354, 356; in dinosaurs (non-avian), 375, 377–378; in early diapsids, 273; in fishlike sarcopterygians, 143; functional significance, 97; in lepospondyls and reptiliomorphs, 239, 243; in lizards, 254–255; in mammals and mammaliaforms, 399, 401–402, 438; in marine reptiles, 293, 303, 307; in non-mammal synapsids, 420, 422–423, 426, 428; in parareptiles, 268, 271; in placoderms, 97; in pterosaurs, 362, 364; in snakes, 294–296; in stem tetrapods, 195; in tetrapodomorphs, 146, 178, 180; in tetrapods, 156, 159, 170–171; in tuatara, 265; in turtles, 283–286. *See also* masseter muscle; temporalis muscle

aetosaurs, 355
Afrotheria, 464–465. *See also* Mammalia
air dam, 133–134
air sacs in archosaurs, 316, 340–341, 374, 384. *See also* lungs
Alectura lathami (Australian brush turkey), 330–331
Alligator mississippiensis (American alligator), 316–325
Allosaurus, 377–378
American alligator. *See Alligator mississippiensis*
Amia calva. See actinopterygians
Amniota: adaptations, 226–237; transition from earlier tetrapods, 237–247. *See also* Mammalia; Reptilia; Synapsida
amniotic egg, 226, 228–229; allantois, 226, 230; amnion, 226, 229; chorion, 226, 230; yolk sac, 226, 230. *See also* placenta
Amphibia: adaptations, 200–203; amphibamids, 219–222, 224; caecilians, 213–214; dissorophoids, 219; early amphibians, 217–219; eryopids, 223–224; frogs and toads, 205–211; Lissamphibia, 202–205; lissamphibian development, 214–217; as a model for early tetrapods, 157, 160–161, 168, 173, 176; and lepospondyls, 202; salamanders and newts, 157, 160–161, 168, 173, 176, 211–213; stereospondyls, 223–224; temnospondyls as, 202
amphioxus. *See* lancelets
anapsid, 232. *See also* Reptilia
angular. *See* hearing
ankle: in amniotes, 233–234; in amphibians, 205, 210, 212, 220–223; in basal tetrapods, 191; in birds, 335, 337; in crocodylians and kin, 323–324, 357; in dinosaurs (non-avian), 376, 380; in early amniotes, 245–246; in early archosaurs, 353; in early diapsids, 273–275; in lepospondyls and reptiliomorphs, 237, 239–242, 244; in lizards, 252, 261; in mammals and mammaliaforms, 417–418, 441, 454, 456–457, 467; in marine reptiles, 304, 308; in non-mammal synapsids, 423; in ornithodirans, 358–360, 387; in parareptiles, 272; in pterosaurs, 368; in tetrapods, 163–164; in

tuatara, 266; in turtles, 281. *See also* astragalus; calcaneum
anteaters, 464–465; *Myrmecophaga tridactyla* (giant anteater), 464
anthracosaurs. *See* reptiliomorphs
antorbital fenestra of archosaurs, 316
Appalachian Mountains, 175
Archaeopteryx, 380–381, 387
Arctognathus, 426
armadillos, 435, 464–465; *Tolypeutes matacus* (three-banded armadillo), 464
articular: in amniotes, 232; functional significance, 120; and jaw opening in pterosaurs, 365; in mammals, 398, 402, 423, 425, 429, 432–439; in osteichthyes, 120; in placoderms, 94; retroarticular process, 255, 302–303, 319, 328, 426; and synovial joint, 254; in tetrapods, 158, 170, 176. *See also* hearing; jaws
Artiodactyla. *See* Cetartiodactyla
astragalus: in amniotes, 233–234; astragalocalcaneum in lizards, 260–261; astragalocalcaneum of tuatara, 265–266; astragalus-like elements in lepospondyls and reptiliomorphs, 238, 242, 244; in birds (fused), 332, 337; in crocodylians, 321, 323–324; in dinosaurs (non-avian), 376; in early amniotes, 244–247; in early archosaurs, 353; in early diapsids, 273–275; in mammals and mammaliaforms, 417, 433, 441, 457, 467; in non-mammal synapsids, 426; in ornithodirans, 358–359; in turtles, 287. *See also* calcaneum
Atlantogenata, 455, 464–465, 469. *See also* Mammalia
atlas vertebra. *See* vertebrae
Australian brush turkey (*Alectura lathami*), 330–331
axis vertebra. *See* vertebrae

*B*alanerpeton, 217–219, 224
basicranial muscle: adaptation for eye blinking, 167, 206; in fishlike sarcopterygians, 142–143; functional significance, 142
bats. *See* Chiroptera (bats)
Benthosuchus, 223
best dinosaur ever (*Aardonyx*), 383
bichir. *See Polypterus* (bichir)

bipedal posture: in archosaurs, 350–352; in crocodylians and kin, 354–355, 357; in dinosaurs (non-avian), 373, 375–376, 380, 383, 387; in lizards, 264–265, 351; in mammals, 463, 466, 471–472; in parareptiles, 272. *See also* center of mass; parasagittal gait and posture

birds: evolution from theropod dinosaurs (non-avian), 379–383; evolution of modern lineages, 341–342; feathers, 337–340; gizzard, 340; heart, 340; respiratory adaptations, 340–341; skeletal adaptations for flight, 327–328, 331–334. *See also* dinosaurs (non-avian); pterosaurs; Reptilia

bone: apatite, 45, 51, 54, 56–57, 59, 115; dermal, 58, 100, 115, 219, 277, 285, 322, 352, 434; in early vertebrates, 49; endochondral, 58, 99, 115; evolution, 53–54; fossilization of, 17–21; long bone growth, 444, 448–450; properties, 51–54; and scurvy, 52

boom mic, 305

Boreoeutheria, 464–465. *See also* Euarchontoglires; Laurasiatheria; Mammalia

Bothriolepis, 98–99

bowfin (*Amia calva*), 127–128

brain: in actinopterygians, 130–132; in amphibians, 208–209; in birds, 327–328, 330–331; in chondrichthyes, 100, 102; in crocodylians, 320–321; in dinosaurs (non-avian), 377–378, 383; in early vertebrates, 46–48; functional units, 43; in hagfish and lampreys, 40, 44–45; in jawed vertebrates, 72; in lizards, 257–259; in mammals, 404–406, 446, 452–454, 470–473; in 'ostracoderms', 59, 66–68; in pterosaurs, 365–366; small brains, 132; and the spinal cord, 38

braincase: in acanthodians, 116; in actinopterygians, 119–120, 125, 127–128; in amniotes, 232; in amphibians, 208–209; in basal tetrapods, 186; in birds, 328–329; in chondrichthyes, 100, 102; in crocodylians and kin, 317, 356; in dinosaurs (non-avian), 375, 378, 387; in early archosaurs, 352; in early diapsids, 273; in early vertebrates, 46, 48; in fishlike sarcopterygians, 139, 142; functional significance, 45; in hagfish and lampreys, 44–45; in jawed vertebrates, 72; in lepidosauromorphs, 291; in lizards, 255; in mammals, 404, 418, 439, 452–453, 471–472; in marine reptiles, 292, 301, 303, 306; and neural crest cells, 45–46; in non-mammal synapsids, 422, 428; in 'ostracoderms', 59, 66; in parareptiles, 271; in pterosaurs, 364–365, 370; in reptiliomorphs, 241;

in snakes, 294, 297; in stem tetrapods, 195; and temporal fenestrae, 232; in tetrapodomorphs, 146, 148, 178; in tetrapods, 158, 176; in turtles, 283–284. *See also* cranial (skull) kinesis; skull

brontotheres, 466

buoyancy: in chondrichthyes, 105; in coelacanths, 142; as definition, 81; a force acting on vertebrates, 81–82; loss of on land, 157–158; and swim bladder in actinopterygians, 121, 134; and turtle evolution, 285. *See also* lungs

buttocks, human, 471–472. *See also* gluteal muscles

caecilians. *See* Amphibia

calcaneum: in amniotes, 233–234; astragalocalcaneum of lizards, 260–261; astragalocalcaneum of tuatara, 265–266; in birds (fused), 332, 337; calcaneum-like elements in lepospondyls and reptiliomorphs, 238, 244; in crocodylians and kin, 321, 323–324, 355–357; in dinosaurs (non avian), 376; in early amniotes, 244–247; in early archosaurs, 353; in early diapsids, 273–275; in mammals and mammaliaforms, 408, 414, 417–418, 433, 441, 467; in non-mammal synapsids, 426, 429, 431; in ornithodirans, 358–359; in turtles, 287. *See also* astragalus

calcaneus, 417

Camarasaurus, 384

Cambrian Explosion, 47, 49

camera F-number: functional significance, 309; in ichthyosaurs, 308–309. *See also* eyes; sclerotic ring

canine. *See* teeth

Carcharocles megalodon (Megalodon), 110. *See also* fairy tale

carpals/carpus. *See* wrist

cartilage: in early vertebrates, 48; and endochondral bone, 55, 58, 448–450; properties, 55

Castorocauda, 446, 455

cats, 395–418, 421, 428, 446–447, 453, 458, 469. *See also* Mammalia

caudal vertebrae: in amniotes, 233; in amphibians, 212–213, 218, 221–222; in basal tetrapods, 187, 190–191; in birds, 332, 336–337; in crocodylians and kin, 321, 354–357; in dinosaurs (non-avian), 373, 375–376, 379–381, 383–384; in early amniotes, 246; in early archosaurs, 352; in lepospondyls and reptiliomorphs, 238, 241; in lizards, 260; in mammals and mammaliaforms, 408, 411, 416, 418, 440, 447, 454, 457, 461, 463, 469–470; in marine reptiles, 292, 302, 304, 306–307; in non-mammal synapsids,

422, 426, 429–430; in ornithodirans, 358, 360; in parareptiles, 271–272; in pterosaurs, 362–364, 366, 370–371; in snakes, 298; in tetrapods, 160, 162, 176; in tuatara, 266; in turtles, 282–283, 286. *See also* urostyle of frogs

caudofemoralis muscles: in crocodylians and kin, 323–324, 354–356; in dinosaurs (non-avian), 376; in early archosaurs, 352; functional significance, 260–261; in lizards, 260–261; loss and reduction in mammals, 411, 416; in ornithodirans, 358. *See also* femur

center of mass, 82; base of support and, 344–347; and bipedal posture in humans, 471–472; and buoyancy, 82, 134; and flexed knees of birds, 336–337; and hips in archosaurs, 351; in mammals, 411; in pterosaurs, 368; and stability of sprawled posture, 261; in *Stegosaurus,* 386. *See also* buoyancy; parasagittal gait and posture

centrum, 104, 146, 161, 191, 202, 233, 238, 246, 259, 282. *See also* vertebrae

cervical vertebrae: in amphibians, 205, 209, 212, 218; in basal tetrapods, 187, 190; in birds, 331–332; in crocodylians, 321; in dinosaurs (non-avian), 374, 376, 378–379, 381, 384, 387; in early amniotes, 245; evolution of the atlas and axis in amniotes, 232–233; evolution of the atlas and axis in mammals, 407, 409; functional significance in tetrapods, 160, 162, 176; in lepospondyls and reptiliomorphs, 238–239, 241; in lizards, 259; in mammals and mammaliaforms, 407–408, 418, 439, 465, 468; in marine reptiles, 303; in non-mammal synapsids, 429; in ornithodirans, 358, 360; in pterosaurs, 362, 366, 370, 372; in turtles, 282–283, 285–286. *See also* vertebrae

Cetacea, 467–469; *Pakicetus,* 467; *Tursiops truncatus* (bottlenose dolphin), 467

Cetartiodactyla, 466–469

chalicotheres, 466; *Moropus,* 466

Charlie the Unicorn, 303

Cheirolepis, 114, 124–125, 130

Chiroptera (bats), 465–466; *Icaronycteris,* 465; *Onychonycteris,* 465

choanae: in amphibians, 203, 207; in basal tetrapods, 186, 190; in crocodylians, 317–318; functional significance, 140; in mammals, 400–401; in snakes, 298; in tetrapodomorphs, 146–147, 152; in tetrapods, 169. *See also* secondary palate

chondrichthyes, 78, 90, 99–112

chordates: defining traits, 35–39; and deuterostomes, 33–34; and vertebrates, 39

Cladoselache, 106–107, 109

claspers: in chrondrichthyes, 104, 107–108, 111; functional significance, 98; in placoderms, 98

clavicle: in amphibians, 210, 219–220, 223; in basal tetrapods, 187–188; in birds as the furcula, 332–333; in dinosaurs (non-avian), 379–380; in early amniotes, 246; functional significance, 121; in lepospondyls and reptiliomorphs, 230, 240, 243; in lizards, 260–261; loss in crocodylians, 321, 322; loss in pterosaurs, 367; in non-mammal synapsids, 423, 426, 429–430; in stem tetrapods, 196; in tetrapodomorphs, 148, 150, 178, 182–183; in tetrapods, 161–162, 164, 176; in tuatara, 265; in turtles, 277, 286; variation in mammals and mammaliaforms, 408, 412, 415, 433, 440, 465, 468. See also cleithrum; interclavicle

cleidoic egg. See amniotic egg

cleithrum: in amphibians, 219–220, 223; in basal tetrapods, 187–188; functional significance, 121; in reptiliomorphs, 241, 243; in stem tetrapods, 196; in tetrapodomorphs, 148, 150, 178, 182–183; in tetrapods, 161–162, 164, 176. See also clavicle; interclavicle

climate change (human-aided), 472

Climatius, 116–117

cloaca: in amniotes, 229–230, 236; in amphibians, 214; in chondrichthyes, 104; in crocodylians, 325; functional significance, 104; in hagfish and lampreys, 40; in turtles, 289

cloacal breathing, 289. See also cloaca; turtles

coelacanth. See fishlike sarcopterygians

coffee mugs, 345

cold-bloodedness. See ectothermy ('cold-bloodedness')

Compagopiscis, 94, 96

concertina motion: in burrowing lizards (amphisbaenians), 264; in caecilians, 213; functional significance, 213; in snakes, 299

cones. See eyes

Coniophis, 300–301

conodonts, 56–57, 73

continental drift. See plate tectonics

coqui frogs (Eleutherodactylus coqui): and amniote evolution, 231, 247; development, 230–231

coracoid: articulation with sternum in amphibians, 210, 212–213; articulation with the sternum in crocodylians, 322; articulation with the sternum in lizards, 260; articulation with sternum in tetrapods, 161; functional

significance, 83–84; in mammals (reduction and loss), 412, 458–459; in marine reptiles, 304; metacoracoid, 234; in monotreme mammals, 412, 454; procoracoid, 234; in pterosaurs, 367; and supracoracoideus, 332–333, 458–459

coracomandibularis muscle: evolution in tetrapods into fleshy tongue, 170–171; functional significance, 93. See also tongue

cranial (skull) kinesis: in actinopterygians, 114, 120–121, 123–124, 126–127, 129–130; in birds, 328–329; in chondrichthyes, 101–102, 109; in dinosaurs (non-avian), 377; in lepidosauromorphs, 291–292; in lizards, 253–255; loss in mammals, 398–399; in marine reptiles, 292, 301–302; in snakes, 265, 293–295, 301; in tuatara and sphenodontians, 266, 292; in turtles, 284. See also jaws

cranial nerves: in chrondrichthyes, 100–101; in early vertebrates, 46–48; functional significance, 42–43; in hagfish and lampreys, 40, 45; and infrared light detection in snakes, 297; and jaw muscle evolution in mammals, 439; and neural crest cells, 45–46; in 'ostracoderms', 59, 63, 66–67; in tetrapods, 167, 171

Cretaceous-Paleogene extinction, 373, 387, 445, 456

Crocodylia: behavior, 320–321; evolution of, 325–326; feeding, 317–320; gizzard, 324; heart, 324–325; locomotion, 322–324; respiration, 322–323. See also Pseudosuchia; Reptilia

crocodyliforms and crocodylimorphs, 355–357

cruciate ligaments. See knee

Crurotarsi. See Pseudosuchia

Cryptoclidus, 303

Darwinopterus, 370–371

deep time, 13–17

Deinonychus, 379–381, 387

Deinosuchus, 326

deltopectoral crest. See humerus

Dendrerpeton, 218–219, 224

density: definition, 81; of vertebrates, 81–82; of water and air, 97, 158

dentary: in actinopterygians, 125–130; in amniotes, 232; in amphibians, 205, 207, 211–212, 214, 217, 220, 222–223; in archosaurs, 316, 352; in basal tetrapods, 190; in birds, 328; in crocodylians and kin, 317, 355–356; in dinosaurs (non-avian), 375, 377, 379, 381; in early amniotes, 244; in early diapsids, 273; in fishlike sarcopterygians, 139–140; functional significance, 119; in lepospondyls and reptiliomorphs, 239–240, 242; in

lizards, 255; in mammals and mammaliaforms, 398–403, 418, 432–439, 454–455, 468; in marine reptiles, 293, 301–303, 306–307; in non-mammal synapsids, 423, 426, 428–429; in ornithodirans, 358; in parareptiles, 271; in pterosaurs, 362, 364–365; in snakes, 293–295, 298, 300; in stem tetrapods, 195; in tetrapodomorphs, 146–147, 178, 180; in tetrapods, 156, 158, 165, 176; in tuatara, 265–266; in turtles, 283, 285–286, 288. See also predentary

depressor mandibulae muscle: in amphibians, 212, 214; in birds, 328; in crocodylians, 319; in dinosaurs (non-avian), 377; functional significance, 171; in lizards, 255; and mammals (evolution into posterior digastric), 402, 438–439; in marine reptiles, 303; in non-mammal synapsids, 426; in tetrapods, 171

deuterostomes, 32–34

Devonian: evolution of tetrapods, 174–176; evolutionary significance of, 91–92

Diadectes, 242–244

diaphragm muscle: diaphragm-like mesentery in crocodylians and kin, 322, 356; diaphragm-like mesentery in turtles, 279–281; true diaphragm in mammals, 409–411, 430, 440. See also lungs

diaphysis. See bone

diapsids, 232, 272–275, 287, 288–289. See also Reptilia

dicynodonts, 423, 425–427, 431, 455

Didelphis (American/Virginian opossum), 434, 436, 456–461

digastric muscle of mammals. See depressor mandibulae muscle; intermandibularis muscle

digitigrade, 337, 358–360, 375, 384, 417–418. See also feet

digits: in amniotes, 233; in amphibians, 202, 210, 220, 224; in basal tetrapods, 184–186, 188–192; in birds, 337; in climbing geckos, 264; in crocodylians and kin, 323, 355; in dinosaurs (non-avian), 375–376, 378 379, 385; in early amniotes, 246; in early diapsids, 274; functional significance, 163–164, 176; in lizards, 260–261; in mammals and mammaliaforms, 417–418, 441, 463–465, 469; in marine reptiles, 293, 302, 304, 307–308; in non-mammal synapsids, 414, 423, 430–431; in ornithodirans, 358–360; in parareptiles, 272; in pterosaurs, 362, 367–369, 387; in reptiliomorphs, 240, 242, 244; in snakes, 300; in stem tetrapods, 195–197; in tuatara, 266; in turtles, 287

Diictodon, 425

Dimetrodon, 420, 422–424, 432–433, 436, 450, 455

dinosaur extinction. *See* Cretaceous-Paleogene extinction

dinosaurs (non-avian): adaptations, 373–374; evolution of birds from theropods, 379–383; extinction, 387–388, 445; ornithischians, 386–387; saurischians, 374–386; sauropods and kin, 383–386; theropods, 374–383. *See also* birds; Reptilia

diphycercal tail, 139, 141, 143, 146, 148, 152. *See also* fishlike sarcopterygians; tetrapodomorphs

Diplodocus, 384

Dipterus, 145

distal, 138

Doleserpeton, 219–220, 224

Doliodus, 106–107

dorsal fin. *See* fins

dorsal ventricular ridge (DVR). *See* brain

dorsal vertebrae: in amphibians, 205, 209–210, 212, 219, 222–223; in basal tetrapods, 187, 190; in birds, 331–332; in crocodylians, 321; in dinosaurs (non-avian), 384; in early amniotes, 245; in amniotes, 233; functional significance and in tetrapods, 160, 162, 176; in lepospondyls and reptiliomorphs, 238–239, 241; in lizards, 259–260; in non-mammal synapsids, 422–424, 429; in ornithodirans, 358, 360; in parareptiles, 271; in pterosaurs, 366, 371–372; in snakes, 298, 300; thoracic and lumbar vertebrae of mammals and mammaliaforms, 406–408, 410, 414–415, 418, 433, 439–440, 456, 465, 468; in tuatara, 265; in turtles, 283, 286

drag, 86–88

driving cat (yes, you read that correctly), 346. *See also* cats; center of mass; dynamic stability; static stability

duck-billed platypus. See *Ornithorhynchus anatinus* (duck–billed platypus)

Dunkleosteus, 94, 96–98

dynamic stability, 345–347

ear drum. *See* tympanum (ear drum)

ears. *See* hearing; pinnae

Eastmanosteus, 94, 96

ectothermy ('cold-bloodedness'): and bone histology, 449–450; and parietal eye/pineal organ in tetrapods, 173, 259, 266, 271; physiological significance, 109. *See also* parietal eye; pineal eye (pineal gland, epiphysis)

ectotympanic. *See* hearing

Elaphe guttata guttata (corn snake), 294

elbow: in amphibians, 210, 220; in archosaurs, 350; in basal tetrapods, 188, 191; in birds, 334, 338; in dinosaurs (non-avian), 376; functional significance, 149; in lepospondyls and

reptiliomorphs, 238, 242; in lizards, 260, 263; in mammals and mammaliaforms, 412–413, 418, 441, 465, 469–470; in marine reptiles, 301–302; in non-mammal synapsids, 430; in stem tetrapods, 196; in tetrapodomorphs, 149, 182; in tetrapods, 163; in turtles, 277, 280–281, 283, 285. *See also* pronation; radio-ulna of frogs; radius; supination; ulna

elephant: in Afrotheria, 464–465; author almost crushed by, 392–394; loss of clavicles in, 412, 415; you can't keep warm with one, 398

Eleutherodactylus coqui. See coqui frogs (*Eleutherodactylus coqui*)

endostyle, 36

endothermy ('warm-bloodedness'): in birds, 327, 330, 340; and bone histology, 449; in chondrichthyes, 109; in dinosaurs (non-avian), 450; in mammals, 398, 404, 450; in non-mammal synapsids, 428, 450; physiological significance, 109; in pterosaurs, 450. *See also* parietal eye; pineal eye (pineal gland, epiphysis)

Eocaecilia, 222

Eomaia, 455, 462–463

epaxial muscles: and back support in tetrapods, 160; functional significance, 96; and head movements, 181, 321; and jaw opening, 96–97, 318–319

epicondyles. *See* humerus

epiphysis. *See* bone; pineal eye (pineal gland, epiphysis)

epipubic bones, 461–462. *See also* Mammalia

Eryops, 223–224

Estemmenosuchus, 426, 447

Euarchontoglires, 455, 465, 469–472. *See also* Mammalia

Eudibamus, 272

eugeneodont tooth whorl. *See* chondrichthyes

Eunotosaurus, 286–287

Euparkeria, 351–352

Euramerica, 175, 180, 185

Eusthenopteron, 146–152

Eutheria. *See* Mammalia

evolution: definition, 9–10; and natural selection, 10–12

evolution and functional significance in tetrapods, 160, 162, 176

Extant Phylogenetic Bracket (EPB) approach, 28. *See also* functional morphology

eyes: in acanthodians, 116–117; in actinopterygians, 125, 127; in amphibians, 204–207, 211, 213–214, 217–220, 223; in basal tetrapods, 186, 190; binocular vision, 205–207, 211, 378, 470; in birds, 328–330; in chondrichthyes, 100, 106, 110–112; cones, 256–257, 329, 446; in

conodonts, 56–57; cornea, 168, 204, 296; in crocodylians and kin, 317, 354, 356; development from the diencephalon, 43; eyelids, 167, 217–220, 296; 352; in fishlike sarcopterygians, 142; in hagfish and lampreys, 40; iris, 168, 204, 296, 307; lens shape and focus, 168, 204, 296–297, 307–310, 329; in lizards, 256–257, 259, 264; in mammals and mammaliaforms, 403, 405–407, 445–446, 454, 470; in marine reptiles, 306–310; nictitating membrane, 167; in non-mammal synapsids, 431; in 'ostracoderms', 59–60, 62–63, 66–68; in placoderms, 92, 98–99; presence in early vertebrates, 47–49; in pterosaurs, 365; in reptiliomorphs, 239; retina, 168, 204, 256–257, 329, 406–407, 446; rods, 256–257, 406, 446; in snakes, 295–297, 300–301; tapetum lucidum, 407, 446; in tetrapodomorphs, 146–147, 178, 180–181, 183; in tetrapods, 167–168, 171. *See also* parietal eye, pineal eye (pineal gland, epiphysis)

fairy tale, extant Megalodon as: 110. *Watch* Megalodon: the Monster Shark Lives

Falcatus, 108–109

feathers: anatomy and function, 337–340; development and evolution, 235, 380–382; for flight, 331–334, 337; as trait of birds, 327

feet: in amniotes, 233–234; in amphibians, 205, 210, 212, 215, 219, 221; in basal tetrapods, 186, 189, 192; in birds, 336–337; in crocodylians and kin, 323–324, 355, 357; development from fins, 184–185; in dinosaurs (non-avian), 375–376, 379, 384–386; in early amniotes, 245–247; in early archosaurs, 353; in early diapsids, 273–274; functional significance in tetrapods, 163–164; in lepospondyls and reptiliomorphs, 238, 240, 242, 244; in lizards, 261; in mammals and mammaliaforms, 413–414, 416–418, 441, 456–457, 461, 463, 465, 470–471; in marine reptiles, 304, 308; in non-mammal synapsids, 423, 431; in ornithodirans, 358–360; in pterosaurs, 368, 372; in stem tetrapods, 195–197; in tuatara, 265–266; in turtles, 281–282, 284–286

Felis catus (domestic cat). *See* cats

femur: in amphibians, 205, 210, 212, 220, 222–223; in basal tetrapods, 189, 191; in birds, 328, 332, 335–336; in crocodylians and kin, 321, 323–324, 354–355, 357; in dinosaurs (non-avian), 376–378, 385; in early amniotes, 245; in early archosaurs, 352–353; in early diapsids, 273; in fishlike sarcopterygians, 139; fourth trochanter of, 261, 324, 352, 355, 358, 376; functional significance, 138; greater trochanter of, 416, 423, 430, 433, 441; in lepospondyls and reptiliomorphs, 238, 240, 242, 244; lesser trochanter of, 416; in lizards, 260–261, 263–264; in mammals and mammaliaforms, 414, 416–417, 441, 446, 471; in marine reptiles, 293, 303–304, 308; in non-mammal synapsids, 423, 426, 430; in ornithodirans, 358–360; in parareptiles, 268; in pterosaurs, 362, 368, 372; in stem tetrapods, 196; in tetrapodomorphs, 149–152, 185; in tetrapods, 161–164; in tuatara, 265; in turtles, 279, 281, 283, 286, 288

fibula: adaptations for tree climbing in mammals, 457; in amniotes, 233–234; in amphibians, 210, 212, 220, 222–223; in basal tetrapods, 189, 191; in birds, 332, 337; in crocodylians, 321, 323; in dinosaurs (non-avian), 376–378; in early amniotes, 245; in early archosaurs, 348; in early diapsids, 273–274; functional significance, 149; in lepospondyls and reptiliomorphs, 238–242, 244; in lizards, 260–261, 263; in mammals and mammaliaforms, 417, 441, 457, 467; in marine reptiles, 308; in non-mammal synapsids, 430; in ornithodirans, 358–360; in parareptiles, 268; in pterosaurs, 362, 368, 371; in stem tetrapods, 196; in tetrapodomorphs, 149–150, 152; in tetrapods, 163–164; in tuatara, 265; in turtles, 281, 283, 286, 288. See also ankle; tibia; tibiofibula of frogs; tibiotarsus of birds

fingers. See digits

fins: in acanthodians, 116; in actinopterygians, 118, 121, 123–124, 126–128, 132–134; in chondrichthyes, 103–104, 107, 110–112; development of fins and limbs, 184–185; in early vertebrates, 46, 48; in fishlike sarcopterygians, 137–139, 141–144, 148–149, 151–152; functional significance, 46; in hagfish and lampreys, 46; in jawed vertebrates, 82–84; in 'ostracoderms', 59–60, 63, 65–66; in placoderms, 93, 98; in tetrapodomorphs, 180, 182, 192; in tetrapods, 160–161, 176, 185; as trait, 24–25; as a trait defining 'fish', 74. See also heterocercal tail

fishlike sarcopterygians: adaptations, 136–138; coelacanth, 141–144, 152; lungfish, 138–143, 148, 152; onychodontiforms, 143–144, 152; porolepiforms, 144–145, 152. See also Eusthenopteron; tetrapodomorphs

Flehmen response, 405. See also pheromones; vomeronasal organ

flowering plants (angiosperms), 193, 388

fluid flow, 61

F-number. See camera F-number

foramen magnum, 156, 159, 238, 407, 429, 471–472; functional significance, 156, 159. See also occipital bones and condyles

forebrain: in birds, 330; and dorsal ventricular ridge (DVR) in reptiles, 257; functional significance, 43; and limbic system in mammals, 453; and optic lobes in actinopterygians, 130; in 'ostracoderms', 66. See also brain

fossils: definition, 17; fossilization, 19–21

fourth trochanter. See femur

frogamander (Gerobatrachus), 221–222

frogfish, 151–152

frogs and toads. See Amphibia

functional morphology: biomechanical analysis, 26–27; definition, 26; and Extant Phylogenetic Bracket (EPB), 28; form-function analysis, 26–27

genes: definition, 11; developmental versus 'house-keeping', 11–12; and natural selection, 11–12

geologic laws, 14–15

Gephyrosaurus, 290, 292

Gerobatrachus, 221–222

gills: in acanthodians, 116–117; in actinopterygians, 120, 122; in amphibians, 201, 204, 215–219; in chondrichthyes, 103, 109–111; in conodonts, 56; in early vertebrates, 46, 48–49; in fishlike sarcopterygians, 138, 140–141; functional significance, 42; gill rakers, 78; in hagfish and lampreys, 42; loss of in amniotes, 229; morphology in jawed vertebrates, 74; and the operculum, 116, 120; in 'ostracoderms', 59–60, 62–63; in placoderms, 93; and the spiracle, 76, 167; in tetrapodomorphs, 148, 181, 186–188, 190–192; in tetrapods, 159, 167, 169, 171–173, 176

gizzard. See stomach

glenoid: in actinopterygians, 121; in amphibians, 210; in basal tetrapods, 187–188; in birds, 333; in crocodylians, 321; in dinosaurs (non-avian), 376; functional significance, 121; in mammals and mammaliaforms, 412, 440, 469; in non-mammal synapsids, 423, 430; in placoderms, 93; in pterosaurs, 367; in tetrapodomorphs, 149, 182; in tetrapods, 162, 164; in turtles, 281. See also coracoid; scapula; scapulocoracoid

glottis: in amphibians, 203; functional significance, 203; modifications in crocodylians, 317–318; modifications in mammals, 400; modifications in

snakes, 294, 298. See also lungs; trachea

gluteal muscles, 189, 210, 241, 416, 429–430, 440–441, 471–472

Glyptemys, 282–283

Gogosardina, 114, 124–125

Gondwana, 175, 463

grass, 388, 466–467, 472

greater trochanter. See femur

Guiyu, 143, 152

gular bones, 119, 124–125, 127–128, 146, 148, 156, 181; loss in tetrapods, 156, 159, 169, 176. See also gills

Hadrocodium, 452–453

hagfish, 39–46

Haikouichthys, 47–49

hair: development and evolution, 235, 428, 442, 446–448, 451–453, 455, 463; as trait of mammals, 397–398, 418. See also Mammalia

hands: in amphibians, 202, 205, 210–212, 220–221, 224; in basal tetrapods, 186–189, 191–192; in birds, 332–334, 340; in crocodylians, 323; development from fins, 184–185; in dinosaurs (non-avian), 374–376, 379–381, 383–386; in early amniotes, 246; in early archosaurs, 353; in early diapsids, 274; in lepospondyls and reptiliomorphs, 238–243; in lizards, 261; in mammals and mammaliaforms, 412–414, 440–441, 463, 465, 470, 473; in marine reptiles, 304, 308; in non-mammal synapsids, 423, 430; in ornithodirans, 358, 360; in parareptiles, 272; in pterosaurs, 367–368; in stem tetrapods, 196; in tetrapods, 163; in turtles, 280–286

hearing: in air, tetrapod adaptations for, 164–167; in amphibians, 207–208, 211, 219; angular (ectotympanic), 433–436, 468; and articular (malleus) 432–437; in birds, 328; in crocodylians, 320; in early amniotes, 244; in early tetrapods, 190, 195; in lepospondyls and reptiliomorphs, 238, 241; in lizards, 256; in mammals, 405–406, 432–433, 435–437, 446, 454, 466, 468; in parareptiles, 271; and quadrate (incus), 432–437; in snakes, 297; and stapes (hyomandibula), 166, 190, 195, 207–208, 211, 219, 238, 241, 244, 256, 271, 284–285, 297, 320, 328, 406, 423, 432–438; in tuatara, 266; in turtles, 284–285; in water. See also lateral line system; operculum

heart: in actinopterygians, 140; in amphibians, 204, 216; in birds, 340; in chordates, 34, 42; in crocodylians, 324–325; in dinosaurs (non-avian), 28; double-pump hearts, 28, 140–141, 171–172, 324–325, 340, 398; in early vertebrates, 47; in fishlike sarcopterygians, 140–141, 148; in hagfish and lampreys, 42; in mammals, 398; single-pump hearts, 43, 140; in tetrapods, 171–172

heterocercal tail: in acanthodians, 117; in actinopterygians, 118, 124, 126, 145; in chondrichthyes, 103–104, 107–108; in fishlike sarcopterygians, 145; functional significance, 64–66; in jawed vertebrates, 83; in 'ostracoderms', 63–66; in placoderms, 93–94, 98–99

hindbrain: in actinopterygians, 130; functional significance, 43; in 'ostracoderms', 67. *See also* brain

homocercal tail fin, 118, 124, 127–128, 151, 306. *See also* heterocercal tail

Homo sapiens (humans), 470–473

horses, 466–467; *Eohippus,* 466; *Equus* (modern horse), 466

HOX genes: snake vertebral column, 300; transition from fin to limb, 184–185

human evolution, 470–473. *See also* Mammalia

humerus: in amphibians, 205, 210, 212, 220, 222–223; in basal tetrapods, 187–188, 191; in birds, 333–334; in crocodylians, 321, 323; deltopectoral crest of, 148–149, 196, 223, 241, 322, 362, 367, 376, 430, 440, 459; in dinosaurs (non-avian), 376, 385; in early amniotes, 245; in early diapsids, 273; epicondyles of, 178, 187–188, 191, 223, 240, 242, 440; in fishlike sarcoptyergians, 143; functional significance, 136, 138–139; in lepospondyls and reptiliomorphs, 240–242, 244; in lizards, 260–261; in mammals and mammaliaforms, 412–413, 415, 440, 459; in marine reptiles, 293, 303–304, 307–308; in non-mammal synapsids, 423, 426, 430; in ornithodirans, 358; in parareptiles, 268; in pterosaurs, 362, 367, 369, 372; in stem tetrapods, 196; in tetrapodomorphs, 148–152, 178, 182–183, 185; in tetrapods, 161–164; in tuatara, 265; in turtles, 279–281, 283, 286

Hybodus, 109

Hylonomus, 244–245

hyoid arch: and breathing in amphibians, 203; development, 76; functional significance, 76; and hearing, 166; and jaw opening in mammals, 438–439; in jawed vertebrates, 72, 76–79; and the spiracle, 77; and tongue projection in amphibians, 207, 216. *See also* hearing

hyomandibula: functional significance, 76; and jaw opening in sharks, 101–102. *See also* hearing; hyoid arch; spiracle

hypothesis: definition, 23; and phylogeny, 23–26

ichthyosaurs, 306–310. *See also* eyes

Ichthyosaurus, 307

Ichthyostega, 189–192

ilium: in amphibians, 205, 210, 213, 219–223; in basal tetrapods, 188–189, 191; in birds, 334; in crocodylians, 322; in dinosaurs (non-avian), 374, 376; in early amniotes, 246; functional significance, 84; and gluteal muscles, 189, 210, 241, 416, 430, 440, 471–472; and jumping in frogs, 210; in lepospondyls and reptiliomorphs, 238, 241, 243; in lizards, 260–261; in mammals and mammaliaforms, 408, 412, 440, 471–472; in non-mammal synapsids, 426, 429–430; in ornithodirans, 358; in pterosaurs, 362; in stem tetrapods, 196; in tetrapodomorphs, 150; in tetrapods, 162, 164; in turtles, 283, 286

incisors. *See* teeth

incus. *See* hearing

infraspinatus muscle. *See* supracoracoideus muscle

interclavicle: in amphibians, 220; in basal tetrapods, 188; in crocodylians, 321–322; in early amniotes, 246; in lepospondyls, 238; in lizards, 260; in mammals and mammaliaforms, 433, 440; in non-mammal synapsids, 426, 429–430; in parareptiles, 271; in tuatara, 265; in turtles, 286. *See also* clavicle; cleithrum

intermandibularis muscle: in actinopterygians, 125; in amphibians, 212; and anterior digastric in mammals, 438–439; in chondrichthyes, 171; functional significance, 121; in tetrapods, 171. *See also* depressor mandibulae muscle

interpterygoid vacuities. *See* pterygoid bones

intracranial joint, 142–144, 146, 152, 167, 176, 180. *See also* basicranial muscle

iris. *See* camera F-number; eyes

ischium: in amphibians, 205, 210, 213, 219–220, 222–223; in basal tetrapods, 189, 191; in birds, 332, 334; in crocodylians, 321–322; in dinosaurs (non-avian), 374, 376; in early amniotes, 246; functional significance, 84; and jumping in frogs, 210; in lepospondyls and reptiliomorphs, 238, 241, 243; in lizards, 260–261; in mammals and mammaliaforms, 408, 412, 433, 440; in non-mammal synapsids, 423, 430; in ornithodirans, 358; in pterosaurs, 362; in stem tetrapods, 196; in tetrapods, 150, 162,

164; and thyroid fenestra in lizards, 261; in turtles, 283, 286

Ischnacanthus, 116–117

jaws: in acanthodians, 116; in actinopterygians, 119–121, 123–125, 126–130, 134; in amphibians, 205, 207, 212, 214; in archosaurs, 314, 316, 352; in basal tetrapods, 186; in birds, 328–329; in chondrichthyes, 101–102, 105–106, 111–112; in crocodylians and kin, 317–320, 323, 325–326, 353–354, 356; dentary-squamosal jaw joint of mammals, 398–399; in dinosaurs (non-avian), 375, 378–379, 381, 384, 386; in early amniotes, 244–245; evolution of, 72, 74–77; in fishlike sarcopterygians, 139–140, 143, 146; functional significance, 74; and gizzard, 324; and herbivory, 27, 242–243, 402; in lepospondyls and reptiliomorphs, 240–241; in lizards, 254–255, 264; in mammals and mammaliaforms, 399–401, 403, 433–434, 438–439, 441, 448, 450–452, 455–456, 458, 466–467, 469; in marine reptiles, 293, 303, 305–307; in non-mammal synapsids, 420, 422–423, 428, 431; in parareptiles, 268, 270–271; in placoderms, 92–94, 96–98; in pterosaurs, 362, 364–365, 372; quadrate-articular jaw joint, 372; in snakes, 294–297; and stapes, 165–166; and stomach, 79–80; and teeth, 77; in tetrapodomorphs, 146, 178, 180, 183; in tetrapods, 156, 158–159; in tuatara, 265–266; in turtles, 283–284, 286, 288. *See also* hearing

jugal: in basal tetrapods, 186; in birds, 328; in early amniotes, 244; functional significance, 156, 159; in lepidosaurs and lepidosauromorphs, 290–292; in mammals, 399, 401–402; in non-mammal synapsids, 426; in parareptiles, 271; in reptiliomorphs, 241, 243; in stem tetrapods, 195; in tetrapodomorphs, 180. *See also* zygomatic arch of mammals

Juramaia, 446, 462

kangaroo, 459, 461, 463; *Macropus,* 461

kidneys, 45, 106, 234–237

knee: in amphibians, 210; in basal tetrapods, 191; in birds, 334–337; in crocodylians and kin, 353; cruciate ligaments, 163, 417; in dinosaurs (non-avian), 376; in early diapsids, 273; functional significance, 149; in lizards, 261, 263–264; in mammals and mammaliaforms, 408, 414, 416–417, 441, 471; in marine reptiles, 301–302; menisci, 163; in non-mammal synapsids, 426, 430; in ornithodirans, 358; in pterosaurs, 369; in reptiliomorphs, 240–241; in stem tetrapods, 196; in tetrapodomorphs, 149; in tetrapods, 163; in turtles, 277. *See also* fibula; tibia; tibo-fibula in frogs; tibiotarsus in birds

K/T extinction. *See* Cretaceous-Paleogene extinction

lacrimal bone and duct, 147, 156, 167, 171, 217–220, 255, 328, 377

lamprey, 39–46

lancelets, 34–35

lateral line system: in acanthodians, 117; in actinopterygians, 121, 127–128; in amphibians, 208–209; in early vertebrates, 46, 48; in fishlike sarcopterygians, 139, 148; functional significance, 44–45; in hagfish and lampreys, 44–45; in 'ostracoderms', 59, 67; in placoderms, 93; in tetrapodomorphs, 186–187, 189, 195, 198; in tetrapods, 164–165. *See also* midbrain

Laurasiatheria, 455, 465–469

law. *See* geologic laws

Lemur catta (ring-tailed lemur), 470

Lepidosauria and lepidosauromorpha. *See* Reptilia

lepospondyls: as Amphibia, 202, 222; as outgroup to amniotes, 237–238. *See also* Amniota; Amphibia

lesser trochanter. *See* femur

levator scapulae muscle. *See* operculum

levers: calcaneum as, 355–357, 408, 417–418, 429, 431, 441; caudofemoralis muscles and femur in dinosaurs (non-avian), 376; and cnemial crest of birds, 336; definition, 94; and inner ear ossicles in mammals, 435, 437; and jaw mechanical advantage, 27, 94–96, 101–102, 242, 254–255, 302–303, 318–319, 400; and mammalian epipubic bones, 462; and olecranon process, 191, 272, 412; and pterosaur rib processes, 367; and snake skull kinesis, 294, 296; and teleost fish skull kinesis, 114, 128–130; and uncinate processes, 181, 226, 332; and urostyle of frogs, 211

Liopleurodon, 303

live birth: in mammals and mammaliaforms, 395–397, 453–454, 460; in placoderms, 98; in reptiles, 252, 307–308. *See also* amniotic egg; placenta

lizards: anatomy, 252–265; relationships, 264–265; and snakes, 265. *See also* Reptilia; snakes; tuatara (*Sphenodon*)

llamas, 467

long bone growth. *See* bone

Loxodonta africana. *See* elephant

lumbar vertebrae. *See* dorsal vertebrae

lungfish. *See* fishlike sarcopterygians

lungs: in actinopterygians, 125; in amniotes, 230, 233; in amphibians, 201, 203–204, 215–216; in basal tetrapods, 190–192, 194; in birds, 340, 341; for buoyancy, 82; chondrichthyes' lack of, 105; in crocodylians, 322, 324–325; and diving in sauropod dinosaurs (non-avian), 385; and elephants' semiaquatic ancestry, 464; in fishlike sarcopterygians, 138, 140–142; functional significance, 121–122; in mammals, 409–411; origin and development, 122; in *Polypterus* (bichir), 118, 121–123; in snakes, 298; as swim bladder, 123, 134; in tetrapodomorphs, 148; in tetrapods, 159, 169–170, 172–173; as trait of osteichthyes, 121; in turtles, 279–281

Lystrosaurus, 423, 425–426

malleus. *See* hearing

Mammalia: and amniotes, 227; anatomy, 395–419; brain in, 452–454; chewing and, 437–439; growth, 448–450; hair, 428, 446–448, 451–453; hearing, 432–437; milk production, 448–454; and mammaliaforms, 432–442, 455; and modern mammals, 454–472; and non-mammal Synapsida, 227, 421–432, 455. *See also* Reptilia

manatees, 464; *Trichechus*, 464

manus. *See* hands

Marasuchus, 357–359

masseter muscle, 399, 401–403, 426, 428–429, 433, 437–438, 467. *See also* adductor mandibulae; temporalis muscle

maxilla and premaxilla: in actinopterygians, 125–130; in amniotes, 232; in amphibians, 205, 207, 211–212, 214, 217, 220, 222–223; in archosaurs, 352; in basal tetrapods, 190; in birds, 328; in crocodylians and kin, 317, 355–356; in dinosaurs (non-avian), 375, 377, 379, 381; in early amniotes, 244; in early diapsids, 273; in fishlike sarcopterygians, 139–140; functional significance, 119; in lepospondyls and reptiliomorphs, 239–240, 242; in lizards, 255; in mammals and mammaliaforms, 399–401, 405, 433, 439, 468; in marine reptiles, 293, 301–303, 306–307; in non-mammal synapsids, 423, 426, 428–429; in ornithodirans, 358; in parareptiles, 271; in pterosaurs, 362, 364–365; in snakes, 293, 296, 300; in stem tetrapods, 195; in tetrapodomorphs, 146–147, 178, 180; in tetrapods, 156, 158, 165, 176; in tuatara, 265–266; in turtles, 283, 285–286, 288

Meckel's cartilage: in acanthodians, 116–117; in actinopterygians, 119, 126; in chondrichthyes, 101–102, 106–107, 109; functional significance, 72, 75–77; in mammals, 434–436; in placoderms, 94, 96; in tetrapods, 166. *See also* gills; hearing; jaws; Mammalia; pharyngeal arches

Megalodon (*Carcharocles megalodon*), 110

Megazostrodon ('*Morganucodon*'), 433, 439–441, 450. See also *Morganucodon*

Meleagris gallopavo (turkey), 327–332, 341

menisci, 163

mesonychids, 467, 469; *Andrewsarchus*, 467

metamorphosis: in amphibians, 173, 205, 209, 214–217, 219–221; in sea squirts, 34, 38–39

metaphysis. *See* bone

Metatheria. *See* Mammalia

metronome. *See* pendulum mechanics

midbrain: in actinopterygians, 130; in amphibians, 209; and balance, 45; in birds, 330; functional significance, 43–45; and hearing, 44–45, 405–406; and lateral line system, 44; in mammals, 405, 446; in 'ostracoderms', 67; in pterosaurs, 365; and vision, 43, 67, 130, 320, 330, 405–406, 446. *See also* brain

molars. *See* teeth

Monodelphis domestica (laboratory opossum), 434–436

Monotremata. *See* Mammalia

Morganucodon, 432–441, 446, 450–455

mosasaurs, 292–293

Mosasaurus, 293

multituberculates. *See* Mammalia

myomeres, 35–37, 39–40, 46–49, 60, 63, 84, 105, 128, 215, 277

natural selection: definition, 10; and genes, 11–12. *See also* evolution

neck: lack of in fishes, 121; in tetrapodomorphs and basal tetrapods, 178, 181–183, 187; in tetrapods, 156, 159–160, 162, 172. *See also* cervical vertebrae

nerve cord. *See* spinal cord

nerve tube, dorsal hollow. *See* spinal cord

neural crest cells: and conodonts, 57; functional significance, 45–46; and jaws, 74; and teeth, 77; as trait, 45–46; and turtle shell development, 277–278

nictitating membrane. *See* eyes

notochord: in acanthodians, 116; in actinopterygians, 121, 124–125, 132; in chondrichthyes, 103–104, 107, 109; in conodonts, 56; in early vertebrates,

47–48; in fishlike sarcopterygians, 139; functional significance, 36–37, 39; in hagfish and lampreys, 40, 46; morphology in jawed vertebrates, 84; morphology in tetrapods, 160; nucleus pulposus, 161, 409; in 'ostracoderms', 64; in placoderms, 93; in tetrapodomorphs, 146, 148, 181, 183; in tetrapods, 161; as trait, 36, 39; and vertebral column, 46

nucleus pulposus. *See* notochord

occipital bones and condyles: in amniotes, 232; in early amniotes, 245–246, 275; functional significance, 156, 159; in lepospondyls, 238; in mammals, 399, 401, 407, 409, 429; in reptiliomorphs, 239; in tetrapods, 156, 159

Odontochelys, 288–289

odor detection, 43–44; in actinopterygians, 130; in basal tetrapods, 190; in birds, 331; and choanae, 140; in chondrichthyes, 100–102; in crocodylians, 320; in dinosaurs (non-avian), 378; and lacrimal duct, 147; in lizards, 257–259; in mammals, 403, 405, 445–446, 454; in 'ostracoderms', 67; in snakes, 298; in tetrapodomorphs, 140, 147; in tetrapods, 171. *See also* pheromones; vomeronasal organ

olecranon process of ulna. *See* ulna

Onychodus, 143–144

operculum: in acanthodians, 117; in actinopterygians, 119–120, 122, 125–129; in amphibian larvae, 215; as amphibian organ of sound transduction, 208, 211; functional significance, 116; in tetrapodomorphs, 181

opossum. See *Didelphis* (American/Virginian opossum); *Monodelphis domestica* (laboratory opossum)

Ornithorhynchus anatinus (duck–billed platypus), 73, 227, 412–413, 435, 440, 451, 454–456

osteichthyes: adaptations, 115, 118. *See also* actinopterygians

'ostracoderms', 57–68, 73, 80, 83, 91, 98, 102, 116

otic notch. *See* hearing; otic region

otic region: in acanthodians, 117; in amphibians, 207–208, 218–220, 223; in basal tetrapods, 190; in chondrichthyes, 101; in crocodylians and kin, 317, 320, 356; in early amniotes, 244; in early vertebrates, 47–48; functional significance, 44; in jawed vertebrates, 72, 76; in 'ostracoderms', 67; in reptiliomorphs, 239–241; in stem tetrapods, 195; in tetrapods, 165–166; in turtles, 283–285. *See also* hearing

oxygen: absorption through skin in amphibians, 204; and amniote embryos, 230; and bone cells, 52; and

fossilization, 18–19; gills and surface area, 42; and evolution of tetrapods in shallow water, 122, 183, 193; and lactic acid, 54; muscle fatigue, 54; levels during Carboniferous, 175, 193–194; and operculum in fishes, 116. *See also* heart; lungs

paddlefish. *See* actinopterygians

palatopterygoid arch: in actinopterygians, 119–120, 124–125, 127–128; in fishlike sarcopterygians, 143; functional significance, 119–120; in tetrapodomorphs, 146. *See also* pterygoid bones

palatopterygoid teeth in snakes, 294

palatoquadrate: in acanthodians, 116–117; in actinopterygians, 119, 126; in chondrichthyes, 101–102, 106–107, 109; functional significance, 72, 75, 77; in placoderms, 94; in tetrapods, 165–166. *See also* gills; jaws; Meckel's cartilage; pharyngeal arches; pharyngeal slits and pouches; pharynx

Paleothyris, 244–247

Paliguana, 290

pangea, 22, 175, 193, 251, 427, 447

Pan troglodytes (chimpanzee), 471

Papio ursinus (baboon), 470

Pappochelys, 289

Paramys, 469–470

Parareptilia. *See* Reptilia

parasagittal gait and posture: in birds, 334, 336, 342; in crocodylians and kin, 355; in dinosaurs (non-avian), 373, 376; functional significance, 334, 346–350; in mammals and mammaliaforms, 394, 412, 440–441; in non-mammal synapsids, 426, 430; in ornithodirans, 358, 360; in pterosaurs, 368

Parasuchus, 353–354

parietal eye, 44, 173, 178, 186, 259, 266, 271, 273, 320, 330. *See also* eyes; pineal eye (pineal gland, epiphysis)

patella. *See* sesamoids

pectoral girdle. *See* coracoid; scapula; scapulocoracoid; sternum

pectoralis muscle: in birds, 332–333; functional significance, 148–149; in mammals, 415, 458, 465; in turtles, 279–281

Pederpes, 194–197

pelvic girdle. *See* ilium; ischium; pubis

pendulum mechanics, 346–348, 350

Perissodactyla, 466–467

Permian extinction. *See* Permo-Triassic extinction

Permo-Triassic extinction, 251–252, 269, 425–427

pes/pedes. *See* feet

Petrolacosaurus, 272–275, 290

pharyngeal arches: in acanthodians, 116–117; in actinopterygians, 127; in amphibians, 215–216, 218; in chondrichthyes, 103, 107; in early vertebrates, 46–49; functional significance, 42; in hagfish and lampreys, 40, 42, 45; in jawed vertebrates, 72, 74–78; in placoderms, 93; in tetrapodomorphs, 148; in tetrapods, 169. *See also* gills; jaws; Meckel's cartilage; palatoquadrate; pharyngeal slits and pouches

pharyngeal slits and pouches: in amphibians, 215; in chondrichthyes, 103, 111; in chordates, 32, 35–36; evolution in jawed vertebrates, 72, 74, 78; functional significance, 35–36; in hagfish and lampreys, 40, 42, 76; in 'ostracoderms', 60, 62–63; presence in early vertebrates, 46–49, 68. *See also* gills; spiracle

pharynx: functional significance, 35–36; and gills, 35, 42

pheromones, 67, 147–148; in lizards, 257; in mammals and mammaliaforms, 405, 439, 454; in salamanders, 212; in snakes, 298; in tetrapodomorphs, 147–148

phylogeny, 23–26

phytosaurs, 353–354

pineal eye (pineal gland, epiphysis), 40, 43–44, 47, 59–60, 63, 67, 99–101, 173, 176, 186, 190, 259, 320, 330–331, 405. *See also* eyes; parietal eye

pinnae, 397, 437, 446, 466. *See also* hearing; Mammalia

placenta: of reptiles, 252, 307–308; of therian mammals, 395–396. *See also* amniotic egg; live birth

placoderms, 92–99

placodonts, 287, 302

plantarflexion, 418, 431, 457. *See also* feet

plantigrade, 418. *See also* feet

plants: coal forests and tetrapods, 193–194; colonization of land and tetrapods, 174–176; evolution of herbivory, 242–244; and plate tectonics, 22; and ungulate mammal teeth, 467

plate tectonics, 21–23; and vertebrate evolution, 22–23

Plesiadapis, 469–470

plesiosaurs, 302–305

Plotosaurus, 293

Polypterus (bichir), 114, 118–125, 127, 129, 138–140, 143, 145, 148, 181–182

Poposuchus, 354–355

post-anal tail, 32, 38–39, 46

Postosuchus, 354

predentary, 374, 386–387. *See also* dentary

premaxilla. *See* maxilla and premaxilla

premolars. See teeth
primates, 469–472
Pristerodon, 423
Procolophon, 271–272
Proganochelys, 285–286
pronation, 281, 413, 440, 470. See also radio-ulna of frogs; radius; ulna
Prosalirus, 221–222
Proterosuchus, 349
Protosuchus, 355–357
proximal, 138
Pseudosuchia: adaptations, 353; evolution, 353–357; and Crocodylia, 355–357. See also Crocodylia; Reptilia
Pteranodon, 371–372
Pterodactylus, 363–370
pterosaurs: adaptations for flight, 362–363, 366–370, 372; hair-like body covering (pycnofibers), 369–370; posture, 368, 372–373; reproduction, 371–373; size, 371–373
pterygoid bones: in crocodylians, 317; in dinosaurs (non-avian), 377–378; in early amniotes, 244; in early diapsids, 273; functional significance, 156, 159; and interpterygoid vacuities of amphibians, 205–206, 211, 213, 222, 224; in lizards, 255; in non-mammal synapsids, 422; and the palatopterygoid teeth of snakes, 294; in parareptiles, 268, 271; in repitiliomorphs, 240–241; in snakes, 293, 295; in tetrapods, 156, 159, 176; in tuatara, 265
pterygoid muscles: in actinopterygians, 120–121, 125; in amphibians, 206, 219; in basal tetrapods, 186; in crocodylians, 318–319; in dinosaurs (non-avian), 375, 377–378; in early diapsids, 273; functional significance, 120; and jaw opening in mammals, 439; in lizards, 254–255; in mammals and mammaliaforms, 402, 439; in non-mammal synapsids, 422, 426; in parareptiles, 271; in reptiliomorphs, 240–241, 243; in stem tetrapods, 195; in tetrapodomorphs, 146, 180; in tetrapods, 156, 159, 170, 176
Ptilodus, 456
pubis: in amphibians, 210, 213, 219–220, 223; in basal tetrapods, 188–189, 191; and bird evolution, 380–381; in birds, 332, 334; in crocodylians and kin, 321–322, 356; in dinosaurs (non-avian), 374, 376, 379–381; in early amniotes, 246; functional significance, 84; and jumping in frogs, 210; in lepospondyls and reptiliomorphs, 238, 241, 243; in lizards, 260–261; in mammals and mammaliaforms, 408, 412, 440; in non-mammal synapsids, 423, 430; in ornithodirans, 358; and respiration in crocodylians, 322; in stem tetrapods,

196; in tetrapods, 150, 162, 164; and thyroid fenestra in lizards, 261; in turtles, 283, 286. See also epipubic bones
pulley mechanics, 262–263
pycnofibers. See pterosaurs
pygostyle of birds, 332, 383

quadrate: in amniotes, 232; in crocodylians, 318; and cranial (skull) kinesis in birds, 329; and cranial (skull) kinesis in lepidosauromorphs, 291; and cranial (skull) kinesis in lizards, 254–255; and cranial (skull) kinesis in snakes, 293, 295; functional significance, 120; and hearing in snakes, 297; and jaw opening in pterosaurs, 365; in mammals, 398, 402, 423, 425, 429, 432–439; and synovial joint, 254; in tetrapods, 158, 170, 176; as trait of osteichthyes, 120; and tympanic membrane, 256, 271, 284–285. See also hearing; jaws
quadratojugal. See jugal
Quetzalcoatlus, 363, 372–373

radio-ulna of frogs, 205, 210–211, 222. See also pronation; radius; supination; ulna
radius: in amphibians, 205, 210–212, 220, 222–223; in basal tetrapods, 187–189; in birds, 332–334; in crocodylians, 321, 323; in dinosaurs (non-avian), 376, 383; in early amniotes, 245; in early diapsids, 273; functional significance, 149; in lepospondyls and reptiliomorphs, 239–242, 244; in lizards, 260–261; in mammals and mammaliaforms, 408, 412–413, 415, 419, 433, 440–441, 463, 468, 470; in marine reptiles, 308; in non-mammal synapsids, 423, 426, 430; in ornithodirans, 358; in parareptiles, 268, 272; in pterosaurs, 362, 367; in stem tetrapods, 196; in tetrapodomorphs, 149–150, 152, 178, 182–183; in tetrapods, 163–164; in tuatara, 265; in turtles, 281, 283, 286, 288. See also ulna
ramp mechanics, 262
rats, 395–418, 431, 440–441, 446–447, 457. See also Euarchontoglires; Mammalia
Rattus norvegicus. See rats
relative dating, 14–16. See also absolute dating
Reptilia: and amniotes, 227, 275; Archosauria, 314–316, 344–353; Diapsida, 231–232, 272; Ichthyosauria, 306–310; Lepidosauria and lepidosauromorphs, 252–266, 291–301; Ornithodira, 357–359; Parareptilia, 268–272; Sauropterygia, 301–305; skin, 235; and Synapsida, 275;

temporal fenestrae, 232; and tetrapods, 198. See also birds; Crocodylia; dinosaurs (non-avian); Pseudosuchia; pterosaurs; Synapsida; tuatara (Sphenodon); turtles
reptiliomorphs: anthracosaurs, 239; carnivorous forms, 239–241; evolution of herbivory, 242–244; as outgroup to amniotes, 237, 239. See also Amniota
retina. See eyes
Rhamphorhynchus, 362–371
rhinos, 466; Ceratotherium simum (white rhino), 466
ribs: in actinopterygians, 121, 127, 132; in amniotes, 233; in amphibians, 202–205, 212, 218–220, 222–223; in basal tetrapods, 186–187, 189–192; in birds, 332, 341; bucket-handle motion of, 169–170; cranial rib of lungfish, 139; in crocodylians and kin, 321, 356; in dinosaurs (non-avian), 376; in early amniotes, 245–247; in early diapsids, 273, 275; in lizards, 259–260; in mammals and mammaliaforms, 408, 410, 412, 415, 418, 440, 468; in marine reptiles, 304; in non-mammal synapsids, 425–426, 429–430; in ornithodirans, 358, 360; in parareptiles, 271; in pterosaurs, 366–367, 370; in reptiliomorphs, 239, 241, 243; in snakes, 298, 300; in stem tetrapods, 195–196; in tetrapodomorphs, 146, 178, 181–183; in tetrapods, 160–162, 169–170, 176, 197–198; in tuatara, 265–266; in turtles, 276–280, 282–283, 286–288
Rocky Mountains, 310
rods. See eyes
rotator cuff muscles. See coracoid; scapula; scapulocoracoid; supracoracoideus muscle
Rugosodon, 456–457

sacrum: in amniotes, 233; in amphibians, 205, 210, 212, 220, 222; in basal tetrapods, 189–190; in birds, 331–332; in crocodylians, 321; in dinosaurs (non-avian), 376, 384; in early amniotes, 245–246; in lepospondyls and reptiliomorphs, 238, 241; in lizards, 259–260; in mammals and mammaliaforms, 411, 440; in marine reptiles, 307; in ornithodirans, 358; in parareptiles, 271; in pterosaurs, 362, 366, 371; in snakes, 298; in stem tetrapods, 196; in tetrapods, 160, 162; in tuatara, 265
salamanders and newts. See Amphibia
salivary glands, 169; modified into toxin glands in snakes, 265, 295
Saltoposuchus, 355–356
Sanajeh indicus, 301
scales: in acanthodians, 114, 116–117; in actinopterygians, 114, 118–119,

121–123, 125–128; in amphibians, 203–204, 213–214, 218, 223; in ancestors of mammals, 447; in chondrichthyes, 105–106; development in amniotes, 235; in early diapsids, 274; feathers do not develop from, 382; in fishlike sarcopterygians, 139, 141, 143, 145; in lepospondyls and reptiliomorphs, 239; in lizards, 252, 257, 264; in marine reptiles, 293; in 'ostracoderms', 58–60, 62–64; in placoderms, 92, 99; in snakes, 294, 296–297, 299–300; teeth derived from, 77; in tetrapodomorphs, 178, 180–181; as trait in lepidosaurs, 310
scapula: in bats, 465; functional significance, 83–84; and hearing in amphibians, 208; in mammals, 411–412, 414–415, 418, 440, 454, 458; and rotator cuff in mammals, 459; in tetrapods, 161. See also scapulocoracoid
scapulocoracoid: in acanthodians, 116; in actinopterygians, 119, 121, 125, 127–128; in amphibians, 210, 212–213, 219–220, 222–223; in basal tetrapods, 188; in birds, 333; in chondrichthyes, 103–104, 107, 109; in crocodylians, 322; in dinosaurs (non-avian), 376; in early amniotes, 246; in fishlike sarcopterygians, 136; in jawed vertebrates, 84; in lepospondyls and reptiliomorphs, 238–241, 243; in lizards, 260; in mammals and mammaliaforms, 433, 456; in marine reptiles, 303–304, 307; in non-mammal synapsids, 423, 426, 429–430, 440; in ornithodirans, 358; in parareptiles, 271; in placoderms, 93, 96–97; in pterosaurs, 362, 367; in tetrapodomorphs, 148–150, 178, 182–183, 187; in tetrapods, 161, 176; in tuatara, 265; in turtles, 276, 278–281, 283, 286, 288. See also coracoid; scapula
science, 9
sclerotic ring: in acanthodians, 116–117; in actinopterygians, 125; in amphibians, 217; in basal tetrapods, 186; in birds, 329; in dinosaurs (non-avian), 445; in early archosaurs, 352; functional significance, 92; in marine reptiles, 307, 309; in placoderms, 92–93; in pterosaurs, 365; in reptiliomorphs, 239; in tetrapodomorphs, 146; in turtles, 283. See also camera F-number
scutes: in amphibians, 204, 219, 223; in crocodylians and kin, 321–322, 354–356; in dinosaurs (non-avian), 386; in early archosaurs, 352; functional significance in tetrapods, 204; in sturgeon, 126; and turtle shells, 277, 285–287
sea squirts. See tunicates

secondary palate: in crocodylians and kin, 317–319, 356; in mammals and mammaliaforms, 401, 439, 461. See also choanae
semicircular canals, 45, 67, 341, 365
serratus ventralis muscle: in mammals, 414–415; in turtles, 278–281
sesamoids: in birds, 336; cartilago transiliens of crocodylians, 318; functional significance, 260, 262–263; in lizards, 260, 262–263; patella of mammals, 408, 416, 441
Seymouria, 240–242
sharks. See chondrichthyes
Sinodelphys, 462–463
skull: in actinopterygians, 114, 118–121, 123–130, 133; in amniotes, 231–232; in amphibians, 204–205, 207, 211–214, 217–224; in archosaurs, 314–316, 352; in basal tetrapods, 186–187, 189–190; in birds, 328–330, 341; in chondrichthyes, 100–102, 106–107, 109; in crocodylians and kin, 317–321, 326, 354–357; development from dermal bone, 58; in dinosaurs (non-avian), 375, 377–381, 383–384, 386; in early amniotes, 244–245, 247; in early archosaurs, 351; in early diapsids, 272–273, 275; in fishlike sarcopterygians, 139–140, 142, 144–145; in jawed vertebrates, 72, 75; in lepidosauromorphs, 290–292; in lepospondyls and reptiliomorphs, 237–242; in lizards, 253–255, 257; in mammals and mammaliaforms, 398–399, 402–404, 409, 418, 432–434, 450, 453–457, 461, 463, 466–470; in marine reptiles, 292, 302–303, 306–307; in non-mammal synapsids, 420, 422–423, 425–426, 428–429, 431; in ornithodirans, 357; in 'ostracoderms', 63, 66; in parareptiles, 268, 270–271; in placoderms, 92–94, 96–98; in pterosaurs, 362–366, 370–372; in snakes, 265, 293–297, 300–301; in stem tetrapods, 195; in synapsids, 422; in tetrapodomorphs, 146–148, 178, 180–182; in tetrapods, 156, 158–159, 161, 164–166, 173–174, 176; in tuatara, 265–266; in turtles, 283–284, 286, 288–289. See also braincase
slithering in snakes, 299
sloths, 464–465; Bradypus variegatus (three-toed sloth), 464
smelling. See odor detection
snakes: anatomy, 293–300; evolution, 300–301; ingestion of prey, 293–296; relationship between jaw kinesis and deafness, 297. See also lizards; Reptilia
somites, 277
Sphenodon. See tuatara (Sphenodon)

spinal cord: in actinopterygians, 121; in amniotes, 233; in amphibians, 212; in basal tetrapods, 190; in chondrichthyes, 103–105; in conodonts, 56; in early vertebrates, 46–48; functional significance, 37–39; in hagfish and lampreys, 45–46; in jawed vertebrates, 72, 84; in mammals, 403, 407, 472; in 'ostracoderms', 58; in tetrapodomorphs, 148; in tetrapods, 156, 159, 161
spiny echidna, 227, 412–413, 451, 454–455
spiracle: in amphibian larvae, 215; in bichir fish, 119, 122; in chondrichthyes, 103, 106, 111–112; and Eustachian (auditory) tube, 77, 166–167; functional significance, 76–77; and hyomandibula, 181, 195, 197; in jawed vertebrates, 72, 76–77; in sturgeon, 126; in tetrapodomorphs and basal tetrapods, 148, 178, 181, 183, 186. See also hearing; hyomandibula; jaws
splitter, 133–134
spoiler, 3
spring: definition, 409; dorsal vertebrae as, 408–409; furcula as, 333; mammalian shoulder sling as, 415; turtle guts as, 280
Squalus acanthias, 100–106
squamosal: in amniotes, 232; in amphibians, 205, 207, 217, 219; in basal tetrapods, 186; in dinosaurs (non-avian), 387; in early amniotes, 244; functional significance, 146; in lepidosaurs, 290; in lizards, 254–255; in mammals and mammaliaforms, 395, 398–401, 418, 432, 434–437, 439, 454–455, 468; in non-mammal synapsids, 423, 425; in reptiliomorphs, 240, 243; in snakes, 293–295; in stem tetrapods, 195; in tetrapodomorphs, 146, 180; in tetrapods, 156, 159, 176; in turtles, 284. See also adductor mandibulae; masseter muscle; temporalis muscle
stapes. See hearing
static stability, 345–346
Stegosaurus, 386
stereospondyls. See Amphibia
sternum: in amniotes, 233; in amphibians, 210, 212, 218; in birds, 332–333, 336, 341; in crocodylians, 322; in early amniotes, 246; in lizards, 259–260; in mammals and mammaliaforms, 408, 412, 414–415, 440, 454, 458–459, 465; in non-mammal synapsids, 423, 430; in ornithodirans, 358; in pterosaurs, 362, 367; in tetrapods, 161–162; in tuatara, 265; in turtles, 277–278
stomach: and jaws, 79–80, 88; vs. gizzard, 324, 340; in ruminants, 467
sturgeon. See actinopterygians

suckling. *See* Mammalia

supination, 413, 440, 470. *See also* radio-ulna of frogs; radius; ulna

supracoracoideus muscle: in birds, 332–333; evolution into rotator cuff in mammals, 458–459; original function in tetrapods, 458. *See also* birds; Mammalia

supraspinatus muscle. *See* supracoracoideus muscle

surface area: and digestion, 79; and gills (and turkey brine), 40–42; in gills of operculum-bearing fishes, 116; and heat loss in mammals, 398; and lungs, 169; and mammal brain expansion, 404; and turbinate bones, 403; why gills don't work in air, 169

swim bladder. *See* lungs

synapsid, 232. *See also* Mammalia; Synapsida

Synapsida: and amniotes, 227, 275; non-mammal cynodonts, 427–431; non-mammal synapsids, 422–427; non-mammal therapsids, 425–427; origins, 421–422; and Reptilia, 252; temporal fenestrae, 232; and tetrapods, 198. *See also* Mammalia; Reptilia

Tachyglossus. *See* spiny echidna

talus, 417. *See also* astragalus

tarsals/tarsus. *See* ankle

Tawa, 375–376

tear duct. *See* lacrimal bone and duct

teeth: in acanthodians, 116–117; in actinopterygians, 119–120, 123, 127, 129; in amphibians, 202–203, 207, 211, 215, 217, 220–222; in birds, 328; in chondrichthyes, 102, 106–107, 109–112; in crocodylians and kin, 318–320, 325–326, 353–355; in dinosaurs (non-avian), 375, 377–378, 383–384; in early amniotes, 244–245; in early archosaurs, 352; in early diapsids, 273; in fishlike sarcopterygians, 142–143; and gizzard, 324, 340; in jawed vertebrates, 77–79; in lizards, 255; in mammals and mammaliaforms, 399–403, 437–439, 441, 448, 450–452, 454, 456–458, 461, 463, 465–467; in marine reptiles, 292, 302–303, 306–307; in non-mammal synapsids, 422, 424–425, 428, 431; in ornithodirans, 358; in parareptiles, 271; in placoderms, 93, 98; properties, 56; in pterosaurs, 364; in reptiliomorphs, 241–243; in snakes, 293–294, 297, 300–301; in stem tetrapods, 186, 190, 195; and suckling, 451–452; in tetrapodomorphs, 146, 180–181; in tuatara, 266; in turtles, 283, 285, 288

teleosts. *See* actinopterygians

temporal fenestrae, 232. *See also* Amniota; Reptilia; Synapsida

temporalis muscle, 399, 401–403, 428–429, 437–438. *See also* adductor mandibulae

Tetrapoda: adaptations, 156–176; transition from water to land, 178–198. *See also* Amniota; Amphibia; Reptilia; Synapsida

tetrapodomorphs: fishlike, 141, 145–153; transitional, 157, 180–184. See also *Eusthenopteron*; *Tiktaalik*

Tetrapodophis, 301

theory, 10

Theria. *See* Mammalia

thoracic vertebrae. *See* dorsal vertebrae

Thrinaxodon, 128–432, 434–436, 439, 446, 448, 450–451, 453, 455

tibia: adaptations for tree climbing in mammals, 457; in amniotes, 233–234; in amphibians, 210, 212, 220, 222–223; in basal tetrapods, 189, 191; in birds, 332, 337; cnemial crest of, 240, 244, 332, 336, 376; in crocodylians, 321, 323; in dinosaurs (non-avian), 376–378; in early amniotes, 245; in early archosaurs, 348; in early diapsids, 273–274; functional significance, 149; in lepospondyls and reptiliomorphs, 238–242, 244; in lizards, 260–261, 263; in mammals and mammaliaforms, 417, 441, 457, 467; in marine reptiles, 308; in non-mammal synapsids, 430; in ornithodirans, 358–360; in parareptiles, 268; in pterosaurs, 362, 368; in stem tetrapods, 196; in tetrapodomorphs, 149–150, 152; in tetrapods, 163–164; in tuatara, 265; in turtles, 281, 283, 286, 288. *See also* ankle; fibula; tibio-fibula of frogs; tibiotarsus of birds

tibio-fibula of frogs, 205, 210, 222. *See also* fibula; tibia

tibiotarsus of birds, 273–274, 332, 335–337. *See also* ankle; fibula; tibia

Tiktaalik, 178, 180–184

Titanoboa, 301

toes. *See* digits

tongue: adaptations in crocodylians, 317; bifurcated tongue in lizards, 257–258, 264; and Flehmen response in mammals, 405; functional significance in jawed vertebrates, 78–79; and hyoid arch, 76–77; in lampreys, 39–40; muscular, fleshy tongue in tetrapods, 170; and odor detection, 171, 257, 264, 294, 298; in snakes, 294, 298; tongue projection in amphibians, 206–207, 215–216. *See also* hyoid arch

trachea, 169, 203, 207, 281, 298, 322, 340, 400, 410. *See also* glottis; lungs

traits: definition, 23; derived, 25; polarity, 25–26; primitive, 25

transversus abdominus muscle, 169, 203, 279–280

Triadobatrachus, 221–222

tribosphenic molars, 455, 458

triceps muscle, 191, 238, 240, 242, 271–272, 412, 440

Triceratops, 386–387

tuatara (*Sphenodon*), 265–267, 290–292. *See also* Reptilia

Tuditanus, 238–239

tunicates, 34

turbinate bones, 403–404

turbulence. *See* drag

turkey. See *Meleagris gallopavo* (turkey)

turtles, 269–270, 275–289; evolutionary relationships, 269–270, 275–276, 286–289; as prey for crocodylians, 326; shell development, 276–280. *See also* diapsids; neural crest cells; Reptilia

tympanic membrane. *See* tympanum (ear drum)

tympanum (ear drum): in amphibians, 205, 207–208, 211, 222–223; functional significance, 165; in lizards, 256; in mammals, 406, 419, 432–437, 468; in parareptiles, 271; in reptiliomorphs, 239, 241; in tetrapods, 165–166; in turtles, 284–285. *See also* hearing; spiracle

Tyrannosaurus, 358, 377–379, 394

ulna: in amphibians, 205, 210–212, 220, 222–223; in basal tetrapods, 187–189, 191; in birds, 332–334, 338; in crocodylians, 321, 323; in dinosaurs (non-avian), 376, 379–381, 383; in early amniotes, 245; in early diapsids, 273; functional significance, 149; in lepospondyls and reptiliomorphs, 238–242, 244; in lizards, 260–261; in mammals and mammaliaforms, 408, 412–413, 415, 418, 433, 440, 463, 468, 470; in marine reptiles, 308; in non-mammal synapsids, 423, 426, 430, olecranon process of, 191, 238, 240, 242, 244, 268, 271–272, 408, 412, 429, 440; in ornithodirans, 358; in parareptiles, 268, 271–272; in pterosaurs, 362, 367; in stem tetrapods, 196; in tetrapodomorphs, 149–150, 152, 178, 182–183; in tetrapods, 163–164; in tuatara, 265; in turtles, 281, 283, 286, 288. *See also* pronation; radio-ulna of frogs; radius; supination

uncinate processes: in birds, 332; similar processes in pterosaurs, 367; similar processes in *Tiktaalik*, 181; in tuatara, 265

ungual, 414, 418, 466

ungulates. *See* Cetartiodactyla, Perissodactyla

unguligrade, 418, 466. *See also* feet; Mammalia

urinary bladder, 230, 236–237. *See also* amniotic egg

urostyle of frogs: evolution, 211, 221–222; functional significance, 205, 209–211. *See also* caudal vertebrae

Velociraptor, 381

vertebrae: in acanthodians, 116; in actinopteryigians, 121, 125, 127–128, 132, 134; in amphibians, 202, 205, 209–213, 218–219, 221–223; in birds, 328, 331–332, 341; in chondrichthyes, 103–104; in conodonts, 57; in crocodylians and kin, 321, 355; in dinosaurs (non-avian), 374, 376, 378, 384; in early amniotes, 245–246; in early diapsids, 273; in early vertebrates, 47–48; in fishlike sarcopterygians, 139; functional significance, 46; in hagfish and lampreys, 40, 46; in lepospondyls and reptilimorphs, 238–239, 241–243; in lizards, 259–260, 292–293; in mammals and mammaliaforms, 407–411, 414, 418, 433, 439–440, 456–457, 465, 468, 472; in marine reptiles, 292–293, 303–304, 307; morphology in amniotes, 232–233; morphology in jawed vertebrates, 72, 84; morphology in tetrapods, 159–163, 170, 176; in non-mammal synapsids, 422–423, 429; in ornithodirans, 358, 360; in 'ostracoderms' 59; in parareptiles, 271; in pterosaurs, 363, 366–367, 370–372; in snakes, 296, 298–301; in tetrapodomorphs, 148, 181, 183, 187, 189–192, 195, 197–198; as trait, 24; in tuatara, 265; in turtles, 278, 282–283, 285–286. *See also* caudal vertebrae; cervical vertebrae; dorsal vertebrae; sacrum

vertebrates: and chordates, 39; defining traits, 39–40, 42–46; and deuterostomes, 33–34

vomeronasal organ: in amphibians, 212; functional significance, 146–147; in lizards, 257–259, 264; in mammals and mammaliaforms, 401, 405, 439; in snakes, 298; in tetrapodomorphs, 146–148

warm-bloodedness. *See* endothermy

Westlothiana, 239–240

Whatcheeria, 197

wrist: in amphibians, 205, 210–211, 220, 222–223; in basal tetrapods, 187–188, 191; in birds, 334, 338; in crocodylians and kin, 323, 355–356; in dinosaurs (non-avian), 379; in early amniotes, 246; in lepospondyls and reptiliomorphs, 238, 240, 242, 244; in lizards, 261; in mammals and mammaliaforms, 398, 414, 440–441, 465, 468; in marine reptiles, 302, 304, 307–308; in non-mammal synapsids, 423, 430; in ornithodirans, 358; in parareptiles, 272; in pterosaurs, 367, 370, 372; in tetrapods, 163–164; in turtles, 281

Xenarthra, 464–465

Zaglossus. *See* spiny echidna

zygapophyses, 159–162, 176, 187, 195, 241, 259, 283, 298–299, 303, 465. *See also* caudal vertebrae; cervical vertebrae; dorsal vertebrae; vertebrae

zygomatic arch of mammals, 401, 426–429, 433. *See also* jugal

This book was designed by Jamison Cockerham at Indiana University Press, set in type by Jamie McKee at MacKey Composition, and printed by Sheridan Books, Inc.

The fonts are Electra, designed by William A. Dwiggins in 1935, Frutiger, designed by Adrian Frutiger in 1975, and Futura, designed by Paul Renner in 1927. All were published by Adobe Systems Incorporated.

MATTHEW F. BONNAN is a vertebrate paleobiologist and Associate Professor of Biology at Stockton University in New Jersey. Since his elementary school days he has been fascinated by dinosaurs and reconstructing life of the past. An Illinois native (but a recent New Jersey transplant), Bonnan received his A.S. in earth sciences from the College of DuPage (Glen Ellyn, Illinois), his B.S. in geological sciences from the University of Illinois at Chicago, and his Ph.D. in biological sciences from Northern Illinois University. His research focuses on the evolution of locomotion in sauropod dinosaurs and the functional morphology of forelimb posture in reptiles, birds, and mammals. Bonnan helped discover and describe three new species of dinosaurs with colleagues in South Africa, and has overseen nearly fifty undergraduate and graduate student research projects on vertebrate functional morphology and evolution. He currently lives in Hammonton, New Jersey, with his wife and fellow academic, Jess Bonnan-White, their two children, Quinn and Max, three cats, and a small dog.